Gijsbertus de With

Structure, Deformation, and Integrity of Materials

Related Titles

Baltes, H., Brand, O., Fedder, G. K., Hierold, C., Korvink, J. G., Tabata, O., Löhe, D., Haußelt, J. (eds.)

Microengineering of Metals and Ceramics

Part I: Design, Tooling, and Injection Molding

2005

ISBN 3-527-31208-0

Zehetbauer, M., Valiev, R. Z. (eds.)

Nanomaterials by Severe Plastic Deformation

2004

ISBN 3-527-30659-5

Trebin, H.-R. (ed.)

Quasicrystals

Structure and Physical Properties

2003

ISBN 3-527-40399-X

Riedel, R. (ed.)

Handbook of Ceramic Hard Materials

2000

ISBN 3-527-29972-6

Meyers, M. A., Armstrong, R. W., Kirchner, H. O. K. (eds.)

Mechanics and Materials

Fundamentals and Linkages

1999

ISBN 0-471-24317-5

Gijsbertus de With

Structure, Deformation, and Integrity of Materials

Volume I: Fundamentals and Elasticity

WILEY-VCH

WILEY-VCH Verlag GmbH & Co. KGaA

The author

Prof. Dr. Gijsbertus de With
Eindhoven University of Technology
Department of Chemical Engineering
Den Dolech 2
5600 MB Eindhoven
The Netherlands

G.d.With@tue.nl

Library of Congress Card No.:
applied for

British Library Cataloguing-in-Publication Data
A catalogue record for this book is available from the British Library.

Bibliographic information published by Die Deutsche Bibliothek
Die Deutsche Bibliothek lists this publication in the Deutsche Nationalbibliografie; detailed bibliographic data is available in the Internet at <http://dnb.ddb.de>.

Printed in the Federal Republic of Germany

Printed on acid-free paper

Printing Strauss GmbH, Mörlenbach

Bookbinding Litges & Dopf GmbH, Heppenheim

ISBN-13: 978-3-527-31426-3

ISBN-10: 3-527-31426-1

no solider wisdom than that which is acquired in struggling against trouble.

Simon L. Altmann, *Icons and Symmetries*, Clarendon Press, Oxford, 1992.

Preface

For many processes and applications of materials a basic knowledge of their mechanical behaviour is a must. This is obviously not only true for materials with a primarily structural function but also for those materials for which the primary function is an electrical, dielectrical, magnetic or optical one and which are frequently (and wrongly) known as functional materials. Although many books on the mechanical behaviour of materials exist, a few drawbacks are generally present. On the one hand, the treatment is either too limited or too extensive and on the other hand the emphasis is typically too much on one type of materials, that is either on metals, on polymers or on inorganics. Moreover, the relation between the behaviour at the atomic, microstructural and macroscopic level is generally poorly developed. In this book a basic but as far as possible self-contained and integrated treatment of the mechanical behaviour of materials and their simplest applications is presented. We try to avoid the drawbacks mentioned by giving an approximately equal weight to the three material categories at a sound basic level. This does imply that not all topics can be treated and a certain initial acquaintance with materials science is probably an advantage for the reader. Meanwhile we try not to forget the need for somewhat more advanced discussion on several topics. Hopefully the proper balance is implemented by the two-level presentation: the 'basic' sections for all students and the 'advanced', or more properly 'intermediate', sections labelled with an asterisk for those who wish to deepen their knowledge.

A particular feature of this book is the attempt to give a basic but balanced presentation of the various aspects relevant at the micro- (atomic or molecular), meso- (microstructural or morphological) and macro-scale (bulk material properties and behaviour) for polymers, metals and inorganics. Another, also quite important aspect is that, wherever useful, the thermodynamic aspects are emphasised. We realise that this approach is not customary but we are convinced that this will make access to the more advanced literature easier. To that purpose we present in the Overview (part I) an introduction and an outline of constitutive behaviour. Part II describes the Fundamentals. It contains some mathematical preliminaries and the essentials of the continuum theory of kinematics, kinetics and thermodynamics. Also a summary of atomic and structural tools is given in this part. The latter has been incorporated to be somewhat self-contained. The remaining chapters discuss several topics in more detail and have been divided into various parts, namely Elasticity (part III), Plasticity (part IV), Visco-elasticity (part V) and Fracture (part VI). Of course, it is quite impossible to deal with every aspect and therefore we have limited ourselves, apart from the essentials of each of these topics, mainly to similarities and differences between the type of materials and their thermodynamic and structural background.

The whole of topics presented is conveniently described as mechanics of materials: it describes the thermomechanical behaviour of materials itself with applications to elementary structures and processes. With respect to the latter aspect the book by A.H. Cottrell, *The mechanical properties of matter*, has been an enlightening example. Unfortunately, in the past the term 'mechanics of materials', sometimes also called 'strength of materials', was claimed to denote the description of the deformation of beams, plates and other structures, given the constitutive behaviour of the material. The material aspect thus only appears on the phenomenological level. This area would rightfully have been called 'mechanics of structures' or 'structural mechanics'. The somewhat different description 'mechanical behaviour of materials' is not entirely adequate for this book since that title suggests that the treatment is essentially mechanical. Moreover, the application of the mechanical behaviour to simple structures and the explanation of the behaviour in structural terms, which is considered as essential in our approach, are not incorporated. The title reflects our final choice! For brevity I generally refer to the field as *thermomechanics*.

Since I realised that my style of writing is compact, I introduced in several chapters some panels. These panels do not interrupt the line of the discussion but can be read as an aside, which put a certain topic in perspective, either from a pragmatic, a context or a historical viewpoint. With respect to history I restricted myself to two or so panels per chapter with a short biography of eminent scientists. Most of the information is taken from the books written by Timoshenko[a], Struik[b], Nye[c], Love[d], Tanner and Walters[e] and Cahn[f]. Another useful source was the book by Hoddeson et al.[g]. The choice of the short biographies is arbitrary and it is likely that other authors will make another choice. It must be said that history is not always kind to people and certain topics or subjects are not known by the name of their first discoverer. I refrained from critical remarks in this respect. For those interested in these aspects the very detailed works[h] of Clifford Ambrose Truesdell[i] might be useful. Contrary to many textbooks, reference is made to the literature, generally to original presentations, other textbooks and reviews. On the other hand, only incidental reference is made to experimental methods.

After having treated the phenomenological equations, the applications of the theory, and the structural aspects of elasticity, plasticity, visco-elasticity and fracture, in the last chapter, I tried to provide a personal view and perspective on the whole of thermomechanics. I hope that the remarks made will be useful to many, although I am quite sure that it does not cover these areas to the satisfaction of everybody.

The essential ingredients of these notes were already contained in a course on the mechanical behaviour of materials at the Department of Chemical Engineering and Chemistry at Eindhoven University of Technology, which I took over some 10 years ago. The overall set-up as given here has been evolved in the last few years in which hopefully both the balance in topics and their presentation is improved. I am obliged to my students and instructors who have followed and used this course and provided

---

[a] Timoshenko, S.P. (1953), *History of strength of materials*, McGraw-Hill, New York (see also Dover, 1983).

[b] Struik, D.J. (1948), *A concise history of mathematics*, Dover, 1948.

[c] Nye, M.J. (1996), *Before big science*, Harvard University Press, Cambridge, MA.

[d] Love, A.E.H. (1927), *A treatise on the mathematical theory of elasticity*, 4th ed., Cambridge University Press, Cambridge (see also Dover, 1944).

[e] Tanner, R.I. and Walters, K. (1998), *Rheology: an historical perspective*, Elsevier, Amsterdam.

[f] Cahn, R.W. (2001), *The coming of materials science*, Pergamon, Amsterdam.

[g] Hoddeson, L., Braun, E., Teichmann, J. and Weart, S. (1992), *Out of the crystal maze*, Oxford University Press, New York.

[h] Truesdell, C.A. (1968), *Essays in the history of mechanics,* Springer, Berlin, and Truesdell, C.A. (1980), *The tragicomical history of thermodynamics 1822-1854*, Springer, Berlin.

[i] Clifford Ambrose Truesdell (1919-2000). American scientist, who played a highly instrumental role in the development of so-called rational mechanics and thermodynamics. His main criticism on the development of the thermodynamics is the continuous and complete mixing up of constitutive behaviour and basic laws as compared to the more or less separation of these aspects in mechanics.

many useful remarks. In particular I want to thank my colleague Dr. Paul. G. Th. van der Varst for the careful reading of and commenting on many parts of the manuscript and many discussions on almost all of the topics covered, which I always enjoyed and which made clearer to me a great number of aspects. Hopefully this led to an improvement in the presentation. It has been said before that authors do not finish their manuscript but abandon it. After the experience of writing this book I recognise that sentiment. My greatest indebtedness is to my wife who I 'abandoned' for many hours and days. Without her patience the book would never have been finished.

Obviously, the border between various classical disciplines is fading out nowadays. It is therefore hoped that these notes are not only useful for the original target audience, chemists and chemical engineers, but also for materials scientists, mechanical engineers, physicists and the like. Finally, I fear, the text will not be free of errors. They are my responsibility. Any comments, corrections or indications of omissions will be appreciated.

G. de With, July 2005

Acknowledgements

Figures

Figures 1.01, 1.08, 8.06, 15.04 and 15.06 are reprinted by permission of the Addison-Wesley Publishing Company, San Francisco, CA.

Figure 11.15 is reprinted by permission of the American Physical Society, College Park, MD.

Figure 22.14 is reprinted by permission of ASM Int., Metals Park, Ohio.

Figures 12.15 and 12.16 are reprinted by permission of the ASTM, Philadelpia.

Figures 14.11 and 14.12 are reprinted by permission of the Akademie-Verlag, Berlin.

Figures 8.21, 11.13, 12.07, 12.22, 23.06, E-3, E-4, E-5, E-7, E-8, E-9, E-10, E-11, E-12 and E-13 are reprinted by permission of the Cambridge University Press, Cambridge.

Figures 1.09, 8.01, 17.09 and 24.14 are reprinted by permission of Chapman & Hall, now CRC Press, Baco Raton.

Figure 22.17 is reprinted by permission of CIRP, Paris.

Figures 1.03, 1.05, 1.11, 1.12, 1.13, 1.14, 8.07, 8.33, 8.34, 15.02, 15.03, 15.13, 15.19, 15.22, 15.23b, 15.32, 15.33, 15.35, 15.36, 16.09, 16.17, 20.05, 20.06, 22.21, 23.07, 23.08 are reprinted by permission of Elsevier, Oxford.

Figure 21.04 is reprinted by permission of Marcel Dekker Inc., now CRC Press, Baco Raton.

Figures: 11.08 and 25-A are reprinted by permission of the Materials Research Society, Warrendale, PA.

Figures 1.15, 11.04, D-2, D-3, 15.09, 15.24, 15.31 and 16.14 are reprinted by permission of McGraw-Hill, New York.

Figures 14.17, 20.01, 20.02 and 20.03 are reprinted by permission of the McMillan Company, Basingstoke, Hampshire.

Figures 7.01, 7.02, 8.12, 8.13, 8.27, 8.28, 11.05 11.10, 11.11, 14.03, 15.28 and 15.30 are reprinted by permission of Oxford University Press, Oxford.

Figures 19.03, 19.04 and 19.05 are reprinted by permission of Prentice-Hall, now Pearson Education, Upper Saddle River, NJ.

Figures 24.16 and 24.17 are reprinted by permission of Scientific American, New York.

Figures 8.17, 20.07, 20.08, 20.09, 20.10, 20.11, 20.12, 20.13, 21.01, 21.11, 21.12, 21.13, E-14 and E-15 are reprinted by permission of Springer, Berlin.

Figure 2.12 are reprinted by permission of Syracuse University Press, Syracuse., New York.

Figure 22.24 is reprinted by permission of the American Ceramic Society, Westerville, Ohio.

Figures 22.15, 22.16, 22.18 and 22.19 are reprinted by permission of NIST, Gaithersburg, MD.

Figures 2.05, 7.03, 8.15, 8.16, 8.18, 11.16, 15.18, 15.20, 15.23a, 16.02, 16.12, 17.01, 17.04, 23.04 and 23.14 are reprinted by permission of Wiley, Chichester, UK.

All portraits have been reproduced from various websites by permission of the copyright holders. Wiley-VCH and the author have attempted to trace the copyright holders of all material reproduced in this publication and apologise to copyright holders if permission to publish in this form has not been obtained.

Cover

Scales: an artist's impression of the length scales aspects in thermomechanics. Martijn de With, 2005.

Contents Volume I

10. Elasticity of structures

11. Molecular basis of elasticity

12. Microstructural aspects of elasticity

Index

Contents Volume II

Part IV: Plasticity

13. Continuum plasticity

19. Applications of visco-elasticity theory

20. Structural aspects of visco-elasticity

Part V: Fracture

21. Continuum fracture

22. Applications of fracture theory

23. Structural aspects of fracture

List of important symbols and abbreviations

Φ dissipation function, Airy stress function, wave function, potential energy
Ξ grand partition function
Ψ wavefunction
Ω external potential energy

α constant
α_{ij} thermal expansion tensor
β constant, kT
γ (engineering) strain, shear strain
δ_{ij} Kronecker delta
ε strain, small scalar, energy
ε_{ij} strain tensor
φ specific dissipation function
ϕ (pair) potential energy
λ Lamé constant, stretch
μ Lamé constant, shear modulus
π (second) Piola-Kirchoff stress
ρ density, radius of curvature
σ (true) stress
σ_{ij} Cauchy stress tensor
τ shear stress
ν Poisson's ratio
ω frequency

C right Cauchy-Green tensor
D left Cauchy-Green tensor
E Euler strain tensor
F deformation gradient
L Lagrange strain tensor
Q generalised force
O zero tensor
0 zero vector
I unit tensor
1 unit vector

a acceleration, generalised displacement
b body force, Burgers vector
d rate of deformation
e unit vector
f force
g reciprocal lattice vector
m, n outer normal vector
l angular momentum, dislocation line
p linear momentum, generalised momentum
q torque
r direct lattice vector, material co-ordinate
s (first) Piola-Kirchhoff stress tensor, shear stress vector
t stress vector, traction
u displacement
v velocity
x spatial co-ordinate

A area, fatigue parameter, generalised force
$A_{(i)}$ basic invariant of A_{ij}
A_{I} principal value of A_{ij}
C constant
C_{ij} elastic stiffness constants
C_{ijkl} elastic stiffness constants
E Young's modulus, energy
F Helmholtz energy, fatigue limit, force
G Gibbs energy, shear modulus, strain energy release rate
H enthalpy, Hamilton function
I moment of inertia
$J_{(i)}$ invariant
K bulk modulus, reaction constant
L power, length, Lagrange function
M moment, orientation factor
N_{A} Avogadro's number
P porosity, probability, power
Q partition function, charge
Q_i component of generalised force
R gas constant, radius, fatigue parameter, fracture energy
S entropy, strength

S_{ij}	elastic compliance constants
S_{ijkl}	elastic compliance constants
T	kinetic energy, temperature
U	(internal) energy
V	potential energy, volume
W	work, strain energy
Y	uniaxial yield strength
Z	section modulus, density of states, co-ordination number, partition functi

a	generalised displacement
a_i	component of acceleration
b	length of Burgers vector
b_i	component of body force
c	constant, inverse spring constant (compliance)
d_{ij}	rate of deformation
e	strain
e_{ijk}	alternator
f	(volume) fraction, specific Helmholtz energy, force
g	specific Gibbs energy
k	Boltzmann's constant, yield strength in shear, spring constant (stiffness)
l	length
m	Weibull modulus, Mie constant, Schmid factor, mass
n	Mie constant, material constant
p	pressure, plastic constraint factor
p_i	component of (generalised) momentum
q_i	component of (generalised) co-ordinate
s	specific entropy, (engineering) stress
t	time
t_i	component of traction
u,v,w	displacement in x,y,z directions
v	volume
w	strain energy density
z	single particle partition function

BCC	body centered cubic
CRSS	critically resolved shear stress
FCC	face centred cubic
FEM	finite element method
HCP	hexagonal closed packed
LEFM	linear elastic fracture mechanics
PCVP	principle of complementary virtual power
PMPE	principle of minimum potential energy
PVP	principle of virtual power
PVW	principle of virtual work
SDG	small displacement gradient
SFE	stacking fault energy
SIF	stress intensity factor
SSY	small scale yielding

$\cong$	approximately equal
$\equiv$	identical
$\sim$	proportional to
$\Leftrightarrow$	corresponds with
(hkl)	specific plane
$\{hkl\}$	set of planes
$[hkl]$	specific direction
$\langle hkl\rangle$	set of directions

1

Introduction

For virtually any structure or process the mechanical behaviour of the materials involved is relevant as elaborated somewhat in the panel 'The importance of MSE'. Either the deformation (strain) or the force (stress) is a controlling parameter. Both quantities depend on the nature of the loads (purely mechanical, thermo-mechanical, electrical, ...), the geometry of the structure or the piece of material and, of course, the thermo-mechanical properties of the materials. Thermomechanics of materials, as defined in this book, deals with the thermomechanical behaviour of solid materials and the application to simple structures and processes. The aim of this chapter is to briefly overview the approach that is to be followed, but for that purpose it is useful to first review briefly the various materials whose thermomechanical behaviour is to be discussed. We deal with inorganics, metals, polymers and composites. Thereafter we present some considerations on length scales and the nature of the continuum, followed by an outline of the topics that will be discussed.

1.1 Inorganics

Inorganics or ceramics are materials that contain either ionically bonded atoms (Fig. 1.1) or covalently bonded atoms[a] (Fig. 1.2). In ionic bonding, positively and negatively charged ions attract each other non-directionally. The cohesive energy ranges from 600 to 1500 kJ/mol. In covalent bonding strong directional bonds are present between the atoms due to shared electrons. The cohesive energy ranges from 300 to 700 kJ/mol. In both cases generally a solid arises with high melting point and corresponding high stiffness and low ductility[b]. For these materials the electrical

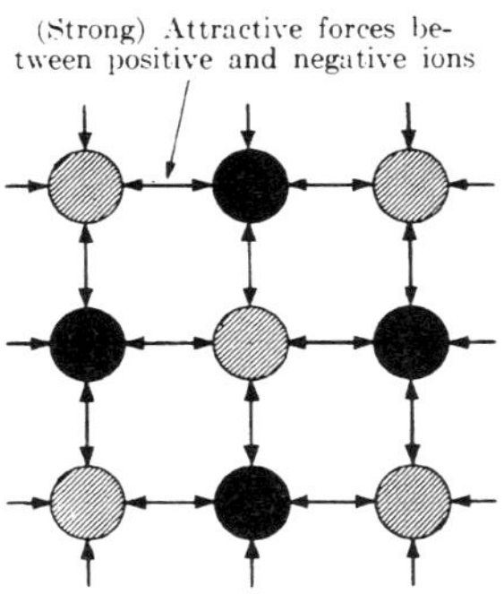

Fig. 1.1: Schematic of an ionically bonded material.

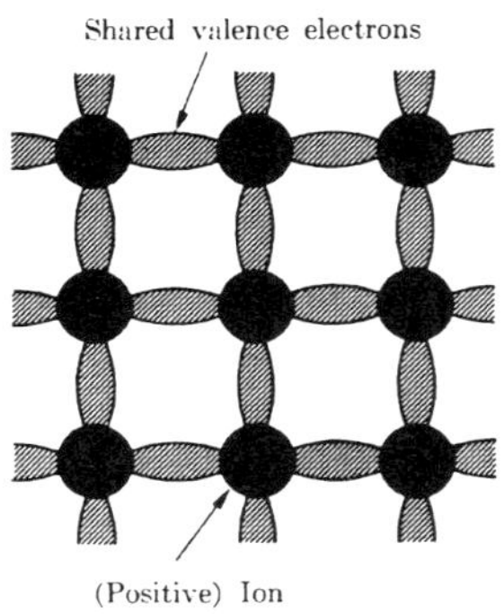

Fig. 1.2: Schematic of a covalently bonded material.

[a] Of course, this classification is one of extremes and a gradual transition between them exists. From the modern point of view bonding ought to be described by quantum mechanics. Nevertheless, the classification is useful and thus frequently used.

conductivity is normally low while the thermal conductivity can vary considerably. Generally two classes can be distinguished: crystalline and amorphous materials.

The importance of MSE

Many of the achievements of Materials Science and Engineering (MSE) are unnoticed. Nevertheless MSE is quite important in modern society, both in daily life and in high-tech applications. MSE was initially dealing mainly with metals and therefore used to be called metallurgy. To mention just a few examples: Cr-based super alloys are indispensable materials for turbine blades in the high-temperature environment of power supply stations and light-weight alloys based on Al and Mg have become a necessity for aircraft. Without stainless steel, as used for knifes and other utensils, daily life in the kitchen would be less easy. However, the importance of polymers and ceramics has increased considerably in the last decades. Today many household objects are made of polymers, e.g. the casings of electronic and household appliances or disposable utensils or children's toys. In fact they are often composites based on a polymer matrix reinforced with inorganic particles. More high tech are the electrically conductive polymers, which are being engineered today. The use of ceramics is usually less noted but nevertheless important. For each electronic chip (still largely from Si, an inorganic material itself), two or so capacitors are required which are most of the time made of a ferroelectric ceramic. Other examples are ceramic magnets in TVs and PC monitors, the high purity optical glass fibres used for telecommunication and the classical example of a spark plug.

Crystalline solids further can be divided into single crystalline or polycrystalline materials. In both single and polycrystalline materials a regularly ordered structure exists at an atomic scale. This structure is maintained, at least in principle, throughout the whole material in a *single crystalline* material, while in a *polycrystalline* material regions of different crystallographic orientations exist. These regions are referred to as *grains* and the boundaries between them as *grain boundaries* (Fig. 1.3). X-ray diffraction clearly reveals the long-range atomic order of these materials. In *amorphous* solids there is no long-range order (Fig. 1.4) although the local co-ordination of a specific atom in the amorphous state may not be that different from the co-ordination of the same type of atom in the corresponding crystalline state (if it exists).

Crystalline solids generally show a distinct melting point. Below the melting point the crystalline structure is present while above the melting point an amorphous, liquid structure arises. Despite the long-range order, various defects may be present in single crystal materials. They can be divided into *point defects* (interstitials, vacancies, substitutional atoms), *line defects* (dislocations), *planar defects* (stacking faults) and *volume defects* (inclusions, pores) (Fig. 1.5). In addition to grain boundaries the same range of defects as mentioned for single crystals occurs in polycrystalline solids. Here

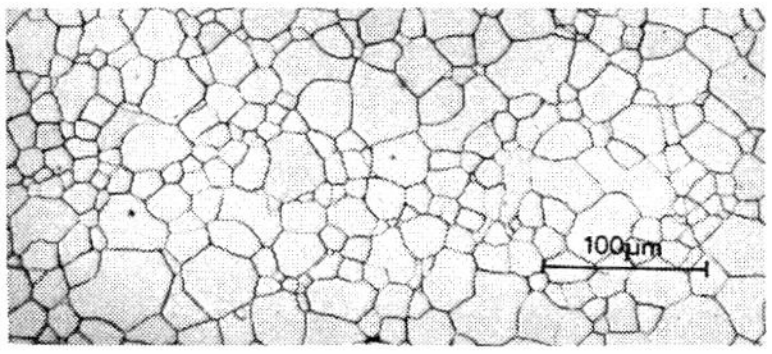

Fig. 1.3: Polycrystalline material.

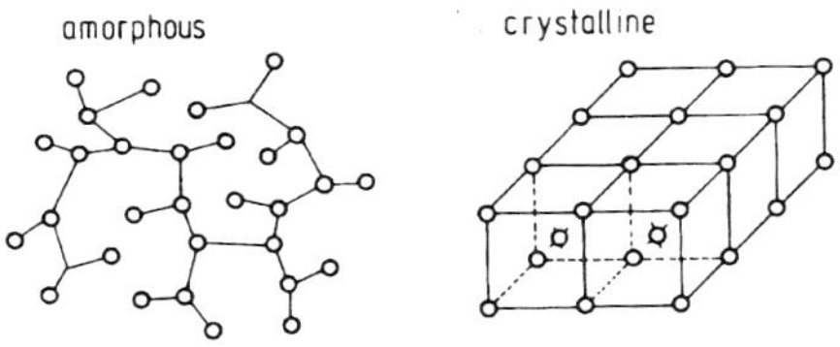

Fig. 1.4: Amorphous and crystalline structure.

---

[b] A number of concepts and ideas are used in this chapter. Most of them will be readdressed later in this book. For the moment we accept their significance as obvious.

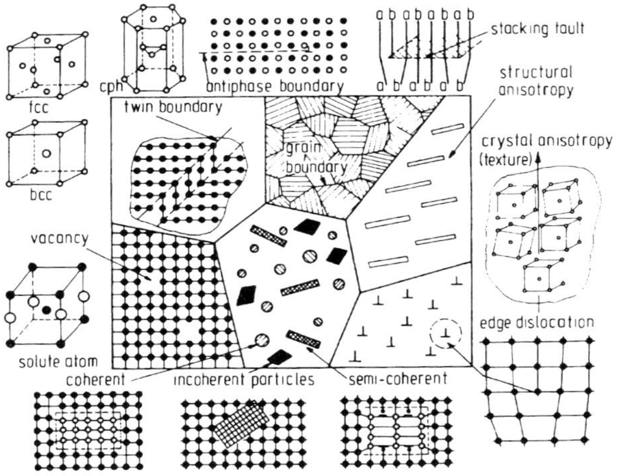

Fig. 1.5: Overview of the various microstructural features.

the average grain size and grain size distribution frequently play an important role as well. The influence of some of these defects on the thermomechanical behaviour will be discussed in later chapters.

Although high purity is often pursued, *impurities* nevertheless are present. These impurities may dissolve in the lattice, if present below the solubility limit, or precipitate, if present in amounts above the solubility limit, sometimes forming new phases. These precipitates also have a pronounced influence on the thermomechanical behaviour. In polycrystalline materials impurities may segregate on the grain boundaries leading to either amorphous or crystalline *grain boundary phases*. These grain boundary phases are often quite important in the mechanical behaviour. Obviously, if sufficient amount of this impurity material is present, the new phases, if they crystallise, can form new grains in the material. If the new phase remains amorphous, it is generally referred to as a *glassy second phase*. If the foreign compounds are added on purpose, e.g. to improve either processing or properties, the indication 'impurity' is usually replaced by *additive* or *dopant*. The type, structure and number of phases; the number, geometric appearance (size, shape, etc.) and topological arrangement of the individual phase regions and their interfaces and the type, structure and geometry of lattice defects define what is called the *microstructure*[c] (see also Fig. 1.5).

Various types of compositions for polycrystalline inorganics exist. The first and most important are the oxides, which show mainly ionic bonding. Silicates as used in bricks, porcelain, etc. are well known. For more advanced applications, mainly of mechanical nature, alumina or Al_2O_3 is the working horse but a large variety of oxides have found their use in a multitude of applications. Some examples are MgO and $Al_2O_3{\cdot}SiO_2$ for refractory applications, $BaTiO_3$ for dielectric applications and MnZn-Fe_2O_4 for magnetic applications. By varying the grain size and composition the properties can be varied over a certain range so as to make them suitable for different applications. Carbides, the most important of which is SiC, are used for abrasive and

[c] Exner, H.E. (1983), *Qualitative and quantitative surface microscopy*, page 581 in *Physical metallurgy*, 3rd ed., R.W. Cahn and P. Haasen, eds., North-Holland, Amsterdam.

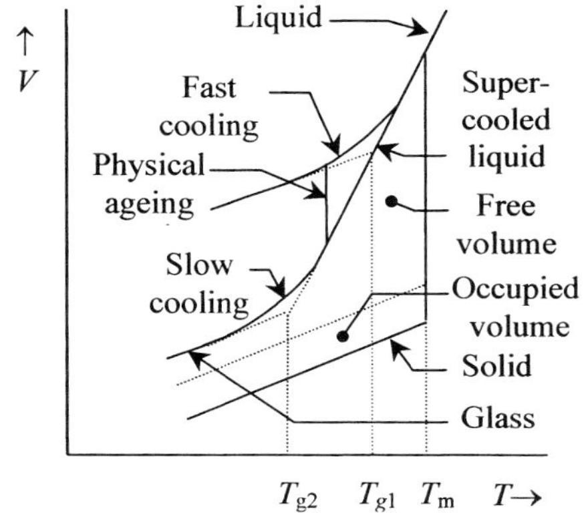

Fig. 1.6: The (idealised) change in specific volume for the glass transition at cooling rate 1 > cooling rate 2 leading to glass transitions temperatures T_{g1} and T_{g2}. Also shown is the normal melting behaviour at melting point T_m.

high-temperature applications. Nitrides, of which the most important representative is Si_3N_4, are gaining importance, particularly in engineering applications for high temperature but also e.g. as sensor tip material in atomic force microscopy (AFM). For these materials the bonding is mainly covalent. Other examples of important carbides and nitrides are WC, TiC, B_4C, AlN and TiN. Furthermore a wide range of sulphides, selenides, tellurides, etc. are (being) engineered for specific applications.

Amorphous inorganics are usually called *glasses*. Glasses do not show clear melting but do show within a certain temperature range a gradual transition from elastic to viscous behaviour with increasing temperature. In this temperature range the material behaviour becomes increasingly more time dependent with temperature. The *glass transition temperature* T_g, located approximately in the middle of that temperature range, characterises the transition and the behaviour is referred to as *visco-elasticity*. For inorganic glasses the glass transition temperature typically ranges from 500 °C to 1000 °C. Unlike during melting, where several properties change abruptly at the melting point, properties for a glass change gradually from one regime to another in the glass transition region, as illustrated for the specific volume in Fig. 1.6. The glass transition temperature T_g is usually determined by the extrapolation of the linear behaviour of the liquid-like and glass-like regions. An important characteristic is that T_g depends on the cooling rate employed. The slower the cooling rate, the lower T_g is obtained. This effect is ascribed to the *free volume*, i.e. the empty space between the molecules. At high cooling rate only limited relaxation of the glass structure can occur before the temperature has decreased so far that further relaxation is very slow. At lower cooling rate more relaxation can take place, thus continuing the liquid-like regime to lower temperature and hence leading to a smaller free volume and therefore a lower T_g. A glass cooled at a high rate can relax slowly in the glass-like regime (at sufficiently high temperature) to a branch associated with a lower cooling rate, a process generally known as *physical ageing*. This is in contrast to *chemical ageing* where a slow chemical reaction modifies the chemical constitution. This effect is relatively unimportant for inorganic glasses but may be considerable for polymers.

For the majority of inorganic glasses silica or SiO_2 is the basic component. Silica itself has a low thermal expansion coefficient and highly homogeneous network

Fig. 1.7: The microstructure of a machinable glass-ceramic.

structure in which the SiO_4 tetrahedra are connected to each other with all corners shared. Other *network forming oxides* are e.g. GeO_2 and B_2O_3 while some others, e.g. Al_2O_3 and Bi_2O_3, are forming glasses only in the presence of other, network forming oxides. By modification of the network through addition of oxides such as Na_2O, K_2O, CaO and Al_2O_3, the *network modifying oxides*, the properties can be varied over a wide range. The full connectivity of the SiO_4 tetrahedra is lost and charge compensation is provided by the other cations. This modification applies to the static properties such as density, hardness, thermal expansion coefficient and refractive index as well as the transport properties such as electrical resistance and diffusion coefficient. In a number of cases the structure becomes inhomogeneous through phase separation, which can be used for strengthening glasses to a considerable extent.

A relatively new class of materials is *glass-ceramics*. They are partially crystallised glasses with a significant remaining volume fraction of amorphous materials. Typically these materials are made by conventional glass technology resulting in a glass, which is partially crystallised by a controlled heat treatment. This crystallisation is induced by the presence of a seed. The near-net shape fabrication option is a definite advantage. The final properties can be rather different depending on the microstructure. We quote two examples. Glass-ceramics with a large number of dispersed crystals can be strong due to internal stress and are used e.g. for household applications. Highly crystallised glass-ceramics with plate-like crystals can be easily machinable and used e.g. for the production of prototype items (Fig. 1.7).

1.2 Metals

Metals are solids in which the bonding between the ions is collectively provided by the electrons (Fig. 1.8). The cohesive energy ranges from 100 to 800 kJ/mol. Depending on the type, metals can show a high melting point and corresponding high hardness and low ductility (e.g. Mo, W) or a relatively low melting point with associated lower hardness and higher ductility (e.g. Cu, Al). Due to the collective electrons, metals typically show a good electrical conductivity and a good thermal conductivity. Like inorganics, metals may be single crystalline or polycrystalline and the same type of defects as in inorganics can occur. Of the 83 metal-like elements 15 have the FCC, 25 the HCP, 15 the BCC, 11 another and 17 an unknown structure.

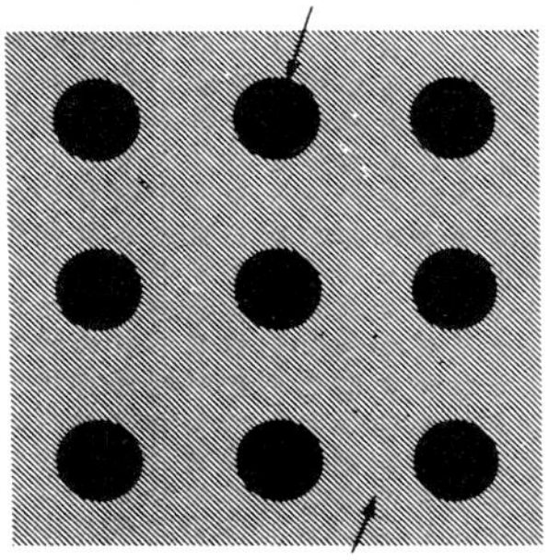

Fig. 1.8: Schematic of a metallically bonded material.

Adding other elements to a base metal is generally called *alloying*. As in the case of inorganics the additions can dissolve in the matrix, segregate at the grain boundaries or lead to new phases. It should be noted that the solubility of additions is typically much higher in the case of metals as compared with crystalline inorganics. By alloying not only the chemical composition but also the microstructure can be influenced significantly leading to widely varying properties for one and the same base metal. Particularly the strength can be increased, usually at the cost of lower ductility.

A typical feature of metals is the possibility of changing the microstructure by a heat treatment below the melting point. A particularly frequently used process is *annealing*, i.e. holding at high temperature for some time. The change in the microstructure is reflected in the properties of the metal; e.g. by annealing and slow cooling the yield strength can often be lowered while after annealing and quenching the yield strength can be increased.

As an example we discuss in some detail Fe, by far the most important metal, mainly used for structural applications. *Ferrite* (or α-Fe with a BCC structure,

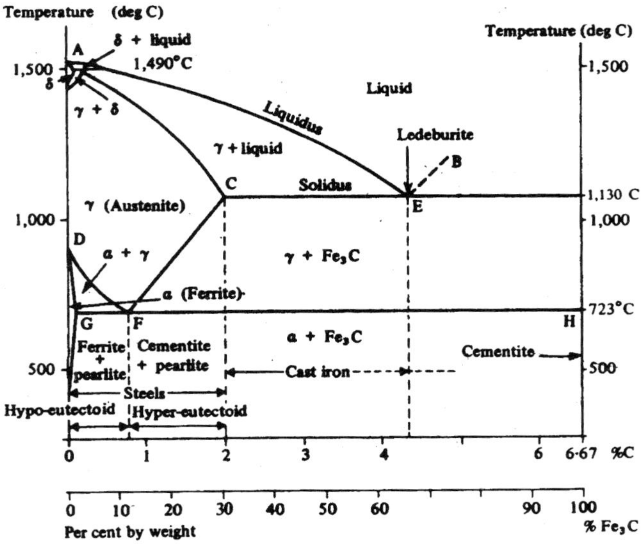

Fig. 1.9: The phase diagram of Fe and C in the Fe-rich region.

magnetic) transforms upon heating to *austenite* (or γ-Fe with a FCC structure, non-magnetic) at 912 °C, which on its turn transforms to δ-ferrite (again a BCC structure) at 1394 °C. The most important alloying element for Fe is C, which is dissolved interstitially in the Fe-matrix. The corresponding phase diagram is shown in Fig. 1.9. The C atoms dissolve only to a limited extent in α-Fe (maximum solubility 0.022 wt% at 727 °C) due to the small size of the interstitial holes of the BCC lattice. In FCC γ-Fe the interstitial holes are much larger and consequently much more C can dissolve (maximum solubility 2.14 wt% at 1147 °C). The phase diagram Fe-C contains an eutectic at 0.76 wt% C between α-Fe and Fe_3C (*cementite*) and one at 4.3 wt% C between γ-Fe and graphite at 1147 °C.

For alloys containing < 0.008 wt% C we denote the metal by *iron*. Its structure is ferritic. In the range between 0.008 and 2.14 wt% C, we denote the alloy as *steel* but typically less than about 1 wt% C is used. The structure is typically a mixture of ferrite and cementite.

If, apart from C, only a small amount of Mn is present, the steel is known as a *plane carbon steel*. At low carbon content we have *low carbon steel* (~ 0.25 wt% C), which is ductile, tough and easily machinable. Typical yield strength values[d] are about 275 MPa while the fracture strength ranges from about 450 to 700 MPa. It is generally used for structural elements with low relative cost. If elements such as Cu, V, Ni and Mo are added to low carbon steel, one speaks of *high strength low alloy* (HSLA) steel. The fracture strength is typically > 600 MPa.

Increasing the C-content results in *medium carbon steel* (typical range 0.25 to 0.6 wt% C; typical alloying elements Ni, Cr, Mo). They are harder and stronger but have lower ductility than the low C steels and are used e.g. for machine parts and railway wheels. By further increasing the C-content results in *high carbon steel* (range 0.6 to 1.4 wt% C; typical alloying elements W, Mo, Cr, V). They are the least ductile of the steels but the hardest and strongest. Applications are e.g. tools and dies.

Apart from the ferritic structure and austenitic structure, several other structures can be present. Slow cooling the 0.76 wt% eutectic composition results in a full *pearlite* structure, consisting of alternating layers of α-Fe and Fe_3C, owing its name to the pearl-like appearance. Moderate cooling results in full *bainite*, again α-Fe and Fe_3C but with the Fe_3C in a needle or plate-like shape. This structure changes after

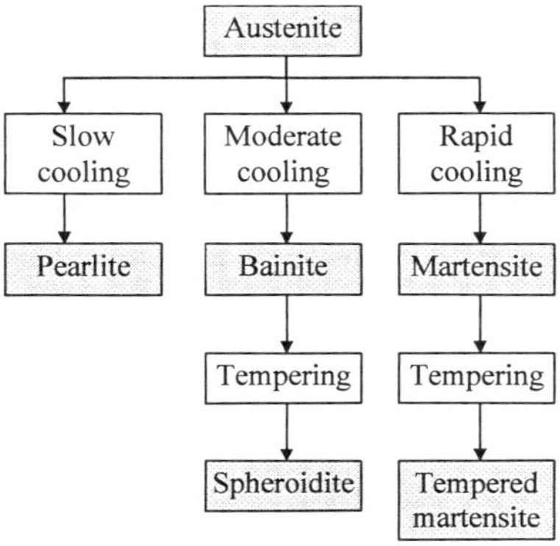

Fig. 1.10: Transformations in steel.

[d] An overview of the various mechanical properties is given in Chapter 2.

tempering for 18 to 24 h at about 700 °C to *spheroidite*, named after the spheroical shape of Fe_3C. Finally rapid cooling of the eutectic composition results in *martensite*. In this case the carbon atoms cannot precipitate and remain dissolved in a tetragonally distorted BCC structure, resulting in a high internal stress and blocking of the slip planes and thus in a high hardness. Moderate tempering (between 250 and 650 °C) results in *tempered martensite*, consisting of α-Fe and Fe_3C. For a C-content somewhat different from 0.76 wt%, the structure still contains pearlite. However, for a C-content < 0.76 wt% islands of Fe_3C are present while for a C-content > 0.76 wt% the structure contains islands of Fe. In Fig. 1.10 a scheme of these transformations is shown.

Alloys with 2.14 wt% up to 6.70 wt% C (corresponding to Fe_3C) are referred to as *cast irons*. Typically they contain 3.0 to 4.5 wt% C and since the melting point by this amount of carbon is considerably decreased (to 1100-1300 °C), these alloys can be cast relatively easily, and hence their name. The large content of C results in graphite inclusions in the alloy. If Fe also contains 1.0 to 3.0 wt% Si, which is often the case, the graphite precipitates as flakes. The Fe structure is ferritic or pearlitic. The material, known as *gray cast iron*, is weak and brittle, due to the shape of the (flake-like) precipitates.

Adding a small amount of Mg or Ce results in sphere-like graphite inclusions and the material is called *nodular cast iron*. It is typically applied in machine parts. If, however, less than 1 wt% Si is present, the C precipitates almost fully as Fe_3C and results in *white iron*, which is extremely hard, difficult to machine and applied e.g. in rollers. Long tempering in neutral atmosphere at 800 to 900 °C results in much more ductile material, called *malleable iron*.

Finally as an example of high alloy steels, we mention the *stainless steels*, typically containing > 11 wt% Cr. They are highly corrosion resistant but relatively soft. They may be ferritic, austenitic or martensitic with corresponding mechanical properties. Important applications not only can be found in food handling, storage and processing, e.g. utensils, containers, kitchen sinks, dough mixers and sterilisation equipment, but also in forging tools.

The main drawbacks of ferrous alloys are their relative high density (about 7.8 g/cm^3), relatively low electrical conductivity and susceptibility to corrosion. Therefore other alloys are in use. We mention briefly the Cu, Al, Mg and Ti alloys.

Most of the Cu alloys can only be strengthened by cold working. Cu-Be alloys form an exception: they can be hardened by precipitation hardening. An important category of alloys forms the Cu-Zn alloys with FCC structure or *brasses*, with < 35 wt% Zn. These alloys are soft and ductile and can easily be cold worked but all have a high density, about 8.9 g/cm^3. Applications are e.g. coins and small parts. Another category is the Cu-Sn alloys or *bronzes*. They are somewhat stronger and are used in bushings and bearings. The yield strength ranges from about 100 to 500 MPa and the fracture strength from about 200 to 600 MPa.

Al and its alloys all have a low density, about 2.7 g/cm^3. The structure is FCC and the materials are generally ductile to relatively low temperature. A major limitation is the low melting point, about 660 °C. Applications are e.g. beverage cans and engine parts. The yield strength ranges typically from 50 to 400 MPa, depending on the type and amount of alloying elements. The fracture strength shows a wide range, say from 100 to 600 MPa, again depending on the alloying conditions.

Mg and its alloys have an extremely low density, about 1.7 g/cm^3. The structure is HCP and results in soft and corrosion sensitive metals. Since their yield strength is close to their fracture strength, these alloys are difficult to deform.

Ti and its alloys have a relatively low density, about 4.5 g/cm^3, but a high melting point of about 1670 °C. The alloys are typically quite strong, up to 1400 MPa, yet ductile. Applications are e.g. again machine parts and also implants.

If sufficiently rapidly cooled from the melt (splat cooling), some metals can solidify in the amorphous state. Typical cooling rates are e.g. for Al alloys 10^6 K/s. These amorphous metals or *metal glasses* can have quite different properties from their crystalline counterparts. In particular, ductility and wear resistance are influenced. It is also possible to produce compositions that cannot be produced in crystalline alloys, sometimes having extraordinary properties. Some alloys are e.g. applied as soft magnets. However, upon thermal processing these alloys usually become microcrystalline again.

The fact that a wide range of microstructures can be obtained and that the final microstructure is largely independent of the original microstructure, is one of the most useful features of metals.

1.3 Polymers

Polymers consist of long molecular chains of covalently bonded atoms. Typically the molecule is constructed from a set of repeating units, the *monomers*. To a good approximation the energy of a single molecule can be estimated by adding bond energies (Table 1.1). In solid polymers bonding occurs via secondary interactions, such as van der Waals interactions and hydrogen bonds, and via cross-linking between the chains[e]. The *van der Waals interaction* is the attraction from charge distribution fluctuations in different molecules mutually influencing each other. The energy of this interaction ranges from 10 to 40 kJ/mol for polymers[f]. In a number of cases also hydrogen bonding is present with similar binding energy. *Hydrogen bonds* arise from the bonding of a hydrogen atom to two other atoms, either symmetrically A–H···A or asymmetrically A–H···B. These bonding types result in rather soft materials with low melting temperature. By joining the chains at points along their length with a chemical bond a *cross-linked structure* (Fig. 1.11) arises, leading to somewhat harder materials showing no melting. Typically the electrical as well as thermal conductivity of polymers is low.

Table 1.1: Bond energy U_{bon} and bond length d for various bonds.

Bond	U_{bon} (eV)	d (Å)	Bond	U_{bon} (eV)	d (Å)
C–H	4.3	1.08	C–Cl	2.8	1.76
C–C	3.6	1.54	C–Si	3.1	1.93
C=C	6.3	1.35	Si–H	3.0	1.45
C≡C	8.7	1.21	Si–Si	1.8	2.34
C–O	3.6	1.43	Si–F	5.6	1.81
C=O	7.6	1.22	Si–Cl	3.7	2.16
C–F	5.0	1.36	Si–O	3.8	1.83

Guy, A.G. (1976), *Essentials of materials science*, McGraw-Hill, New York. 1 eV = 96.48 kJ/mol.

[e] Using this description an oxide glass can be considered as an inorganic polymer.
[f] For small molecules such as N_2, CH_4 and CCl_4 the range is often quoted as 1 to 10 kJ/mol.

Generally two classes of polymers can be distinguished: *addition* (or *chain-grown*) *polymers* and *condensation* (or *step-grown*) *polymers*. Members of the first class are made by initialising a molecule by a catalyst to provide it with an activated end site via opening of a double bond and then growing the molecule to a chain by addition of a monomer until growth is terminated, either by exhaustion of the monomer supply or via a side reaction. At any time there are essentially only monomers and growing polymer chains present. The number of the latter is always low. Possibly the simplest example is polyethylene, which consists of long chains of a $-[CH_2-CH_2]-$ repeating unit. The monomer is ethylene, $CH_2{=}CH_2$. If the monomer is modified to $CH_2{=}CHX$, where X represents a certain chemical group, the polymers are called vinyl polymers. If X is a methyl, phenyl or chloride group the resulting polymer is indicated by polypropylene, polystyrene or polyvinylchloride, respectively. Linear members of the second class are made by reacting bifunctional molecules with the elimination of a low molar mass condensation product, e.g. water. At any moment the mixture contains growing chains and water. The number of reactive groups decreases with increasing chain length. The reaction between a suitable organic dicarboxylic acid and a diol yields a *polyester*, e.g. polyethylene terephthalate (Dacron), made from ethylene glycol and terephthalic acid. Similarly, a *polyamide* can be the condensation product of a dicarboxylic acid and a diamine, e.g. nylon 66, made from adipic acid and hexamethylene diamine.

The chemical structure of the chains is complicated somewhat by isomerism. A simple example of chemical isomerism is provided by the vinyl polymers for which one may have *head-to-head* ($-CH_2-CHX-CHX-CH_2-$) or *head-to-tail* ($-CH_2-CHX-CH_2-CHX-$) addition. A somewhat more complex case involves steric isomerism. Consider again the case of vinyl polymers in which a side group is added to every alternate carbon atom. If the groups are all added in an identical way, we obtain an *isotactic* polymer (Fig. 1.11). If on the other hand there is an inversion for each monomer unit, we obtain a *syndiotactic* polymer. Finally, an irregular addition sequence leads to an *atactic* polymer.

Each sample of polymer will consist of molecular chains of varying length and consequently of varying molecular weight. The molecular weight distribution is important for many properties. One can distinguish between the number average M_n and weight average M_w, defined by

$$M_n = \frac{\sum_i N_i M_i}{\sum_i N_i} \quad \text{and} \quad M_w = \frac{\sum_i (N_i M_i) M_i}{\sum_i N_i M_i} \tag{1.1}$$

respectively, where N_i is the number of molecules with molecular weight M_i and the summation is over all molecular weights. The weight average is always larger than the number average and in fact $(M_w/M_n) - 1$ represents the relative variance of the number distribution and therefore the width of that distribution.

Like in inorganics and metals, mixtures of various kinds are possible. A *blend* is a mixture of two or more polymers. In a *graft* a chain of a second polymer is attached to the base polymer. If in the main chain a chemical combination exists between two monomers [A] and [B] the material is a *copolymer*. In the latter case we distinguish between a *block* copolymer, where the monomer A is followed first by a sequence of other monomers A like AAA and subsequently by a series of B monomers, and *random* copolymers, where there is no long sequence of A and B monomers.

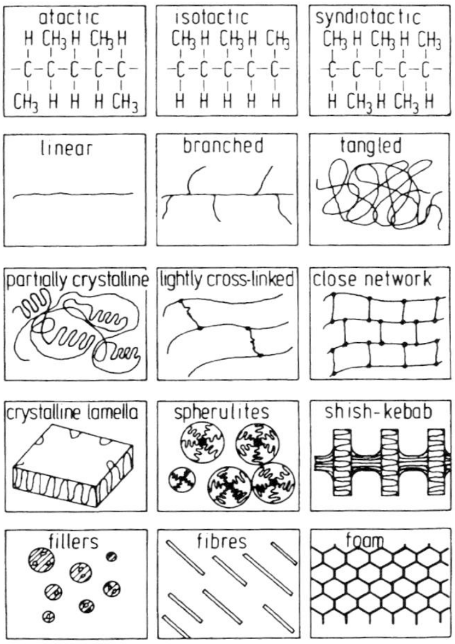

Fig. 1.11: Microstructural features of polymers.

The structure of polymers is described by the geometrical structure of the chain itself and of the arrangement of the chains with respect to each other. The geometrical structure of the chains themselves is largely determined by rotational isomerism. Although single bonds rotate relatively easily along the bond axis, this rotation is somewhat hindered. For a unit like –CHY–CHX– the preferred orientations (orientations with the lowest energy) are *gauche*, i.e. with the groups X and Y rotated through 60° along the bond, and *trans*, i.e. with the groups X and Y rotated through 180° along the bond. For a unit like –CHY=CHX– the preferred orientations are *cis*, with the groups X and Y on the same side of the double bond, and *trans*, with the groups X and Y on different sides of the double bond. In this case the rotational barrier is considerably higher.

Considering the arrangement of the chains with respect to each other, there are four relevant aspects: *entanglement*, *cross-linking*, *preferential orientation* and *crystallinity*.

Considering the first aspect, we note that the individual chains, having a relatively large internal flexibility, form *coils*, the size of which depends on temperature and on whether the chain is in a melt or in solution. In the latter case the size also depends on the solvent and the temperature. Like in inorganics and metals, polymers in the molten state are amorphous and the individual chains are entangled, i.e. the chains get mixed up and are difficult to unravel since at various positions a kind of knots, *entanglements*, are formed. A crude analogy is that of a bowl of wriggling spaghetti with length to diameter ratios of 10^4 or more. The number of entanglements per molecule increases with increasing molecular weight.

As indicated before, in many cases the individual chains are chemically bonded to each other at points along their length to make a *cross-linked structure* (Fig. 1.11). Heavily cross-linked polymers (*thermosets*) are relatively difficult to deform, even at

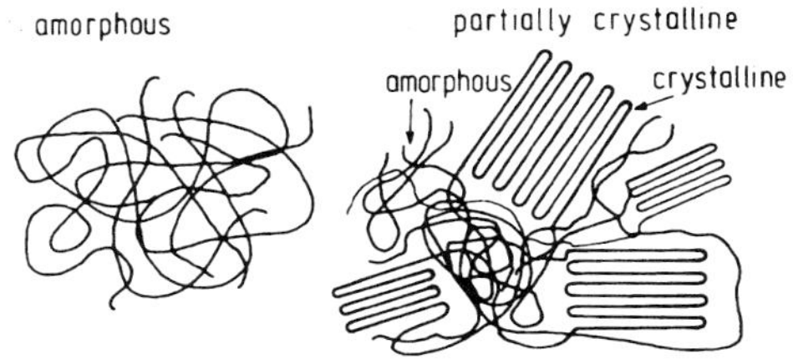

Fig. 1.12: Amorphous structure, orientation and crystallinity in polymers.

elevated temperature. A less rigorous way to connect molecules to each other is via *chain branching* where a secondary chain initiates from a point on the main chain. This leads to a more entangled structure as compared with linear polymers and thus these materials are more difficult to deform. Lightly branched and linear polymers (*thermoplastics*) are relatively easily deformed at elevated temperature since thermal motion in combination with mechanical load can change the entanglement structure relatively easily.

Many polymers, in particular atactic polymers, random copolymers and highly branched polymers, when cooled down from the molten state remain in the disordered or *amorphous* state (Fig. 1.12). Obviously complete random organisation is impossible in view of the covalent bonds between the atoms in the chain. Generally this slight orientational preference is non-detectable by X-ray diffraction. If a polymer is stretched, the molecules may be preferentially aligned along the stretch direction, so that the structure shows some more orientation. This orientation is still not detectable by X-ray diffraction, but can be detected possibly by optical means. Such a structure is called *oriented amorphous*. Further stretching will lead to a strong *preferential orientation*, also detectable by X-ray diffraction.

Finally we note that polymers, in particular those with a more regular chain structure, when cooled down sufficiently slowly from the melt, also can crystallise. Similar to polycrystalline inorganics and metals, these crystallised polymers are macroscopically isotropic but microscopically non-homogeneous. Generally these polymers are not completely crystallised but only partially: alternating regions of order (crystallites) and disorder (amorphous regions) exist (Fig. 1.12). The crystallites generally have the shape of *lamellae*, several tens of micrometres of lateral dimensions and about 10 to 20 nm in thickness. The chains are folded and the large surfaces of the lamellae contain the folds. In the crystallites not only different chains align but also a single chain participates in several lamellae. The crossover of chains, chain ends and defects within the chain largely collect in the amorphous regions between the lamellae. The end-to-end distance of the individual molecules remains largely preserved in melt-grown lamellae, so that the entanglement density is also largely preserved. In solution-grown lamellae, on the other hand, the end-to-end distance has decreased considerably as compared to the value in solution leading to a much lower entanglement density. The lamellae orient typically more or less similarly forming *stacks* with amorphous material in between. These stacks in turn form superstructures of which the *spherulite* (Fig. 1.13), in which stacks emanate radially from a certain nucleus, is the most important. The collective of features detected by microscopic means is referred to as *morphology*, a characteristic comparable to microstructure (although the noun morphology originally was a synonym for shape).

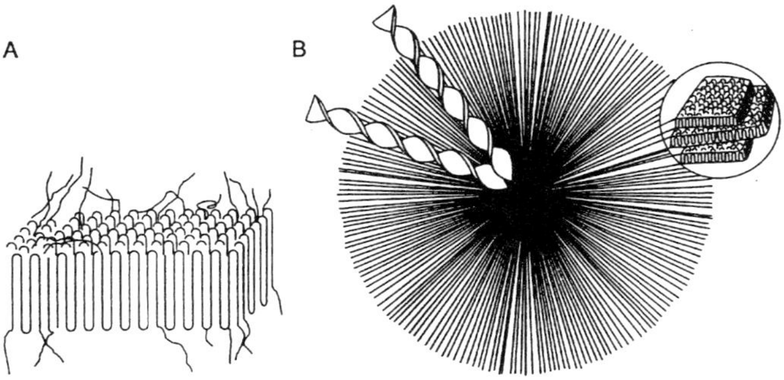

Fig. 1.13: The structure of a lamella and a spherulite.

Depending on the degree of crystallinity polymers may or may not show a clear melting point. Therefore depending on the temperature, a polymer may exhibit elastic/brittle behaviour, visco-elastic behaviour or viscous behaviour, similar to inorganic glasses. Amorphous polymers do show a glass transition temperature T_g rather than a melting point. The slower the cooling rate, the lower T_g is obtained. Apart from cooling rate, T_g also depends on chemical structure, molecular weight, branching and cross-linking. The more flexible the chains, the lower T_g and bulky and/or polar side groups tend to increase T_g. Since there is generally more free volume associated with the chain ends than with the chain middle, the glass transition temperature increases with increasing molecular mass M. The behaviour is approximately described by the Fox[g] equation

$$T_g = T_{g\infty} - \frac{k_1}{M} \tag{1.2}$$

where $T_{g\infty}$ indicates the T_g for very high M and k_1 is an empirical constant. Similarly, a small amount of branching reduces T_g while a large amount restricts mobility and therefore increases T_g. Cross-linking increases the density implying that the free volume decreases. Hence cross-linking increases T_g and the behaviour is approximately described by

$$T_g = T_{g\infty} - \frac{k_1}{M} + \frac{k_2}{M_{sub}} \tag{1.3}$$

where M_{sub} is the molecular mass of the sub-chains between the cross-links and k_2 is another empirical constant. A non-crystalline polymer above T_g behaves like a viscous liquid. Flow can take place since the chains can slide along each other. Cross-linking can stop this process and in this way one obtains *rubbers* (or *elastomers*). In this state the material can be extended many times its own length and will return upon unloading rapidly to its original shape. The basic material here is natural rubber, consisting of *cis*-isoprene, which crystallises but with difficulty. Cross-linking or as the jargon reads *vulcanisation*, was originally done by sulphur but nowadays usually with peroxides. The most well-known synthetic rubber is a random copolymer of styrene and butadiene (SBR), often reinforced with particles such as carbon black and used e.g. in vehicle tyres.

---

[g] Fox, T.G. (1956), Bull. Am. Phys. Soc. **1**, 123.

1.4 Composites

Ever since men used materials, composites (a combination of more than one kind of material) have been used. Typically a composite implies a material consisting of the matrix material with dispersed particles or fibres or a laminate (Fig. 1.14). Nowadays composites with a matrix of an inorganic material, a metal or a polymer are utilised, polymeric matrices being the most frequently used.

Polymer matrices are often made of the relatively inexpensive polyesters. For better matrix properties, but also higher cost, epoxy resins are used, while for high temperature application polyimide resins are applied. Polymers are often mixed with other materials in the shape of fibres or particles. For polymeric matrices in particular the stiffness and strength are improved by using inorganic fibres, for which often glass is used. This may lead to anisotropic behaviour if the fibres are aligned, resulting in excellent strength in one direction but low strength in the perpendicular direction. In addition to an increase in strength the resistance towards creep is improved. Other reinforcement materials used are carbon and aramide fibres. Both can be introduced as strands or as woven mats. To avoid anisotropy, relatively short fibres should be randomly distributed. However, improvement in properties is then more limited. Particles are also used to improve the polymer behaviour. In the latter case inorganic and rubber particles are used. While the former are used to increase stiffness and to reduce visco-elastic behaviour (and for economic reasons), the latter are used to improve the impact resistance.

Metal matrices are typically combined with inorganic particles. For metal matrices the first goal usually is to improve the creep behaviour meanwhile also improving stiffness and wear resistance. The most well-known example is so-called *hard metal*, a dispersion of 70 to 90 vol% WC (or other carbide) particles in a Co matrix. This material is quite hard (due to the WC particles) but still tougher than monolithic WC (due to the Co) and is applied in cutting tools. A more recent example is provided by Al_2O_3 or SiC particles in Al alloys as used in engines and ThO_2 particles in Ni alloys for high-temperature applications. It should be realised that precipitation strengthened metals are actually composites with particles of diameter 0.01 to 0.1 μm generated *in situ* by the thermal treatment applied.

Inorganic matrices are mixed with other inorganic or metal particles. Metal inclusions are used to improve ductility and thereby toughness. As an example we

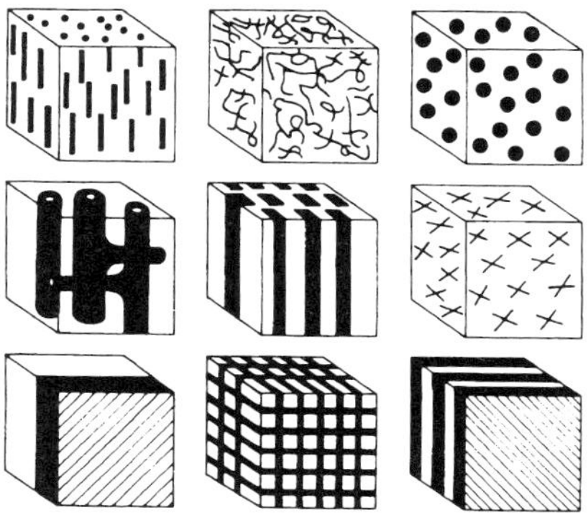

Fig. 1.14: A schematic representation of the microstructural options for composites.

mention silver-hydroxyapatite composites for biomedical applications. The addition of other inorganics can improve strength (and to some extent toughness) as well as wear resistance. A classic example is concrete, a composite of relatively large inorganic particles, e.g. gravel, in an inorganic matrix, typically cement and sand. The coarse particles act as a filler material to reduce the cost. The so-called *transformation toughened* materials provide an example of a more modern class of composite materials. Small particles of partially stabilised ZrO_2 are dispersed in a matrix of e.g. Al_2O_3 or ZrO_2 itself, leading to better toughness and/or strength. Fibres are used as well, the most frequently used material being SiC often in a polycrystalline SiC matrix. Another example is provided by C fibres in a polycrystalline C matrix, the so-called carbon-carbon composites.

Laminates are made from polymers, metals and inorganic materials. A well-known example of a laminate is plywood. Other examples are provided by polyester impregnated fabrics as used e.g. in small ships, hardened glass sandwiching PVB (polyvinylbutyral) as used for wind shield for cars and the Aral laminate, aluminium plates sandwiching layers reinforced with aramide fibres, for structural applications.

In all cases one attempts to improve the matrix material by taking advantage of the properties of the dispersed material, meanwhile avoiding the disadvantages of the dispersant. Obviously, these attempts are not always successful.

1.5 Length scales

The present knowledge of mechanical behaviour of solids can be most conveniently described using three levels.

Consider first the level of structures. A piece of material has a finite size and is used in a structure, which can be loaded by various forces. The typical size of a structure can vary in a wide range: from small components of a few mm or cm, such as a capacitor, a spring or a cup, to medium sized parts of a metre size, such as windows and doors or turbine rotors, to large structures of several tens of metres, such as buildings or ships. At this level the mechanical behaviour of the material, or equivalently its *constitutive behaviour*, is assumed to be known, either by measurement, as was done frequently in the past, or in combination with modelling, which is attempted more and more nowadays. The constitutive behaviour is assumed to be valid at all material points but the state of the material may differ from place to place. Denoting a quantity defined as a continuous function of position and time throughout a given region by a *field*, the state of a structure is described by a set of field quantities, which is subsequently used in continuum, engineering calculations. The associated length scale is referred to as the *macro(-scopic)* level.

On the other hand it has been known many years that matter is composed of atoms that are bonded together. As mentioned these bonds may be primarily covalent, primarily ionic or primarily metallic. In addition we know that the van der Waals bonds are responsible for the bonding between molecules in molecular crystals, such as benzene, and for the so-called secondary interactions in many solid polymers. As briefly indicated in the previous sections, materials also have different microstructures with elements such as grains, their boundaries and defects. The length scale involved is nanometre to micrometre: a typical bond length is 0.15 nm while the size of these grains can vary but is usually between 0.1 and 100 μm. It is the joint domain of solid state and materials science and the associated length scale is referred to in this book as the *micro(-scopic)* level. At this level a material is highly heterogeneous.

Mechanical integrity everywhere

Most of the properties of materials can be divided into the so-called functional properties, i.e. the electrical, dielectrical, magnetical and optical ones, and the structural properties, i.e. those that are associated with the mechanical integrity of materials and structures. The importance of functional properties is relatively easily recognised once mentioned. We just quote here as examples semi-conductor materials as used widely in electronics, ferroelectric materials as applied in capacitors in electronics, ceramic magnets as required in cathode ray tubes (televisions and computer monitors) and power transformers and high purity glass as used for optical fibres. The importance of structural properties is often thought of as being important only for engineering structures. However, all structures, either engineering or otherwise, either large or small, have to endure (thermo-)mechanical loads, either due to production processes or in use. For the semi-conductors mentioned these loads will be of pure mechanical nature during bending. For the capacitors thermal stresses due to soldering occur as well as bending stresses during the dewarping process of mounted components. For magnets thermal stresses occur during mounting as well as in use and a significant tensile load occurs during glass fibre cable positioning and also in use due to bends in the cable. Even this short list of examples will make the importance of the thermomechanical behaviour evident. One of the main problems is that one is often so focused on the functional behaviour that mechanical integrity is largely neglected.

The behaviour just described cannot be ascribed to arbitrary volumes of the material. The grains are bonded together in a polycrystalline material. An amorphous material may also contain inclusions or pores. Therefore the mechanical behaviour of a small representative volume element of the material has to be considered. For a dense polycrystalline material it is a suitable average of the behaviour of the single grains taking into account the connectivity between the various grains and their different crystallographic orientations. This average behaviour is characteristic for the representative volume element and has a certain length scale associated with it. The reason is that on the one hand one wants the representative volume element as small as possible while on the other hand it must have a certain minimum size so that averaging yields reliable values. The length scale involved is typically ten times the grain size in the case of polycrystalline materials. This length scale is referred to in this book as the *meso(-scopic)* level and is the domain of materials science. The representative volume element is also denoted as meso-cell. The behaviour at the meso-level is determined by the micro-phenomena, typically active at a length scale one to three orders of magnitude smaller than the meso-scale. The meso-behaviour on its turn represents the constitutive behaviour of the materials, which is used as field quantities in structures, typically one to three orders of magnitude larger than the meso-scale. The concept is shown schematically in Fig. 1.15.

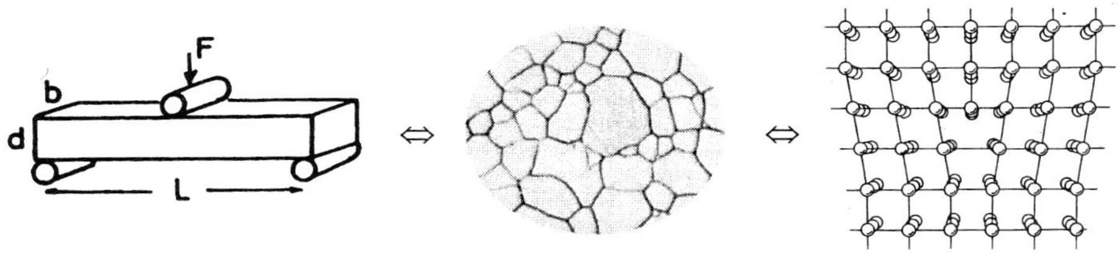

Fig. 1.15: An illustration of the 3M concept showing a simple macro-structure and the underlying meso- and micro-description. While the atoms within each grain are ordered in a definite crystallographic structure with the various associated defects, the grains themselves are joined together to yield the polycrystalline material, which is applied in a simple structure, here a bend bar.

Hence, it is clear that the meso-level determines the macro-level but is in itself determined by the micro-level. Two remarks are appropriate. First, it should be realised that the distinction as presented above is not always made so clearly. Second, the influence of the intermediate meso-level is different for different properties. The relation between processing, microstructure and properties is given schematically in Fig. 1.16[h]. The essence of this relation is the difference between the 'compound' on the one hand and the 'material' on the other hand. The properties of a compound are intrinsic and can hardly be influenced when its composition is fixed. They comprise properties such as crystal structure, thermal expansion coefficient, refractive index and magnetic crystalline anisotropy. The properties of a material are to a large extent extrinsic and can be drastically changed by altering the microstructure through different processing routes. Typical examples are the mechanical properties such as fracture strength and fracture toughness, the permittivity for ferroelectric materials and the permeability for ferromagnetic materials.

In conclusion, usually the macro properties are not only determined by the micro-level but also by the meso-level although in a number of cases the intermediate, meso-level is not separately developed but the macro behaviour is more directly connected with the micro-considerations. This is true for some properties, as mentioned above, and some materials, e.g. in amorphous polymers and glasses. Nevertheless we use the above description, occasionally referred to as the 3M aspects[i] and graphically illustrated in Fig. 1.15, as a useful template.

1.6 The nature of the continuum

At the structural or macro-level a *continuum description* is used where for every material point of the material a response –as predicted by the constitutive behaviour and the accompanying phenomenological parameters– is prescribed. The response may be different for different directions in which case we call the material *anisotropic*. The response may also be different at different locations and in this case we call the material *inhomogeneous*. Single crystals are anisotropic and homogeneous while composites are typically both anisotropic and inhomogeneous. As mentioned previously, the macroscopic properties are a result of the collective behaviour of microstructural and crystallographic features that make up a representative volume or meso-cell. By the way, this often results in an isotropic and homogeneous response. So, although in a continuum description one speaks of material points of infinitely small size, one actually means such a representative volume.

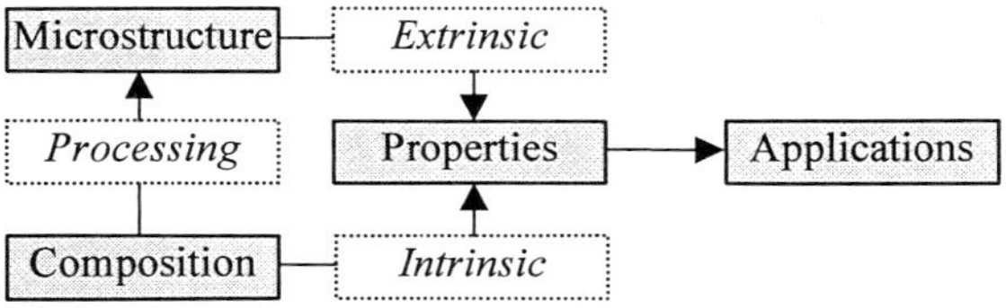

Fig. 1.16: From compound to material. While proper intrinsic compound properties are a prerequisite for obtaining good material properties, the realisation of the ultimate material properties is achieved in the microstructure through proper processing.

---

[h] de With, G. (1996), *Process control in the manufacture of ceramics*, page 27 in Materials Science and Technology, vol. 17A, R.J. Brook, ed., VCH, Weinheim.

[i] 3M is shorthand for micro, meso and macro.

Let us try to be a little more precise and take an arbitrary volume element $\Delta v = \int dV$ at position $\mathbf{x}$. Indicating the integral of the relevant parameter y over this volume by $\Delta y = \int y\, dV$, the space average of y is given by $\Delta y/\Delta v$. If we denote the volume of the meso-cell by Δv^* and the average over the meso-cell by y^*, in continuum theory it is thus assumed that the identification of the physical quantity y^* with the field quantity y is allowed, i.e.

$$y^* \equiv \lim_{\Delta v \to \Delta v^*} (\Delta y/\Delta v) \cong \lim_{\Delta v \to 0} (\Delta y/\Delta v) \equiv y(\mathbf{x},t) \tag{1.4}$$

A similar statement can be made for the time dependence. Indicating the integral of the relevant parameter z over an arbitrary time interval $\Delta t = \int dt$ with $\Delta z = \int z\, dt$, the time average of z is given by $\Delta z/\Delta t$. If the relaxation time of the parameter z is τ, the necessary minimum time for measuring z in the meso-cell is Δt^*, for which it holds that $\Delta t^* >> \tau$. Denoting the average over Δt^* by z^*, the transition

$$z^* \equiv \lim_{\Delta t \to \Delta t^*} (\Delta z/\Delta t) \cong \lim_{\Delta t \to 0} (\Delta z/\Delta t) \equiv z(\mathbf{x},t) \tag{1.5}$$

should be allowed. In later chapters we will discuss these aspects further.

Any field property in continuum theory should be interpreted accordingly. Hence to each continuum point the properties of a meso-cell, centred at the co-ordinates of the continuum point, must be given. These types of points are usually denoted as *particles* in continuum theory (although obviously this indication has quite different interpretations as well) and the continuum consists of a set of these particles. In order for continuum theory to be physically meaningful, the size of the meso-cell should be large in comparison with the underlying microstructural and atomic features of the material but small compared to the length of the macroscopic variation over the structure. Obviously the above implies that continuum theory is useful for the macroscopic description of a structure but has to be supplemented by models, based on (micro)-structural information, or measurements, which yield a prediction for the behaviour of the phenomenological parameters.

1.7 Approach

The approach in this book is based on four considerations:

- Since the thermomechanical behaviour of materials is determined by the behaviour of the atoms or molecules collectively, *thermodynamics*, dealing amongst others with the thermal behaviour of a large number of atoms or molecules, must form an important tool in the study of thermomechanical behaviour. Moreover, in modern literature frequent use is made of thermodynamic descriptions so that a treatment along these lines is also important for further study.
- The different types of bonds mentioned form the origin for distinguishing the three main categories of materials: *metals, polymers* and *inorganics*. It is highly desirable for future engineers and scientists to have a basic knowledge of all these three categories. Therefore attention is divided as evenly as possible between all three categories.
- As discussed previously, three relevant length scales exist in the description of the mechanical behaviour of materials so that the relevant aspects of the *micro-, meso- and macro-behaviour* will be discussed.

- Finally, the treatment of the mechanical behaviour in the literature is frequently either too elementary or too advanced for the readers. Here we chose to discuss the topics chosen at a *sound basic level*. Occasionally we feel that supplementary material is highly useful, either for a deeper understanding or for interest. Therefore we distinguished in the text between sections of what is called the 'basic' text, relevant to all students, and what is called 'advanced', or more properly 'intermediate' text, relevant to those with a more than average interest. The latter sections are indicated with an asterisk.

1.8 Topics

It is impossible to deal with all aspects of mechanical behaviour of materials and for that reason we chose to deal only with solid materials and volume properties. Hence we discuss:

- *Elasticity*. When a material is mechanically loaded it deforms. As long as during unloading the material immediately returns to its original shape, we speak of elastic behaviour. The deformation response can be linear in the load (linear elastic behaviour) or non-linear in the load (non-linear elastic behaviour). While for many inorganics, metals and polymers the elastic behaviour is predominantly linear, the elastic behaviour of rubbers is usually non-linear.
- *Plasticity*. When a certain threshold stress is exceeded during loading, the material not longer returns to its original shape during unloading. We speak of plastic deformation in this case. Here we assume that the permanent deformation is mainly dependent on the exceedance of this threshold level and is independent of the applied deformation rate.
- *Visco-elasticity*. As soon as the deformation rate becomes important, time enters explicitly the description. Upon unloading the material may or may not return to its initial state. In the former case one deals with visco-elasticity (in the strict sense) while the latter case is often denoted as elasto-visco-plasticity or visco-plasticity for short. We deal with them under the header visco-elasticity.
- *Fracture*. Apart from permanent deformation fracture can also occur. This usually occurs if the energy stored in the structure is sufficient to create new surfaces so that the total energy decreases.

Although time may be involved in all these cases, e.g. for most metals a higher strain rate results in a higher yield strength, it can be considered as an ordering parameter rather than a conventional variable (in the sense that this parameter can only increase contrary to a conventional variable which can take any value). In these cases the time effect can be approximately incorporated by using effective properties, e.g. as in rate-independent plasticity and in instantaneous fracture. When time dependence is explicitly present, as in visco-elasticity or delayed fracture, the history of the material plays a role. For each of the four topics mentioned above we will discuss the description at the macro-level first, followed by micro- and meso-considerations.

Finally, it should be said that both permanent deformation and fracture can be considered as failure (if the effect is not wanted, e.g. in a structure with specified size or load-bearing capacity) or as processing (if the effect is desired, e.g. in metal forming or grinding). We will refer to both aspects throughout this book.

1.9 Preview

In order to reach the goals described we have divided the text into six main parts. The first part entails the introduction and an *overview* of the various types of constitutive behaviour, elucidated by means of the uni-axial tensile test, in Chapter 2. This will make the need for a three-dimensional macroscopic description clear and for that description we need tools. Therefore in the second part the *basic tools* and laws[j] of continuum thermomechanics are introduced. First, before dealing with the actual subject, we realise that many practical problems are more or less complex, the result being that the simple mathematics used till now is not sufficient. Therefore we need some additional mathematical tools and a brief overview of useful *mathematics* (without proofs) is given in Chapter 3. Then we discuss the extension of our purely mechanical concepts. The methods for describing motion and deformation, i.e. the more elaborate description of strain, is usually addressed as *kinematics* and is treated in Chapter 4. Similarly, in Chapter 5 the more general discussion on the forces involved in the deformation, i.e. on the stress, is presented. We refer to this as *kinetics*, in order to emphasise that dynamical effects, where inertia becomes important, are only marginally addressed. After that, thermal aspects enter and our approach towards *thermodynamics* using a field description is outlined in Chapter 6. Also at the micro- and meso-level we need some tools. Therefore classical, quantum and statistical mechanics are summarised in Chapter 7, while the structure of and bonding in solids are reviewed in Chapter 8.

The third part describes *elastic deformation.* Here the constitutive behaviour as seen from a macroscopic point of view is treated in Chapter 9 and applications to some structures are discussed in Chapter 10. After that the microscopic (atomic) mechanisms are dealt with in Chapter 11. The introduction of the microstructure leads to the need for a representative volume element or meso-cell. Proper averaging over the meso-cell leads to a mesoscopic description, the resulting expressions of which provide a description of the phenomenological parameters to be used in the macroscopic approach. This is accomplished in Chapter 12.

The fourth part deals with *plastic deformation* in a similar way. First the macroscopic phenomenological treatment is presented in Chapter 13 while the application to structures and processes is given in Chapter 14. Thereafter the microscopic mechanisms are discussed (Chapter 15, part of 17), which again lead after proper averaging over the meso-cell to expressions for the macroscopic parameters (Chapter 16, part of 17). In the case of plasticity, however, the full picture is not so well developed as for elasticity.

Visco-elasticity is treated in the fifth part along the same line. We start with the continuum description (Chapter 18) and discuss subsequently the applications (Chapter 19) and structural aspects (Chapter 20).

Finally, the sixth part discusses *fracture phenomena*. The approach should be clear by now: first, the phenomenological macroscopic treatment (Chapter 21 basics, Chapter 22 application to structures and processes), followed by structural considerations which after introduction of microstructure and proper averaging over

---

[j] It may be useful to recall that generally a law indicates either a general relation valid independent of the type and the state of the material, e.g. Newton's laws in mechanics or the first and second law in thermodynamics, or a particular relation valid for a limited range of states for particular materials, e.g. Hooke's law for solids under limited deformation or Vegard's law for the molar volume of solutions. In the latter case the phrase *auxiliary relations*, or more commonly *constitutive relations*, is also used.

the representative volume element again connects with the phenomenological parameters (Chapter 23). Fatigue and damage are discussed in Chapter 24.

The whole approach is thus based on a proper attention for the 3M aspects. In Chapter 25 we present a rather personal perspective and outlook. Finally, in the appendices some data are collected which are useful in the quantitative application of the material presented.

2

Constitutive behaviour

In this chapter we survey the classes of constitutive behaviour of materials, approximately in the order of increasing complexity using the tensile test. We deal primarily with explicitly time independent behaviour and start with elastic behaviour. After that we discuss plastic behaviour, followed by fracture. A brief overview of explicitly time-dependent phenomena is presented, including viscous and visco-elastic behaviour. Thereafter a short survey of the order of magnitude of the relevant phenomenological parameters is presented. At the end the need for more complete description is addressed, indicating the use of more elaborate thermomechanical tools.

2.1 The tensile test

In order to illustrate the material behaviour we shall make use of the tensile test (Fig. 2.1). In this test a bar of gauge length l_0 and with a (often circular) cross-section A_0 is loaded by a force F. Normally the ends have an enlarged diameter for (better) gripping, which taper smoothly through shoulders to the central uniform gauge length. The test is actually performed by increasing the load F and simultaneously measuring the current length l of the gauge section. With increasing load the length l increases and the current cross-section, denoted by A, becomes smaller. The test may also be conducted in compression, although different shapes are often used in that case. Normally the response of such a specimen is dependent on both the geometry and the materials properties. To be able to discuss materials properties independent of geometry we prefer to represent the results of such a test by the use of stress and strain instead of force and elongation. Here there are several options, which are all used in practice. Whatever definition is used, in a tensile test the stress and strain are constant throughout the gauge length of the specimen, i.e. they are *uniform*. Moreover, the stress points in only one direction, i.e. the stress distribution is *uniaxial*.

Let us first consider two frequently used options for the stress. The *engineering* (or *nominal*) *stress* is defined by

► $s = F/A_0$

while the *natural* (or *true* or *logarithmic*) *stress* is defined by

► $\sigma = F/A$

For the strain comparable options exist. The increment in *engineering* (or *nominal*)

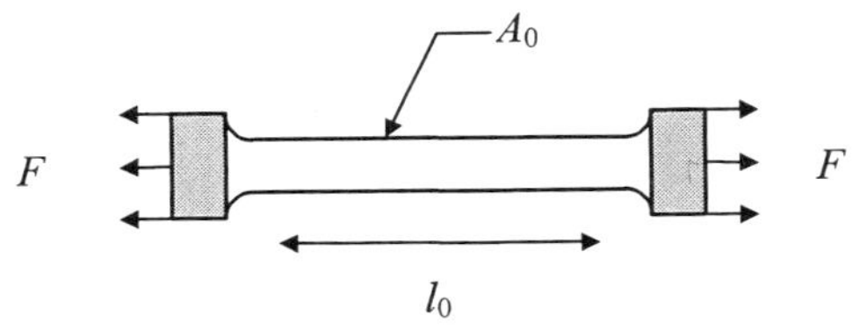

Fig. 2.1: The tensile test.

strain is defined by

► $\mathrm{d}e = \mathrm{d}l/l_0$

while the increment in *natural* (or *true* or *logarithmic*) *strain* is given by

► $\mathrm{d}\varepsilon = \mathrm{d}l/l.$

For the total engineering and natural strain we find, respectively,

► $$e = \int_{l_0}^{l} \mathrm{d}l/l_0 = \frac{l-l_0}{l_0} = \Delta l/l_0 \tag{2.1}$$

► $$\varepsilon = \int_{l_0}^{l} \mathrm{d}l/l = \ln\frac{l}{l_0} = \ln(1+e) \tag{2.2}$$

The total natural strain reduces to the total engineering strain for small values of strain[a]. The engineering strain can also be used to relate the engineering and natural stress and we write

► $$\sigma = \frac{F}{A} = \frac{F}{A_0}\frac{A_0}{A} \cong s\exp(\varepsilon) = s\exp[\ln(1+e)] = s(1+e) \tag{2.3}$$

where use has been made of the incompressibility relation $l/l_0 = A_0/A$, which is only approximately valid.

Problem 2.1

Determine the limiting value of the total engineering strain e below which it is indistinguishable from the total true strain ε, accepting differences up to 5%.

2.2 Elastic behaviour

We now consider the response of a tensile loaded specimen using the engineering quantities (Fig. 2.2). As long as the strain remains smaller than a few percent, the difference in stress and strain definitions used is relatively unimportant. We assume

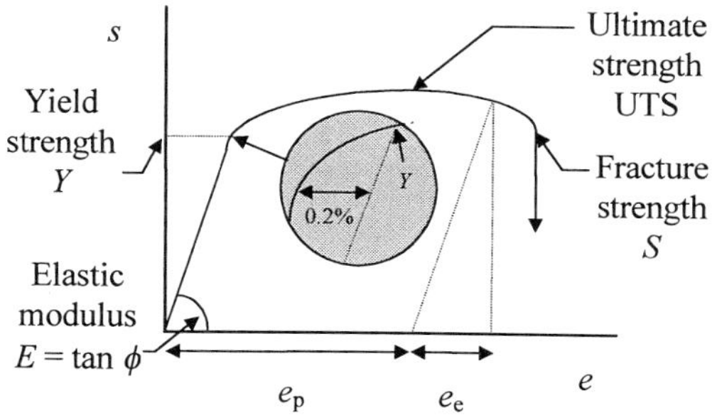

Fig. 2.2: The stress-strain relationship.

[a] There are intrinsic problems with the use of the total natural strain for other than homogeneous deformations. For some details, see Chapter 4.

that the response is *isotropic*, i.e. no preferential orientation exists in the material, and *homogeneous*, i.e. the properties are independent of position in the material. Moreover, for the time being, we neglect all rate effects. Initially, with increasing load, the response is proportional to the load and we obtain a linear relationship between the stress s and the strain e. Upon unloading the original cross-section and length are regained. In other words, there is no permanent deformation. Moreover, there is no *hysteresis* and the material is called (linear) *elastic*. At this stage the deformation is characterised by the *elastic strain*[b] e_e. The slope of the stress-strain curve $s(e)$ in this region is *Young's modulus* E and is given by

$$\blacktriangleright \quad E = \frac{\mathrm{d}s}{\mathrm{d}e} = \frac{s}{e} = \frac{(F/A_0)}{(\Delta l/l_0)} \tag{2.4}$$

At higher loads non-linear, but still elastic behaviour might be present. There are two ways of characterising this situation. One can use the derivative of s with respect to e at a certain load. This is called the *tangent modulus* $E_{\tan}$ and is given by

$$E_{\tan} = \frac{\mathrm{d}s(e)}{\mathrm{d}e} \tag{2.5}$$

Another possibility is to use the *secant modulus* $E_{\sec}$, given by the slope of the line connecting the origin with the stress at the strain considered, i.e.

$$E_{\sec} = \frac{s(e)}{e} \tag{2.6}$$

Obviously for a linear stress-strain relationship $E = E_{\tan} = E_{\sec}$. Since at low strain the difference between s and σ on the one hand and e and ε on the other hand is negligible, the above relations also hold for σ and ε.

For some materials, e.g. rubbers, the unloading curve does not coincide with the loading curve but the original situation is restored upon complete unloading after some time. This behaviour is called *anelasticity*. The energy dissipated is converted to heat in this case.

Thomas Young (1773-1829)
Born in Milverton, Somerset, he started with the study of medicine in 1792, first in London and Edinburgh and later in Göttingen where he received his doctorate in 1796. After his return to England he was admitted as a Fellow Commoner of Emmanuel College, Cambridge. In spite of his great talents, Young did not assert superiority. As early as 1793 one of his papers on the theory of sight was presented to the Royal Society. While at Cambridge in 1798 he became

[b] Although in this chapter the elastic (plastic) strain is indicated by e_e (e_p, see Section 2.3), in later chapters we will denote e_e (e_p) as $e^{(e)}$ ($e^{(p)}$) since subscripts are often used for components, e.g. $e_{ij}^{(e)}$.

interested in the theory of sound. In 1801 he made his famous discovery on the interference of light. In 1802 he was elected as a member of the Royal Society and in the same year installed as a professor of natural philosophy by the Royal Institution, but his lectures were a failure since they generally were too terse. He resigned his professorship in 1803 but continued to be interested in natural philosophy and published his lectures as *A course on natural philosophy and the mechanical arts* in 1807, in which his contributions to mechanics of materials are given. Some of the solutions to important problems given there were completely new in Young's time. However, the work did not gain much attention from engineers because his presentation was always brief and seldom clear. One of the most striking points was Young's estimate of the size of molecules as between two-thousand and ten-thousand millionth of an inch. It should be said that according to Truesdell the naming of the material elasticity constant after him was a serious historical error.

Apart from an elongation of the bar also a decrease in cross-section takes place upon loading. We denote the original radius by r_0 and the current radius by r. For a linear stress-strain relationship the decrease in radius $\Delta r = r - r_0$ is again proportional to the load. It is conventional to call the ratio ν of the relative decrease in radius over the relative increase in length as *Poisson's ratio*

► $$\nu = -\left(\frac{\Delta r}{r_0}\right) \Big/ \left(\frac{\Delta l}{l_0}\right) \tag{2.7}$$

Siméon-Denis Poisson (1781-1840)

Born in Pithiviers near Paris, France, he had in his early childhood no chance to learn more than to read and write due to the poverty of his family. In 1796 he was sent to his uncle in Fontainebleau where he appeared to be so good that in 1798 he was able to pass the entrance examinations of the École Polytechnique. After graduation in 1800 he remained at the school as instructor and had been in charge of the course in calculus in 1806. His original publications in mathematics made him to become a member of the French Academy in 1812. The theory of elasticity attracted Poisson's attention and his principal results were summarised in *Traité de mécanique*, published in 1833, which also contains the equations of motion. He showed that when a body is disturbed it results in dilatational and shear waves. He further made contributions to the theory of plates and seemed to be the first in using trigonometric series for the solution of bend bars. Poisson did not contribute such fundamental ideas as Navier or Cauchy but did solve many problems of practical importance for which he is still recognised.

Problem 2.2

A cylindrical bar with a diameter of 8 mm is loaded elastically by 15 700 N and this results in a diameter reduction of 0.005 mm. Determine Poisson's ratio ν if Young's modulus of the material E = 140 GPa.

Problem 2.3

Show that the value of Poisson's ratio ν of an incompressible material as obtained from a tensile test becomes $\nu = ½$.

Problem 2.4

Consider the deformation of a bar loaded in tension. Discuss briefly what is the effect of anisotropy and inhomogeneity.

2.3 Plastic behaviour

Above a certain stress, the *yield strength*[c] Y, the material behaviour changes qualitatively. The increase in stress with increasing strain generally decreases as compared with the linear region. Moreover, upon unloading the original situation is not regained: a certain permanent deformation, known as *plastic deformation*, remains. During plastic deformation the total volume for most materials is constant, i.e. $A_0l_0 = Al$. Notable exceptions are porous materials. The plastic deformation is characterised by the *plastic strain* $e_p = e - e_e = e - s/E$ (Fig. 2.2), where e represents the total strain. The stress-strain curve resulting from the initial loading into the plastic range is called the *flow curve*. Reloading leads to elastic behaviour until the previously obtained position at the stress-strain curve is reached, after which the plastic deformation increases again. This previously reached point is thus 'remembered' and acts as the 'new' yield strength.

Generally the unloading-loading cycle shows a small amount of hysteresis (Fig. 2.3) but this effect is often neglected. The exact location of the *proportional* or *elastic limit* is virtually impossible to determine. Therefore one usually uses the yield strength that is defined as the value for which a certain 'offset' strain results after unloading (Fig. 2.2). Typical values used are 0.1%, 0.2% and 0.5%. In this case one thus states e.g. the 0.2% *offset yield strength*. For strains larger than the strain associated with Y, the stress may either increase, typically in a non-linear way, or remain approximately constant.

If the yield strength is constant during the deformation process, the behaviour is called *perfectly plastic*. Upon loading and unloading the yield strength also may increase and that behaviour is described as *hardening*. In some cases the hardening is approximately linear if natural stress and strain are used. In that case the plastic behaviour can be characterised by a *hardening modulus* h as defined by the derivative

$$h = \mathrm{d}\sigma/\mathrm{d}\varepsilon \qquad \text{for} \qquad \sigma > Y$$

In other cases the complete flow curve can be approximately described by the *Ramberg-Osgood formula*[d] containing both the elastic strain ε_e and the plastic strain ε_p

$$\varepsilon = \varepsilon_e + \varepsilon_p = \frac{\sigma}{E} + \alpha \frac{Y}{E}\left(\frac{\sigma}{Y}\right)^m \tag{2.8}$$

---

[c] Many authors use yield stress. We will use the designation *stress* in conjunction with applied forces to a material (a field parameter) and denote the critical value of a stress for a certain property of a certain material by *strength* (a material parameter).

[d] Ramberg, W. and Osgood, W.R. (1943), N.A.C.A. TN 902.

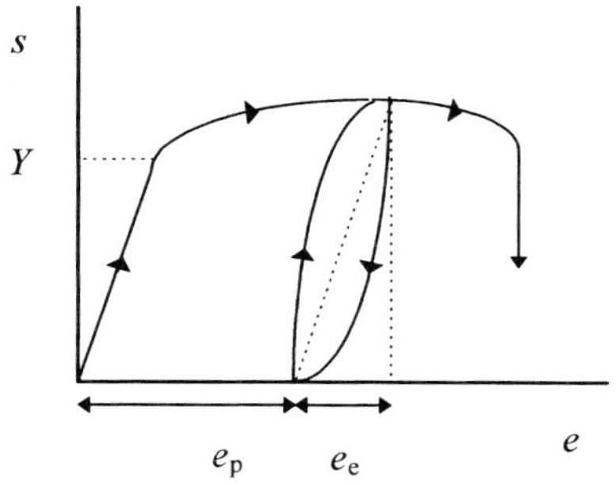

Fig. 2.3: Hysteresis loop during unloading and loading (width largely exaggerated).

where α and m are non-dimensional constants and Y is a reference stress. A value of $\alpha = 3/7$ is often used. If m is large, ε_p remains small until σ approaches Y and increases rapidly when σ exceeds Y so that Y can be considered as an approximate yield strength. In the limit $m\to\infty$ the plastic strain is zero when $\sigma < Y$, indeterminate when $\sigma = Y$ while $\sigma > Y$ cannot occur since it produces an infinite flow. This limiting behaviour is thus the perfectly plastic behaviour. If we assume the elastic part to be negligible, the behaviour is referred to as *rigid-perfectly plastic* or *ideally plastic* for short. The main advantage of the Ramberg-Osgood relation is its flexibility to describe the stress-strain curve (Fig. 2.4), but it cannot be explicitly solved for the stress σ as a function of the strain ε.

In case an explicit expression for σ as a function of ε is required, one often uses

► $$\sigma = Y' + K\varepsilon_p^{\,n} \tag{2.9}$$

originally proposed by Ludwik[e]. Here Y' represents the initial yield strength. The parameters K and n are considered to be material constants. The value for n, generally known as the *strain-hardening exponent*, ranges from about 0.1 to 0.5 for most metals (Table 2.1). The n-value is usually higher in the annealed state. For large plastic deformation the contribution of Y' is sometimes neglected resulting in

$$\sigma = K\varepsilon_p^{\,n} \tag{2.10}$$

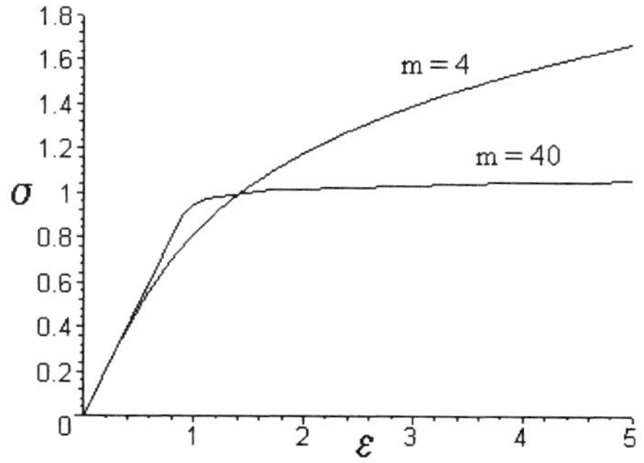

Fig. 2.4: Non-dimensional Ramberg-Osgood plot with $\varepsilon E/Y \to \varepsilon$ and $\sigma/Y \to \sigma$.

[e] Ludwik, P. (1909), *Elemente der technologischen Mechanik*, Julius Springer, Berlin.

Table 2.1: Values for Y', K and n for various alloys at room temperature.

Material	Condition	Y' (MPa)	K (MPa)	n
0.05% C steel	Annealed	210	530	0.26
Al	Annealed	40	180	0.20
Al 2024 alloy	Heat treated	310	690	0.16
Cu	Annealed	60	315	0.54
Brass 70Cu-30Zn	Annealed	80	895	0.49
0.6% C steel	Quenched, 540 °C*	–	1570	0.10
0.6% C steel	Quenched, 700 °C*	–	1230	0.19
Stainless steel 304	Annealed	600	1275	0.45
Alloy steel 4135	Rolled	650	1100	0.15

*: tempering temperature. -: fitted with power law behaviour. Data from Callister (1997).

and referred to as power law behaviour. Note that expression (2.10) has an infinite initial slope since usually $n < 1$. Therefore sometimes an elastic range is added with the initial yield strength Y',

$$\sigma = E\varepsilon \quad \text{for} \quad \varepsilon \le Y'/E \qquad \text{and} \qquad \sigma = Y'\left(\frac{E\varepsilon}{Y'}\right)^n \quad \text{for} \quad \varepsilon \ge Y'/E \tag{2.11}$$

resulting in a discontinuous σ-ε curve. Finally, at large plastic deformation, say $\varepsilon_p > 0.2$, the σ-ε relation has approximately a constant slope, i.e. $n = 1$, for several metals.

In a log-log plot of stress versus strain, power-law behaviour displays as a straight line with slope n. For those cases where this behaviour is not obeyed we still can define a strain hardening exponent by

► $$n(\varepsilon_p) = \frac{d\ln\sigma}{d\ln\varepsilon_p} \tag{2.12}$$

although n is no longer a constant but dependent on the strain ε_p.

Hardening often induces anisotropy in the material. One of the consequences is the *Bauschinger effect*: a previous plastic strain with a certain sign diminishes the material's resistance with respect to the next plastic strain with an opposite sign (Fig. 2.5).

Using the engineering stress and strain in the stress-strain plot, generally a decrease in stress is observed after a certain value (but before fracture, see Fig. 2.2). This value is called the *ultimate tensile strength*, in the literature sometimes indicated by UTS. From the various definitions of stress and strain one easily obtains, very nearly,

$$s = \sigma\exp(-\varepsilon) \qquad \text{and its differential} \qquad ds = (d\sigma - \sigma\, d\varepsilon)\exp(-\varepsilon)$$

The ultimate tensile strength and thus the maximum load occur when

$$ds = 0 \quad \text{or} \quad d\sigma/d\varepsilon = \sigma$$

If the material obeys power law behaviour one can show that this happens when $\varepsilon = n$. Interpreting n according to Eq. (2.12) this relation holds for any constitutive behaviour. Moreover, for metals data in compression and tension differ considerably when plotted in engineering terms. Frequently a single curve results when the data are plotted in terms of the natural stress and strain (Fig. 2.6).

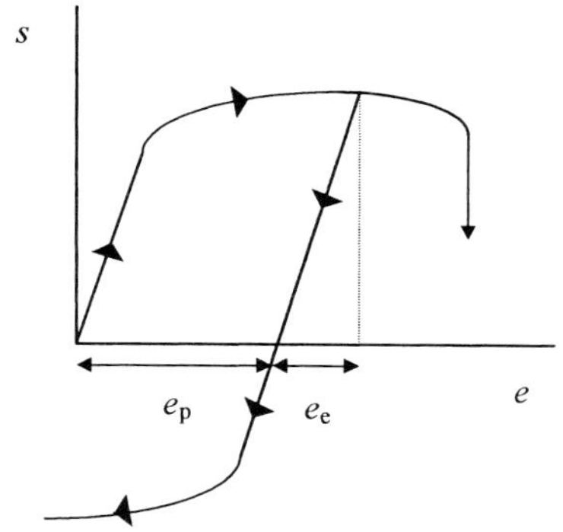

Fig. 2.5: The Bauschinger effect.

Up to the stress associated with the ultimate tensile strength, the deformation is homogeneous. At higher stress *necking* occurs. With necking a local decrease in the diameter of the test specimen is indicated. For (perfectly) plastic materials necking is easily imaginable. Small random deviations in diameter are always present and at the smallest diameter the yield strength is reached first. The material cannot withstand a higher stress and at constant load unstable necking occurs. In such a neck the stress distribution is not uniaxial but multiaxial. Contrary to graphs using the engineering stress and strain, the stress-strain curve continues to rise after necking starts if natural stress and strain are used, at least if the local diameter at the neck is used.

Bridgman[f] has given an approximate expression[g] for the mean axial stress in the z-direction (Fig. 2.7)

$$\bar{\sigma}_{zz} = \int_0^a 2\pi\sigma_{zz}(r)r\,\mathrm{d}r \Big/ \pi a^2 \tag{2.13}$$

in the neck for a cylindrically shaped test specimen. The analysis is based upon the following assumptions:

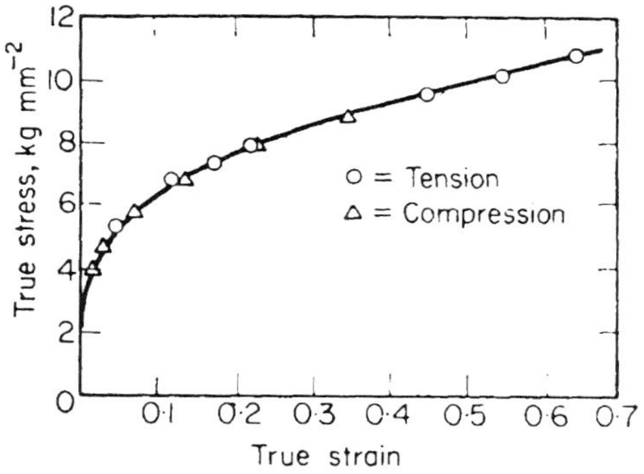

Fig. 2.6: True stress-strain curve for Al, determined from tension and compression tests. Data from Cottrell (1964).

[f] Percy Williams Bridgman (1882-1961). American physicist who received the Nobel Prize for physics in 1946 for his work on high pressure physics.

[g] Bridgman, P.W. (1944), Trans. Am. Soc. Met. **32**, 553.

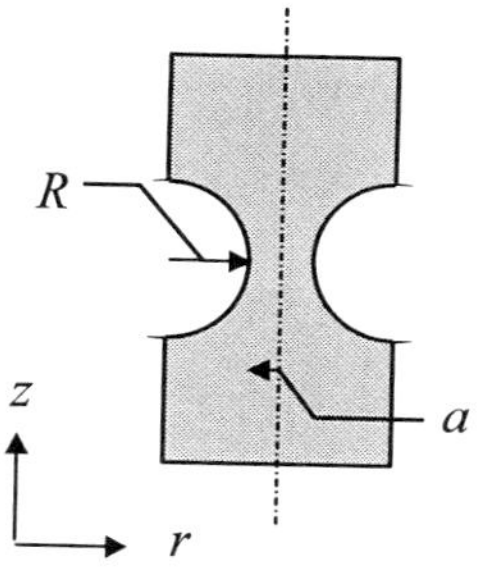

Fig. 2.7: Geometry of the necking region.

- the contour of the neck is approximated by an arc of a circle,
- the cross-section remains circular,
- von Mises criterion for yielding (Chapter 13) applies and
- the strains are constant over the cross-section of the neck.

The final expression reads

$$\frac{\overline{\sigma}_{zz}}{\sigma_{nom}} = \left\{ \left(1+\frac{2R}{a}\right)\left[\ln\left(1+\frac{a}{2R}\right)\right]\right\}^{-1} \tag{2.14}$$

where R represents the radius of the neck contour along the cylinder at the weakened section, a is the radius of the minimum cross-section (Fig. 2.7) and σ_{nom} is the nominal stress. However, the experiments with necking have to be considered with caution since the initial assumptions are not always fulfilled. Moreover, the onset of necking signals the onset of instabilities during the experiment that cannot be controlled if the experiment is performed in load control, that is if the total load F is prescribed as the global parameter[h].

For a stable neck in tension hardening must occur. At the point of instability an increment in strain gives no increment in load,

$$\mathrm{d}F = \mathrm{d}(\sigma A) = A\,\mathrm{d}\sigma + \sigma\,\mathrm{d}A = 0$$

Because plastic deformation occurs essentially at constant volume we also have

$$\mathrm{d}V = \mathrm{d}(Al) = A\,\mathrm{d}l + l\,\mathrm{d}A = 0$$

Combining we obtain

$$\mathrm{d}\sigma/\sigma = -\,\mathrm{d}A/A = \mathrm{d}l/l = \mathrm{d}\varepsilon = \mathrm{d}e\,/(1+e)$$

Instability is thus reached when

► $$\mathrm{d}\sigma/\mathrm{d}\varepsilon = \sigma \quad \text{or} \quad \mathrm{d}\sigma/\mathrm{d}e = \sigma/(1+e) \quad \text{or} \quad \mathrm{d}\ln\sigma/\mathrm{d}\ln\varepsilon = \varepsilon \tag{2.15}$$

If we plot the natural stress σ versus the nominal strain e we can illustrate the instability using the second of these equations. In this plot a line originating at $e = -1$ is drawn tangent to the stress-strain curve, a graphical construction known as *Considère's construction*. The point P where the tangent touches the stress-strain

[h] In compression also a non-homogeneous stress distribution occurs due to friction between a specimen and anvil. For a cylindrical specimen of radius a and height h, the average pressure $\overline{p}$ during yielding is given by $\overline{p} \cong (1+2\mu a/3h)Y$, where μ denotes the friction coefficient between the specimen and anvil.

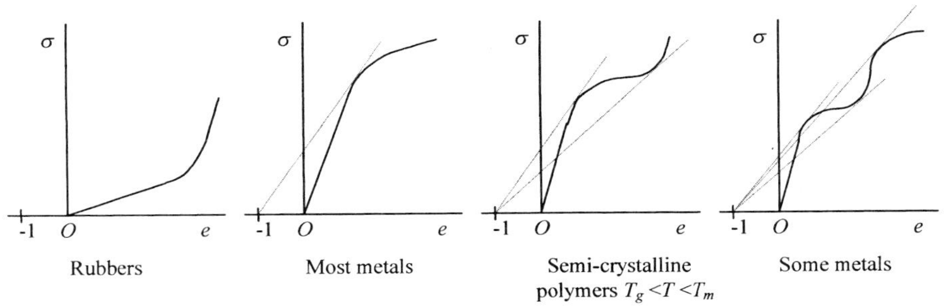

Fig. 2.8: Various stress-strain curves and Considère's construction.

curve represents the instability and the ultimate strength is $\sigma/(1+e)$. In Fig. 2.8 typical experimental stress-strain diagrams are shown with the corresponding Considère construction indicated. Four possibilities arise:

- During the test there is no point for which $d\sigma/de = \sigma/(1+e)$ or $d\sigma/de > \sigma/(1+e)$; a line drawn from $e = -1$ is nowhere tangent to the stress-strain curve. In this case the deformation is homogeneous up to fracture. Most rubbers behave like this.
- During the test there is one point for which $d\sigma/de = \sigma/(1+e)$. Deformation is homogeneous up to a certain strain after which unstable necking occurs, leading to fracture. Most metals behave like this.
- There are two points for which $d\sigma/de = \sigma/(1+e)$. Deformation is stable up to a certain strain after which a stable neck spreads through the specimen followed by further homogeneous deformation until fracture. For metals the deformation is generally limited to a few percent but for crystalline polymers it may be very large and is e.g. important for the production of fibres via a stretching process. It is also possible that a drop in stress occurs, the *yield drop*, and that the material then deforms essentially at constant stress, the *yield plateau* (Fig. 2.9). In that case the initial yield stress is known as the *upper yield point* while the lower value indicating the plateau is known as the *lower yield point*. The stress at this plateau fluctuates slightly because the deformation in this region occurs in a limited number of discrete narrow zones, known as *Lüders bands,* at approximately 45° in the tensile direction. The associated strain is known as *Lüders strain.* Some authors call this particular behaviour the yield point phenomenon or discontinuous yielding and reserve the word yielding for this process. The process occurs

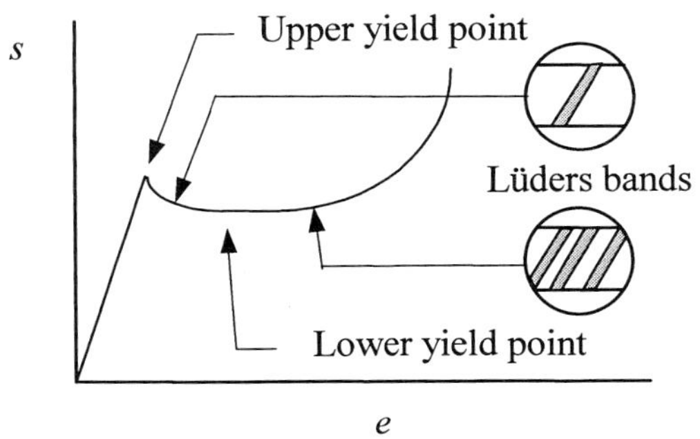

Fig. 2.9: Yield plateau.

primarily in low impurity BCC metals, in particular mild (low carbon) steel, and in some polymers, e.g. polypropylene.

- There are three points for which $d\sigma/de = \sigma/(1+e)$. Similar to the previous case a stable neck is formed followed by homogeneous deformation but this is followed by unstable neck growth, again leading to fracture.

Considère's construction can be used to determine at which strain (stable or unstable) necking occurs. However, it does not explain why instability occurs and for this we need micro and meso considerations.

Problem 2.5

Show that for a power law material the UTS is reached at $\varepsilon_p = n$.

2.4 Fracture behaviour

Deforming still further leads to fracture and the corresponding stress is usually called the *fracture strength S*. The various types of materials can fracture in rather different ways. At low and intermediate temperature most inorganic materials loaded in tension fracture before plastic deformation occurs. This behaviour is called *brittle* (Fig. 2.10). Many metals, on the other hand, do show necking before fracturing. Usually this is accompanied with void nucleation. The final part is often the formation of a lip by shearing. The morphology is called *cone and cup* while the behaviour is addressed as *semi-brittle*. Metals with high ductility, e.g. Pb and Au, and polymers and inorganic glasses at elevated temperature can show failure all the way by necking with virtually 100% area reduction. In this case failure is thus not due to a fracture process at all. It is generally known as *ductile failure*.

Generally fracture behaviour is dependent on the size of the structure. If sufficiently large, every structure behaves macroscopically brittle. To elucidate a bit, we need two length scales. The first is the characteristic size of the process zone, say p, in which the fracture processes ahead of the crack tip occur. The second is the characteristic size of the structure itself, say l. If $l/p >> 1$, fracture is macroscopically brittle.

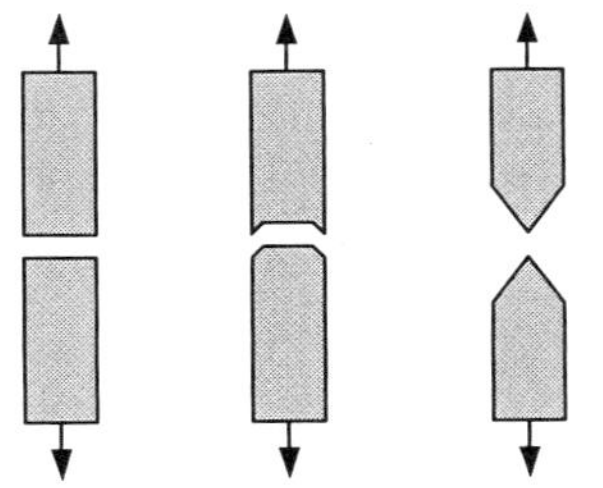

Brittle Semi-brittle Ductile

Fig. 2.10: Morphology of fracture.

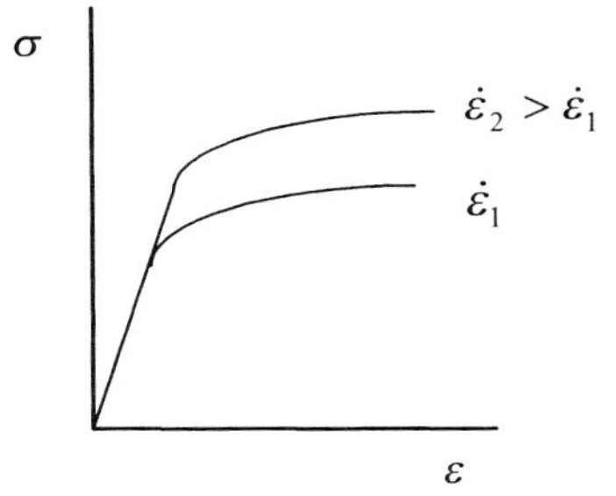

Fig. 2.11: Stress strain curves at deformation rates $\dot{\varepsilon}_1$ and $\dot{\varepsilon}_2 > \dot{\varepsilon}_1$.

2.5 Temperature and rate effects

Many metals show a more or less limited rate-dependent plastic behaviour while polymers generally show more significant rate effects. However, the yield strength and hardening behaviour for a particular metal can depend significantly on the strain rate (see Fig. 2.11). In the same way as the stress-strain dependency often can be described by the power law $\sigma = K\varepsilon_p{}^n$, the stress-strain rate dependency can be described by

$$\sigma = K'\dot{\varepsilon}_p^m \tag{2.16}$$

Here $\dot{\varepsilon}_p = d\varepsilon_p/dt$ denotes the *strain rate* while m is the *strain rate sensitivity exponent*. If this behaviour is followed, a log-log plot of the stress versus the strain rate will provide a straight line with slope m. Again we can generalise the definition of the strain rate sensitivity exponent by using

▶ $$m(\dot{\varepsilon}_p) = \frac{d\ln\sigma}{d\ln\dot{\varepsilon}_p} \tag{2.17}$$

if the power law is not obeyed, that is if a plot of stress versus strain rate is not a straight line. The strain rate effect is responsible for *superplasticity*, meaning that under suitable conditions certain alloys can be elongated without local necking and fracture. This is due to the fact that if a neck begins to form during tensile deformation, the local strain rate increases strongly for these alloys and thus deformation shifts to another position. For superplastic alloys the strain rate sensitivity exponent has a value $m \cong 0.5$, while for normal alloys m is less than 0.2. The actual m-value varies with temperature, structure and strain rate; its maximum value occurs for small grain size and usually at intermediate strain rate.

Justification 2.1*: Superplasticity

In fact the extremum condition $dF[\sigma(\varepsilon),A] = 0$ is insufficient, since stability is governed by rate effects as well. If we take $\sigma = \sigma(\varepsilon,\dot{\varepsilon})$ we obtain

$$d\sigma = \frac{\partial\sigma}{\partial\varepsilon}d\varepsilon + \frac{\partial\sigma}{\partial\dot{\varepsilon}}d\dot{\varepsilon}$$

Further differentiating $d\varepsilon = -dA/A$ with respect to time t, the result is

$$d\dot{\varepsilon} = -\frac{d\dot{A}}{A} + \frac{\dot{A}\,dA}{A^2}$$

Inserting these expressions in $d\sigma$ and dividing by σ results in

$$\frac{d\sigma}{\sigma} = -\frac{\partial \ln\sigma}{\partial \varepsilon}\frac{dA}{A} + \frac{\partial \ln\sigma}{\partial \dot{\varepsilon}}\left(-\frac{d\dot{A}}{A} + \frac{\dot{A}\,dA}{A^2}\right)$$

Solving for $d\dot{A}/dA$, meanwhile using $d\sigma/\sigma = d\varepsilon = -dA/A$ and $\dot{\varepsilon} = -\dot{A}/A$, the result is

$$\frac{d\dot{A}}{dA} = \left(1 - \frac{\partial \ln\sigma}{\partial \varepsilon} - \frac{\partial \ln\sigma}{\partial \ln\dot{\varepsilon}}\right) \Big/ \frac{\partial \ln\sigma}{\partial \dot{\varepsilon}}$$

If for a small fluctuation in diameter dA, $d\dot{A} < 0$, dA will decrease in time and the fluctuation will lead to failure. On the other hand, if $d\dot{A} > 0$, dA will increase and the fluctuation will be stabilised. Onset of instability thus occurs for $d\dot{A} = 0$ or

$$\frac{d\dot{A}}{dA} = 0 = 1 - \frac{\partial \ln\sigma}{\partial \varepsilon} - \frac{\partial \ln\sigma}{\partial \ln\dot{\varepsilon}}$$

Obviously, the strain rate affects the stability behaviour. In the case of a power law material for strain- and strain-rate hardening Eqs. (2.10) and (2.17) can be combined to read $\sigma = K''\varepsilon^n\dot{\varepsilon}^m$. The instability criterion then becomes[i]

$$0 = 1 - (n/\varepsilon_{cri}) - m \qquad \text{or} \qquad \varepsilon_{cri} = n/(1-m)$$

For increasing value of m the value of the critical strain for instability ε_{cri} increases, thus explaining superplastic behaviour. The above expression is also valid for other constitutive laws if we interpret the parameters n and m as $n = \partial \ln\sigma / \partial \ln\varepsilon$ and $m = \partial \ln\sigma / \partial \ln\dot{\varepsilon}$. Finally, we note that for strain-rate independent behaviour ($m = 0$), the expression reduces to the Considère result.

Yield strength and hardening behaviour are also temperature dependent. Also the elastic parameters are temperature dependent, but apart from materials (either organic or inorganic) in the glass transition temperature, not so much on strain rate. Typically the various quantities are exponentially dependent on temperature and their change is described by an *Arrhenius equation*. For any quantity x this relationship reads

$$\blacktriangleright \qquad x = x_0 \exp(-\Delta U / kT) \qquad (2.18)$$

where x_0 is a reference value of the quantity x, ΔU is the so-called *activation energy* and k and T denote Boltzmann's constant and temperature, respectively. Generally the behaviour is limited to a certain temperature range where a particular mechanism is dominant. For other ranges a different mechanism may dominate with different reference value x_0 and activation energy ΔU.

Although at ambient condition the various materials behave more or less in one particular manner, many materials can show the whole or nearly the whole range of

[i] Hart, E.W. (1967), Acta Metall. **15**, 351.

the types of behaviour, dependent on the temperature and strain rate. Take inorganic glasses as an example. At room temperature and below these materials behave nearly ideally elastic, show no plastic or viscous deformation in tension and fracture in a brittle fashion. Increasing the temperature introduces, loosely speaking, some viscosity so that the behaviour becomes *visco-elastic* (rate dependent without a stress threshold as in plasticity). Finally, at sufficiently high temperature the behaviour becomes fully viscous. In this temperature range shaping of e.g. bottles, sheet glass, etc. takes place. A similar change in behaviour can also occur for inorganics and metals. For example, at room temperature polycrystalline aluminium oxide (alumina) behaves elastically and fractures in a brittle way. At sufficiently high temperature, dependent on grain size and purity, the material starts to behave plastically with rate effects comparable to metals (Fig. 2.12[j]). In this case a threshold value (yield strength) is present and the behaviour is called *elasto-visco-plasticity* or *visco-plasticy* for short. At the proper temperature and for sufficiently small grain size alumina can even show superplastic deformation, similar to the behaviour observed in alloys. This effect has been shown for other inorganics as well. Also for polymers this variety in behaviour is present. Although at room temperature their behaviour is often significantly rate dependent, at low temperature they become brittle. Even at room temperature failure is often due to brittle or semi-brittle fracture.

Summarising, the deformation response of a material to a mechanical load can be divided into four categories:

- *Elasticity*, the rate-independent response without hysteresis,
- *Plasticity*, the rate-independent response that shows hysteresis,
- *Visco-elasticity*, the rate-dependent response without equilibrium hysteresis and
- *Visco-plasticity*, the rate-dependent response with equilibrium hysteresis.

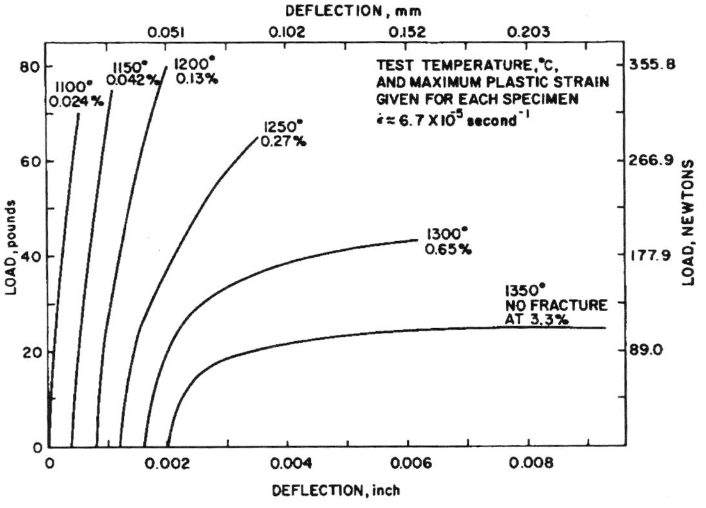

Fig. 2.12: Load-deflection curve of hot-pressed alumina (grain size ~ 1 μm, relative density > 99%, 0.25% MgO as grain growth inhibitor) indicating the brittle-ductile transition. Superplasticity is observed above 1300 °C.

[j] Heuer, A.H., Cannon, R.M. and Tighe, N.J. (1970), page 339 in *Ultrafine-grain ceramics*, J.J. Burke, N.L. Reed, and V. Weiss, eds., Syracuse University Press.

Often the last two categories are taken together under the header visco-elasticity. The fracture response can similarly be divided into three categories:

- *Brittle*, i.e. failure by fracture with no preceding plastic and/or visco-elastic deformation,
- *Semi-brittle*, i.e. failure by fracture with limited preceding plastic and/or visco-elastic deformation and
- *Ductile*, i.e. failure directly by plastic and/or visco-elastic deformation.

However, the microscopic and mesoscopic mechanisms responsible for the behaviour of the various material types are entirely different. Moreover, a single theory dealing with all aspects of thermo-mechanical behaviour, even for one class of materials, becomes immensely complex. Therefore the division in elastic, plastic, visco-elastic and fracture behaviour provides a useful simplification.

2.6 Work and power

Before indicating the order of magnitude for various quantities, let us consider the *work W* done and the *power P* consumed during a tensile test. Evidently

$$\mathrm{d}W = F\mathrm{d}l \qquad \text{and} \qquad W = \int F\,\mathrm{d}l \tag{2.19}$$

If we refer to engineering and natural quantities, respectively, we have

$$W = A_0 l_0 \int \frac{F\,\mathrm{d}l}{A_0\,l_0} = A_0 l_0 \int s\,\mathrm{d}e = V_0 \int s\,\mathrm{d}e \qquad \text{and} \tag{2.20}$$

$$W = \int Al\frac{F\,\mathrm{d}l}{A\,l} = \int Al\sigma\,\mathrm{d}\varepsilon = \int V\sigma\,\mathrm{d}\varepsilon \cong V\int \sigma\,\mathrm{d}\varepsilon \tag{2.21}$$

The last step can be made only if the material is incompressible since then $V = A_0 l_0 \cong Al$ is constant. Incompressibility is usually well obeyed for plastic deformation but not for elastic deformation. For the power $P = \mathrm{d}W/\mathrm{d}t$ one similarly finds

$$► \qquad P = F\frac{\mathrm{d}l}{\mathrm{d}t} = sA_0 l_0 \dot{e} = sV_0\dot{e} = \sigma Al\dot{\varepsilon} = \sigma V\dot{\varepsilon} \tag{2.22}$$

so that $s\dot{e}$ and $\sigma\dot{\varepsilon}$ are the rates of work per unit original and current volume, respectively. In the rate formulation the incompressibility aspect does not arise.

As is well known not all the power delivered to a material is recoverable. In fact the discussion of the tensile test behaviour have shown this already clearly. In the elastic region there is no (or very limited) hysteresis, implying that in a closed stress-strain cycle no (or very little) energy is dissipated. The energy put in one part of the cycle is recoverable in another part and the process is said to be *conservative*. The *resilience* is defined as the capacity of a material to absorb energy when it is deformed elastically, up to the point of yield. It is quantified by the *resilience* U_{res} (Fig. 2.13), defined as the density of strain energy and given by

$$U_{\text{res}} = \int_0^Y \sigma\,\mathrm{d}\varepsilon = \frac{1}{2}Y\varepsilon_Y = \frac{1}{2}Y\frac{Y}{E} = \frac{Y^2}{2E} \tag{2.23}$$

where Y is the uniaxial yield strength, ε_Y is the yield strain, and use has been made of the constitutive behaviour for a linear elastic material. Resilient materials thus have a high yield strength and a low elastic modulus.

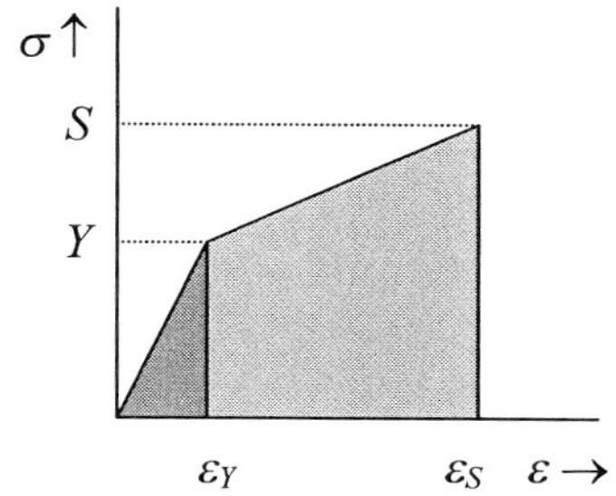

Fig. 2.13: Resilience U_{res} (dark grey) and toughness U_{tou} (dark + light grey).

In the plastic region there is significant hysteresis. The associated energy is converted mainly to heat and for 5 to 15% in internal (strain) energy. Obviously the heat is irrecoverable and the process is therefore called *dissipative*. The *toughness* U_{tou} (Fig. 2.13) is defined as the capacity of a material to absorb energy when it is loaded up to fracture. For a quasi-static tensile test it is the area below the stress-strain curve. Other examples of dissipative processes are provided by visco-elastic deformation and viscous deformation. In the latter two cases all energy spent is 'dissipated' into heat.

Problem 2.6

Show that for material with a stress-strain curve given by $\sigma = C\varepsilon^{1/2}$, the toughness $U_{tou} = 2\varepsilon_{fra}\sigma_{ult}/3$, where ε_{fra} and σ_{ult} denote the fracture strain and ultimate tensile strength, respectively.

2.7 Typical values

Having indicated the typical phenomena that will be encountered in this book, it remains to discuss briefly the order of magnitude of the various phenomenological parameters. Here we distinguish again between polymers, metals and inorganics.

In Table 2.2 typical values for the density ρ, Young's modulus E, yield strength Y and fracture strength S for a number of materials are given. More data can be found in Appendix B. From this table the following observations can be made:

- The density ρ of polymers is in the range of 1 to 2 Mg/m^3, for metals in the range of 4 to 8 Mg/m^3 and for inorganics in the range of 2 to 6 Mg/m^3.
- The elastic modulus E for polymers is in the range of 1 to 10 GPa, for metals in the range of 70 to 400 GPa and for inorganics in the range of 100 to 500 GPa.
- While the yield strength Y of polymers is usually negligible as compared with that for metals for which the range is about 0.1 to 2 GPa, metals are usually soft as compared with inorganics for which the yield strength ranges from about 2 to 7 GPa. However, since most inorganic materials are brittle, the yield strength applies only in compression and is usually estimated from the hardness H (Chapter 14) via the approximate relation $H \cong 3Y$. The fracture strength S is highly determined by processing, as we will discuss later more extensively. Typical strength values for polymers are 50 to 100 MPa. For both metals and inorganics the strength ranges from about 100 to 1000 MPa.

Table 2.2: Characteristic data for various materials.

Material	ρ (Mg/m^3)	E (GPa)	Y (GPa)	S (MPa)
SiO_2	2.2	70	2	50-150
B_4C	2.5	460	7-10	300
Al_2O_3	3.9	400	7	300-400
ZrO_2	5.8	200	4	200-1000
Mg	1.7	42	0.1-0.2	180-230
Al	2.7	70	0.5-1.5	50-150
Fe	7.9	200	1-3	200-600
W	19.4	360	1-2	1000
PE LD (HD)	0.9	0.4-1.0 (1.0-1.3)	9-15 (25-30)	10-30 (35-70)
PS	1.1	3-4	~50	20-30
PMMA	1.2	2-3	50-75	50-80
PTFE	2.2	3-5	~25	15-30

PE polyethylene, LD low density, HD high density, PS polystyrene, PMMA polymethylmetacrylate (perspex), PTFE polytetrafluorethylene (Teflon).

The fracture strength S and yield strength Y can vary considerably depending on the processing of the material.

2.8 Towards the 3D reality of solids

The above description and the associated parameters are convenient for a homogeneous uniaxial stress distribution. However, although in practice tensile loaded bars are used as structural elements, the stress distributions in most mechanically loaded structures are not homogeneous and not uniaxial. Moreover, we have seen that even for a tensile test necking occurs and thus a non-homogeneous stress distribution arises. It is thus clear that we need a more elaborate description of stress and strain for most problems.

Another aspect is the influence of temperature. We have stated that the various phenomenological parameters are temperature dependent. Conventional mechanics does not deal with temperature. Moreover, we have seen that mechanical energy can be partly or wholly converted to heat in dissipative processes. Both aspects lead to the use of thermodynamics. However, the usual approach to thermodynamics is based on homogeneous fluids with the volume V and the pressure p as the only independent and dependent mechanical variables, respectively. Furthermore typically quasi-reversible processes are considered. This description is not sufficient for solids. Solids have a definite shape determined primarily by the material itself in sharp contrast to fluids where the external shape is determined largely by the container. Intrinsically dissipating processes occur frequently, e.g. plastic deformation. On top of that we have already noticed the occurrence of inhomogeneous deformations. Therefore in order to be able to link with mechanics our approach to thermodynamics incorporates

- the notion of a field (to deal with inhomogeneous situations),
- the use of the extended stress and strain definitions as discussed in Chapters 4 and 5 (to deal with more complex mechanical loading) and
- conservative and dissipative forces (to deal with dissipative phenomena).

Exactly in this combined area between thermodynamics and mechanics significant progress has been made. This amalgamation is referred to as *thermomechanics*.

At the end of this introductory chapters two remarks remain. First, rather different presentations of the mechanical behaviour of materials are given by e.g. Courtney (1990), Dowling (1993) and Meyers and Chawla (1999). These authors present the topic from a rather pure material point of view. Second, the tensile test as used here to illustrate the mechanical behaviour of materials is deceptively simple. Lubliner (1990) provides a further brief discussion of the complexities of its nature.

2.9 A note on notation

To be able to fully exploit the aforementioned aspects, some useful mathematics are discussed in Chapter 3. Although in that chapter a rather systematic system of notation for scalars (a, .., B, .., γ, .., Δ, ..), vectors (**a**, ..), tensors (**A**, ..), column matrices ($\boldsymbol{a}$, ..) and square matrices ($\boldsymbol{A}$, ..) is introduced, a strict adherence to this system will frequently lead to a clumsy, complicated and non-conventional notation. To avoid this, deviations from these rules will occur regularly, e.g. if we need the volume integral over a tensor. In this case it is convenient to denote the local tensor by **a** or a_{ij} and its integral by **A** or A_{ij}. However, we stick to the rule that scalars are indicated by italic letters, vectors and tensors by boldface letters and their matrix representations by italic boldface letters. Making distinction in notation between a tensor and its matrix representation deviates from the IUPAC rules[k]. Further we try to adhere to the basic rule of representing a physical quantity by an italic symbol and a label by a regular symbol. This applies to quantities as well as subscripts. A deviation is made for non-specified indices which are shown as italic. Further, in principle, a symbol should be used for one variable only. Strict adherence to this again will lead to excessive use of symbols. Hence a symbol will be used doubly by now and then, e.g. T can denote temperature or kinetic energy. The meaning will be clear from the context. Finally, subscripts can be used as indices or as labels. Indices are given in italics while labels are in regular print. If confusion is likely, labels are given by some more letters, typically the first three letters of the adjective involved, e.g. the initial force f_{ini} instead of f_{i}. On a number of occasions a single capital subscript is used to indicate the (true or alleged) originator of that quantity, e.g. N_{A} for Avogadro's number. An initial or reference state is also sometimes indicated by the subscript 0 (zero), e.g. the initial area A_0. Superscript labels will be given between parentheses to avoid confusion with powers, e.g. the dissipative stress $\sigma_{ij}^{(\mathrm{d})}$.

2.10 Bibliography

Callister, W.D. (1997), *Materials science and engineering*, 4th ed., Wiley, New York.

Cottrell, A.H. (1964), *The mechanical properties of matter*, Wiley, New York.

Courtney, T.H. (1990), *Mechanical behaviour of materials*, McGraw-Hill, Singapore.

Dowling, N.E. (1993), *Mechanical behaviour of materials*, Prentice-Hall, Englewood Cliffs, NJ.

Lubliner, J. (1990), *Plasticity theory*, McMillan, New York.

Meyers, M.A. and Chawla, K.K. (1999), *Mechanical behaviour of materials*, Prentice Hall, Upper Saddle River, NJ.

---

[k] Mills, I., Cvitaš, T., Homann, K., Kallay, N. and Kuchitsu, K. (eds.) (1993), *Quantities, units and symbols in physical chemistry, IUPAC green book*, 2nd ed., Blackwell Science.

3

Mathematical preliminaries

In the description of physical phenomena, we will encounter functions, matrices and determinants. Moreover we will encounter scalars, vectors and tensors. We will also need co-ordinate transformations, some transforms and calculus of variations. In the following we will briefly review these concepts as well as some of the operations between them and some useful general results. We restrict our considerations normally to three-dimensional Euclidean space and Cartesian co-ordinates[a]. In this chapter free use has been made of several books quoted in the bibliography.

3.1 Symbols and conventions

Often we will use quantities with subscripts, e.g. A_{ij}. Upfront we define two frequently used symbols and a useful convention in connection with subscripts.

A convenient symbol is the *Kronecker delta* denoted by δ_{ij}. It is defined by

$$\delta_{ij} = 1 \quad \text{if } i = j \quad \text{and} \quad \delta_{ij} = 0 \text{ if } i \neq j \tag{3.1}$$

We also introduce the *alternator* e_{ijk} for which it holds that

$$\begin{aligned} &e_{ijk} = 1 \text{ for } ijk = 123, 231, 312 \text{ (even permutations[b])}, \\ &e_{ijk} = -1 \text{ for } ijk = 132, 213, 321 \text{ (odd permutations) and} \\ &e_{ijk} = 0 \text{ otherwise, i.e. if any of the three indices are equal.} \end{aligned} \tag{3.2}$$

An alternative expression is given by $e_{ijk} = \frac{1}{2}(i-j)(j-k)(k-i)$.

We introduce also at this point the *summation convention*[c], which states that whenever an index occurs twice in a term in an expression it should be summed over the range of the index. For three-dimensional space the summation is thus over 1, 2 and 3, corresponding to x, y and z. The symbol A_{jj} thus means $A_{11}+A_{22}+A_{33}$ and represents a single equation. On the other hand, the expressions $c_i = a_i + b_i$ and $c_{ij} = a_{ij} + b_{ij}$ represent 3 and 9 expressions, respectively, one for each of the possible indices i and j. A few implications of the summation convention are

$$\begin{aligned} &\delta_{ii} = 3 \\ &e_{ijk}\, e_{imn} = \delta_{jm}\,\delta_{kn} - \delta_{jn}\,\delta_{km} \\ &e_{ijk}\, e_{ijn} = 2\delta_{kn} \\ &e_{ijk}\, e_{ijk} = 6 \end{aligned} \tag{3.3}$$

---

[a] After René Descartes (1596-1650). French philosopher, mathematician and founder of analytical geometry. He invented the co-ordinates, according to his own record on 10 November 1619, idly watching a fly buzzing around in the corner of his room when he suddenly realised that the position of the fly could be represented by three numbers, giving its distance from each of the three walls that met in the corner.

[b] An even (odd) permutation is the result of an even (odd) number of interchanges. The character (even or odd) of a permutation is independent of the order of the interchanges.

[c] Sometimes also called the Einstein (summation) convention.

Further, note the identity

$$\delta_{ij}\ a_i = a_j \tag{3.4}$$

which is frequently used. Since the effect of multiplying a quantity a_i by δ_{ij} is replacing the index i by j, δ_{ij} is also called the *substitution operator*. Applying the Kronecker delta to A_{ij} thus results in $A_{ij}\delta_{ij} = A_{jj}$.

When using the summation convention, a summation index may have to be replaced by another letter in order to avoid appearance more than twice and is usually called a *dummy index*. For instance, substituting $x_i = C_{ip}z_p$ in $y_p = D_{pi}x_i$ would result without change of index in $y_p = D_{pi}C_{ip}z_p$, which is ambiguous. Rewriting $x_i = C_{ip}z_p$ as $x_i = C_{ir}z_r$ yields $y_p = D_{pi}C_{ir}z_r$, which is non-ambiguous. Indices occurring once should occur on the left-hand side as well as the right-hand side of an equation and are called the *free indices* since they can take any of the values 1, 2 or 3.

3.2 Partial derivatives

A function f may be dependent on *variables* x_i and *parameters* p_j. Such a dependence is denoted by $f(x_i;p_j)$. Reference to the parameters is often omitted by writing $f(x_i)$. In practice, reference to the variable is also often omitted by writing just f. For a function f, several derivatives exist. If all variables but one, say x_1, is kept constant during differentiation, the derivative of f with respect to x_1 is called the *partial derivative* and denoted by $(\partial f(x_i)/\partial x_1)_{\mathrm{xi}\neq\mathrm{x1}}$. Once a choice of independent variables is made, there is no need to indicate, as frequently done, which variables are kept constant. Therefore $(\partial f(x_i)/\partial x_1)_{\mathrm{xi}\neq\mathrm{x1}}$ can be indicated without confusion by $\partial f(x_i)/\partial x_1$. Since differentiation with respect to a particular co-ordinate is frequently required, e.g. in $\partial f/\partial x_i$ or $\partial a_{ij}/\partial y$ or $\partial a_{ij}/\partial x_k$, convenient further abbreviations for these expressions are $\partial_i f = f_{,i}$ and $\partial_2 a_{ij} = a_{ij,2}$ and $\partial_k a_{ij} = a_{ij,k}$, respectively. In the latter notation differentiation of f_i with respect to x_j is indicated by an extra subscript j separated by a comma. Note that the index after the comma is counted as the first one (see Section 3.10). The function $f_{,i}$ generally is a function of all variables x_i and, if continuous, may be differentiated again to yield the *second partial derivatives* $f_{,ij} = \partial^2 f/\partial x_i\partial x_j$. It holds that $\partial^2 f/\partial x_i\partial x_j = \partial^2 f/\partial x_j\partial x_i$ or, equivalently, $f_{,ij} = f_{,ji}$.

Example 3.1

For $f(x,y) = x^2y^3$, one simply calculates

$\partial f/\partial x = 2xy^3$	$\partial^2 f/\partial x^2 = 2y^3$	$\partial^2 f/\partial x\partial y = 6xy^2$
$\partial f/\partial y = 3x^2y^2$	$\partial^2 f/\partial y^2 = 6x^2y$	$\partial^2 f/\partial y\partial x = 6xy^2$

In case the independent variables x_i increase by $\mathrm{d}x_i$, the value of the function f at $x_i+\mathrm{d}x_i$ is given by Taylor's expansion

$$f(x_i + \mathrm{d}x_i) = f(x_i) + \frac{\partial f}{\partial x_i}\mathrm{d}x_i + \frac{1}{2!}\frac{\partial^2 f}{\partial x_i \partial x_j}\mathrm{d}x_i\mathrm{d}x_j + \cdots \tag{3.5}$$

One can also write symbolically

$$f(x_i + \mathrm{d}x_i) = \exp(\mathrm{d}x_j \frac{\partial}{\partial x_j}) f(x_i) \tag{3.6}$$

which upon expansion of the exponential in the conventional manner yields Eq. (3.5). Another way is to write

$$f(x_i + \mathrm{d}x_i) = f(x_i) + \mathrm{d}f(x_i) + \frac{1}{2!}\mathrm{d}^2 f(x_i) + \cdots + \frac{1}{n!}\mathrm{d}^n f(x_i) \tag{3.7}$$

where the *first* and *second* (*order*) *differentials* are given by

$$\mathrm{d}f(x_i) = \frac{\partial f}{\partial x_k}\mathrm{d}x_k \qquad \text{and} \qquad \mathrm{d}^2 f(x_i) = \frac{\partial f}{\partial x_k \partial x_l}\mathrm{d}x_k \mathrm{d}x_l \tag{3.8}$$

Generally one can write for the (*order*) *differential*

$$\mathrm{d}^n f(x_i) = \left(\mathrm{d}x_j \frac{\partial}{\partial x_j} \right)^n f(x_i) \tag{3.9}$$

Example 3.2

Consider again the function $f(x,y) = x^2y^3$. Evaluation of $f(2,2)$ yields $4 \cdot 8 = 32$. A first-order estimate for $f(2.1,2.1)$ is provided by

$$f(2.1,2.1) = f(2,2) + (\partial f/\partial x)\mathrm{d}x + (\partial f/\partial y)\mathrm{d}y$$

$$= 32 + 2xy^3\mathrm{d}x + 3x^2y^2\mathrm{d}x = 32 + 2\cdot2\cdot8\cdot0.1 + 3\cdot4\cdot4\cdot0.1 = 40.00$$

The actual value is 40.84.

A function $f(x)$ is *analytic* at $x = c$ if $f(x)$ can be written as (a sum of) Taylor series (with a positive convergence radius). If $f(x)$ is analytic at each point on the open interval I, $f(x)$ is analytic on the interval I. For a function $w(z)$ of a complex variable $z = x+\mathrm{i}y$ ($\mathrm{i} = \sqrt{-1}$) to be analytic, it must satisfy the *Cauchy-Riemann conditions*

$$\frac{\partial u}{\partial x} = \frac{\partial v}{\partial y} \qquad \text{and} \qquad \frac{\partial v}{\partial x} = -\frac{\partial u}{\partial y} \tag{3.10}$$

where $u(x,y) = \mathrm{Re}\, w(z)$ and $v(x,y) = \mathrm{Im}\, w(z)$ denote the real and imaginary parts of w, respectively. Moreover, if $u(x,y)$ and $v(x,y)$ have continuous second derivatives the function $w(z)$ obeys the *Laplace equation* $u_{,ii} = v_{,ii} = 0$ and is said to be *harmonic*.

Example 3.3

Consider the function $w(z) = \mathrm{e}^x \cos y + \mathrm{i}\, \mathrm{e}^x \sin y = \exp(z)$. Then it holds that

$$u = \mathrm{Re}\, w = \mathrm{e}^x \cos y \qquad\qquad v = \mathrm{Im}\, w = \mathrm{e}^x \sin y$$

The derivatives are given by

$\partial u/\partial x = e^x \cos y \qquad \partial v/\partial x = e^x \sin y$

$\partial v/\partial y = e^x \cos y \qquad \partial u/\partial y = -e^x \sin y$

$\partial^2 u/\partial x^2 = e^x \cos y \qquad \partial^2 v/\partial y^2 = -e^x \sin y$

Hence Cauchy-Riemann conditions are satisfied and the function is harmonic.

3.3 Composite, implicit and homogeneous functions

If for a first-order differential $df = (\partial f/\partial x_i)dx_i$ the variables x_i are themselves function of y_j, it holds that

$$dx_i = \frac{\partial x_i}{\partial y_j}dy_j \qquad \text{and} \qquad df = \frac{\partial f}{\partial x_i}\frac{\partial x_i}{\partial y_j}dy_j \tag{3.11}$$

Such a dependency is called a *composite function* and the operation is known as the *chain rule*.

In many cases the variables x_i are not independent, i.e. a relation exists between them meaning that an arbitrary member, say x_1, can be expressed as a function of $x_2, \ldots, x_n$. Often this relation is given in the form $f = f(x_i) =$ constant. In this case the function is called an *implicit function*. Of course, if the equation can be solved, the relevant differentials can be obtained from the solution. The appropriate relations between the differentials can also be obtained by observing that $df = 0$ resulting in

$$df = \frac{\partial f}{\partial x_i}dx_i = \frac{\partial f}{\partial x_1}dx_1 + \frac{\partial f}{\partial x_2}dx_2 + \frac{\partial f}{\partial x_3}dx_3 = 0 \tag{3.12}$$

Assuming that x_1 is the dependent variable and putting $dx_1 = 0$, division by dx_i ($i \neq 1$) yields

$$\left(\frac{\partial f}{\partial x_i}\right)_{xj,x1} + \left(\frac{\partial f}{\partial x_j}\right)_{xi,x1}\left(\frac{\partial x_j}{\partial x_i}\right)_{f,x1} = 0 \quad \text{or}$$
$$\left(\frac{\partial x_j}{\partial x_i}\right)_{f,x1} = -\frac{\left(\partial f/\partial x_i\right)_{xj,x1}}{\left(\partial f/\partial x_j\right)_{xi,x1}} \tag{3.13}$$

Example 3.4

To evaluate the consequences of Eq. (3.13), consider explicitly a function f of the three variables x, y and z, where z is the dependent variable. The first consequence is obtained by taking $x_1 = z$, $x_i = x$ and $x_j = y$. Eq. (3.13) then reads

$$(\partial f/\partial x)_{y,z} + (\partial f/\partial y)_{x,z}(\partial y/\partial x)_{f,z} = 0 \quad \text{or} \quad (\partial y/\partial x)_{f,z} = -(\partial f/\partial x)_{y,z}/(\partial f/\partial y)_{x,z}$$

On the other hand, taking $x_i = y$ and $x_j = x$ for $x_1 = z$ results in

$$(\partial f/\partial y)_{x,z} + (\partial f/\partial x)_{y,z}(\partial x/\partial y)_{f,z} = 0 \quad \text{or} \quad (\partial x/\partial y)_{f,z} = -(\partial f/\partial y)_{x,z}/(\partial f/\partial x)_{y,z}$$

Hence it easily follows that

- $(\partial x/\partial y)_{f,z} = 1/(\partial y/\partial x)_{f,z}$

The second consequence is obtained by cyclic permutation of the variables

$$(\partial f/\partial x)_{y,z} + (\partial f/\partial y)_{z,x}(\partial y/\partial x)_{f,z} = 0$$

$$(\partial f/\partial y)_{z,x} + (\partial f/\partial z)_{x,y}(\partial z/\partial y)_{f,x} = 0$$

$$(\partial f/\partial z)_{x,y} + (\partial f/\partial x)_{y,z}(\partial x/\partial z)_{f,y} = 0$$

resulting, after substitution in each other, in

- $(\partial x/\partial y)_{f,z}(\partial y/\partial z)_{f,x}(\partial z/\partial x)_{f,y} = -1$

The third consequence is obtained if x, y and z are considered to be a composite function of another variable u. If f is constant there is a relation between x, y and z and thus also between $\partial x/\partial u$, $\partial y/\partial u$ and $\partial z/\partial u$. Moreover, $\mathrm{d}f = 0$ and Eq. (3.13) explicitly reads

$$\mathrm{d}f = [(\partial f/\partial x)_{y,z}(\partial x/\partial u)_f + (\partial f/\partial y)_{z,x}(\partial y/\partial u)_f + (\partial f/\partial z)_{x,y}(\partial z/\partial u)_f]\mathrm{d}u = 0$$

Further taking z as constant, independent of u, results in

$$(\partial f/\partial x)_{y,z}(\partial x/\partial u)_{f,z} + (\partial f/\partial y)_{z,x}(\partial y/\partial u)_{f,z} = 0 \quad \text{or}$$

$$(\partial y/\partial u)_{f,z}/(\partial x/\partial u)_{f,z} = -(\partial f/\partial x)_{y,z}/(\partial f/\partial y)_{z,x}$$

Comparing with $(\partial y/\partial x)_{f,z} = -(\partial f/\partial x)_{y,z}(\partial f/\partial y)_{z,x}$ one obtains

- $(\partial y/\partial x)_{f,z} = (\partial y/\partial u)_{f,z}/(\partial x/\partial u)_{f,z}$

The three relations, indicated by •, are frequently used in thermodynamics.

A function $f(x_i)$ is said to be positively *homogeneous* of degree n if for every value of x_i and for every $\lambda > 0$ we have

$$f(\lambda x_i) = \lambda^n f(x_i) \tag{3.14}$$

For such a function *Euler's theorem*

$$x_i(\partial f/\partial x_i) = nf(x_i)$$

applies, which can be proven by differentiation with respect to λ first and taking $\lambda = 1$ afterwards.

Example 3.5

Consider the function $f(x,y) = x^2+xy-y^2$. One easily finds

$$f_{,x} = \partial f/\partial x = 2x+y \quad \text{and} \quad f_{,y} = \partial f/\partial y = x-2y$$

Consequently, $xf_{,x} + yf_{,y} = x(2x+y) + y(x-2y) = 2(x^2+xy-y^2) = 2f$. Hence f is homogeneous of degree 2.

3.4 Extremes and Lagrange multipliers

For obtaining an extreme of a function $f(x_i)$ of n independent variables x_i the first variation δf has to vanish. This leads to

$$\delta f = \frac{\partial f}{\partial x_i}\delta x_i = 0 \tag{3.15}$$

and, since the variables x_i are independent and the variations δx_i are arbitrary, to $\partial f/x_i = 0$ for $i = 1, \ldots, n$. If, however, the extreme of f has to be found when x_i are dependent and satisfy r constraint functions c_j,

$$c_j(x_i) = C_j \qquad (j = 1,\ldots,r \text{ and } r < n) \tag{3.16}$$

where the parameters C_j are constants, the variables x_i must also obey

$$\frac{\partial c_j(x_i)}{\partial x_i}\delta x_i = 0 \qquad (j = 1,\ldots,r) \tag{3.17}$$

Of course, the system can be solved in principle by solving Eq. (3.16) for the independent $n - r$ variables x_i as functions of the others but the procedure is often complex. It can be shown that finding the extreme of f subject to the constraint of Eq. (3.16) is equivalent to finding the extreme of the function g defined by

$$g(x_i, \lambda_j) = f(x_i) - \sum\nolimits_{j=1}^{r} \lambda_j \left[c_j(x_i) - C_j\right] \tag{3.18}$$

where now the original variables x_i and the additional variables λ_j, which are called *Lagrange (undetermined) multipliers*, are to be considered independent. Variation of λ_j leads to Eq. (3.16) and variation of x_i to

$$\frac{\partial f(x_i)}{\partial x_i} - \lambda_j \frac{\partial c_j(x_i)}{\partial x_i} = 0 \tag{3.19}$$

From Eq. (3.19) the values for x_i can be determined. These values are still functions of λ_j but they can be eliminated using Eq. (3.16). In physics, chemistry and materials science the Lagrange multiplier often can be physically interpreted.

Example 3.6

One can ask what is the minimum circumference L of a rectangle given the area A. Denoting the edges by x and y, the circumference is given by $L = 2(x+y)$ while the area is given by $A = xy$. The equations to be solved are

$$\frac{\partial L}{\partial x} - \lambda\frac{\partial A}{\partial x} = 2 - \lambda y = 0 \qquad \Rightarrow \qquad y = \frac{2}{\lambda}$$

$$\frac{\partial L}{\partial y} - \lambda\frac{\partial A}{\partial y} = 2 - \lambda x = 0 \qquad \Rightarrow \qquad x = \frac{2}{\lambda}$$

Hence the solution is $x = y$, $\lambda = 2/\sqrt{A}$ and $\min(L) = 4\sqrt{A}$.

3.5 Legendre transforms

In many problems we meet the demand to interchange between dependent and independent variables. If $f(x_i)$ denotes a function of n variables x_i, we have

$$df = \frac{\partial f}{\partial x_i} dx_i \equiv X_i dx_i \qquad (i = 1,\ldots,n) \tag{3.20}$$

Now we consider the function $g = f - X_1x_1$. For the differential we obtain

$$dg = df - d(X_1x_1) = -x_1 dX_1 + X_j dx_j \qquad (j = 2,\ldots,n) \tag{3.21}$$

and we see that the roles of x_1 and X_1 have been interchanged. Of course, this transformation can be applied to only one variable, to several variables or to all variables. In the last case we use $g = f - X_ix_i$ and obtain $dg = -x_j dX_j$ $(j = 1,\ldots,n)$. This type of transformations is known as *Legendre transformations*. The Legendre transform is often used in thermodynamics. For example, the Gibbs energy with pressure p and temperature T as independent variables and the Helmholtz energy with volume V and temperature T as independent variables are related by a Legendre transform.

Example 3.7

Consider the function $f(x) = ½x^2$. The dependent variable X is given by

$$X = \partial f/\partial x = x$$

which can be solved to yield $x = X$. Therefore the function expressed in the variable X reads $f(X) = ½X^2$. For the transform $g(X)$ one thus obtains

$$g(X) = f(X) - Xx = ½X^2 - X\,X = -½X^2$$

Adrien-Marie Legendre (1752-1833)

Born in a wealthy family in Paris, he was given a top quality education in mathematics and physics at the Collège Mazarin. At the age of 18, Legendre defended his thesis in mathematics and physics there but this was not quite as grand an achievement as it sounds to us today, for this consisted more of a plan of research rather than a complete thesis. With no need for employment to support himself, Legendre lived in Paris and concentrated on research. From 1775 to 1780 he taught with Laplace at the École Militaire. Winning the 1782 prize on projectiles offered by the Berlin Academy launched Legendre on his research career, as he came to the notice of Lagrange, then Director of Mathematics at the Academy in Berlin. Due to his study on the attraction of ellipsoids, leading to the Legendre functions, he was appointed an adjoint in the Académie des Sciences. Over the next few years Legendre published work in a number of areas. In particular, he published papers on celestial mechanics, which contain the Legendre polynomials. Legendre became an associé in 1785 and in 1787 a member of the team to make measurements of the Earth involving a triangulation survey between the Paris and Greenwich observatories. This work resulted in his election to the Royal Society of London in 1787 and also to an important publication, which contains Legendre's theorem on spherical triangles. In 1791 Legendre became a member of the committee of the Académie des Sciences with the task to standardise weights and measures. The committee worked on the metric system. In 1792 he supervised the major task of producing logarithmic and trigonometric tables. He had between 70 and 80 assistants and the work was completed in 1801. In 1794 Legendre published *Eléments de géométrie*, which was the leading elementary text on the topic for around 100 years. Legendre published a book on determining the orbits of comets in 1806 and his major work on elliptic functions appeared in three volumes in 1811, 1817 and 1819.

3.6 Matrices and determinants

A *matrix* is an array of numbers (or functions), represented by an italic boldface uppercase symbol, e.g. $\boldsymbol{A}$, or by italic regular uppercase symbols with indices, e.g. A_{ij}. In full we write

$$\boldsymbol{A} = A_{ij} = \begin{pmatrix} A_{11} & A_{12} & \cdot & A_{1n} \\ A_{21} & & & \cdot \\ \cdot & & & \cdot \\ A_{m1} & \cdot & \cdot & A_{mn} \end{pmatrix} \tag{3.22}$$

The numbers A_{ij} are called the *elements*. The matrix with m rows and n columns is called an $m \times n$ matrix or a matrix of order (m,n). The *transpose* of a matrix, indicated by a superscript T, is formed by interchanging rows and columns. Hence

$$\boldsymbol{A}^{\mathrm{T}} = A_{ji} \tag{3.23}$$

Often we will use square matrices A_{ij} for which $m = n$ (order n). A *column matrix* (or *column* for short) is a matrix for which $n = 1$ and is denoted by a lowercase italic standard symbol with an index, e.g. by a_i, or by a lowercase italic bold symbol, e.g. $\boldsymbol{a}$. A row matrix is a matrix for which $m = 1$ and is the transpose of a column matrix and thus denoted by $(a_i)^{\mathrm{T}}$ or $\boldsymbol{a}^{\mathrm{T}}$.

Two matrices of the same order are *equal* if all their corresponding elements are equal. The *sum* of two matrices $\boldsymbol{A}$ and $\boldsymbol{B}$ of the same order is given by the matrix $\boldsymbol{C}$ whose corresponding elements are the sums of the elements of $\boldsymbol{A}$ and $\boldsymbol{B}$ or

$$\boldsymbol{C} = \boldsymbol{A} + \boldsymbol{B} \qquad \text{or} \qquad C_{ij} = A_{ij} + B_{ij} \tag{3.24}$$

The *product* of two matrices $\boldsymbol{A}$ and $\boldsymbol{B}$ is given by the matrix $\boldsymbol{C}$ whose elements are given by

$$\boldsymbol{C} = \boldsymbol{AB} \qquad \text{or} \qquad C_{ij} = A_{ik} B_{kj} \tag{3.25}$$

representing the *row-into-column* rule. In the last equation explicit use is made of the summation convention, e.g. the index k is summed over 1 to n. Note that, if $\boldsymbol{A}$ represents a matrix of order (k,l) and $\boldsymbol{B}$ a matrix of order (m,n), the product $\boldsymbol{BA}$ is not defined unless $k = n$. For square matrices generally $\boldsymbol{AB} \neq \boldsymbol{BA}$, so that the order must be maintained in any multiplication process. The *transpose* of a *product* $(\boldsymbol{ABC}..)^{\mathrm{T}}$ is given by $(\boldsymbol{ABC}..)^{\mathrm{T}} = ..^{\mathrm{T}}\boldsymbol{C}^{\mathrm{T}}\boldsymbol{B}^{\mathrm{T}}\boldsymbol{A}^{\mathrm{T}}$.

A *real* matrix is a matrix with real elements only. If a real, square matrix $\boldsymbol{A}$ is equal to its transpose

$$\boldsymbol{A} = \boldsymbol{A}^{\mathrm{T}} \tag{3.26}$$

then $\boldsymbol{A}$ is a *symmetric* matrix. A *complex* matrix is a matrix with complex elements. The *complex conjugate* of a matrix $\boldsymbol{A}$ is the matrix $\boldsymbol{A}^*$ formed by the complex conjugate elements of $\boldsymbol{A}$ or

$$\boldsymbol{A}^* = A_{ij}^* \tag{3.27}$$

while the *adjoint*[d] (or conjugate transpose) is defined by the transpose of the complex conjugate

$$(\boldsymbol{A}^*)^\mathrm{T} = A_{ji}^* \tag{3.28}$$

A *Hermitian* matrix obeys the relation

$$\boldsymbol{A} = (\boldsymbol{A}^*)^\mathrm{T} \tag{3.29}$$

so that a symmetric matrix is a real Hermitian matrix. For an *antisymmetric* matrix it holds that

$$\boldsymbol{A} = -(\boldsymbol{A}^\mathrm{T}) \tag{3.30}$$

and thus $\boldsymbol{A}$ has the form

$$\boldsymbol{A} = \begin{pmatrix} 0 & A_{12} & \cdot & A_{1n} \\ -A_{12} & 0 & & \cdot \\ \cdot & & & \cdot \\ -A_{1n} & \cdot & \cdot & 0 \end{pmatrix} \tag{3.31}$$

A *diagonal* matrix has only non-zero entries along the diagonal:

$$\boldsymbol{A} = \begin{pmatrix} A_{11} & 0 & \cdot & 0 \\ 0 & A_{22} & & \cdot \\ \cdot & & \cdot & \cdot \\ 0 & \cdot & \cdot & A_{nn} \end{pmatrix} \tag{3.32}$$

The *unit* matrix $\boldsymbol{I}$ is a diagonal matrix with unit elements:

$$\boldsymbol{I} = \delta_{ij} = \begin{pmatrix} 1 & 0 & \cdot & 0 \\ 0 & 1 & & \cdot \\ \cdot & & \cdot & \cdot \\ 0 & \cdot & \cdot & 1 \end{pmatrix} \tag{3.33}$$

Obviously, $\boldsymbol{IA} = \boldsymbol{AI} = \boldsymbol{A}$, where $\boldsymbol{A}$ is any square matrix of the same order as the unit matrix.

The *determinant* of a square matrix of order n is defined by

$$\det\boldsymbol{A} = |\boldsymbol{A}| = \sum (\pm A_{1i} A_{2j} A_{3k} \cdots A_{np}) \tag{3.34}$$

where the summation is over all permutations of the indices $i, j, k, \cdots, p$. The sign in brackets is positive when the permutation involves an even number of permutations from the initial term $A_{11}A_{22}A_{33}..A_{nn}$ while it is negative for an odd number of permutations.

[d] This definition is according to the usual mathematical convention. In continuum mechanics one also uses the following definition: the adjoint $\boldsymbol{D}^\mathrm{a}$ of a matrix $\boldsymbol{D}$ is uniquely defined by the requirement that for each column matrix $\boldsymbol{v}$ and $\boldsymbol{w}$ it must hold that $(\boldsymbol{Dv})\times(\boldsymbol{Dw}) = \boldsymbol{D}^\mathrm{a}(\boldsymbol{v}\times\boldsymbol{w})$ (Chadwick, P. (1979), *Continuum mechanics, concise theory and problems*, George Allen and Unwin, London).

Example 3.8

For a matrix $\boldsymbol{A}$ of order 3, Eq. (3.34) yields

$$\det\boldsymbol{A} = A_{11}A_{22}A_{33} + A_{12}A_{23}A_{31} + A_{13}A_{21}A_{32} - A_{12}A_{21}A_{33} - A_{11}A_{23}A_{32} - A_{13}A_{22}A_{31}$$

Alternatively, it can be written as $\det\boldsymbol{A} = e_{rst}A_{1r}A_{2s}A_{3t}$.

The determinant of the product $\boldsymbol{AB}$ is given by

$$\det\boldsymbol{AB} = (\det\boldsymbol{A})(\det\boldsymbol{B}) \tag{3.35}$$

Further the determinant of a matrix equals the determinant of its transpose, that is

$$\det\boldsymbol{A} = \det\boldsymbol{A}^{\mathrm{T}} \tag{3.36}$$

The *inverse* of a square matrix $\boldsymbol{A}$ is denoted by $\boldsymbol{A}^{-1}$ and is defined by

$$\boldsymbol{AA}^{-1} = \boldsymbol{I} \tag{3.37}$$

where $\boldsymbol{I}$ is the unit matrix of the same order as $\boldsymbol{A}$. From the above it follows that

$$\boldsymbol{A}^{-1}\boldsymbol{A} = \boldsymbol{I} \tag{3.38}$$

so that a square matrix *commutes* with its inverse. The inverse only exists if $\det\boldsymbol{A} \neq 0$. The inverse of the product $(\boldsymbol{ABC}..)^{-1}$ is given by $(\boldsymbol{ABC}..)^{-1} = ..^{-1}\boldsymbol{C}^{-1}\boldsymbol{B}^{-1}\boldsymbol{A}^{-1}$. The inverse of a transpose is equal to the transpose of the inverse, i.e. $(\boldsymbol{A}^{\mathrm{T}})^{-1} = (\boldsymbol{A}^{-1})^{\mathrm{T}}$, often written as $\boldsymbol{A}^{-\mathrm{T}}$.

The *co-factor* α_{ij} of the element A_{ij} is $(-1)^{i+j}$ times the *minor* θ_{ij}. The latter is the determinant of a matrix obtained by removing row i and column j from the original matrix. The inverse of $\boldsymbol{A}$ is then found from *Cramers's rule*

$$\left(\boldsymbol{A}^{-1}\right)_{ij} = \alpha_{ji} / \det\boldsymbol{A} \tag{3.39}$$

Note the reversal of the element and co-factor indices.

Example 3.9

Consider the matrix $\boldsymbol{A} = A_{ij} = \begin{pmatrix} 1 & 2 \\ 3 & 4 \end{pmatrix}$. The determinant is $\det\boldsymbol{A} = -2$. The co-factors are given by

$$\alpha_{11} = (-1)^{1+1}\,\theta_{11} = (-1)^{1+1}\,4 = 4 \qquad \alpha_{12} = (-1)^{1+2}\,\theta_{12} = (-1)^{1+2}\,3 = -3$$
$$\alpha_{22} = (-1)^{2+2}\,\theta_{22} = (-1)^{2+2}\,1 = 1 \qquad \alpha_{21} = (-1)^{2+1}\,\theta_{21} = (-1)^{2+1}\,2 = -2$$

The elements of the inverse $\boldsymbol{A}^{-1}$ are thus given by

$$[A_{ij}^{-1}] = \begin{pmatrix} \alpha_{11} & \alpha_{21} \\ \alpha_{12} & \alpha_{22} \end{pmatrix} \Big/ \det\boldsymbol{A} = \begin{pmatrix} 4 & -2 \\ -3 & 1 \end{pmatrix} \Big/ -2 = \begin{pmatrix} -2 & 1 \\ 1.5 & -0.5 \end{pmatrix}$$

For a *diagonal* matrix $\boldsymbol{A}$ the inverse is particularly simple and given by

$$\boldsymbol{A}^{-1} = \begin{pmatrix} A_{11}^{-1} & 0 & . & 0 \\ 0 & A_{22}^{-1} & & . \\ . & & . & . \\ 0 & . & . & A_{nn}^{-1} \end{pmatrix} \tag{3.40}$$

For an *orthogonal* matrix it holds that

$$\boldsymbol{A}^{\mathrm{T}} = \boldsymbol{A}^{-1} \qquad \text{or} \qquad \boldsymbol{A}^{\mathrm{T}}\boldsymbol{A} = \boldsymbol{I} \tag{3.41}$$

This implies $(\det \boldsymbol{A})^2 = 1$ or $\det \boldsymbol{A} = \pm 1$. Choosing $\det \boldsymbol{A} = 1$, the matrix $\boldsymbol{A}$ denotes a *proper* orthogonal matrix. Finally, we mention a *unitary* matrix defined by

$$\boldsymbol{A}^{*} = \boldsymbol{A}^{-1} \qquad \text{or} \qquad \boldsymbol{A}^{*}\boldsymbol{A} = \boldsymbol{I} \tag{3.42}$$

3.7 Change of variables

It is also often required to use different independent variables, in particular in integrals. For definiteness consider the case of three 'old' variables x, y and z and three 'new' variables u, v and w. In this case we have

$$u = u(x,y,z) \qquad v = v(x,y,z) \qquad \text{and} \qquad w = w(x,y,z)$$

where the functions u, v and w are continuous and have continuous first derivatives in some region R*. The transformations $u(x,y,z)$, $v(x,y,z)$ and $w(x,y,z)$ are such that a point (x,y,z) corresponding to (u,v,w) in R* lies in a region R and that there is a one-to-one correspondence between the points (u,v,w) and (x,y,z). The Jacobian matrix $J = \partial(x,y,z)/\partial(u,v,w)$ is defined by

$$J = \frac{\partial(x,y,z)}{\partial(u,v,w)} = \begin{pmatrix} \frac{\partial x}{\partial u} & \frac{\partial x}{\partial v} & \frac{\partial x}{\partial w} \\ \frac{\partial y}{\partial u} & \frac{\partial y}{\partial v} & \frac{\partial y}{\partial w} \\ \frac{\partial z}{\partial u} & \frac{\partial z}{\partial v} & \frac{\partial z}{\partial w} \end{pmatrix} \tag{3.43}$$

The determinant, $\det J$, should be either positive or negative throughout the region R*. Consider now the integral

$$I = \int_{\mathrm{R}} F(x,y,z)\,\mathrm{d}x\mathrm{d}y\mathrm{d}z \tag{3.44}$$

over the region R. If the function F is now expressed in u, v and w instead of x, y and z, the integral has to be evaluated over the region R* as

$$I = \int_{\mathrm{R}^*} F(u,v,w)\left|\det J\right|\mathrm{d}u\mathrm{d}v\mathrm{d}w \tag{3.45}$$

where $|\det J|$ denotes the absolute value of the determinant of the Jacobian matrix[e] J. The expression is easily generalised to more variables than 3.

---

[e] In the literature the name Jacobian sometimes indicates the Jacobian determinant instead of the matrix of derivatives. To avoid confusion we use Jacobian matrix and Jacobian determinant explicitly.

Example 3.10

In many cases the use of cylindrical co-ordinates is convenient. Here we consider the Cartesian co-ordinates x_1, x_2 and x_3 as 'new' variables and the cylindrical co-ordinates r, θ and z as 'old' variables. The relations between the Cartesian co-ordinates and the cylindrical co-ordinates (Fig. 3.1) are

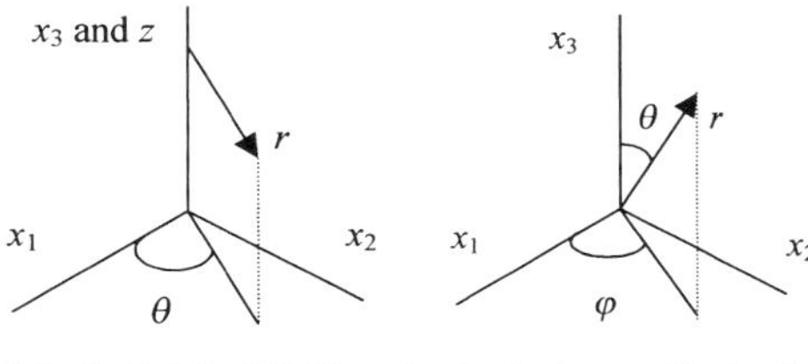

Fig. 3.1: Cylindrical (left) and spherical co-ordinates (right).

$$x_1 = r\cos\theta \qquad x_2 = r\sin\theta \qquad x_3 = z$$

while the inverse equations are given by

$$r = (x_1^2 + x_2^2)^{1/2} \qquad \theta = \tan^{-1}(x_2/x_1) \qquad z = x_3$$

The Jacobian determinant is easily calculated as $|\det J| = r$. Similarly for spherical co-ordinates[f]

$$x_1 = r\cos\varphi\sin\theta \qquad x_2 = r\sin\varphi\sin\theta \qquad x_3 = r\cos\theta$$

and the corresponding inverse equations

$$r = (x_1^2 + x_2^2 + x_3^2)^{1/2} \quad \varphi = \tan^{-1}(x_2/x_1) \qquad \theta = \tan^{-1}[\,(x_1^2 + x_2^2)^{1/2}/x_3]$$

In this case the Jacobian determinant becomes $|\det J| = r^2 \sin\theta$.

3.8 Co-ordinate axes rotations

A co-ordinate axes rotation is frequently required. If we have a Cartesian co-ordinate system $\mathbf{e}_i$ and another system having the same origin as the initial system with $\mathbf{e}_p'$, we can define the *direction cosines* C_{pi} as the cosines of the angles between the new axes $\mathbf{e}_p'$ and the old axes $\mathbf{e}_i$. If P denotes a point with co-ordinates x_i, its co-ordinates in the second system x_i' are given by the projections of the basis $\mathbf{e}_i$ on the basis $\mathbf{e}_p'$ or

$$\begin{aligned} x_1' &= C_{11}x_1 + C_{12}x_2 + C_{13}x_3 \\ x_2' &= C_{21}x_1 + C_{22}x_2 + C_{23}x_3 \\ x_3' &= C_{31}x_1 + C_{32}x_2 + C_{33}x_3 \end{aligned} \tag{3.46}$$

or, more compactly, in index notation and by summation convention $x_p' = C_{pi}x_i$ or in matrix notation $\boldsymbol{x}' = \boldsymbol{C}\boldsymbol{x}$. The inverse relation is given by

[f] Unfortunately in the usual convention for spherical co-ordinates the angle φ corresponds to the angle θ in cylindrical co-ordinates.

$$x_1 = C_{11}x_1' + C_{21}x_2' + C_{31}x_3'$$
$$x_2 = C_{12}x_1' + C_{22}x_2' + C_{32}x_3' \qquad (3.47)$$
$$x_3 = C_{13}x_1' + C_{23}x_2' + C_{33}x_3'$$

or, again more compactly, in index notation and by summation convention $x_i = C_{pi}x_p'$ or in matrix notation by $\boldsymbol{x} = \boldsymbol{C}^{\mathrm{T}}\boldsymbol{x}'$. Substituting $\boldsymbol{x}' = \boldsymbol{Cx}$ in $\boldsymbol{x} = \boldsymbol{C}^{\mathrm{T}}\boldsymbol{x}'$ yields $\boldsymbol{x} = \boldsymbol{CC}^{\mathrm{T}}\boldsymbol{x}$ or $\boldsymbol{CC}^{\mathrm{T}} = \boldsymbol{I}$. The same result can be obtained by substituting $x_r' = C_{ri}x_i$ in $x_i = C_{pi}x_p'$. Since x_p' and x_r' are identical, it follows that $C_{pi}C_{ri} = C_{ip}C_{ir} = \delta_{pr}$. Hence the inverse of the matrix of coefficients of C_{kl}, denoted by $(C_{kl})^{-1}$, is equal to the transpose of C_{kl}, denoted by $(C_{kl})^{\mathrm{T}} = C_{lk}$, which is the definition of an orthogonal matrix. For obvious reasons the matrix $\boldsymbol{C}$ is also called a *rotation* matrix[g].

Example 3.11

Consider a rotation of axes in two-dimensional space of the x_1-axis over an angle θ (Fig. 3.2). The original basis is $x_i = \{x_1, x_2\}$ and the rotated one is $x_i' = \{x_1', x_2'\}$. The direction cosines are given by $C_{11} = \cos\theta$, $C_{12} = \cos(\pi/4-\theta) = \sin\theta$, $C_{21} = \cos(\pi/4+\theta) = -\sin\theta$ and $C_{22} = \cos\theta$. The matrix $\boldsymbol{C}$ thus reads

$$C_{ij} = \begin{pmatrix} \cos\theta & \sin\theta \\ -\sin\theta & \cos\theta \end{pmatrix}$$

For definiteness, consider the point $(x_1, x_2) = (2, 4)$ and a rotation over $\theta = \pi/4$. The new co-ordinates will be

$$x_1' = C_{11}x_1 + C_{12}x_2 = \tfrac{1}{2}\sqrt{2}\times 2 + \tfrac{1}{2}\sqrt{2}\times 4 \qquad \text{and}$$
$$x_2' = C_{21}x_1 + C_{22}x_2 = -\tfrac{1}{2}\sqrt{2}\times 2 + \tfrac{1}{2}\sqrt{2}\times 4$$

yielding $(x_1', x_2') = (3\sqrt{2}, \sqrt{2})$

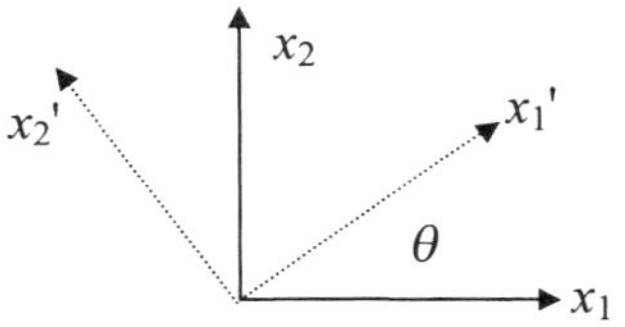

Fig. 3.2: Co-ordinate rotation.

So far we have interpreted the relation $x_p' = C_{pi}x_i$ as a rotation of 'old' co-ordinate axes x_i over a positive (counterclockwise) angle θ to 'new' co-ordinate axes x_p'. However, we can also suppose that there is only one co-ordinate system and that the vector x_i is rotated through an angle θ in the positive direction to give a new vector x_p' in the same co-ordinate system. This is equivalent to a rotation of the co-ordinate system over an angle θ in the *negative* (clockwise) direction. Since any rotation can be

[g] From $\boldsymbol{CC}^{\mathrm{T}} = \boldsymbol{I}$ it follows that $\det \boldsymbol{C} = \pm 1$. We restrict ourselves to proper rotations with $\det \boldsymbol{C} = 1$. If $\det \boldsymbol{C} = -1$, $\boldsymbol{C}$ is an improper rotation matrix representing a rotation combined with a reflection.

described by the matrix $\boldsymbol{C}(\theta)$ where θ is the angle of rotation around a proper axis, as illustrated in Example 3.11, we obtain in this case $\boldsymbol{x}' = \boldsymbol{C}(-\theta)\boldsymbol{x} = \boldsymbol{C}^{\mathrm{T}}(\theta)\boldsymbol{x}$.

Although a scalar has a numerical value independent of the co-ordinate system used, a scalar function is generally expressed by a different function in a different co-ordinate system. The following example illustrates the effect for the same rotation as used in the previous example.

Example 3.12

Consider the function $x_2 = f(x_1) = x_1^2$ in two-dimensional space. For the evaluation of the expression for $x_2' = g(x_1')$ the inverse relations

$$x_1 = C_{11}x_1' + C_{21}x_2'$$
$$x_2 = C_{12}x_1' + C_{22}x_2'$$

are required. Substitution in $x_2 = x_1^2$ yields $C_{11}^2(x_1')^2 + 2C_{11}C_{21}(x_1'x_2') + C_{21}^2(x_2')^2 - C_{12}x_1' - C_{22}x_2' = 0$. Solving for x_2' yields $x_2' = g(x_1')$. For definiteness, consider again the point $(x_1, x_2) = (2, 4)$ and a rotation over $\theta = \pi/4$. While the point $(x_1, x_2) = (2, 4)$ satisfies $x_2 = f(x_1)$, it can easily be verified that $(x_1', x_2') = (3\sqrt{2}, \sqrt{2})$ satisfies $x_2' = g(x_1')$.

3.9 Scalars, vectors and tensors

A *scalar* is an entity with a magnitude. It is denoted by an italic, lowercase or uppercase, Latin or Greek letter, e.g. a, A, γ or Γ.

A *vector* is an entity with a magnitude and direction. It is denoted by a lowercase boldface (Latin) letter[h], e.g. **a**. It can be interpreted as an arrow from a point O (origin) to a point P. Let **a** be this arrow. Its *magnitude* (length), equal to the distance OP, is denoted by $\|\mathbf{a}\|$. A unit vector in the same direction as the vector **a**, here denoted by **e**, has a length of 1. Vectors obey the following rules (Fig. 3.3):

- $\mathbf{c} = \mathbf{a} + \mathbf{b} = \mathbf{b} + \mathbf{a}$ (commutative rule)
- $\mathbf{a} + (\mathbf{b} + \mathbf{d}) = (\mathbf{a} + \mathbf{b}) + \mathbf{d}$ (associative rule)
- $\mathbf{a} + (-\mathbf{a}) = \mathbf{0}$ (zero vector definition)
- $\mathbf{a} = \|\mathbf{a}\|\mathbf{e}$, $\|\mathbf{e}\| = 1$
- $\alpha\mathbf{a} = \alpha\|\mathbf{a}\|\mathbf{e}$

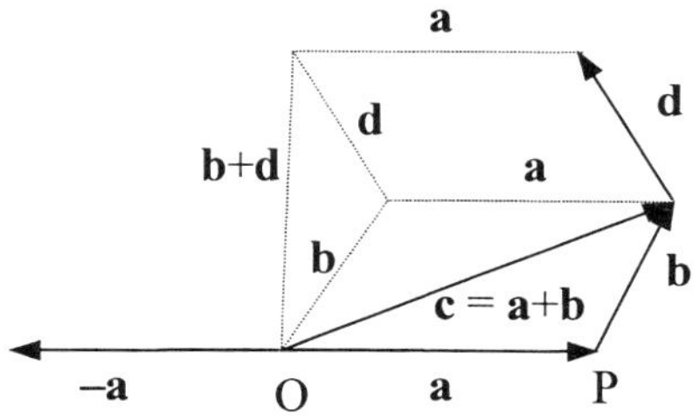

Fig. 3.3: Vector properties.

---

[h] In handwritten text vectors are often indicated by a wavy underscore ($\underset{\sim}{a}$), by underlining $\underline{a}$ or by an arrow above the letter ($\vec{a}$). For consistency with tensors, a wavy underscore is advised.

Various products can be formed using vectors. The *scalar* (or *dot* or *inner*) *product* of two vectors $\mathbf{a}$ and $\mathbf{b}$ yields a scalar and is defined as $\mathbf{a}\cdot\mathbf{b} = \mathbf{b}\cdot\mathbf{a} = \|\mathbf{a}\|\,\|\mathbf{b}\|\cos(\phi)$, where ϕ is the enclosed angle between $\mathbf{a}$ and $\mathbf{b}$. From this definition it follows that $\|\mathbf{a}\| = (\mathbf{a}\cdot\mathbf{a})^{1/2}$. Two vectors $\mathbf{a}$ and $\mathbf{b}$ are orthogonal if $\mathbf{a}\cdot\mathbf{b} = 0$. The scalar product is commutative ($\mathbf{a}\cdot\mathbf{b} = \mathbf{b}\cdot\mathbf{a}$) and distributive ($\mathbf{a}\cdot(\mathbf{b}+\mathbf{c}) = \mathbf{a}\cdot\mathbf{b} + \mathbf{a}\cdot\mathbf{c}$).

The *vector* (or *cross* or *outer*) *product* of $\mathbf{a}$ and $\mathbf{b}$ denotes a vector $\mathbf{c} = \mathbf{a}\times\mathbf{b}$. We define a unit vector $\mathbf{n}$ perpendicular to the plane spanned by $\mathbf{a}$ and $\mathbf{b}$. The sense of $\mathbf{n}$ is right-handed: rotate from $\mathbf{a}$ to $\mathbf{b}$ along the smallest angle and the direction of $\mathbf{n}$ is given by a right-hand screw. It holds that $\mathbf{a}\cdot\mathbf{n} = \mathbf{b}\cdot\mathbf{n} = 0$ and $\|\mathbf{n}\| = 1$. Explicitly, $\mathbf{n} = \mathbf{a}\times\mathbf{b}/\|\mathbf{a}\|\,\|\mathbf{b}\|$. The vector product $\mathbf{c}$ is equal to $\mathbf{c} = \mathbf{a}\times\mathbf{b} = -\mathbf{b}\times\mathbf{a} = \|\mathbf{a}\|\,\|\mathbf{b}\|\sin(\phi)\mathbf{n}$. The length of $\|\mathbf{c}\| = \|\mathbf{a}\|\,\|\mathbf{b}\|\sin(\phi)$ is equal to the area of the parallelogram whose sides are given by $\mathbf{a}$ and $\mathbf{b}$. The vector product is anti-commutative ($\mathbf{a}\times\mathbf{b} = -\mathbf{b}\times\mathbf{a}$) and distributive ($\mathbf{a}\times(\mathbf{b}+\mathbf{c}) = \mathbf{a}\times\mathbf{b} + \mathbf{a}\times\mathbf{c}$). However, it is not associative ($\mathbf{a}\times(\mathbf{b}\times\mathbf{c}) \neq (\mathbf{a}\times\mathbf{b})\times\mathbf{c}$).

The *triple product* is a scalar and given by $d = \mathbf{a}\cdot\mathbf{b}\times\mathbf{c} = \mathbf{a}\times\mathbf{b}\cdot\mathbf{c}$. It yields the volume of the block (or parallelepepid) the edges of which are $\mathbf{a}$, $\mathbf{b}$ and $\mathbf{c}$. Three vectors $\mathbf{a}$, $\mathbf{b}$ and $\mathbf{c}$ are independent if from $\alpha\mathbf{a} + \beta\mathbf{b} + \gamma\mathbf{c} = \mathbf{0}$ it follows that $\alpha = \beta = \gamma = 0$. This is only the case if $\mathbf{a}$, $\mathbf{b}$ and $\mathbf{c}$ are non-coplanar or, equivalently, the product $\mathbf{a}\cdot\mathbf{b}\times\mathbf{c} \neq 0$.

Finally, we need the *tensor* (or *dyadic*) *product* $\mathbf{ab}$. Operating on a vector $\mathbf{c}$, it associates with $\mathbf{c}$ a new vector according to $\mathbf{ab}\cdot\mathbf{c} = \mathbf{a}(\mathbf{b}\cdot\mathbf{c}) = (\mathbf{b}\cdot\mathbf{c})\mathbf{a}$. Note that $\mathbf{ba}$ operating on $\mathbf{c}$ yields $\mathbf{ba}\cdot\mathbf{c} = \mathbf{b}(\mathbf{a}\cdot\mathbf{c}) = (\mathbf{a}\cdot\mathbf{c})\mathbf{b}$. A useful relation involving three vectors using the tensor product is $\mathbf{a}\times(\mathbf{b}\times\mathbf{c}) = (\mathbf{ba} - (\mathbf{a}\cdot\mathbf{b})\mathbf{I})\cdot\mathbf{c}$.

A *tensor* (of rank 2), denoted by an uppercase boldface (Latin) letter[i], e.g. $\mathbf{A}$, is a linear mapping that associates with a vector $\mathbf{a}$ another vector $\mathbf{b}$ according to $\mathbf{b} = \mathbf{A}\cdot\mathbf{a}$. Tensors obey the following rules:

- $\mathbf{C} = \mathbf{A} + \mathbf{B} = \mathbf{B} + \mathbf{A}$ (commutative law)
- $\mathbf{A} + (\mathbf{B} + \mathbf{C}) = (\mathbf{A} + \mathbf{B}) + \mathbf{C}$ (associative law)
- $(\mathbf{A} + \mathbf{B})\cdot\mathbf{u} = \mathbf{A}\cdot\mathbf{u} + \mathbf{B}\cdot\mathbf{u}$ (distributive law)
- $\mathbf{A}\cdot(\alpha\mathbf{u}) = (\alpha\mathbf{A})\cdot\mathbf{u} = \alpha(\mathbf{A}\cdot\mathbf{u})$
- $\mathbf{I}\cdot\mathbf{u} = \mathbf{u}$ ($\mathbf{I}$ unit tensor)
- $\mathbf{A} + (-\mathbf{A}) = \mathbf{O}$ (zero tensor definition)
- $\mathbf{O}\cdot\mathbf{u} = \mathbf{0}$ ($\mathbf{O}$ zero tensor, $\mathbf{0}$ zero vector)

where α is an arbitrary scalar and $\mathbf{u}$ is an arbitrary vector. The simplest example of a tensor is the tensor product of two vectors, e.g. if $\mathbf{A} = \mathbf{bc}$, the vector associated with $\mathbf{a}$ is given by $\mathbf{A}\cdot\mathbf{a} = \mathbf{bc}\cdot\mathbf{a} = (\mathbf{c}\cdot\mathbf{a})\mathbf{b}$.

So far we have discussed vectors and tensors using the *direct notation* only, that is using a symbolism, which represents the quantity without referring to a co-ordinate system. It is convenient though to introduce a co-ordinate system. In this book we will make use primarily of Cartesian co-ordinates, which are a rectangular and rectilinear co-ordinate system with origin O and unit vectors $\mathbf{e}_1$, $\mathbf{e}_2$ and $\mathbf{e}_3$ along the axes. The set $\mathbf{e}_i = \{\mathbf{e}_1, \mathbf{e}_2, \mathbf{e}_3\}$ is called an *orthonormal basis*. It holds that $\mathbf{e}_i\mathbf{e}_j = \delta_{ij}$. The vector OP = $\mathbf{x}$ is called the position of point P. The real numbers x_1, x_2 and x_3, defined uniquely by the relation $\mathbf{x} = x_1\mathbf{e}_1 + x_2\mathbf{e}_2 + x_3\mathbf{e}_3$, are called the (Cartesian) *components* of the vector $\mathbf{x}$. It follows that $x_i = \mathbf{x}\cdot\mathbf{e}_i$ for $i = 1, 2, 3$. Using the components x_i in equations, we use the *index notation*. Using the index notation and the summation convention the scalar product $\mathbf{u}\cdot\mathbf{v}$ can be written as $\mathbf{u}\cdot\mathbf{v} = u_1v_1 + u_2v_2 + u_3v_3 = u_iv_i$. The length of a vector $\mathbf{x}$,

[i] In handwritten text a tensor is often identified by underlining ($\underline{A}$).

$\|\mathbf{x}\| = (\mathbf{x}\cdot\mathbf{x})^{1/2}$, is thus also equal to $(x_1^2 + x_2^2 + x_3^2)^{1/2} = (x_i x_i)^{1/2}$. Sometimes it is also convenient to use *matrix notation*. In this case the components x_i are written collectively as a column matrix $\boldsymbol{x}$. In matrix notation the scalar product $\mathbf{u}\cdot\mathbf{v}$ is written as $\boldsymbol{u}^{\mathrm{T}}\boldsymbol{v}$. The tensor product $\mathbf{ab}$ in matrix notation is given by $\boldsymbol{ab}^{\mathrm{T}}$.

Using the alternator e_{ijk} the relations between the unit vectors can be written as $\mathbf{e}_i\times\mathbf{e}_j = e_{ijk}\mathbf{e}_k$. Similarly, the vector product $\mathbf{c} = \mathbf{a}\times\mathbf{b}$ can alternatively be written as $\mathbf{c} = \mathbf{a}\times\mathbf{b} = e_{ijk}\mathbf{e}_i a_j b_k$. In components this leads to the following expressions:

$$c_1 = a_2b_3 - a_3b_2 \qquad c_2 = a_3b_1 - a_1b_3 \qquad c_3 = a_1b_2 - a_2b_1 \tag{3.48}$$

The triple product $\mathbf{a}\cdot\mathbf{b}\times\mathbf{c}$ in components is given by $e_{ijk}a_ib_jc_k$ while the tensor product $\mathbf{ab}$ is represented by a_ib_j.

If $\mathbf{e}_i = \{\mathbf{e}_1, \mathbf{e}_2, \mathbf{e}_3\}$ is a basis, the tensor products $\mathbf{e}_i\mathbf{e}_j$, $i, j = 1, 2, 3$, form a basis for representing a tensor and we can write $\mathbf{A} = A_{kl}\mathbf{e}_k\mathbf{e}_l$. The nine real numbers A_{kl} are the (Cartesian) components of the tensor $\mathbf{A}$ and are conveniently arranged in a square matrix. It follows that $A_{kl} = \mathbf{e}_k\cdot(\mathbf{A}\cdot\mathbf{e}_l)$, which can be taken as the definition of the components. Applying this definition to the unit tensor, it follows that δ_{kl} are the components of the unit tensor, i.e. $\mathbf{I} = \delta_{kl}\mathbf{e}_k\mathbf{e}_l$. If $\mathbf{v} = \mathbf{A}\cdot\mathbf{u}$, we also have $\mathbf{v} = (A_{kl}\mathbf{e}_k\mathbf{e}_l)\cdot\mathbf{u} = \mathbf{e}_k A_{kl}u_l$. Tensors, like vectors, can form different products. The inner product $\mathbf{A}\cdot\mathbf{B}$ of two tensors (of rank 2) $\mathbf{A}$ and $\mathbf{B}$ yields another tensor of rank 2 and is defined by $(\mathbf{A}\cdot\mathbf{B})\cdot\mathbf{u} = \mathbf{A}\cdot(\mathbf{B}\cdot\mathbf{u}) = A_{kp}B_{pm}u_m$ wherefrom it follows that $(\mathbf{A}\cdot\mathbf{B})_{km} = A_{kp}B_{pm}$, representing conventional matrix multiplication. The expression $\mathbf{A}:\mathbf{B}$ denotes the double inner product, yields a scalar and is given in index notation by $A_{ij}B_{ij}$. Equivalently, $\mathbf{A}:\mathbf{B} = \mathrm{tr}\,\boldsymbol{AB}^{\mathrm{T}} = \mathrm{tr}\,\boldsymbol{A}^{\mathrm{T}}\boldsymbol{B}$. Sometimes another double inner product $\mathbf{A}\cdot\cdot\mathbf{B} = A_{ij}B_{ji}$ is also defined. If one of the tensors $\mathbf{A}$ or $\mathbf{B}$ is symmetric, the difference is obviously immaterial. The tensor product rule of vectors can also be applied to tensors of rank 2. Conforming the notation with vectors the tensor product is denoted by $\mathbf{AB}$, yields a tensor of rank 4 and is denoted by an open uppercase symbol, e.g. $\mathbb{L}$. Equivalent representations are $\mathbb{L} = L_{ijkl} = \mathbf{AB} = A_{ij}B_{kl}$. Similar to the interpretation of a tensor of rank 2 as a mapping of one vector to another, a tensor of rank 4 represents a mapping of a tensor of rank 2 to another tensor of rank 2.

We noticed that the components of a vector $\mathbf{a}$ can be transformed to another Cartesian frame by $a_p' = C_{pi}a_i$ in index notation or $\boldsymbol{a}' = \boldsymbol{Ca}$ in matrix notation. Since a tensor $\mathbf{A}$ of rank 2 can be interpreted as the tensor product of two vectors $\mathbf{b}$ and $\mathbf{c}$, i.e.

$$\mathbf{A} = \mathbf{bc} \qquad \text{or} \qquad A_{ij} = b_ic_j \qquad \text{or} \qquad \boldsymbol{A} = \boldsymbol{bc}^{\mathrm{T}}$$

the transformation rule for the components of a tensor $\mathbf{A}$ obviously is

$$A_{pq}' = b_p'c_q' = C_{pi}b_iC_{qj}c_j = C_{pi}C_{qj}b_ic_j = C_{pi}C_{qj}A_{ij} \qquad \text{or in matrix notation}^{\mathrm{j}}$$

$$\boldsymbol{A}' = \boldsymbol{b}'\boldsymbol{c}'^{\mathrm{T}} = (\boldsymbol{Cb})(\boldsymbol{c}^{\mathrm{T}}\,\boldsymbol{C}^{\mathrm{T}}) = \boldsymbol{CAC}^{\mathrm{T}}$$

Similarly for a tensor $\mathbb{L}$ of rank 4 it holds that

$$L_{pqrs}' = C_{pi}C_{qj}C_{rk}C_{sl}L_{ijkl}$$

If $\mathbf{A}' = \mathbf{A}$ and thus $\boldsymbol{A}' = \boldsymbol{A}$, then $\mathbf{A}$ is an *isotropic* (or *spherical*) *tensor*. Further if the component matrix of a tensor has a property which is not changed by a co-ordinate axes rotation that property is shared by $\boldsymbol{A}'$ and $\boldsymbol{A}$. Such a property is called an

---

[j] Obviously if the transformation is interpreted as a rotation of the tensor instead of the frame, we obtain $\boldsymbol{A}' = \boldsymbol{C}^{\mathrm{T}}\boldsymbol{AC}$. This is the conventional definition of an orthogonal transformation.

invariant. An example is the transpose of a tensor of rank 2: If $A' = CAC^{\mathrm{T}}$, then $A'^{\mathrm{T}} = CA^{\mathrm{T}}C^{\mathrm{T}}$. Consequently we may speak of the transpose $\mathbf{A}^{\mathrm{T}}$ of the tensor $\mathbf{A}$ and we may define the symmetric parts $\mathbf{A}^{(s)}$ and antisymmetric $\mathbf{A}^{(a)}$ parts by

$$\mathbf{A}^{(s)} = (\mathbf{A} + \mathbf{A}^{\mathrm{T}})/2 \qquad \text{and} \qquad \mathbf{A}^{(a)} = (\mathbf{A} - \mathbf{A}^{\mathrm{T}})/2$$

While originally a distinction in terminology is made for a scalar, a vector and a tensor, it is clear that they all transform similarly under a co-ordinate transformation. Therefore a scalar is frequently denoted as a tensor of rank 0 and a vector as a tensor of rank 1. Expressed in components, all tensors obey the same type of transformation rules, e.g. $A_{i..j} = C_{ip} \ldots C_{jq} A_{p\ldots q}$, where the transformation matrix C represents the rotation of the co-ordinate system. Scalars have no index, a vector has one index, a tensor of rank 2 has two indices while a tensor of rank 4 has four indices. Their total transformation matrix contains a product of respectively 0, 1, 2 and 4 individual transformation matrices C_{ij}. Obviously, if we define (Cartesian) tensors as quantities obeying the above transformation rules[k], extension to any order is immediate.

Example 3.13

Consider the vectors **a**, **b** and **c** the matrix representations of which are $\boldsymbol{a}^{\mathrm{T}} = (1,0,0)$, $\boldsymbol{b}^{\mathrm{T}} = (0,1,1)$ and $\boldsymbol{c}^{\mathrm{T}} = (1,1,2)$. Then

$$\mathbf{a} + \mathbf{b} = \boldsymbol{a} + \boldsymbol{b} = (1,1,1)^{\mathrm{T}}$$

$$(\mathbf{b} \times \mathbf{c})^{\mathrm{T}} = (b_2c_3 - b_3c_2,\ b_3c_1 - b_1c_3,\ b_1c_2 - b_2c_1)$$
$$= (1\cdot 2 - 1\cdot 1,\ 1\cdot 1 - 0\cdot 2,\ 0\cdot 1 - 1\cdot 1) = \begin{pmatrix}1 & 1 & -1\end{pmatrix}$$

$$\|\mathbf{c}\| = (\boldsymbol{c}^{\mathrm{T}}\boldsymbol{c})^{1/2} = \left[\begin{pmatrix}1 & 1 & 2\end{pmatrix}\begin{pmatrix}1\\1\\2\end{pmatrix}\right]^{1/2} = \sqrt{6} \qquad \mathbf{b}\cdot\mathbf{c} = \boldsymbol{b}^{\mathrm{T}}\boldsymbol{c} = \begin{pmatrix}0 & 1 & 1\end{pmatrix}\begin{pmatrix}1\\1\\2\end{pmatrix} = 3$$

$$\mathbf{ab} = \boldsymbol{ab}^{\mathrm{T}} = \begin{pmatrix}1\\0\\0\end{pmatrix}\begin{pmatrix}0 & 1 & 1\end{pmatrix} = \begin{pmatrix}0 & 1 & 1\\0 & 0 & 0\\0 & 0 & 0\end{pmatrix} \qquad \mathbf{a}\cdot\mathbf{b}\times\mathbf{c} = \begin{pmatrix}1 & 0 & 0\end{pmatrix}\begin{pmatrix}1\\1\\-1\end{pmatrix} = 1$$

$$\mathbf{ab}\cdot\mathbf{c} = \begin{pmatrix}0 & 1 & 1\\0 & 0 & 0\\0 & 0 & 0\end{pmatrix}\begin{pmatrix}1\\1\\2\end{pmatrix} = \begin{pmatrix}3\\0\\0\end{pmatrix} = 3\mathbf{a}$$

---

[k] This transformation rule is only appropriate for proper rotations of the axes. For improper rotations, which involve a reflection and change of handedness of the co-ordinate system, there are two possibilities. If the rule applies we call the tensor *polar*; if an additional change of sign occurs for an improper rotation we call the tensor *axial*. Hence generally $L_{i..j} = (\det \mathbf{C})^p\, C_{ik}\ldots C_{jl}L_{k..l}$, where $p = 0$ for a polar tensor and $p = 1$ for an axial tensor. Since $\det \mathbf{C} = 1$ for a proper and $\det \mathbf{C} = -1$ for an improper rotation, this results in an extra change of sign for an axial tensor under improper rotation. It follows that the inner and outer product of two polar or two axial tensors is polar while the product of a polar and an axial tensor is axial. The permutation tensor e_{ijk} is axial since $e_{123} = 1$ for both right-handed and left-handed systems. Hence the vector product of two polar vectors is axial. If one restricts oneself to right-handed systems the distinction is irrelevant.

Finally, we have to mention that, like scalars, vectors and tensors can be a function of one or more variables x_k. The appropriate notation is $f(x_k)$, $a_i(x_k)$ and $A_{ij}(x_k)$ or, equivalently, $f(\mathbf{x})$, $\mathbf{a}(\mathbf{x})$ and $\mathbf{A}(\mathbf{x})$. If x_k represent the co-ordinates, $f(\mathbf{x})$, $\mathbf{a}(\mathbf{x})$ and $\mathbf{A}(\mathbf{x})$ are referred to as a scalar, vector and tensor field, respectively.

3.10 Tensor analysis

In this section we consider various differential operators, the divergence theorem and their representation in cylindrical and spherical co-ordinates. Consider the tensor a_i as a typical representative of tensors of rank 1 and take the partial derivatives $\partial_j a_i = \partial a_i/\partial x_j$. Such a derivative transforms like a tensor of rank 2, is called the *gradient* and denoted by grad a_i or in direct notation as grad $\mathbf{a}$ or $\nabla\mathbf{a}$. The gradient can operate on any tensor thereby increasing its rank by 1. An alternative and frequently used notation for $\partial_j a_i = \partial a_i/\partial x_j$ is $a_{i,j}$. In the comma notation the index j is counted as the first one, conform $\nabla\mathbf{a}$, the reason being that this rule avoids many transpose operations. The notation $\nabla\mathbf{a}$ is not universal though and various authors write $(\nabla\mathbf{a})^{\mathrm{T}}$.

If two indices are the same, a summation is implied, thereby decreasing the rank of a tensor by 2. The process is known as *contraction*. If we apply contraction to a second-rank tensor A_{ij} we calculate A_{ii} and the result is known as the *trace*. In direct notation the trace is written as tr $\mathbf{A}$. Contraction also applies to the gradient. The expression $a_{i,i}$ yields a scalar, is called *divergence* and denoted in direct notation by div $\mathbf{a}$ or $\nabla\cdot\mathbf{a}$.

Another important operator is *curl* (or *rot* as abbreviation of rotation). The curl of a vector $\mathbf{a}$, in direct notation written as curl $\mathbf{a}$ or as $\nabla\times\mathbf{a}$, is a vector with components $e_{ijk}a_{k,j}$ (note again the order of the indices). It is defined for 3D space only.

The last operator we mention is the *Laplace operator*. It can act on a scalar a or on a vector $\mathbf{a}$, is denoted by ∇^2 or Δ and defined by $\nabla^2 a = \Delta a = a_{,ii}$ or $\nabla^2\mathbf{a} = \Delta\mathbf{a} = a_{j,ii}$.

Example 3.14

Assume a vector field $\mathbf{a}(\mathbf{x})$ represented by $\boldsymbol{a}^{\mathrm{T}}(\boldsymbol{x}) = \left(3x^2+2y,\quad x+z^2,\quad x+y^2\right)$. Then

$$a_{i,j} = \nabla\mathbf{a} = \begin{pmatrix}\partial/\partial x\\ \partial/\partial y\\ \partial/\partial z\end{pmatrix}\left(3x^2+2y,\ x+z^2,\ x+y^2\right) = \begin{pmatrix}6x & 1 & 1\\ 2 & 0 & 2y\\ 0 & 2z & 0\end{pmatrix}$$

$$\nabla\times\mathbf{a} = \mathrm{e}_{ijk}\mathbf{e}_i a_{k,j} = \begin{pmatrix}2y-2z\\ 0-1\\ 1-2\end{pmatrix} = \begin{pmatrix}2(y-z)\\ -1\\ -1\end{pmatrix} \qquad \Delta\mathbf{a} = a_{j,ii} = \begin{pmatrix}6\\ 2\\ 2\end{pmatrix}$$ a

nd

$$\nabla\cdot\mathbf{a} = a_{i,i} = 6x+0+0 = 6x$$

Next we introduce the divergence theorem. Therefore we consider a region of volume V with a piecewise smooth surface S on which a single-valued tensor field $\mathbf{A}$ or A_{ij} is defined. The body may be convex or non-convex. The components of the exterior normal vector $\mathbf{n}$ of S are denoted by n_i. The *divergence theorem* or the *theorem of Gauss* states that

$$\int A_{ij,j}\,\mathrm{d}V = \int A_{ij}n_j\,\mathrm{d}S \qquad \text{or equivalently} \qquad \int \nabla\cdot\mathbf{A}\,\mathrm{d}V = \int \mathbf{A}\cdot\mathbf{n}\,\mathrm{d}S \tag{3.49}$$

The divergence theorem connects a volume integral to a surface integral and is mainly used in theoretical work. The divergence theorem can be applied to a variety of integrands. Applied to a scalar a, a vector $\mathbf{a}$ or a tensor $\varepsilon_{ijk}a_{k,j}$ we obtain in direct notation

$$\int \nabla a\,\mathrm{d}V = \int \mathbf{n}a\,\mathrm{d}S \quad \int \nabla\cdot\mathbf{a}\,\mathrm{d}V = \int \mathbf{n}\cdot\mathbf{a}\,\mathrm{d}S \quad \text{and} \quad \int \mathrm{curl}\,\mathbf{a}\,\mathrm{d}V = \int \mathbf{n}\times\mathbf{a}\,\mathrm{d}S \tag{3.50}$$

From the divergence theorem we can derive *Stokes' theorem*

$$\int n_i\varepsilon_{ijk}a_{k,j}\,\mathrm{d}S = \int a_i\,\mathrm{d}r_i \qquad \text{or} \qquad \int \mathbf{n}\cdot\mathrm{curl}\,\mathbf{a}\,\mathrm{d}S = \int \mathbf{a}\cdot\mathrm{d}\mathbf{r} \tag{3.51}$$

The theorem connects a surface and line integral and implies that the surface integral is the same for all surfaces bounded by the same curve.

The above operations can also be performed in other co-ordinate systems. Often one considers systems where the base vectors *locally* still form an orthogonal basis, although the orientation may differ through space. These systems are normally addressed as *orthogonal curvilinear co-ordinates*. Cylindrical and spherical co-ordinates form examples with practical importance. Using the relations of Example 3.10 one can show that the unit vectors for cylindrical co-ordinates are

$$\mathbf{e}_r = \mathbf{e}_1\cos\theta + \mathbf{e}_2\sin\theta \qquad \mathbf{e}_\theta = -\mathbf{e}_1\sin\theta + \mathbf{e}_2\cos\theta \qquad \mathbf{e}_z = \mathbf{e}_3$$

so that the only non-zero derivatives are

$$\mathrm{d}\mathbf{e}_r/\mathrm{d}\theta = \mathbf{e}_\theta \qquad \mathrm{d}\mathbf{e}_\theta/\mathrm{d}\theta = -\mathbf{e}_r$$

Using the chain rule for partial derivatives one may show that the gradient operator becomes

$$\nabla = \mathbf{e}_r\frac{\partial}{\partial r} + \mathbf{e}_\theta\frac{1}{r}\frac{\partial}{\partial\theta} + \mathbf{e}_z\frac{\partial}{\partial z} \tag{3.52}$$

The divergence of a vector $\mathbf{a}$ becomes

$$\nabla\cdot\mathbf{a} = \frac{\partial a_r}{\partial r} + \frac{1}{r}\left(a_r + \frac{\partial a_\theta}{\partial\theta}\right) + \frac{\partial a_z}{\partial z} \tag{3.53}$$

while the Laplace operator acting on a scalar a is expressed by

$$\nabla^2 a = \frac{\partial^2 a}{\partial r^2} + \frac{1}{r}\frac{\partial a}{\partial r} + \frac{1}{r^2}\frac{\partial^2 a}{\partial\theta^2} + \frac{\partial^2 a}{\partial z^2} \tag{3.54}$$

Using again the relations of Example 3.10 one can show that the unit vectors for spherical co-ordinates are

$$\mathbf{e}_r = (\mathbf{e}_1\cos\varphi + \mathbf{e}_2\sin\varphi)\sin\theta + \mathbf{e}_3\cos\theta$$

$$\mathbf{e}_\theta = -\mathbf{e}_1\sin\varphi + \mathbf{e}_2\cos\varphi$$

$$\mathbf{e}_\varphi = (\mathbf{e}_1\cos\varphi + \mathbf{e}_2\sin\varphi)\cos\theta - \mathbf{e}_3\sin\theta$$

so that the only non-zero derivatives are

$d\mathbf{e}_r/d\theta = \mathbf{e}_\theta$ $\qquad$ $d\mathbf{e}_\theta/d\theta = -\mathbf{e}_r$

$d\mathbf{e}_r/d\varphi = \mathbf{e}_\varphi \sin\theta$ $\qquad$ $d\mathbf{e}_\theta/d\varphi = \mathbf{e}_\varphi \cos\theta$ $\qquad$ $d\mathbf{e}_\varphi/d\varphi = -\mathbf{e}_r \sin\theta - \mathbf{e}_\varphi \cos\theta$

The gradient operator becomes

$$\nabla = \mathbf{e}_r \frac{\partial}{\partial r} + \mathbf{e}_\theta \frac{1}{r}\frac{\partial}{\partial \theta} + \mathbf{e}_\varphi \frac{1}{r\sin\theta}\frac{\partial}{\partial \varphi} \tag{3.55}$$

The divergence of a vector **a** becomes

$$\nabla \cdot \mathbf{a} = \frac{\partial a_r}{\partial r} + \frac{2a_r}{r} + \frac{1}{r}\frac{\partial a_\theta}{\partial \theta} + \frac{\cot\theta}{r} a_\theta + \frac{1}{r\sin\theta}\frac{\partial a_\varphi}{\partial \varphi} \tag{3.56}$$

while the Laplace operator acting on a scalar a is expressed by

$$\nabla^2 a = \frac{\partial^2 a}{\partial r^2} + \frac{2}{r}\frac{\partial a}{\partial r} + \frac{1}{r^2}\frac{\partial^2 a}{\partial \theta^2} + \frac{\cot\theta}{r^2}\frac{\partial a}{\partial \theta} + \frac{1}{r^2\sin^2\theta}\frac{\partial^2 a}{\partial \varphi^2} \tag{3.57}$$

Johann Carl Friedrich Gauss (1777-1855)

Born in Brunswick, Germany, a son of a father who did not believe in formal education and a mother who encouraged her son in his studies and took considerable pride in his (later) achievements until she died at the age of 97. Fortunately, at the age of 15 the Duke of Brunswick became his patron and helped him to enter Brunswick College and 3 years later to enter the University of Göttingen. He decided only to proceed with mathematics and not with philosophy after he invented the method of least-squares (a decade before Legendre) and a method of constructing a polygon, with the number of sides not being a multiple of 2, 3 or 5, using a compass and ruler, a problem for which the solution was sought for roughly 2000 years. Thereafter he went to the University of Helmstädt where he received his doctorate degree in 1798, proving for the first time that a polynomial of degree n had exactly n roots, a problem which Euler, Newton and Lagrange could not prove. In 1807 he became professor of mathematics at Göttingen University where he remained until his death. He made many contributions to physics, e.g. in electricity and astronomy, but is most well known for his contributions to mathematics, in particular to algebra (complex numbers), geometry, matrix theory (Gauss-Jordan elimination), numerical analysis (Gauss quadrature). Gauss' greatest single publication is his *Disquisitiones arithmeticae*, a work of fundamental importance in the modern theory of numbers. His successors were P.G.L. Dirichlet (1805-1859), G.F. Riemann (1826-1866) and A. Clebsch (1833-1872).

3.11 The eigenvalue problem

Consider a square matrix $\boldsymbol{A}$ of order n, a column matrix $\boldsymbol{e}$ and a scalar λ. One can ask whether it is possible to find column matrices $\boldsymbol{e}$ which through the mapping $\boldsymbol{Ae}$ are mapped upon themselves, e.g. produce the column matrices $\lambda\boldsymbol{e}$. In formula

$$\boldsymbol{Ae} = \lambda\boldsymbol{e} \qquad \text{or} \qquad (\boldsymbol{A} - \lambda\boldsymbol{I})\boldsymbol{e} = 0 \tag{3.58}$$

where $\boldsymbol{I}$ is the unit matrix. The scalar λ is called the *eigenvalue* while the column(s) $\boldsymbol{e}$ are called the *eigenvector*(*s*). This equation can be solved using det $(\boldsymbol{A} - \lambda\boldsymbol{I}) = 0$ or

$$\begin{vmatrix} A_{11} - \lambda & A_{12} & \cdot & A_{1n} \\ A_{21} & A_{22} - \lambda & & \cdot \\ \cdot & & & \cdot \\ A_{n1} & \cdot & \cdot & A_{nn} - \lambda \end{vmatrix} = 0 \tag{3.59}$$

because otherwise the matrix $\boldsymbol{A} - \lambda\boldsymbol{I}$ is regular and Eq. (3.58) has only the (trivial) solution $\boldsymbol{e} = 0$. Expanding the determinant for a matrix of order 3 yields the *characteristic equation*

$$\lambda^3 - J_{(1)}\lambda^2 - J_{(2)}\lambda - J_{(3)} = 0 \tag{3.60}$$

where the (*principal*) *invariants* $J_{(1)}$, $J_{(2)}$ and $J_{(3)}$ are given by

$$J_{(1)} = A_{11} + A_{22} + A_{33} = A_{ii} = \text{tr}\boldsymbol{A}$$

$$\begin{aligned} J_{(2)} &= A_{12}^2 + A_{23}^2 + A_{31}^2 - A_{11}A_{22} - A_{22}A_{33} - A_{33}A_{11} \\ &= \left(A_{ij}A_{ji} - A_{ii}A_{jj}\right)/2 = [\text{tr}(\boldsymbol{A}^2) - (\text{tr}\,\boldsymbol{A})^2]/2 \end{aligned} \tag{3.61}$$

$$J_{(3)} = \det\boldsymbol{A} = \left(2A_{ij}A_{jk}A_{ki} - 3A_{ij}A_{ji}A_{kk} + A_{ii}A_{jj}A_{kk}\right)/6$$

The various J's are called invariants since their numerical value is independent of the co-ordinate system chosen. The roots of the characteristic equation are the eigenvalues λ_1, λ_2 and λ_3 and these values are also independent of the co-ordinate system. It can be shown that for a Hermitian matrix the eigenvalues are real. Backsubstituting λ_1, λ_2 and λ_3 in the eigenvalue equation yields the eigenvectors $\boldsymbol{e}_1$, $\boldsymbol{e}_2$ and $\boldsymbol{e}_3$, respectively. These eigenvectors are indeterminate to the extent of a scalar constant, which can be used to normalise the eigenvectors. Moreover, they are also mutually orthogonal or can be chosen to be so. Hence we can write

$$\mathbf{e}_i \cdot \mathbf{e}_j = \boldsymbol{e}_i^{\mathrm{T}}\boldsymbol{e}_j = \delta_{ij} \tag{3.62}$$

Putting the eigenvectors $\boldsymbol{e}_i$ as columns of a matrix $\boldsymbol{E}$, the complete set of equations is given by $\boldsymbol{E}^{\mathrm{T}}\boldsymbol{AE} = \boldsymbol{\Lambda}$, where $\boldsymbol{\Lambda}$ is a diagonal matrix with λ_1, λ_2 and λ_3 as elements. The original question of the mapping is equivalent to whether it is possible to choose a co-ordinate system in such a way that the matrix $\boldsymbol{A}$ becomes a diagonal matrix.

Example 3.15

Consider the matrix $\boldsymbol{A} = A_{ij} = \begin{pmatrix} 1 & 0 & 0 \\ 0 & 0 & a \\ 0 & a & 0 \end{pmatrix}$ with $a \neq 0$. The characteristic equation is

$$\left|\begin{pmatrix}1&0&0\\0&0&a\\0&a&0\end{pmatrix}-\lambda\begin{pmatrix}1&0&0\\0&1&0\\0&0&1\end{pmatrix}\right|=0 \quad \text{or} \quad \begin{vmatrix}1-\lambda&0&0\\0&-\lambda&a\\0&a&-\lambda\end{vmatrix}=0$$

Evaluation yields $(1-\lambda)(\lambda^2-a^2)=0$. The eigenvalues (roots) are $\lambda_1=1$, $\lambda_2=a$ and $\lambda_3=-a$. Substitution of $\lambda_1=1$ in the eigenvalue equation results in

$$\begin{pmatrix}1-\lambda&0&0\\0&-\lambda&a\\0&a&-\lambda\end{pmatrix}\begin{pmatrix}e_{11}\\e_{21}\\e_{31}\end{pmatrix}=\begin{pmatrix}0\\0\\0\end{pmatrix} \quad \text{or} \quad \begin{pmatrix}0&0&0\\0&-1&a\\0&a&-1\end{pmatrix}\begin{pmatrix}e_{11}\\e_{21}\\e_{31}\end{pmatrix}=\begin{pmatrix}0\\-e_{21}+ae_{31}\\ae_{21}-e_{31}\end{pmatrix}=\begin{pmatrix}0\\0\\0\end{pmatrix}$$

Consequently, we may take $e_{21}=e_{31}=0$. Since the eigenvectors have to be normalised the magnitude $\|\mathbf{e}\|=e_{11}^2+e_{21}^2+e_{31}^2=1$, $e_{11}=1$. Substitution of $\lambda_2=a$ results in

$$\begin{pmatrix}1-a&0&0\\0&-a&a\\0&a&-a\end{pmatrix}\begin{pmatrix}e_{12}\\e_{22}\\e_{32}\end{pmatrix}=\begin{pmatrix}(1-a)e_{12}\\-ae_{22}+ae_{32}\\ae_{22}-ae_{32}\end{pmatrix}=\begin{pmatrix}0\\0\\0\end{pmatrix}$$

Hence, if $(1-a)\neq 0$, $e_{12}=0$. Further, since $a\neq 0$, $e_{22}=e_{32}$. Normalising yields $e_{22}=e_{32}=\tfrac{1}{2}\sqrt{2}$ (or $-\tfrac{1}{2}\sqrt{2}$). Substitution of $\lambda_3=-a$ results in

$$\begin{pmatrix}1+a&0&0\\0&a&a\\0&a&a\end{pmatrix}\begin{pmatrix}e_{13}\\e_{23}\\e_{33}\end{pmatrix}=\begin{pmatrix}(1+a)e_{13}\\ae_{23}+ae_{33}\\ae_{23}+ae_{33}\end{pmatrix}=\begin{pmatrix}0\\0\\0\end{pmatrix}$$

Hence, if $(1+a)\neq 0$, $e_{13}=0$. Further, since $a\neq 0$, $e_{23}=-e_{33}$. Normalising yields $e_{23}=-e_{33}=\tfrac{1}{2}\sqrt{2}$ (or $-\tfrac{1}{2}\sqrt{2}$). Collecting results leads to

$$\begin{pmatrix}1&0&0\\0&\tfrac{1}{2}\sqrt{2}&\tfrac{1}{2}\sqrt{2}\\0&\tfrac{1}{2}\sqrt{2}&-\tfrac{1}{2}\sqrt{2}\end{pmatrix}\begin{pmatrix}1&0&0\\0&0&a\\0&a&0\end{pmatrix}\begin{pmatrix}1&0&0\\0&\tfrac{1}{2}\sqrt{2}&\tfrac{1}{2}\sqrt{2}\\0&\tfrac{1}{2}\sqrt{2}&-\tfrac{1}{2}\sqrt{2}\end{pmatrix}=\begin{pmatrix}1&0&0\\0&a&0\\0&0&-a\end{pmatrix}$$

The eigenvalues are also known as *principal values* and the co-ordinate system in which the matrix is diagonal is known as the *principal axes system*. The principal values are also often indicated by the matrix symbol and a Roman numerical subscript, e.g. A_I, A_II and A_III. In the principal axis system the invariants become

$$\begin{aligned}J_{(1)}&=\lambda_1+\lambda_2+\lambda_3=A_\mathrm{I}+A_\mathrm{II}+A_\mathrm{III}\\J_{(2)}&=\lambda_1\lambda_2+\lambda_2\lambda_3+\lambda_3\lambda_1=A_\mathrm{I}A_\mathrm{II}+A_\mathrm{II}A_\mathrm{III}+A_\mathrm{III}A_\mathrm{I}\\J_{(3)}&=\lambda_1\lambda_2\lambda_3=A_\mathrm{I}A_\mathrm{II}A_\mathrm{III}\end{aligned}\tag{3.63}$$

In the above the invariants are expressed in terms of the principal values λ_1, λ_2 and λ_3, Eq. (3.63), or in terms of combinations of the elements A_{ij} of the matrix $\boldsymbol{A}$, Eq. (3.61). We can also write them in terms of the *basic invariants* $A_{(1)}$, $A_{(2)}$ and $A_{(3)}$ defined by

$$A_{(1)}=A_{ii}=\mathrm{tr}\boldsymbol{A} \qquad A_{(2)}=A_{ij}A_{ji}=\mathrm{tr}\boldsymbol{A}^2 \qquad A_{(3)}=A_{ij}A_{jk}A_{ki}=\mathrm{tr}\boldsymbol{A}^3 \tag{3.64}$$

or, equivalently, in principal values

$$A_{(1)} = \lambda_1 + \lambda_2 + \lambda_3 \qquad A_{(2)} = \lambda_1^2 + \lambda_2^2 + \lambda_3^2 \qquad A_{(3)} = \lambda_1^3 + \lambda_2^3 + \lambda_3^3 \tag{3.65}$$

This results in

$$J_{(1)} = A_{ii} = A_{(1)} \tag{3.66}$$

$$J_{(2)} = \left(A_{ij}A_{ji} - A_{ii}A_{jj}\right)/2 = \left(A_{(2)} - A_{(1)}^2\right)/2 \tag{3.67}$$

$$\begin{aligned} J_{(3)} &= \det\boldsymbol{A} = \left(2A_{ij}A_{jk}A_{ki} - 3A_{ij}A_{ji}A_{kk} + A_{ii}A_{jj}A_{kk}\right)/6 \\ &= \left(2A_{(3)} - 3A_{(2)}A_{(1)} + A_{(1)}^3\right)/6 \end{aligned} \tag{3.68}$$

In principal axes the matrix $\boldsymbol{A} = A_{ij}$ is represented by

$$\boldsymbol{A} = \begin{pmatrix} A_{\mathrm{I}} & 0 & 0 \\ 0 & A_{\mathrm{II}} & 0 \\ 0 & 0 & A_{\mathrm{III}} \end{pmatrix} \tag{3.69}$$

Powers of $\boldsymbol{A}$, in index notation represented by $A_{ik}A_{kj}$, $A_{ik}A_{kl}A_{lj}$, etc., are in principal axes space given by

$$\boldsymbol{A}^n = \begin{pmatrix} A_{\mathrm{I}}^n & 0 & 0 \\ 0 & A_{\mathrm{II}}^n & 0 \\ 0 & 0 & A_{\mathrm{III}}^n \end{pmatrix} \tag{3.70}$$

Since the principal values A_{I}, A_{II} and A_{III} satisfy the characteristic equation (3.60) we have

$$A_{\mathrm{I}}^3 = J_{(1)}A_{\mathrm{I}}^2 + J_{(2)}A_{\mathrm{I}} + J_{(3)}$$

Similar equations can be written for A_{II} and A_{III}. Collectively one writes

$$\boldsymbol{A}^3 = J_{(1)}\boldsymbol{A}^2 + J_{(2)}\boldsymbol{A} + J_{(3)}\boldsymbol{I} \tag{3.71}$$

generally referred to as the *Hamilton-Cayley equation*. In components it reads

$$A_{ik}A_{kl}A_{lj} = J_{(1)}A_{ip}A_{pj} + J_{(2)}A_{ij} + J_{(3)}\delta_{ij} \tag{3.72}$$

Using this equation it is possible to express any power of $\boldsymbol{A}$ in terms of $\boldsymbol{A}^2$, $\boldsymbol{A}$ and $\boldsymbol{I}$. Therefore any power series $\boldsymbol{B} = a\boldsymbol{I} + b\boldsymbol{A} + c\boldsymbol{A}^2 + d\boldsymbol{A}^3 + \cdots$ can be reduced to

$$\boldsymbol{B} = \alpha\boldsymbol{I} + \beta\boldsymbol{A} + \gamma\boldsymbol{A}^2 \tag{3.73}$$

where α, β and γ are power series in $J_{(1)}$, $J_{(2)}$ and $J_{(3)}$.

Example 3.16

Consider the fourth power $\boldsymbol{A}^4$. This term can be written as

$$\begin{aligned} \boldsymbol{A}\cdot\boldsymbol{A}^3 &= \boldsymbol{A}[J_{(1)}\boldsymbol{A}^2 + J_{(2)}\boldsymbol{A} + J_{(3)}\boldsymbol{I}] = J_{(1)}\boldsymbol{A}^3 + J_{(2)}\boldsymbol{A}^2 + J_{(3)}\boldsymbol{A} \\ &= J_{(1)}[\,J_{(1)}\boldsymbol{A}^2 + J_{(2)}\boldsymbol{A} + J_{(3)}\boldsymbol{I}] + J_{(2)}\boldsymbol{A}^2 + J_{(3)}\boldsymbol{A} \\ &= (J_{(1)}^2 + J_{(2)})\boldsymbol{A}^2 + (J_{(1)}J_{(2)} + J_{(3)})\boldsymbol{A} + J_{(1)}J_{(3)}\boldsymbol{I} \end{aligned}$$

3.12 Decompositions

For a complex scalar z two representations exist based on an additive and multiplicative decomposition, respectively. To be specific we can write $z = x+iy$ or $z = r\exp(i\theta)$, where $i = \sqrt{-1}$. In analogy with the complex scalar, a matrix can be decomposed in an additive way and a multiplicative way.

Any matrix $\boldsymbol{A}$ can be written as the sum of a *symmetric* part $A_{(ij)}$ and an *anti-symmetric* part $A_{[ij]}$, defined by

$$A_{(ij)} = (A_{ij}+A_{ji})/2 \qquad \text{and} \qquad A_{[ij]} = (A_{ij}-A_{ji})/2 \tag{3.74}$$

The additive decomposition then is

$$A_{ij} = A_{(ij)} + A_{[ij]} \tag{3.75}$$

Equivalently,

$$A_{(ij)} = (A+A^{\mathrm{T}})/2 \qquad \text{and} \qquad A_{[ij]} = (A-A^{\mathrm{T}})/2 \tag{3.76}$$

It follows easily that for two matrices $\boldsymbol{A}$ and $\boldsymbol{B}$ the product $A_{(ij)}B_{[ij]}$ vanishes and that

$$A_{ij}B_{ij} = \left(A_{(ij)} + A_{[ij]}\right)\left(B_{(ij)} + B_{[ij]}\right) = A_{(ij)}B_{(ij)} + A_{[ij]}B_{[ij]} \tag{3.77}$$

Example 3.17

If we have a matrix $A_{ij} = \begin{pmatrix} 1 & 2 & 3 \\ 4 & 5 & 6 \\ 7 & 8 & 9 \end{pmatrix}$, the symmetric and asymmetric parts are $A_{(ij)} = \begin{pmatrix} 1 & 3 & 5 \\ 3 & 5 & 7 \\ 5 & 7 & 9 \end{pmatrix}$ and $A_{[ij]} = \begin{pmatrix} 0 & -1 & -2 \\ 1 & 0 & -1 \\ 2 & 1 & 0 \end{pmatrix}$, respectively. Similarly for the matrix $B_{ij} = \begin{pmatrix} 1 & 2 & 2 \\ 2 & 4 & 4 \\ 4 & 8 & 8 \end{pmatrix}$ $B_{(ij)} = \begin{pmatrix} 1 & 2 & 3 \\ 2 & 4 & 6 \\ 3 & 6 & 8 \end{pmatrix}$ and $B_{[ij]} = \begin{pmatrix} 0 & 0 & -1 \\ 0 & 0 & -2 \\ 1 & 2 & 0 \end{pmatrix}$. The product $A_{ij}B_{ij} =$ 211 while $A_{(ij)}B_{(ij)} = 219$ and $A_{[ij]}B_{[ij]} = -8$.

For another way to decompose a symmetric matrix in an additive way, we need the *trace* of a matrix $\boldsymbol{A}$, defined by

$$\mathrm{tr}\,\boldsymbol{A} = A_{ii} = A_{ij}\delta_{ij} \tag{3.78}$$

and the *deviator* $\boldsymbol{A}'$, defined by

$$A_{ij}' = A_{ij} - (A_{kk}\delta_{ij})/3 = \boldsymbol{A} - \mathrm{tr}(\boldsymbol{A})\boldsymbol{I}/3 \tag{3.79}$$

The term $\mathrm{tr}(\boldsymbol{A})\boldsymbol{I}/3$ denotes the *isotropic part* or *spherical symmetric part* of the matrix $\boldsymbol{A}$. Solving the deviator expression for A_{ij} yields $A_{ij} = (A_{kk}\delta_{ij})/3 + A_{ij}'$ from which it is clear that every symmetric matrix can be considered as the sum of an isotropic part and a deviator. Note that the double inner product of two tensors **A** and **B**, represented by matrices $\boldsymbol{A}$ and $\boldsymbol{B}$, when decomposed into their isotropic and deviatoric parts, can be written as

$$A_{ij}B_{ij} = \left(\frac{A_{kk}\delta_{ij}}{3} + A_{ij}'\right)\left(\frac{B_{ll}\delta_{ij}}{3} + B_{ij}'\right) = \frac{A_{kk}B_{ll}}{3} + A_{ij}'B_{ij}' \tag{3.80}$$

The quantity $\mathrm{tr}(\boldsymbol{A})/3 = A_{kk}/3$ is sometimes addressed as the mean value of the diagonal components of the matrix A_{ij} and is indicated by A (or A_m if confusion can arise). The matrix A_{ij} therefore can also be written as $A_{ij} = A_{ij}' + A\delta_{ij}$. This convention is in particular used with the stress matrix σ_{ij} and the strain matrix ε_{ij}.

Example 3.18

If we have a matrix $A_{ij} = \begin{pmatrix} 1 & 2 & 3 \\ 2 & 3 & 4 \\ 3 & 4 & 5 \end{pmatrix}$, the isotropic and deviatoric parts are $\frac{1}{3}A_{kk}\delta_{ij} = \begin{pmatrix} 3 & 0 & 0 \\ 0 & 3 & 0 \\ 0 & 0 & 3 \end{pmatrix}$ and $A_{ij}' = \begin{pmatrix} -2 & 2 & 3 \\ 2 & 0 & 4 \\ 3 & 4 & 2 \end{pmatrix}$, respectively. Similarly for the matrix $B_{ij} = \begin{pmatrix} 1 & 1 & 3 \\ 1 & 3 & 5 \\ 3 & 5 & 5 \end{pmatrix}$, the isotropic part becomes $\frac{1}{3}B_{kk}\delta_{ij} = \begin{pmatrix} 3 & 0 & 0 \\ 0 & 3 & 0 \\ 0 & 0 & 3 \end{pmatrix}$ while the deviatoric part is given by $B_{ij}' = \begin{pmatrix} -2 & 1 & 3 \\ 1 & 0 & 5 \\ 3 & 5 & 2 \end{pmatrix}$. The product $(A_{kk}B_{ll})/3 = 27$ while $A_{ij}'B_{ij}' = 70$ resulting in $A_{ij}B_{ij} = 97$.

Similar to the full symmetric matrix $\boldsymbol{A}$, the invariants and basic invariants of the deviator $\boldsymbol{A}'$ can be calculated. Note that, since the reduction to the deviator removes only the spherical symmetrical part from $\boldsymbol{A}$, the principal axes of $\boldsymbol{A}'$ are the same as for $\boldsymbol{A}$ while the principal values are given by $\lambda_\mathrm{I}' = \lambda_\mathrm{I} - A_{ii}/3$. Obviously $J_{(1)}' = A_{(1)}' = \mathrm{tr}\,\boldsymbol{A}' = A_{ii}' = A_{ii} - A_{kk}\delta_{ii}/3 = 0$. The second invariants are given by

$$\begin{aligned} A_{(2)}' &= \mathrm{tr}\,\boldsymbol{A}'^2 = \mathrm{tr}[\mathbf{A} - \mathrm{tr}(\mathbf{A})\mathbf{I}/3][\mathbf{A} - \mathrm{tr}(\mathbf{A})\mathbf{I}/3] \\ &= \mathrm{tr}\,\boldsymbol{A}^2 - \frac{1}{3}\mathrm{tr}^2\boldsymbol{A} = A_{(2)} - \frac{1}{3}A_{(1)}^2 \end{aligned} \tag{3.81}$$

$$J_{(2)}' = \frac{1}{2}\left(A_{(2)}' - A_{(1)}'^2\right) = \frac{1}{2}A_{(2)}' = \frac{1}{2}\left(\mathrm{tr}\boldsymbol{A}^2 - \frac{1}{3}\mathrm{tr}^2\boldsymbol{A}\right) \tag{3.82}$$

$$\begin{aligned} J_{(2)}' &= \frac{1}{6}\left[(A_{11} - A_{22})^2 + (A_{22} - A_{33})^2 + (A_{33} - A_{11})^2\right] \\ &\qquad + A_{12}^2 + A_{23}^2 + A_{31}^2 \\ &= \frac{1}{6}\left[(A_\mathrm{I} - A_\mathrm{II})^2 + (A_\mathrm{II} - A_\mathrm{III})^2 + (A_\mathrm{III} - A_\mathrm{I})^2\right] \end{aligned} \tag{3.83}$$

while the third basic invariant reduces to

$$A_{(3)}' = A_{(3)} - A_{(2)}A_{(1)} + \frac{2}{9}A_{(1)}^3 \tag{3.84}$$

Using a multiplicative decomposition, a matrix $\boldsymbol{A}$ can be uniquely decomposed into

$\boldsymbol{A} = \boldsymbol{RU}$ and $\boldsymbol{A} = \boldsymbol{VR}$

where $\boldsymbol{R}$ is an orthogonal matrix; hence $\boldsymbol{RR}^{\mathrm{T}} = \boldsymbol{R}^{\mathrm{T}}\boldsymbol{R} = \boldsymbol{I}$, and $\boldsymbol{U}$ and $\boldsymbol{V}$ are symmetric and positive definite matrices, $\boldsymbol{U} = \boldsymbol{U}^{\mathrm{T}}$ and $\boldsymbol{V} = \boldsymbol{V}^{\mathrm{T}}$. These decompositions, conventionally denoted as right and left *polar decompositions*, can only be realised if $\det \boldsymbol{A} \neq 0$. The actual decompositions are generally quite tedious but fortunately in practice we frequently only make use of the fact that the decomposition always exists.

3.13 Some special functions*

A convenient function is the *Dirac (delta) function* $\delta(x)$, in one dimension defined by

$$\int_{-a}^{a}\delta(x)f(x)\,\mathrm{d}x = f(0) \qquad \text{or} \qquad \int_{-a}^{a}\delta(x-t)f(t)\,\mathrm{d}t = f(x) \tag{3.85}$$

where $a = \infty$ is included and which selects the value of a function f at the value of variable x from an integral expression. A generalisation to n dimensions is immediate. Alternatively $\delta(x)$ is defined by

$$\delta(x-t) = \infty \ \text{ if } x = t \qquad \text{and} \qquad 0 \text{ otherwise,} \qquad \int_{-a}^{a}\delta(x-t)\,\mathrm{d}t = 1 \tag{3.86}$$

Some properties of the delta function are

$$\delta(-x) = \delta(x) \quad x\delta(x) = 0 \quad \delta(ax) = a^{-1}\delta(x) \ \ (a>0) \ \text{ and}$$
$$\delta(x^2 - a^2) = \tfrac{1}{2}a^{-1}[\delta(x-a) + \delta(x+a)]$$

Related is the *Heaviside (step) function* $h(x)$, defined by

$$h(x-t) = 0 \ \text{ if } x < t \qquad \text{and} \qquad h(x-t) = 1 \ \text{ if } x > t \tag{3.87}$$

For $x = t$, the conventions $h = 0$, $h = \tfrac{1}{2}$ and $h = 1$ are used by various authors. The step function can be considered as the integral of the delta function.

3.14 Calculus of variations*

In many cases variational principles are used. One of the chief problems is to find a function for which some given integral is an extremum. The solution is provided by the calculus of variations. We treat the problem essentially as one-dimensional.

Suppose we wish to find a path $x = x(t)$ between two given values $x(t_1)$ and $x(t_2)$ such that the *functional*[1] $J = \int_{t_1}^{t_2} f(x,\dot{x},t)\,\mathrm{d}t$ of some function $f(x,\dot{x},t)$ with $\dot{x} = \mathrm{d}x/\mathrm{d}t$ is an extremum. Let us assume that $x_0(t)$ is the solution we are looking for. Other possible curves close to $x_0(t)$ are written as $x(t,\alpha) = x_0(t) + \alpha\eta(t)$, where $\eta(t)$ is any function that satisfies $\eta(t_1) = \eta(t_2) = 0$. Using such a representation, the integral J becomes a function of α,

$$J(\alpha) = \int_{t_1}^{t_2} f\left[x(t,\alpha),\dot{x}(t,\alpha),t\right]\mathrm{d}t \tag{3.88}$$

and the condition for obtaining the extremum is $(\mathrm{d}J/\mathrm{d}\alpha)_{\alpha=0} = 0$. We obtain

[1] A function maps a number on a number. A functional maps a function on a number.

$$\frac{\partial J}{\partial \alpha} = \int_{t_1}^{t_2} \left(\frac{\partial f}{\partial x}\frac{\partial x}{\partial \alpha} + \frac{\partial f}{\partial \dot{x}}\frac{\partial \dot{x}}{\partial \alpha} \right) \mathrm{d}t = \int_{t_1}^{t_2} \left(\frac{\partial f}{\partial x}\eta + \frac{\partial f}{\partial \dot{x}}\dot{\eta} \right) \mathrm{d}t \tag{3.89}$$

Through integration by parts the second term of the integral evaluates to

$$\int_{t_1}^{t_2} \frac{\partial f}{\partial \dot{x}}\dot{\eta}\,\mathrm{d}t = \frac{\partial f}{\partial \dot{x}}\eta\bigg|_{t_1}^{t_2} - \int_{x_1}^{x_2} \eta \frac{\mathrm{d}}{\mathrm{d}t}\frac{\partial f}{\partial \dot{x}}\,\mathrm{d}t \tag{3.90}$$

At t_1 and t_2, $\eta(t) = \partial x/\partial \alpha$ vanishes and we obtain for Eq. (3.89)

$$\frac{\partial J}{\partial \alpha} = \int_{t_1}^{t_2} \left(\frac{\partial f}{\partial x} - \frac{\mathrm{d}}{\mathrm{d}t}\frac{\partial f}{\partial \dot{x}} \right) \eta\,\mathrm{d}t$$

If we define the *variations* $\delta J = (\mathrm{d}J/\mathrm{d}\alpha)_{\alpha=0}\mathrm{d}\alpha$ and $\delta x = (\mathrm{d}x/\mathrm{d}\alpha)_{\alpha=0}\mathrm{d}\alpha$, we find

$$\delta J = \int_{t_1}^{t_2} \left(\frac{\partial f}{\partial x} - \frac{\mathrm{d}}{\mathrm{d}t}\frac{\partial f}{\partial \dot{x}} \right) \delta x\,\mathrm{d}t = \left[\int_{t_1}^{t_2} \left(\frac{\partial f}{\partial x} - \frac{\mathrm{d}}{\mathrm{d}t}\frac{\partial f}{\partial \dot{x}} \right) \eta\,\mathrm{d}t \right] \mathrm{d}\alpha = 0 \tag{3.91}$$

and since η must be arbitrary

$$\frac{\partial f}{\partial x} - \frac{\mathrm{d}}{\mathrm{d}t}\frac{\partial f}{\partial \dot{x}} = 0 \tag{3.92}$$

Once this so-called *Euler condition* is fulfilled an extremum is obtained. It should be noted that this extremum is not necessarily a minimum. The extension to more than one variable is evident. Finally, we note that in case the variations at the boundaries do not vanish, i.e. the values of η are not prescribed, the boundary term evaluates, instead of to zero, to

$$\left[\frac{\partial f}{\partial \dot{x}}\eta \right]_{t_1}^{t_2} \tag{3.93}$$

If we now require $\delta J = 0$ we obtain in addition to Eq. (3.92) also the boundary condition $\partial f/\partial \dot{x} = 0$ at $t = t_1$ and $t = t_2$.

Example 3.19

Let us calculate the shortest distance between two points in a plane. An element of an arc length in a plane is

$$\mathrm{d}s = \sqrt{\mathrm{d}x^2 + \mathrm{d}y^2}$$

and the total length of any curve between two points 1 and 2 is

$$I = \int_1^2 \mathrm{d}s = \int_{x_1}^{x_2} f(\dot{y},x)\,\mathrm{d}x = \int_{x_1}^{x_2} \sqrt{1 + \left(\frac{\mathrm{d}y}{\mathrm{d}x}\right)^2}\,\mathrm{d}x \qquad \text{where} \qquad \dot{y} = \frac{\partial y}{\partial x}$$

The condition that the curve is the shortest path is

$$\frac{\partial f}{\partial y} - \frac{\mathrm{d}}{\mathrm{d}x}\frac{\partial f}{\partial \dot{y}} = 0$$

Since $\dfrac{\partial f}{\partial y}=0$ and $\dfrac{\partial f}{\partial \dot{y}}=\dot{y}\big/\sqrt{1+\dot{y}^2}$, we have $\dfrac{\mathrm{d}}{\mathrm{d}x}\left(\dfrac{\dot{y}}{\sqrt{1+\dot{y}^2}}\right)=0$ or $\dfrac{\dot{y}}{\sqrt{1+\dot{y}^2}}=c$, where c is a constant. This solution holds if $\dot{y}=a$ where a is given by $a = c/(1+c^2)^{1/2}$. Obviously this is the equation for a straight line $y = ax + b$, where b is another constant of integration. The constants a and b are determined by the condition that the curve should go through (x_1,y_1) and (x_2,y_2).

3.15 Laplace and Fourier transforms*

In many cases transforms of some kind are useful to ease the solution of a problem. In particular, the Laplace and Fourier[m] transforms are useful in this respect. The *Laplace transform* of a function $f(t)$, defined by

$$L[f(t)]=\hat{f}(s)=\int_0^\infty f(t)\exp(-st)\,\mathrm{d}t \tag{3.94}$$

transforms $f(t)$ into $\hat{f}(s)$ where s may be real or complex. The operation is linear, i.e.

$$L[c_1f_1(t)+c_2f_2(t)]=c_1\hat{f}_1(s)+c_2\hat{f}_2(s) \tag{3.95}$$

The product of two Laplace transforms $L[f(t)]$ and $L[g(t)]$ equals the transform of the convolution of the functions $f(t)$ and $g(t)$

$$L[f(t)]L[g(t)]=L[\int_0^t f(t-\lambda)g(\lambda)\,\mathrm{d}\lambda]=L[\int_0^t f(\lambda)g(t-\lambda)\,\mathrm{d}\lambda] \tag{3.96}$$

This theorem is generally known as the *convolution theorem*. Since the Laplace transform has the property

$$L[\frac{\mathrm{d}f(t)}{\mathrm{d}t}]=s\hat{f}(s)-f(0) \tag{3.97}$$

it can transform differential equations in t to algebraic equations in s. Generalisation to higher derivatives is straightforward and reads

$$L[\frac{\mathrm{d}^nf(t)}{\mathrm{d}t^n}]=s^n\hat{f}(s)-s^{n-1}f(0)-\cdots-sf^{n-2}(0)-f^{n-1}(0) \tag{3.98}$$

Similarly for integration it is found that

$$L[\int_0^t f(u)\,\mathrm{d}u]=\frac{1}{s}\hat{f}(s) \tag{3.99}$$

Some useful transforms are given in Table 3.1.

The *Fourier transform* of a function $f(t)$ is defined by

$$F[f(t)]=\tilde{f}(\omega)=N^{(-)}\int_{-\infty}^{+\infty} f(t)\exp(-\mathrm{i}\omega t)\,\mathrm{d}t \tag{3.100}$$

---

[m] Joseph Fourier (1768-1830), French mathematician, interested in the application of mathematics to physics and mechanics and considered as the founder of mathematical physics. He is well known for his development of the idea that 'every' function can be developed in a trigonometric series and for the theory of heat conduction as described in his book *Théorie analytique du chaleur* (1822).

Table 3.1: Laplace transform pairs.

$f(t)$	$\hat{f}(s)$	$f(t)$	$\hat{f}(s)$
1	$1/s$	$t^n, n>-1$	$\Gamma(n+1)/s^{n+1}$
A	a/s	$\exp(-at)$	$1/(s+a)$
$h(t)$	$1/s$	$t^n \exp(-at)$, $n=0,1,\cdots$	$n!/(s+a)^{n+1}$
$h(t-a)$	$\exp(-as)/s$	$\sin at$	$a/(s^2+a^2)$
$\delta(t)$	1	$\cos at$	$s/(s^2+a^2)$
$\delta(t-a)$	$\exp(-as)$	$\sinh at$	$a/(s^2-a^2)$
T	$1/s^2$	$\cosh at$	$s/(s^2-a^2)$

Its inverse is

$$F^{-1}[\tilde{f}(\omega)] = f(t) = N^{(+)}\int_{-\infty}^{+\infty} \tilde{f}(\omega)\exp(\mathrm{i}\omega t)\,\mathrm{d}\omega \tag{3.101}$$

The normalisation constants $N^{(-)}$ and $N^{(+)}$ in front of the integrals can take any value as long as their product is $(2\pi)^{-1}$. In solid state physics the convention $N^{(-)}=1$ and $N^{(+)}=(2\pi)^{-1}$ is frequently used and we will do likewise. If $N^{(-)} = N^{(+)} = (2\pi)^{-1/2}$ is taken the transform is called symmetric. Similar to the Laplace transform, the Fourier transform is a linear operation for which the convolution theorem holds.

Since for the delta function $\delta(t)$ it holds that

$$\tilde{\delta}(\omega) = \int_{-\infty}^{+\infty}\delta(t)\exp(-\mathrm{i}\omega t)\,\mathrm{d}t = 1 \tag{3.102}$$

we have as a representation of the delta function

$$\delta(t) = (2\pi)^{-1}\int_{-\infty}^{+\infty}\exp(\mathrm{i}\omega t)\,\mathrm{d}\omega \tag{3.103}$$

Similarly for the three-dimensional delta function $\delta(\mathbf{t})$ we have

$$\tilde{\delta}(\boldsymbol{\omega}) = \int_{-\infty}^{+\infty}\delta(\mathbf{t})\exp(-\mathrm{i}\boldsymbol{\omega}\cdot\mathbf{t})\,\mathrm{d}\mathbf{t} = 1 \quad \text{and} \quad \delta(\mathbf{t}) = (2\pi)^{-3}\int_{-\infty}^{+\infty}\exp(\mathrm{i}\boldsymbol{\omega}\cdot\mathbf{t})\,\mathrm{d}\boldsymbol{\omega} \tag{3.104}$$

Finally, we note that by the Gauss theorem applied to a sphere with radius r

$$\int\nabla^2\left(\frac{1}{r}\right)\mathrm{d}V = \int\nabla\left(\frac{1}{r}\right)\cdot\mathbf{n}\,\mathrm{d}S = -\int\frac{\mathbf{r}\cdot\mathbf{n}}{r^3}\,\mathrm{d}S = -4\pi \quad \text{or} \quad \nabla^2\left(\frac{1}{r}\right) = -4\pi\delta(\mathbf{r})$$

since $\nabla^2(1/r) = 0$ for $r \neq 0$ and $\nabla^2(1/r) = \infty$ for $r = 0$. Therefore we have

$$\nabla^2 t^{-1} = \nabla^2 F^{-1}\left[F[t^{-1}]\right] = \nabla^2(2\pi)^{-3}\int F[t^{-1}]\exp(\mathrm{i}\boldsymbol{\omega}.\mathbf{t})\,\mathrm{d}\boldsymbol{\omega} =$$

$$(2\pi)^{-3}\int F[t^{-1}](-\omega^2)\exp(\mathrm{i}\boldsymbol{\omega}.\mathbf{t})\,\mathrm{d}\boldsymbol{\omega} = -4\pi(2\pi)^{-3}\int\exp(\mathrm{i}\boldsymbol{\omega}\cdot\mathbf{t})\,\mathrm{d}\boldsymbol{\omega} \qquad \text{or}$$

$$F[t^{-1}] = 4\pi/\omega^2 \tag{3.105}$$

Applying the inverse transform we obtain $F^{-1}\left[F[t^{-1}]\right]=F^{-1}\left[4\pi/\omega^2\right]$ or

$$\frac{1}{t}=\frac{1}{2\pi^2}\int_{-\infty}^{+\infty}\frac{1}{\omega^2}\exp(\mathrm{i}\boldsymbol{\omega}\cdot\mathbf{t})\,\mathrm{d}\boldsymbol{\omega} \tag{3.106}$$

as the representation of the function $1/t$.

Pierre-Simon de Laplace (1749-1827)

Born at Beaumont-en-Auge, Normandy, he became from a pupil an usher in the school at Beaumont but, having procured a letter of introduction to D'Alembert, he went to Paris. A paper on the principles of mechanics excited D'Alembert's interest, and on his recommendation a place in the military school was offered to him. In the next 17 years, 1771-1787, he produced much of his original work in astronomy followed by several papers on points in the integral calculus, finite differences and differential equations. During the years 1784-1787 he produced some memoirs of exceptional power. Prominent among these is one read in 1784, and reprinted in the third volume of the *Méchanique céleste* (1799-1825), in which he completely determined the attraction of a spheroid on a particle outside it. This is memorable for the introduction into analysis of spherical harmonics or Laplace's coefficients, as also for the development of the use of the potential – a name first given by Green in 1828. He was also active in probability theory, which he summarised in his book *Théorie analytique des probabilités* (1812). His work on celestial mechanics and probability was also introduced via less technical expositions, namely *Exposition du système du monde* (1796) and *Essai philosophique des probabilités* (1814). In the times of turmoil in which he lived he refrained from any political statements and changed easily of political conviction, which made that Napoleon as well as Louis XVIII praised him. Well-known is the answer he gave to Napoleon when teasing Laplace with the remark that God nowhere appeared in his books: *Sire, je n'avais besoin de cette hypothèse.*

3.16 Bibliography

Adams, R.A. (1995), *Calculus*, 3rd ed., Addison-Wesley, Don Mills, Ontario.

Jeffreys, H. and Jeffreys, B.S. (1972), *Methods of mathematical physics*, Cambridge University Press, Cambridge.

Kreyszig, E. (1988), *Advanced engineering mathematics*, 6th ed., Wiley, New York.

Teodosiu, C. (1982), *Elastic models of crystal defects*, Springer, Berlin.

Ziegler, H. (1983), *An introduction to thermomechanics*, 2nd ed., North-Holland, Amsterdam.

4

Kinematics

Kinematics describes the motion and deformation of a continuous body with respect to a reference frame without paying attention to the origin of this motion and deformation or to the nature of the body. To this purpose it is useful to distinguish between material and spatial co-ordinates. Only in case the deformation is small, the difference between the two descriptions vanishes and this reduced description is referred to as the small displacement gradient or infinitesimal strain approximation. The strain tensor is introduced and its physical interpretation discussed. Compatibility is briefly treated. This chapter also describes the deformation using the formal relations between the material and spatial descriptions. The physical interpretation of the resulting strain tensors is revisited. To other presentations is referred to in the reference list. Free use has been made of the books quoted in the bibliography.

4.1 Material and spatial description

An adequate description of the mechanical state of solids requires more parameters than for a fluid where the volume as an extensive kinematical parameter generally is sufficient to describe the mechanical state. Essential is that a solid has a definite shape, contrary to a fluid, and can withstand shear deformation. A general motion of a body is composed of a global and a local part. The global part of the motion, also called *rigid body motion*, is the one where the distance between any two particles of the body does not change at all. The local part is the one where the distance between (at least) two particles does change. The local part of the motion leads to so-called deformation because it leads to changes in the shape and/or volume of the body whereas the global motion leads only to a different position and/or orientation in space (Kuiken, 1994; Maugin, 1992).

For the description of the deformation and motion of a continuous body we use a right-handed Cartesian axes co-ordinate system fixed in space with unit vectors $\mathbf{e}_1$, $\mathbf{e}_2$ and $\mathbf{e}_3$. Fixed points in this *reference co-ordinate system* are known as *spatial points* or just *points*. It is the co-ordinate system used by an observer at rest. The continuous body itself is supposed to be the assembly of *material points* or *particles*, the collective motion of which describes the deformation. To eliminate confusion as far as possible from now on spatial points will be referred to as points and material points as particles. Note that these particles are mathematical entities, as discussed in Chapter 1. To such a particle the properties averaged over a representative volume element or meso-cell centred at the co-ordinates of the particle can be ascribed. It is convenient to introduce another co-ordinate system, parallely oriented to the reference system that accompanies the particle considered during its motion. This *accompanying co-ordinate system* is the system used by an observer moving along with the particle.

Suppose that a particle P of the body is located at point $\mathbf{r}$ at time t_0. The configuration of all the particles at time t_0 forms the undeformed, *reference configuration*. At a (later) time t, conventionally known as the *current configuration*, the particle considered will be at point $\mathbf{x}$. The mapping $\mathbf{x} = \mathbf{x}(\mathbf{r},t)$ for all particles describes the motion of the body. It is known as the *Lagrangian* or *material description* and considers the position **x** as a function of time for one and the same particle P, labelled with **r** in the reference configuration. Here we use the compact notation where the dependent variable and function are denoted by the same symbol.

It is assumed that the mapping $\mathbf{x} = \mathbf{x}(\mathbf{r},t)$ is invertible. This results in $\mathbf{r} = \mathbf{r}(\mathbf{x},t)$, referred to as the *Eulerian* or *spatial description*. This description considers the occupation of a point **r** by a particle which was in the reference configuration at position **x**. This implies that at different times different particles are considered. The components of **r** are called the *material co-ordinates* and the components of **x** the *spatial co-ordinates*.

A simple analogy of these two descriptions is provided by the motion of traffic. If you are a driver (moving observer), you move with your car (particle) and you will describe the motion in Lagrangian terms. On the other hand, if you are a policeman (observer at rest), you are fixed in space, you will see the cars pass and describe the motion in Eulerian terms.

For solids for many formal calculations the Lagrange description is the most useful while for fluids the Euler description is more often used. If the deformations are small enough, the difference between the two descriptions becomes negligible. In Sections 4.2 and 4.3 a description of small deformations is presented in which the distinction of spatial and material co-ordinates is negligible. Those interested in a more complete description should consult later sections.

Joseph Louis Lagrange (1736-1813)

Born and educated in Turin, Italy, he became professor at the Turin Academy of Sciences at the age of 19. In 1766 he moved to Berlin (on invitation of the Prussian King Frederick the Great) and from 1787 onwards he worked at the École Polytechnique in Paris, France. In 1788 his famous book *Mécanique analytique* appeared. At first no printer could be found who would publish the book, but Legendre at last persuaded a Paris firm to undertake it. In this book he laid down the law of virtual work and from that one fundamental principle by the aid of the calculus of variations deduces the whole of mechanics, both of solids and fluids. The object of the book is to show that the subject is implicitly included in a single principle and to give general formulae from which any particular result can be obtained. The method of generalized co-ordinates by which he obtained this result is perhaps the most brilliant result of his analysis. In the preface he emphasized that *On ne trouvera point des figures dans cette ouvrage, seulement des operations algébriques*, so that hundred years after the appearance of Newton's *Principia* mechanics had become largely analytical. Lagrange is probably most well known for his attempts to make calculus mathematically rigorous. Other important contributions are to the calculus of variations, the theory of ordinary and partial differential equations, numerical analysis and algebra. He played a major role in the introduction of the metric system.

4.2 Small displacement gradient deformations

In this section we assume that the deformations are small, so differences between spatial and material co-ordinates can be neglected. Consequently a simple derivation of strain is possible. The result is known as the *small displacement gradient approximation* and is the limiting situation from a more complete description.

To describe the relevant deformations, consider a mechanical system in a (reference) state at time t_0 with two embedded points P and Q having an infinitesimal distance d**x** (Fig. 4.1). Point P is at the position **x** relative to some origin O while point Q is at the point **x**+d**x** (Lubliner, 1990, Teodosiu, 1982).

Consider also another state with new positions for P and Q, denoted by P′ and Q′, respectively, at a later time t. If the displacement for point P (i.e. the vector PP′) is denoted by **u**(**x**), the displacement for Q is **u**(**x**+d**x**). The vector P′Q′, denoted by d**x**′, is obtained by noting that (Fig. 4.1)

$$\mathbf{u}(\mathbf{x})+\mathrm{d}\mathbf{x}' = \mathrm{d}\mathbf{x}+\mathbf{u}(\mathbf{x}+\mathrm{d}\mathbf{x}) \qquad \text{or} \qquad \mathrm{d}\mathbf{x}' = \mathrm{d}\mathbf{x}+\mathbf{u}(\mathbf{x}+\mathrm{d}\mathbf{x})-\mathbf{u}(\mathbf{x}).$$

If we expand **u**(**x**+d**x**) in a Taylor series around **x**, we find that

$$\mathbf{u}(\mathbf{x}+\mathrm{d}\mathbf{x}) = \mathbf{u}(\mathbf{x}) + \nabla\mathbf{u}(\mathbf{x})\cdot\mathrm{d}\mathbf{x} + \cdots$$

For small deformations higher order terms can be neglected. If we do so, the above equation for d**x**′ can be conveniently written as

$$\mathrm{d}\mathbf{x}' = \mathrm{d}\mathbf{x} + \mathbf{U}(\mathbf{x})\cdot\mathrm{d}\mathbf{x} = \left[\mathbf{I} + \mathbf{U}(\mathbf{x})\right]\cdot\mathrm{d}\mathbf{x} \tag{4.1}$$

with **I** the unit tensor and **U**(**x**) the displacement gradient matrix written explicitly as

$$\nabla\mathbf{u}(\mathbf{x}) = \mathbf{U}(\mathbf{x}) = \begin{pmatrix} \dfrac{\partial u_1(\mathbf{x})}{\partial x_1} & \dfrac{\partial u_1(\mathbf{x})}{\partial x_2} & \dfrac{\partial u_1(\mathbf{x})}{\partial x_3} \\ \dfrac{\partial u_2(\mathbf{x})}{\partial x_1} & \dfrac{\partial u_2(\mathbf{x})}{\partial x_2} & \dfrac{\partial u_2(\mathbf{x})}{\partial x_3} \\ \dfrac{\partial u_3(\mathbf{x})}{\partial x_1} & \dfrac{\partial u_3(\mathbf{x})}{\partial x_2} & \dfrac{\partial u_3(\mathbf{x})}{\partial x_3} \end{pmatrix} \tag{4.2}$$

Remember that in $\nabla\mathbf{u} = u_{i,j}$ the index j is counted as the first one. Eq. (4.1) gives the relative position (d**x**′) of two points (P’ and Q’) in any state in terms of their relative positions (d**x**) in the reference state (P and Q).

Expression (4.1) actually needs a little refinement since rigid body translations and rotations are clearly irrelevant to the description of deformation but are present in expression (4.1). A rigid body motion leaves the distance between particles of a body unchanged. The general expression for a small rigid body displacement is

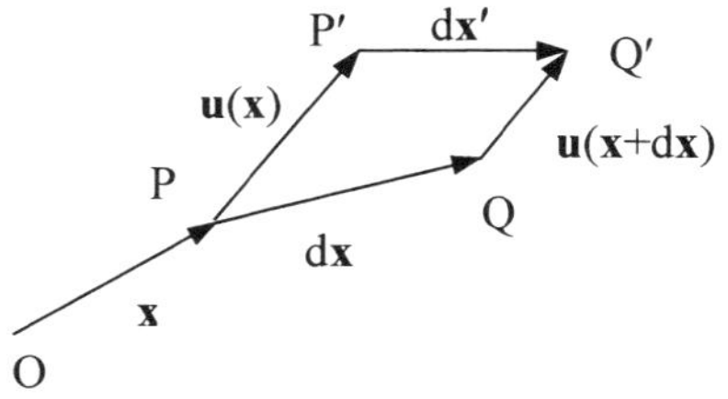

Fig. 4.1: Geometry of deformation.

$$\mathbf{u}(\mathbf{x}) = \mathbf{c} + \boldsymbol{\omega} \times \mathbf{x} \tag{4.3}$$

where **c** is a position-independent vector and $\boldsymbol{\omega} = \omega\mathbf{n}$ with $\omega = ||\boldsymbol{\omega}||$ the angle of rotation for an axis through the origin around a vector **n**. Both **c** and **ω** are possibly time dependent. The first term describes the rigid body translation while the second represents a small rotation.

Consider first rigid body translation only. In that case all points of the body translate with the same amount $\mathbf{u}(\mathbf{x}) = \mathbf{c}$, independent of **x**, and the tensor $\mathbf{U}(\mathbf{x})$ is identically zero. Hence the rigid body translation is properly taken care of because **c** does not contribute to **u**.

If we consider rigid body rotation only, the displacement of a point **x** of the body is described by $\mathbf{u}(\mathbf{x}) = \boldsymbol{\omega}\times\mathbf{x}$. In components $u_i = e_{ijk}\omega_j x_k$ and hence $U_{ij} = \partial u_i/\partial x_j = u_{i,j}$ is

$$U_{ij} = \begin{pmatrix} 0 & -\omega_3 & \omega_2 \\ \omega_3 & 0 & -\omega_1 \\ -\omega_2 & \omega_1 & 0 \end{pmatrix} \tag{4.4}$$

Hence a small rotation leads to an anti-symmetric matrix U_{ij}. Conversely, it can be shown that a constant anti-symmetric matrix $\boldsymbol{A}$ ($||\boldsymbol{A}|| << 1$) describes a small rotation. Since any matrix can be decomposed into a symmetric and an anti-symmetric matrix via

$$U_{ij} = \left(U_{ij} + U_{ji}\right)/2 + \left(U_{ij} - U_{ji}\right)/2 = U_{(ij)} + U_{[ij]} \tag{4.5}$$

we conclude that $U_{(ij)}$, that is the symmetric part of U_{ij}, describes the deformation and $U_{[ij]}$, that is the anti-symmetric part of U_{ij}, describes the rotation. From now on we will denote $U_{(ij)}$ by ε_{ij} (or in direct notation by $\boldsymbol{\varepsilon}$) and refer to it as the *strain* tensor. Note that the use of the symbol $\boldsymbol{\varepsilon}$ deviates from our 'tensor = uppercase (bold)' rule but the notation is conventional. Because $\varepsilon_{ij} = \varepsilon_{ji}$ only six of the nine terms are independent and the strain tensor is symmetrical.

A different way to reach the same conclusion is by considering the change in length for a line element d**x** at position **x**. Clearly, a rigid body rotation should leave the length of d**x** invariant whereas a deformation changes d**x**. The square of the length after deformation is

$$||\,\mathrm{d}\mathbf{x}'\,||^2 = \mathrm{d}\mathbf{x}'\cdot\mathrm{d}\mathbf{x}' = \mathrm{d}\mathbf{x}\cdot[\mathbf{I}+\mathbf{U}]^{\mathrm{T}}\cdot[\mathbf{I}+\mathbf{U}]\cdot\mathrm{d}\mathbf{x} \tag{4.6}$$

After expansion to first order in **U**, one obtains directly

$$||\,\mathrm{d}\mathbf{x}'\,||^2 = \mathrm{d}\mathbf{x}\cdot\left[\mathbf{I}+\mathbf{U}+\mathbf{U}^{\mathrm{T}}\right]\cdot\mathrm{d}\mathbf{x} \tag{4.7}$$

with $\mathbf{U}+\mathbf{U}^{\mathrm{T}} = 2\boldsymbol{\varepsilon}$, or equivalently

► $$\varepsilon_{ij} = U_{(ij)} = u_{(i,j)} \tag{4.8}$$

Example 4.1

Consider a displacement field given by $\begin{pmatrix} u_1 \\ u_2 \\ u_3 \end{pmatrix} = \begin{pmatrix} x^2+y^2 \\ y^2+z^2 \\ z^2+x^2 \end{pmatrix}$. The displacement gradient U_{ij} is then $U_{ij} = \begin{pmatrix} 2x & 2y & 0 \\ 0 & 2y & 2z \\ 2x & 0 & 2z \end{pmatrix}$ so that the strain $\varepsilon_{ij} = U_{(ij)} = u_{(i,j)}$ is

represented by the matrix $\varepsilon_{ij} = \begin{pmatrix} 2x & y & x \\ y & 2y & z \\ x & z & 2z \end{pmatrix}$.

Before we discuss in the next section the physical interpretation of the strain tensor we present first some definitions we need frequently in the rest of this book. A strain is called *homogeneous* when the components ε_{ij} are constant, independent of the co-ordinates. In that case the displacement is linear with $\mathbf{x}$. There are only three macroscopically homogeneous deformation modes (Fig. 4.2). The first mode is *uniaxial tension* or *compression*, e.g. along the x_1-axis described by

$$\varepsilon_{ij} = \begin{pmatrix} \varepsilon_{11} & 0 & 0 \\ 0 & -\nu\varepsilon_{11} & 0 \\ 0 & 0 & -\nu\varepsilon_{11} \end{pmatrix} \tag{4.9}$$

where ν represents Poisson's ratio, a material constant (see Chapter 9). Although the deformation is homogeneous the strain is not uni-axial. The second mode is *simple shear*, e.g. in the x_1-x_2 plane, described by

$$\varepsilon_{ij} = \begin{pmatrix} 0 & \varepsilon_{12} & 0 \\ \varepsilon_{21} & 0 & 0 \\ 0 & 0 & 0 \end{pmatrix} \tag{4.10}$$

and the third is (*pure*) *tension* or *compression* given by

$$\varepsilon_{ij} = \begin{pmatrix} \varepsilon_{11} & 0 & 0 \\ 0 & \varepsilon_{22} & 0 \\ 0 & 0 & \varepsilon_{33} \end{pmatrix} \tag{4.11}$$

If in the latter case $\varepsilon_{11} = \varepsilon_{22} = \varepsilon_{33}$, one obtains *isotropic* (or *hydrostatic*) *tension* or *compression.* Another special case of strain is *plane strain* for which it holds that in one direction a component of the strain tensor and its derivatives with respect to that direction are zero, e.g. $\varepsilon_{i3} = 0$ and $\varepsilon_{ij,3} = 0$.

The components such as ε_{11}, ... are called the *normal strains* and the components ε_{12}, ... the *shear strains*. Obviously the shear strains are symmetric, i.e. $\varepsilon_{ij} = \varepsilon_{ji}$. Hence all the properties of a symmetric tensor, as discussed in Chapter 3, are applicable. In

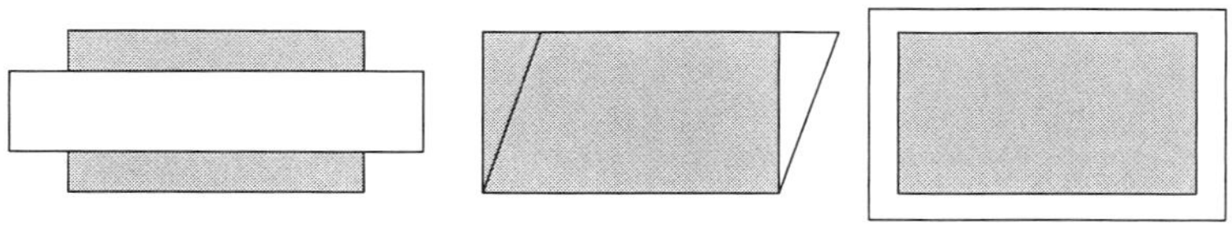

Fig. 4.2: Uniaxial length change, simple shear and isotropic volume change. The grey and white areas represent the original shape and deformed shape, respectively.

particular the tensor can be brought at principal axes. In that case the diagonal elements ε_{11}, ε_{22} and ε_{33} are the only non-zero elements of the strain tensor ε_{ij}. These strains are called the *principal strains* (or *values*) and the corresponding axes the *principal axes*. An equivalent terminology is *characteristic values* and *eigenvalues*.

Using the decomposition of a symmetric tensor in an isotropic part and a deviator, we can write $\varepsilon_{ij} = (\varepsilon_{kk}\delta_{ij})/3+\varepsilon_{ij}'$, where the term $(\varepsilon_{kk}\delta_{ij})/3$ represents the *isotropic strain* and ε_{ij}' represents the *deviatoric strain*. While the latter represents the change in shape (at constant volume), the former is related to the change in volume (at constant shape). In fact the trace of the strain tensor, $\varepsilon_{ii} = \nabla\cdot\mathbf{u}$, equals the *dilatation*, defined by $\Delta V/V_0$, where ΔV is the volume change and V_0 is initial volume (see Section 4.3). We will use the isotropic and deviatoric part again in the separation of the elastic energy in a volume-dependent and shape-dependent part and in the description of plasticity.

Apart from the displacement u_i or strain ε_{ij}, we will also need the strain rate $\dot{\varepsilon}_{ij}$. In the present approximation $\dot{\varepsilon}_{ij} = \dot{u}_{(i,j)} \cong v_{(i,j)}$, where $v_{(i,j)}$ is the symmetrised velocity gradient. The latter is frequently also called *rate of deformation* and sometimes indicated by D_{ij}. In the literature the symbol d_{ij} (deviating from the 'tensor = uppercase rule') is also often used and we will do likewise.

Leonhard Euler (1707-1783)

Born in the vicinity of Basel, Switzerland, he received his master's degree at the University of Basel at the age of 16. He published his first scientific paper before he was 20. He moved to St. Petersburg, Russia in 1727 as an associate of the 1725 founded Russian Academy of Sciences. He became a member of the department of physics in 1730 and when Daniel Bernoulli left St. Petersburg in 1733, he took his place as the head of the mathematics department. Here he also wrote his famous book on mechanics *Mechanica sive motus scientia analytica exposita* (1736), in which he used analytical methods instead of the geometrical methods as employed by Newton. In 1741 he accepted a post offered by King Frederick the Great at the Prussian Academy of Sciences in Berlin. While in Berlin he published the first book on variational calculus entitled *Methodus inveniendi lineas curves* ...(1744) and wrote *Introductio in Analysi Infinitorum* (1748), *Institutiones calculi differentialis* (1755) and *Institutiones calculi integralis* (1770), books that guided mathematicians for many years. After the death of Maupertuis he was also in charge of the Academy, resulting in a great deal of executive work. In 1766 he returned to St. Petersburg on invitation of Empress Catherina II, where he could focus again on science. Euler made amongst others important contributions to mechanics, variational, differential and integral calculus and left us with a large number of papers (more than 400 during the period 1766-1783 and in total 886), in spite that he lost one eye in 1735 and later approached complete blindness due to a cataract. Forty years after his death the Russian Academy was still publishing papers from his legacy.

Problem 4.1

The displacement field of a body (a set of material points) is described by

$$\mathbf{u} = \alpha \begin{pmatrix} 2x^2 + xy \\ y^2 \\ 0 \end{pmatrix}$$

where α is a constant. Calculate at the point (1,1,0)
a) the displacement gradient matrix and the associated strain field,
b) the isotropic part, the deviatoric part and the dilatation of the strain field and
c) the principal values of the strain field and the associated eigenvectors.

Problem 4.2

Consider the displacement field $\mathbf{u} = \begin{pmatrix} -\gamma\, y + \beta\, z \\ \gamma\, x - \alpha\, z \\ -\beta\, x + \alpha\, y \end{pmatrix}$ where α, β and γ denote constants. Determine
a) the strain matrix,
b) for which values of α, β and γ this description is useful and
c) the volume change.

Problem 4.3

The co-ordinates $\mathbf{r}'$ of a body after deformation are given by $\mathbf{r}' = \mathbf{F}\cdot\mathbf{r}$, where **F** is given by $\mathbf{F} = \alpha \begin{pmatrix} x & x & x \\ y & y & y \\ z & z & z \end{pmatrix}$. Here x, y and z denote the components of the initial co-ordinates r and α is a constant. Calculate
a) the displacement gradient field and the associated strain field,
b) the isotropic and the deviatoric part of the strain field,
c) the principal values of the strain field and the associated eigenvectors,
d)*for which values of α this description is valid, given that the boundaries of the body are given by $|x| \le 1$, $|y| \le 1$ and $|z| \le 1$ and
e)*the strain rate given that $x = at$, $y = bt$ and $z = ct$, where t denotes the time and a, b and c are constants.

4.3 Physical interpretation

The strain tensor ε describes the deformation of continuous media for small values of the displacement gradient $\nabla\mathbf{u}(\mathbf{x})$. Let us consider in turn the interpretation of the diagonal elements ε_{ii} and the off-diagonal elements ε_{ij} (Ziegler, 1983).

Consider for definiteness the diagonal element ε_{11} or ε_{xx} of the strain tensor. If we take $\mathrm{d}\mathbf{x}$ to be of length $\mathrm{d}l$ along the x-axis,

$$\mathrm{d}\mathbf{x} = \mathrm{d}l\, \mathbf{e}_1 \tag{4.12}$$

then the length of $\mathrm{d}\mathbf{x}'$ squared, $\|\mathrm{d}\mathbf{x}'\|^2$, is given by

$$\|\,\mathrm{d}\mathbf{x}'\,\|^2 = \mathrm{d}l\mathbf{e}_1 \cdot (\mathbf{I} + 2\boldsymbol{\varepsilon}) \cdot \mathrm{d}l\mathbf{e}_1 = (1 + 2\varepsilon_{11})(\mathrm{d}l)^2 \tag{4.13}$$

For small α, $(1+2\alpha)^{1/2} \cong 1+\alpha$, so that

$$||\mathrm{d}\mathbf{x}'|| = \mathrm{d}l\,(1+\varepsilon_{11}) \tag{4.14}$$

Similar expressions can be derived for ε_{22} and ε_{33}. Thus the diagonal components of $\boldsymbol{\varepsilon}$ represent the relative elongation along the co-ordinate axes. They are counted positive for tension and negative for compression along these axes.

The diagonal components can be used to calculate the volume change. Consider a rectangular block with edges l_1, l_2 and l_3 ($V_0 = l_1 l_2 l_3$) parallel to the co-ordinate axes. When the block is infinitesimally deformed, the lengths of the edges change to $(1+\varepsilon_{11})l_1$, $(1+\varepsilon_{22})l_2$ and $(1+\varepsilon_{33})l_3$, respectively. To first order the volume becomes $(1+\varepsilon_{11})(1+\varepsilon_{22})(1+\varepsilon_{33})V_0 \cong (1+\varepsilon_{ii})V_0 = (1+\nabla\cdot\mathbf{u})V_0$. The relative change in volume $\Delta V/V_0$ is

► $$\Delta V/V_0 = \nabla\cdot\mathbf{u} = \mathrm{tr}(\boldsymbol{\varepsilon}) = \varepsilon_{ii} \tag{4.15}$$

For the relative elongation in any direction with unit direction vector $\mathbf{n}$, $\mathrm{d}\mathbf{x} = \mathrm{d}l\,\mathbf{n}$ and

$$||\,\mathrm{d}\mathbf{x}'\,||^2 = \mathrm{d}l\,\mathbf{n}\cdot(\mathbf{I}+2\boldsymbol{\varepsilon})\cdot\mathrm{d}l\,\mathbf{n} = \mathrm{d}l^2(1+2\mathbf{n}\cdot\boldsymbol{\varepsilon}\cdot\mathbf{n}) \tag{4.16}$$

The length $||\mathrm{d}\mathbf{x}'||$ yields to first order in the strain components

$$||\mathrm{d}\mathbf{x}'|| = \mathrm{d}l\left[1+\begin{pmatrix} n_1 & n_2 & n_3\end{pmatrix}\begin{pmatrix}\varepsilon_{11} & \varepsilon_{12} & \varepsilon_{13}\\ \varepsilon_{12} & \varepsilon_{22} & \varepsilon_{23}\\ \varepsilon_{23} & \varepsilon_{13} & \varepsilon_{33}\end{pmatrix}\begin{pmatrix}n_1\\ n_2\\ n_3\end{pmatrix}\right] \tag{4.17}$$

Evaluation of the relative length change $\in(\mathbf{n}) = (||\mathrm{d}\mathbf{x}'||-||\mathrm{d}\mathbf{x}||)/||\mathrm{d}\mathbf{x}|| = ||\mathrm{d}\mathbf{x}'||/||\mathrm{d}\mathbf{x}||-1$ along the unit direction vector $\mathbf{n}$, also referred to as *unit length extension*, results in

► $$\in(\mathbf{n}) = \mathbf{n}\cdot\boldsymbol{\varepsilon}\cdot\mathbf{n} = n_i\varepsilon_{ij}n_j \quad \text{or} \quad \in(\mathbf{n}) = \begin{pmatrix} n_1 & n_2 & n_3\end{pmatrix}\begin{pmatrix}\varepsilon_{11} & \varepsilon_{12} & \varepsilon_{13}\\ \varepsilon_{21} & \varepsilon_{22} & \varepsilon_{23}\\ \varepsilon_{31} & \varepsilon_{32} & \varepsilon_{33}\end{pmatrix}\begin{pmatrix}n_1\\ n_2\\ n_3\end{pmatrix} \tag{4.18}$$

Example 4.2

Assume a strain matrix $\varepsilon_{ij} = \begin{pmatrix}0.02 & 0.01 & 0\\ 0.01 & 0.03 & 0\\ 0 & 0 & 0\end{pmatrix}$. To evaluate the relative length change in the $n = (1/\sqrt{2})[1,1,0]$ direction, we calculate

$$\in(\mathbf{n}) = \frac{1}{\sqrt{2}}\begin{pmatrix}1 & 1 & 0\end{pmatrix}\begin{pmatrix}0.02 & 0.01 & 0\\ 0.01 & 0.03 & 0\\ 0 & 0 & 0\end{pmatrix}\frac{1}{\sqrt{2}}\begin{pmatrix}1\\ 1\\ 0\end{pmatrix} = \frac{1}{2}\begin{pmatrix}1 & 1 & 0\end{pmatrix}\begin{pmatrix}0.03\\ 0.04\\ 0\end{pmatrix} = 0.035$$

Consequently the length increase in the [1,1,0] direction is 3.5%. The relative change in volume ε_{ii} is evidently $\varepsilon_{11}+\varepsilon_{22}+\varepsilon_{33} = 0.02+0.03+0 = 0.05$.

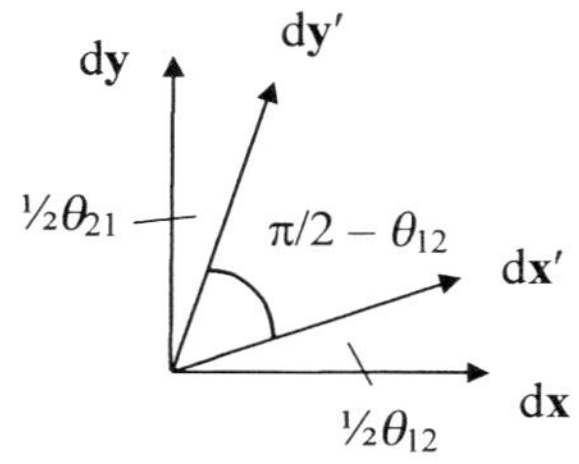

Fig. 4.3: Angle change upon deformation.

Let us now interpret the off-diagonal element ε_{12}. Therefore, we have to consider two elements $d\mathbf{x} = dl\ \mathbf{e}_1$ and $d\mathbf{y} = dl\ \mathbf{e}_2$ directed along the x- and y-axes in the reference state (Fig. 4.3). Obviously in the reference state it holds that $d\mathbf{x}\cdot d\mathbf{y} = 0$, in consonance with an enclosed angle of $\pi/2$ between the vectors $\mathbf{e}_1$ and $\mathbf{e}_2$. In the strained state this angle will change to $\pi/2 - \theta_{12}$. This change can be calculated from the change in the vectors $dl\ \mathbf{e}_1$ and $dl\ \mathbf{e}_2$ given by the expression

$$d\mathbf{x}' = dl\mathbf{e}_1 + dl\boldsymbol{\varepsilon}.\mathbf{e}_1 = dl[(1+\varepsilon_{11})\mathbf{e}_1 + \varepsilon_{12}\mathbf{e}_2 + \varepsilon_{13}\mathbf{e}_3] \quad \text{and} \tag{4.19}$$

$$d\mathbf{y}' = dl\,\mathbf{e}_2 + dl\,\boldsymbol{\varepsilon}.\mathbf{e}_2 = dl\left[\varepsilon_{21}\mathbf{e}_1 + (1+\varepsilon_{22})\mathbf{e}_2 + \varepsilon_{23}\mathbf{e}_3\right] \tag{4.20}$$

The cosine of the angle ($\pi/2-\theta_{12}$) between the two elements is the scalar product of $d\mathbf{x}'$ and $d\mathbf{y}'$ divided by $\|d\mathbf{x}'\|\cdot\|d\mathbf{y}'\|$. To first order in the strain components we obtain

$$\cos(\pi/2-\theta_{12}) = \varepsilon_{21} + \varepsilon_{12} = 2\varepsilon_{12} = 2\varepsilon_{21} \tag{4.21}$$

Since θ_{12} is small, to first order in θ_{12}, it holds that $\cos(\pi/2-\theta_{12}) = \sin(\theta_{12}) \cong \theta_{12} = 2\varepsilon_{12} = 2\varepsilon_{21}$. Similar expressions can be derived for ε_{23} and ε_{13}. Thus the off-diagonal components of $\boldsymbol{\varepsilon}$ represent half the change in the angle between two elements initially along the co-ordinate axes.

Similar to the unit length extension $\in(\mathbf{n})$ for an arbitrary unit vector $\mathbf{n}$, half the change in the angle $\theta(\mathbf{n},\mathbf{m})$ between two general unit direction vectors $\mathbf{n}$ and $\mathbf{m}$ can be obtained. The cosine of the angle $\phi_{\text{fin}}(\mathbf{n},\mathbf{m})$ between $\mathbf{n}$ and $\mathbf{m}$ after deformation is, to first order in the strain components, given by

$$\begin{aligned}\cos\phi_{\text{fin}} &= \begin{pmatrix} n_1 & n_2 & n_3 \end{pmatrix} \begin{pmatrix} 1+2\varepsilon_{11} & 2\varepsilon_{12} & 2\varepsilon_{13} \\ 2\varepsilon_{12} & 1+2\varepsilon_{22} & 2\varepsilon_{23} \\ 2\varepsilon_{31} & 2\varepsilon_{32} & 1+2\varepsilon_{33} \end{pmatrix} \begin{pmatrix} m_1 \\ m_2 \\ m_3 \end{pmatrix} \\ &= \mathbf{n}\cdot\mathbf{m} + 2n_i\varepsilon_{ij}m_j = \cos\phi_{\text{ori}} + 2n_i\varepsilon_{ij}m_j \end{aligned} \tag{4.22}$$

Since the change in ϕ is small, we can write to first order

$$\cos\phi_{\text{fin}} \cong \cos\phi_{\text{ori}} + (\phi_{\text{fin}} - \phi_{\text{ori}})\sin\phi_{\text{ori}}$$

Consequently $\phi_{\text{fin}} - \phi_{\text{ori}} \equiv \theta = 2n_i\varepsilon_{ij}m_j/\sin\phi_{\text{ori}}$. If in the original configuration the directions $\mathbf{n}$ and $\mathbf{m}$ are perpendicular we have $\sin\phi_{\text{ori}} = 1$, so that ($\mathbf{n}\cdot\mathbf{m} = 0$)

► $$\frac{\theta(\mathbf{n},\mathbf{m})}{2} = \mathbf{n}\cdot\boldsymbol{\varepsilon}\cdot\mathbf{m} = n_i\varepsilon_{ij}m_j = \begin{pmatrix} n_1 & n_2 & n_3 \end{pmatrix} \begin{pmatrix} \varepsilon_{11} & \varepsilon_{12} & \varepsilon_{13} \\ \varepsilon_{21} & \varepsilon_{22} & \varepsilon_{23} \\ \varepsilon_{31} & \varepsilon_{32} & \varepsilon_{33} \end{pmatrix} \begin{pmatrix} m_1 \\ m_2 \\ m_3 \end{pmatrix} \tag{4.23}$$

Example 4.3

Assume again the strain matrix $\varepsilon_{ij} = \begin{pmatrix} 0.02 & 0.01 & 0 \\ 0.01 & 0.03 & 0 \\ 0 & 0 & 0 \end{pmatrix}$. To evaluate the angle change between the $n = [1,0,0]$ and $m = [0,1,0]$ directions, we calculate

$$\frac{\boldsymbol{\theta}(\mathbf{n},\mathbf{m})}{2} = \begin{pmatrix} 1 & 0 & 0 \end{pmatrix} \begin{pmatrix} 0.02 & 0.01 & 0 \\ 0.01 & 0.03 & 0 \\ 0 & 0 & 0 \end{pmatrix} \begin{pmatrix} 0 \\ 1 \\ 0 \end{pmatrix} = \begin{pmatrix} 1 & 0 & 0 \end{pmatrix} \begin{pmatrix} 0.01 \\ 0.03 \\ 0 \end{pmatrix} = 0.01$$

Consequently, the angle change $\theta\,(\mathbf{n},\mathbf{m}) = 0.02$ rad $\cong 1.1°$.

Summarising, for a deformation the unit extension of a line element, Eq. (4.18), and half the change in the angle between two orthogonal line elements, Eq. (4.23), in direct notation are given by

$$\in(\mathbf{n}) = \mathbf{n}\cdot\boldsymbol{\varepsilon}\cdot\mathbf{n} \qquad \text{and} \qquad \theta(\mathbf{n},\mathbf{m})/2 = \mathbf{n}\cdot\boldsymbol{\varepsilon}\cdot\mathbf{m} \tag{4.24}$$

In the small displacement gradient approximation the deformations are thus completely defined by the strain tensor. Finally, we note that since the angle of shear $\theta(\mathbf{n},\mathbf{m})$ is equal to $2\mathbf{n}\cdot\boldsymbol{\varepsilon}\cdot\mathbf{m}$, frequently also the *conventional* or *engineering shear strains* $\gamma_{ij} = 2\varepsilon_{ij}$ $(i{\neq}j)$ are introduced. The ε_{ij} denote the *tensorial shear strains*.

Problem 4.4

Derive Eq. (4.22).

Problem 4.5

For the deformation as described in Problem 4.1 calculate the length change in the $\langle 011\rangle$ direction and the angle change between the $\langle 001\rangle$ and $\langle 010\rangle$ directions.

Problem 4.6

The strain matrix at point P in a body is given by

$$\boldsymbol{\varepsilon} = \begin{pmatrix} 2\times10^{-4} & 0 & 0 \\ 0 & 10^{-4} & 0 \\ 0 & 0 & 0 \end{pmatrix}$$

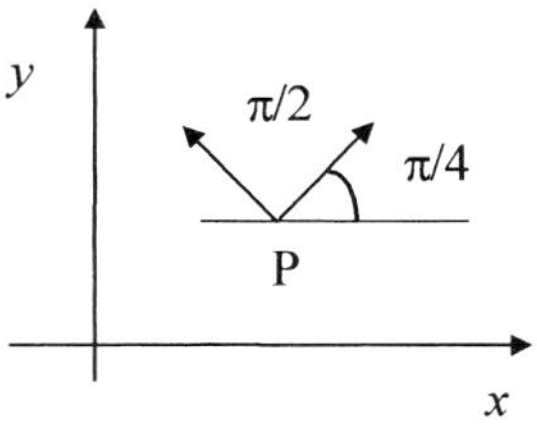

Consider the directions at point P as indicated in the accompanying figure. Calculate the change in the angle between these two directions.

4.4 Strain in cylindrical and spherical co-ordinates

In many problems Cartesian co-ordinates are not the most suitable, e.g. for cylindrical holes cylindrical co-ordinates are a logical choice and with spherical holes spherical co-ordinates can be used. To that purpose we need to express the strain components in these co-ordinates. Taking the symmetric part of $\nabla\mathbf{u}$ as given in Section 4.2, we obtain for the strain in cylindrical co-ordinates (Lubliner, 1990)

$$\begin{aligned}
&\varepsilon_{rr} = \frac{\partial u_r}{\partial r} \qquad &&\varepsilon_{\theta\theta} = \frac{u_r}{r} + \frac{1}{r}\frac{\partial u_\theta}{\partial \theta} \qquad &&\varepsilon_{zz} = \frac{\partial u_z}{\partial z} \\
&\gamma_{r\theta} = \frac{\partial u_\theta}{\partial r} + \frac{1}{r}\frac{\partial u_r}{\partial \theta} - \frac{u_\theta}{r} \qquad &&\gamma_{rz} = \frac{\partial u_z}{\partial r} + \frac{\partial u_r}{\partial z} \qquad &&\gamma_{\theta z} = \frac{1}{r}\frac{\partial u_z}{\partial \theta} + \frac{\partial u_\theta}{\partial z}
\end{aligned} \tag{4.25}$$

In spherical co-ordinates we find for the strain

$$\begin{aligned}
&\varepsilon_{rr} = \frac{\partial u_r}{\partial r} \qquad &&\varepsilon_{\theta\theta} = \frac{1}{r}\frac{\partial u_\theta}{\partial \theta} + \frac{u_r}{r} \\
&\varepsilon_{\varphi\varphi} = \frac{1}{r\sin\theta}\frac{\partial u_\varphi}{\partial \varphi} + \frac{u_r}{r} + \frac{\cot\theta}{r}u_\theta \qquad &&\gamma_{r\theta} = \frac{\partial u_\theta}{\partial r} + \frac{1}{r}\frac{\partial u_r}{\partial \theta} - \frac{u_\theta}{r} \\
&\gamma_{r\varphi} = \frac{\partial u_\varphi}{\partial r} + \frac{1}{r\sin\theta}\frac{\partial u_r}{\partial \varphi} - \frac{u_\varphi}{r} \qquad &&\gamma_{\theta\varphi} = \frac{1}{r}\frac{\partial u_\varphi}{\partial \theta} + \frac{1}{r\sin\theta}\frac{\partial u_\theta}{\partial \varphi} - \frac{\cot\theta}{r}u_\varphi
\end{aligned} \tag{4.26}$$

The corresponding expressions can be obtained for other co-ordinate systems.

4.5 Material derivatives and integrals*

The distinction between the spatial and material co-ordinates may be ignored in the small displacement gradient approximation. There remains, however, a difference in some types of derivatives that we need (Ziegler, 1983).

Consider a tensor which is, deviating from our 'tensor = uppercase bold' rule, here indicated by **a** or a_{ij}. The *local change* of that tensor a_{ij} in the time element dt, i.e. its increment at a certain spatial point P during dt, is given by $(\partial a_{ij}/\partial t)\mathrm{d}t$, where $\partial a_{ij}/\partial t$ is the partial derivative with respect to time. Occasionally we will write $\partial a_{ij}/\partial t = a_{ij,0}$, i.e. using the standard notation for partial differentiation but using a subscript number 0 for time. To avoid confusion, the letter o is not used as an index.

The *instantaneous distribution* of any tensor a_{ij} in the vicinity of the material point P is described by its gradient and thus given by $\partial a_{ij}/\partial x_k = a_{ij,k}$. For an observer moving with the accompanying co-ordinate system, displaced $\mathrm{d}x_k$ in the time interval dt, the change in a_{ij} is given by $a_{ij,0}\,\mathrm{d}t + a_{ij,k}\,\mathrm{d}x_k$. For that observer $\mathrm{d}x_k = v_k\,\mathrm{d}t$, where v_k denote the components of the velocity **v**. The *material derivative* is defined by

► $$\frac{\mathrm{d}a_{ij}}{\mathrm{d}t} \equiv \dot{a}_{ij} = a_{ij,0} + a_{ij,k}v_k \qquad \text{or} \qquad \frac{\mathrm{d}\mathbf{a}}{\mathrm{d}t} \equiv \dot{\mathbf{a}} = \frac{\partial \mathbf{a}}{\partial t} + \nabla\mathbf{a}\cdot\mathbf{v} \tag{4.27}$$

The first term denotes the *local derivative* and the second the *convective derivative*.

Consider now quantities defined as volume integrals over a certain material region, that is to say a volume element always enclosing the same set of particles. A general expression would be

$$C_{ij} = \int c_{ij}(x_k, t)\, \mathrm{d}V \tag{4.28}$$

where for the moment C_{ij} is used for the integral over the tensor c_{ij}. The volume V is bounded by the surface A and the outer normal is indicated by n_k (Fig. 4.4). Since the deformation is continuous, the particles at the surface A at time t are displaced to A' at time $t' = t+\mathrm{d}t$. A material point located originally at x_j moves to $x_j' = x_j+v_j\,\mathrm{d}t$ at time $t' = t+\mathrm{d}t$. The change in C_{ij} is

$$\mathrm{d}C_{ij} = \dot{C}_{ij}\,\mathrm{d}t = \int c_{ij}(x_k', t')\, \mathrm{d}V' - \int c_{ij}(x_k, t)\, \mathrm{d}V \tag{4.29}$$

where V' is the volume at time t'. The material points contained in both V and V' lead to the change $c_{ij,0}\,\mathrm{d}t\,\mathrm{d}V$. The volume elements inside V' but not in V contribute to the volume $\mathrm{d}V = v_k\,\mathrm{d}t\,n_k\,\mathrm{d}A$ and thus contribute to the integral $c_{ij}v_k\,\mathrm{d}t\,n_k\,\mathrm{d}A$. The same contribution arises from the volume elements inside V but not in V'. The total result is thus in index and direct notation, respectively,

$$\dot{C}_{ij} = \int c_{ij,0}\, \mathrm{d}V + \int c_{ij} v_k n_k\, \mathrm{d}A \quad \text{or} \quad \dot{\mathbf{C}} = \int \frac{\partial \mathbf{c}}{\partial t} \mathrm{d}V + \int \mathbf{c}(\mathbf{v}\cdot\mathbf{n})\mathrm{d}A \tag{4.30}$$

where the first and second terms on the right-hand side are the *accumulation* and *flux term*, respectively. Using the divergence theorem, Eq. (4.30) may be written as

$$\blacktriangleright \quad \begin{aligned} \dot{C}_{ij} &= \int \left[c_{ij,0} + (c_{ij}v_k)_{,k}\right] \mathrm{d}V \\ &= \int \left[c_{ij,0} + c_{ij,k}v_k + c_{ij}v_{k,k}\right] \mathrm{d}V = \int \left[\dot{c}_{ij} + c_{ij}v_{k,k}\right] \mathrm{d}V \end{aligned} \tag{4.31}$$

If a conservation law for the quantity $\mathbf{C}$ is valid, $\mathrm{d}\mathbf{C}/\mathrm{d}t = 0$. Since the volume element is arbitrary, not only the integral but also the integrand must be zero, resulting in

$$c_{ij,0} + (c_{ij}v_k)_{,k} = 0 \quad \text{or} \quad \dot{c}_{ij} + c_{ij}v_{k,k} = 0 \tag{4.32}$$

valid for an observer at rest and an observer moving with the accompanying system, respectively.

An important example for which a conservation law applies is the mass

$$m = \int \rho(x_j, t)\, \mathrm{d}V \tag{4.33}$$

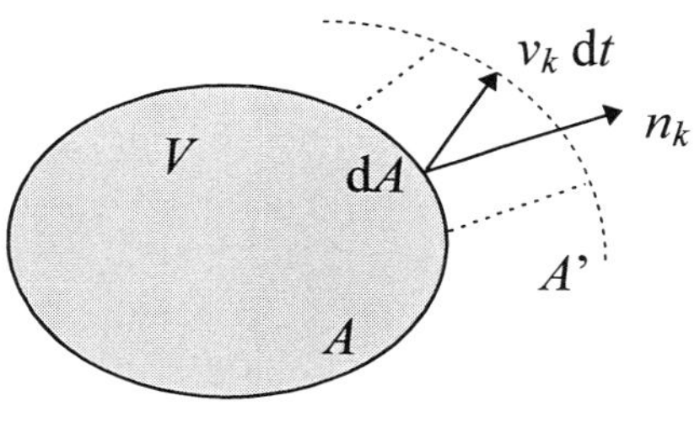

Fig. 4.4: Material volume in motion.

defined as the integral over the *density*, i.e. the mass per unit volume, $\rho = \mathrm{d}m/\mathrm{d}V$. Here we obtain

$$\dot{m} = \left(\int \rho \mathrm{d}V \right)^{\bullet} = \frac{\mathrm{d}}{\mathrm{d}t} \int \rho \mathrm{d}V = 0 \tag{4.34}$$

resulting in

$$\dot{m} = \int \frac{\partial \rho}{\partial t} \mathrm{d}V + \int \rho(\mathbf{v} \cdot \mathbf{n}) \mathrm{d}A = \frac{\partial}{\partial t} \int \rho \mathrm{d}V + \int \rho(\mathbf{v} \cdot \mathbf{n}) \mathrm{d}A \tag{4.35}$$

After applying the divergence theorem, the local relations in index and direct notation become

$$\rho_{,0} + (\rho v_k)_{,k} = 0 \qquad \text{and} \qquad \frac{\partial \rho}{\partial t} + \operatorname{div}(\rho \mathbf{v}) = 0 \tag{4.36}$$

$$\dot{\rho} + \rho v_{k,k} = 0 \qquad \text{and} \qquad \dot{\rho} + \rho \operatorname{div} \mathbf{v} = 0 \tag{4.37}$$

again valid for an observer at rest and an observer moving with the accompanying system, respectively. These equations are referred to as the *continuity equations*.

A direct implication is that for a quantity where one of the factors in the integrand is the density, e.g.

$$C_{ij} = \int \rho \, c_{ij} \, \mathrm{d}V \tag{4.38}$$

its material derivative is

$$\mathrm{d}C_{ij} / \mathrm{d}t = \dot{C}_{ij} = \int \rho \, \dot{c}_{ij} \, \mathrm{d}V \tag{4.39}$$

so that now the material derivative of C_{ij} equals the integral over the density ρ and the material derivative $\mathrm{d}c_{ij}/\mathrm{d}t$.

4.6 Compatibility*

Although each displacement field $\mathbf{u}$ defines a strain $\varepsilon_{ij} = u_{(i,j)}$, the reverse is not automatically true. Given a strain tensor ε_{ij}', it is not always possible to find an associated displacement field, that is to say a vector field $\mathbf{u}$ such that $\varepsilon_{ij}' = (\nabla \mathbf{u})_{ij}$. However, when it is possible the strain tensor is said to be *compatible*. Obviously, compatibility is closely related to integrability. The compatibility equations can be obtained by eliminating the displacements from Eq. (4.8). As an example consider

$$\varepsilon_{11} = \frac{\partial u_1}{\partial x_1} \qquad \varepsilon_{22} = \frac{\partial u_2}{\partial x_2} \qquad 2\varepsilon_{12} = \frac{\partial u_1}{\partial x_2} + \frac{\partial u_2}{\partial x_1}$$

Taking the second derivative of ε_{11} with respect to x_2, the second derivative of ε_{22} with respect to x_1 and the second derivative of ε_{12} once with respect to x_1 and once with respect to x_2, we obtain

$$\varepsilon_{11,22} = \frac{\partial^3 u_1}{\partial x_1 \partial^2 x_2} \qquad \varepsilon_{22,11} = \frac{\partial^3 u_2}{\partial x_2 \partial^2 x_1} \qquad 2\varepsilon_{12,12} = \frac{\partial^3 u_1}{\partial x_1 \partial^2 x_2} + \frac{\partial^3 u_2}{\partial^2 x_1 \partial x_2}$$

Substituting the first two equations in the third yields

$$\varepsilon_{11,22} + \varepsilon_{22,11} - 2\varepsilon_{12,12} = 0 \tag{4.40}$$

For plane strain conditions with, say $\varepsilon_{i3} = 0$, Eq. (4.40) is sufficient. It can be shown that the only displacement field compatible with plane strain is one of *plane displacement*: $u_1 = u_1(x_1,x_2)$ and $u_2 = u_2(x_1,x_2)$. Carrying out a similar operation for each of the components of the strain tensor ε_{ij} and collecting the results one can obtain

► $$\varepsilon_{ij,kl} + \varepsilon_{kl,ij} - \varepsilon_{ik,jl} - \varepsilon_{jl,ik} = 0 \tag{4.41}$$

Without proof we quote the alternative expression

$$e_{ijk} e_{mln} \varepsilon_{km,jn} = 0 \tag{4.42}$$

The left-hand side, which is a second-order tensor, is known as the *incompatibility tensor*. If this equation is satisfied, there is no incompatibility for a simply connected region.

4.7 General deformations*

Before we proceed we expand a little on material derivatives. As we have seen in Section 4.5, the time derivative of any quantity ϕ in the material description is called the *material time derivative*, denoted by $\dot{\phi} = \mathrm{d}\phi/\mathrm{d}t$. In fact it is the partial time derivative with the material co-ordinates $\mathbf{r}$ kept constant. In particular, the *material velocity* $\mathbf{v}$ and the *material acceleration* $\mathbf{a}$ are given by (Kuiken, 1994)

$$\mathbf{v}(\mathbf{x},t) = \dot{\mathbf{x}}(\mathbf{x},t) = \frac{\mathrm{d}}{\mathrm{d}t}\mathbf{x}(\mathbf{r},t) \qquad \text{and} \qquad \mathbf{a} = \dot{\mathbf{v}} = \frac{\mathrm{d}\mathbf{v}}{\mathrm{d}t} = \frac{\mathrm{d}^2\mathbf{x}}{\mathrm{d}t^2} \tag{4.43}$$

respectively. For the material time derivative of any quantity ϕ one finds

$$\dot{\phi}(\mathbf{x},t) = \frac{\mathrm{d}}{\mathrm{d}t}\phi(\mathbf{x},\mathrm{t}) = \frac{\partial\phi}{\partial t} + \frac{\partial\phi}{\partial\mathbf{x}}\cdot\frac{\mathrm{d}\mathbf{x}}{\mathrm{d}t} = \frac{\partial\phi}{\partial t} + \mathbf{v}\cdot\frac{\partial\phi}{\partial\mathbf{x}} \tag{4.44}$$

The first term is the *local derivative* and is the change observed by an observer at constant position $\mathbf{x}$ while the second term is called the *convective derivative*.

Returning to the topic of general deformations, let us consider a particle in the reference configuration. A line element at $\mathbf{r}$ is given by $\mathrm{d}\mathbf{r}$. At a (later) time t, the particle considered will be at point $\mathbf{x}$. In the deformed configuration the corresponding line element is given by $\mathrm{d}\mathbf{x}$. The deformation is described by

$$\mathrm{d}\mathbf{x} = \mathbf{F}\cdot\mathrm{d}\mathbf{r} \qquad \text{or} \qquad \mathbf{F} = \frac{\partial\,\mathbf{x}(\mathbf{r})}{\partial\,\mathbf{r}} \tag{4.45}$$

where the tensor $\mathbf{F}$ is called the *deformation gradient*. The squared length of the line element $\mathrm{d}\mathbf{x}$ can be calculated from

$$\mathrm{d}\mathbf{x}\cdot\mathrm{d}\mathbf{x} = \mathrm{d}\mathbf{r}\cdot\mathbf{F}^{\mathrm{T}}\cdot\mathbf{F}\cdot\mathrm{d}\mathbf{r} = \mathrm{d}\mathbf{r}\cdot\mathbf{C}\cdot\mathrm{d}\mathbf{r} \qquad \text{and} \qquad \mathbf{C} = \mathbf{F}^{\mathrm{T}}\cdot\mathbf{F} = \left(\frac{\partial\mathbf{x}}{\partial\mathbf{r}}\right)^{\mathrm{T}}\cdot\left(\frac{\partial\mathbf{x}}{\partial\mathbf{r}}\right) \tag{4.46}$$

where $\mathbf{C}$, the *right Cauchy-Green tensor*, is a measure of the pure deformation.

Example 4.4

The tensor **F** can, according to the polar decomposition theorem, be written as **F** = **R**·**U**, where **U** is a symmetric tensor, the *right stretch tensor*, describing the deformation and **R** is an orthogonal tensor, describing the rotation. Therefore

$$\mathbf{F}^{\mathrm{T}} = \mathbf{U}^{\mathrm{T}}\cdot\mathbf{R}^{\mathrm{T}} \quad \text{and} \quad \mathbf{F}^{\mathrm{T}}\cdot\mathbf{F} = \mathbf{U}^{\mathrm{T}}\cdot\mathbf{R}^{\mathrm{T}}\cdot\mathbf{R}\cdot\mathbf{U} = \mathbf{U}^{\mathrm{T}}\cdot\mathbf{U}$$

Consequently the right Cauchy-Green **C** tensor describes pure deformation.

Consider next the deformation of a line element d**r** that in the reference state has a length $\mathrm{d}l_0$ and a (fixed) direction **m**. After deformation the length has changed to dl and the direction to **n**. So the relations

$$\mathrm{d}\mathbf{r} = \mathrm{d}l_0\,\mathbf{m} \qquad \mathrm{d}\mathbf{x} = \mathrm{d}l\,\mathbf{n} \qquad \text{and} \qquad \mathrm{d}\mathbf{x} = \mathbf{F}\cdot\mathrm{d}\mathbf{r}$$

can also be written as

$$\mathrm{d}l\,\mathbf{n} = \mathbf{F}\cdot\mathrm{d}l_0\,\mathbf{m} \qquad \text{or} \qquad \lambda\mathbf{n} = \mathbf{F}\cdot\mathbf{m}$$

where $\lambda = \mathrm{d}l/\mathrm{d}l_0$. The quantity λ denotes the *stretch* and describes the elongation ratio of the vectors d**x** and d**r**. Note that the stretch must be positive. It follows that

► $$\lambda^2 = \mathbf{m}\cdot\mathbf{F}^{\mathrm{T}}\cdot\mathbf{F}\cdot\mathbf{m} = \mathbf{m}\cdot\mathbf{C}\cdot\mathbf{m} \tag{4.47}$$

This kind of expression represents an ellipsoid for which the length of a vector in the direction **m** from the centre of the ellipsoid to a particular point at the surface is given by $\lambda^2(\mathbf{m})$. Such an ellipsoid is called a *representation ellipsoid.*

Let us consider how the stretch varies with time. The time derivative of λ^2 is given by (Ziegler, 1983)

$$\left(\lambda^2\right)^{\bullet} = 2\lambda(\dot{\lambda}) = \mathbf{m}\cdot\dot{\mathbf{C}}\cdot\mathbf{m} = \mathbf{m}\cdot\left(\dot{\mathbf{F}}^{\mathrm{T}}\cdot\mathbf{F}+\mathbf{F}^{\mathrm{T}}\cdot\dot{\mathbf{F}}\right)\cdot\mathbf{m} \tag{4.48}$$

because **m** has a fixed direction. Using

$$\dot{\mathbf{F}}^{\mathrm{T}}\cdot\mathbf{F} = \frac{\mathrm{d}}{\mathrm{d}t}\left(\frac{\partial\mathbf{x}}{\partial\mathbf{r}}\right)^{\mathrm{T}}\cdot\frac{\partial\mathbf{x}}{\partial\mathbf{r}} = \left(\frac{\partial\mathbf{v}}{\partial\mathbf{r}}\right)^{\mathrm{T}}\cdot\frac{\partial\mathbf{x}}{\partial\mathbf{r}} = \left(\frac{\partial\mathbf{x}}{\partial\mathbf{r}}\right)^{\mathrm{T}}\cdot\left(\frac{\partial\mathbf{v}}{\partial\mathbf{x}}\right)^{\mathrm{T}}\cdot\frac{\partial\mathbf{x}}{\partial\mathbf{r}} = \mathbf{F}^{\mathrm{T}}\cdot\left(\frac{\partial\mathbf{v}}{\partial\mathbf{x}}\right)^{\mathrm{T}}\cdot\mathbf{F} \tag{4.49}$$

and its transpose yields together with $\lambda\mathbf{n} = \mathbf{F}\cdot\mathbf{m}$

► $$\dot{\lambda}/\lambda = \tfrac{1}{2}\mathbf{n}\cdot\left[\left(\frac{\partial\mathbf{v}}{\partial\mathbf{x}}\right)^{\mathrm{T}}+\left(\frac{\partial\mathbf{v}}{\partial\mathbf{x}}\right)\right]\cdot\mathbf{n} = \mathbf{n}\cdot\mathbf{d}\cdot\mathbf{n} \tag{4.50}$$

so that, maybe somewhat surprising, the relative rate of elongation $\dot{\lambda}/\lambda$ can be expressed entirely in the spatial co-ordinates **x**. The tensor **d** is known as the *rate of deformation*. In the spatial description the deformation is thus exactly given by **d**.

Let us again consider a particle in the reference configuration. In the material description the *displacement* vector **u** can be introduced by (Kuiken, 1994)

$$\mathbf{x}(\mathbf{r}) = \mathbf{r}+\mathbf{u}(\mathbf{r})-\mathbf{b} \tag{4.51}$$

where **u** is considered as a function of the material co-ordinates **r**. The vector **b** is a constant vector, specifying the origin $\mathrm{O_s}$ of the spatial co-ordinate system relative to the origin O_m of the material co-ordinate system. Normally $\mathbf{b} = \mathbf{0}$ is taken. In that case in the reference configuration the material co-ordinates $\mathbf{r}$ and spatial co-ordinates $\mathbf{x}$ coincide. If for all particles $\mathbf{r}$ the displacement $\mathbf{u}(\mathbf{r})$ is known, a complete material description of the deformation is given. Taking $\mathbf{b} = \mathbf{0}$ and using $\mathbf{u}(\mathbf{r})$, it follows that

$$\mathbf{F} = \mathbf{I} + \frac{\partial \mathbf{u}(\mathbf{r})}{\partial \mathbf{r}} \tag{4.52}$$

where $\mathbf{I}$ denotes the unit tensor and the derivative denotes the *material displacement gradient*. The right Cauchy-Green tensor $\mathbf{C} = \mathbf{F}^T \cdot \mathbf{F}$ tensor thus can also be written as

$$\mathbf{C} = \mathbf{I} + \left[\left(\frac{\partial \mathbf{u}}{\partial \mathbf{r}}\right)^T + \left(\frac{\partial \mathbf{u}}{\partial \mathbf{r}}\right)\right] + \left(\frac{\partial \mathbf{u}}{\partial \mathbf{r}}\right)^T \cdot \left(\frac{\partial \mathbf{u}}{\partial \mathbf{r}}\right) \tag{4.53}$$

We can calculate the increase in length of the line element from

$$d\mathbf{x} \cdot d\mathbf{x} - d\mathbf{r} \cdot d\mathbf{r} = d\mathbf{r} \cdot \mathbf{F}^T \cdot \mathbf{F} \cdot d\mathbf{r} - d\mathbf{r} \cdot d\mathbf{r} = d\mathbf{r} \cdot \mathbf{C} \cdot d\mathbf{r} - d\mathbf{r} \cdot d\mathbf{r} = d\mathbf{r} \cdot 2\mathbf{L} \cdot d\mathbf{r} \tag{4.54}$$

where $\mathbf{L} = ½(\mathbf{C} - \mathbf{I})$ denotes the *Lagrange strain tensor*. The expression for $\mathbf{L}$, given by

► $$\mathbf{L} = \tfrac{1}{2}\left[\left(\frac{\partial \mathbf{u}}{\partial \mathbf{r}}\right) + \left(\frac{\partial \mathbf{u}}{\partial \mathbf{r}}\right)^T\right] + \tfrac{1}{2}\left(\frac{\partial \mathbf{u}}{\partial \mathbf{r}}\right)^T \cdot \left(\frac{\partial \mathbf{u}}{\partial \mathbf{r}}\right) \equiv \boldsymbol{\varepsilon} + \tfrac{1}{2}\left(\frac{\partial \mathbf{u}}{\partial \mathbf{r}}\right)^T \cdot \left(\frac{\partial \mathbf{u}}{\partial \mathbf{r}}\right) \tag{4.55}$$

is easily obtained. If quantities of second order can be neglected, $\mathbf{L} \cong \boldsymbol{\varepsilon}$. The quantity $\boldsymbol{\varepsilon}$ denote the (Cauchy) *small displacement gradient* (SDG) *strain tensor*, although in the literature it is frequently called the *infinitesimal strain tensor*. We will refer to $\boldsymbol{\varepsilon}$ as the strain tensor. Note that the use of the symbol $\boldsymbol{\varepsilon}$ deviates from our 'tensor = uppercase (bold)' rule but the notation is conventional.

Example 4.5

If we use the spatial description, we similarly can write $\mathbf{u}(\mathbf{x}) = \mathbf{x} - \mathbf{r}(\mathbf{x})$, where $\mathbf{u}$ and $\mathbf{r}$ are now considered as functions of $\mathbf{x}$. Differentiation yields

$$\partial\mathbf{u}/\partial\mathbf{x} = \mathbf{I} - \partial\mathbf{r}/\partial\mathbf{x} = \mathbf{I} - \mathbf{F}^{-1} \quad \text{or} \quad \mathbf{F}^{-1} = \mathbf{I} - \partial\mathbf{u}/\partial\mathbf{x}$$

Since

$$\partial\mathbf{u}/\partial\mathbf{x} = \partial\mathbf{u}/\partial\mathbf{r}\cdot\partial\mathbf{r}/\partial\mathbf{x} = \partial\mathbf{u}/\partial\mathbf{r}\cdot\mathbf{F}^{-1} = \partial\mathbf{u}/\partial\mathbf{r}\cdot(\mathbf{I} - \partial\mathbf{u}/\partial\mathbf{x}) \cong \partial\mathbf{u}/\partial\mathbf{r}$$

to first order, we have $\partial\mathbf{u}(\mathbf{x})/\partial\mathbf{x} = \partial\mathbf{u}(\mathbf{r})/\partial\mathbf{r}$, implying that when $||\nabla\mathbf{u}(\mathbf{r})|| << 1$ is valid, it is immaterial whether material or spatial co-ordinates are used.

The expression for $\mathbf{u}(\mathbf{x})$ can be obtained from $\mathbf{u}(\mathbf{r})$ and vice versa although the actual evaluation can be quite complicated. It must be admitted that in the spatial description the use of the displacement is somewhat artificial and the deformation is most easily described by the velocity field. Example 4.6 nevertheless illustrates for an easy case the indifference of using material or spatial co-ordinates in the SDG approximation.

Let us now consider another consequence of the small displacement gradient approximation. From the definition of **L**, **C** and the SDG approximation, one obtains

$$\mathbf{C} = \mathbf{I} + 2\mathbf{L} \cong \mathbf{I} + 2\boldsymbol{\varepsilon} \tag{4.56}$$

To obtain the stretch λ we note that $\mathrm{d}\mathbf{x} = (\mathbf{I}+\boldsymbol{\varepsilon}+\boldsymbol{\omega})\cdot\mathrm{d}\mathbf{r} = (\mathbf{I}+\boldsymbol{\varepsilon})\cdot\mathrm{d}\mathbf{r}$, where the rotation $\boldsymbol{\omega}$ in the last expression is neglected since it does not lead to length changes. Hence

$$\lambda = \frac{\|\mathrm{d}\mathbf{x}\|}{\|\mathrm{d}\mathbf{r}\|} = \left[\frac{\mathrm{d}\mathbf{r}\cdot(\mathbf{I}+2\boldsymbol{\varepsilon})\cdot\mathrm{d}\mathbf{r}}{\mathrm{d}\mathbf{r}\cdot\mathrm{d}\mathbf{r}}\right]^{1/2} \cong 1 + \frac{\mathrm{d}\mathbf{r}\cdot\boldsymbol{\varepsilon}\cdot\mathrm{d}\mathbf{r}}{\mathrm{d}\mathbf{r}\cdot\mathrm{d}\mathbf{r}} \tag{4.57}$$

Consequently the elongation of a line element $\mathrm{d}\mathbf{r}$ is described by the strain $\boldsymbol{\varepsilon}$.

Example 4.6

Consider the 2D displacement field $u_1(\mathbf{r}) = \alpha r_2$, $u_2(\mathbf{r}) = \alpha r_1$ with α a number and r_1 and r_2 the components of the material position vector $\mathbf{r}$. The spatial coordinates x_1 and x_2 are $x_1 = r_1+u_1$ and $x_2 = r_2+u_2$, respectively. Hence

$$\begin{pmatrix} x_1 \\ x_2 \end{pmatrix} = \begin{pmatrix} 1 & \alpha \\ \alpha & 1 \end{pmatrix}\begin{pmatrix} r_1 \\ r_2 \end{pmatrix} \quad \text{or in direct notation} \quad \mathbf{x} = \mathbf{A}\cdot\mathbf{r}$$

Inversion yields $\mathbf{r} = \mathbf{A}^{-1}\cdot\mathbf{x}$ or in full

$$\begin{pmatrix} r_1 \\ r_2 \end{pmatrix} = \frac{1}{1-\alpha^2}\begin{pmatrix} 1 & -\alpha \\ -\alpha & 1 \end{pmatrix}\begin{pmatrix} x_1 \\ x_2 \end{pmatrix} \quad \text{leading to} \quad \begin{pmatrix} r_1 \\ r_2 \end{pmatrix} = \begin{pmatrix} 1 & -\alpha \\ -\alpha & 1 \end{pmatrix}\begin{pmatrix} x_1 \\ x_2 \end{pmatrix}$$

to first order ($\alpha << 1$). Therefore $r_1 = x_1 - u_1(\mathbf{x})$ and $r_2 = x_2 - u_2(\mathbf{x})$ with $u_1 = \alpha x_2$ and $u_2 = \alpha x_1$. If we assume that the approximation $\|\nabla\mathbf{u}(\mathbf{r})\| << 1$ is valid, it is immaterial whether during differentiation points are considered as spatial or as material points: In both cases $(\nabla\mathbf{u})_{11} = (\nabla\mathbf{u})_{22} = 0$ and $(\nabla\mathbf{u})_{12} = (\nabla\mathbf{u})_{21} = \alpha$.

The strain rate $\mathrm{d}\boldsymbol{\varepsilon}/\mathrm{d}t$ is related to the deformation rate in the SDG approximation

$$2\dot{\boldsymbol{\varepsilon}} = \frac{\mathrm{d}}{\mathrm{d}t}\left[\left(\frac{\partial\mathbf{u}}{\partial\mathbf{r}}\right)^{\mathrm{T}} + \left(\frac{\partial\mathbf{u}}{\partial\mathbf{r}}\right)\right] \cong \frac{\mathrm{d}}{\mathrm{d}t}\left[\left(\frac{\partial\mathbf{u}}{\partial\mathbf{x}}\right)^{\mathrm{T}} + \left(\frac{\partial\mathbf{u}}{\partial\mathbf{x}}\right)\right] = \left[\left(\frac{\partial\mathbf{v}}{\partial\mathbf{x}}\right)^{\mathrm{T}} + \left(\frac{\partial\mathbf{v}}{\partial\mathbf{x}}\right)\right] = 2\mathbf{d} \tag{4.58}$$

In the SDG approximation $\mathrm{d}\boldsymbol{\varepsilon}/\mathrm{d}t$ and $\mathbf{d}$ are thus identical. Hence the strain can be calculated as the integral of $\mathrm{d}\boldsymbol{\varepsilon}/\mathrm{d}t$, which is equal to the integral of $\mathbf{d}$, and thus

$$\boldsymbol{\varepsilon} = \int\dot{\boldsymbol{\varepsilon}}\,\mathrm{d}t = \int\mathbf{d}\,\mathrm{d}t \tag{4.59}$$

So far we have described deformation by the SDG strain. Under certain circumstances another measure of deformation is useful. Taking logarithms of Eq. (4.47) we obtain

$$\ln\lambda = \tfrac{1}{2}\ln(\mathbf{m}\cdot\mathbf{C}\cdot\mathbf{m}) = \tfrac{1}{2}\ln(1+2\mathbf{m}\cdot\mathbf{L}\cdot\mathbf{m}) \cong \mathbf{m}\cdot\mathbf{L}\cdot\mathbf{m} \cong \mathbf{m}\cdot\boldsymbol{\varepsilon}\cdot\mathbf{m} \tag{4.60}$$

where $\mathbf{m}$ is a unit vector and the third step can be made if $\|\mathbf{m}\cdot\mathbf{L}\cdot\mathbf{m}\| << \tfrac{1}{2}$. Since the stretch $\lambda = \mathrm{d}l/\mathrm{d}l_0$ represents the elongation ratio of a line of length $\mathrm{d}l_0$, the quantity $\ln\lambda$ is known as the *logarithmic strain*. If $\mathrm{d}l - \mathrm{d}l_0$ is small with respect to $\mathrm{d}l_0$, we have

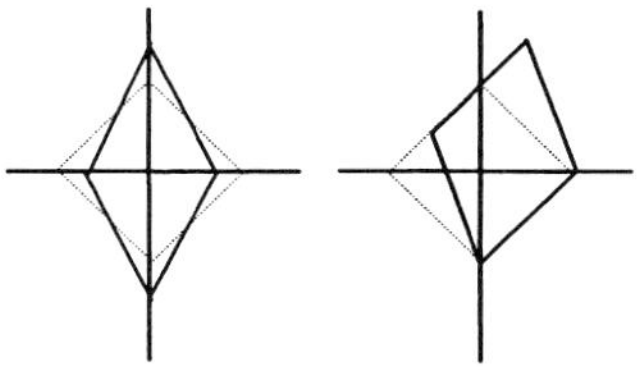

Fig. 4.5: Two modes of shear. The original shape is indicated by - - -.

$$\ln\frac{\mathrm{d}l}{\mathrm{d}l_0} = \ln\left(\frac{\mathrm{d}l-\mathrm{d}l_0}{\mathrm{d}l_0}+1\right) \cong \frac{\mathrm{d}l-\mathrm{d}l_0}{\mathrm{d}l_0} = \lambda - 1 = \|\mathbf{m}\cdot\boldsymbol{\varepsilon}\cdot\mathbf{m}\| \tag{4.61}$$

The logarithmic strain, also known as *natural strain*, thus reduces to the SDG strain in the considered direction **m** for small length increase. From this it also follows that a description using the SDG strain is only valid if the directions of the principal values of the strain tensor do not rotate significantly in the time element considered. There are only three deformations for which this is rigorously true: isotropic volume change, uniaxial length change and pure shear. Note that simple shear is a combination of pure shear and rotation (Fig. 4.5).

Problem 4.7

Show that, using the *left Cauchy-Green tensor* $\mathbf{D} = \mathbf{F}^{-\mathrm{T}}\cdot\mathbf{F}^{-1}$, the *Euler strain* **E**, defined according to $\mathrm{d}\mathbf{x}\cdot\mathrm{d}\mathbf{x}-\mathrm{d}\mathbf{r}\cdot\mathrm{d}\mathbf{r} = 2\mathrm{d}\mathbf{x}\cdot\mathbf{E}\cdot\mathrm{d}\mathbf{x}$, is given by $\mathbf{E} = ½(\mathbf{I}-\mathbf{D})$ or

$$\mathbf{E} = ½\left[\left(\frac{\partial\mathbf{u}}{\partial\mathbf{x}}\right)+\left(\frac{\partial\mathbf{u}}{\partial\mathbf{x}}\right)^{\mathrm{T}}\right] - ½\left(\frac{\partial\mathbf{u}}{\partial\mathbf{x}}\right)^{\mathrm{T}}\cdot\left(\frac{\partial\mathbf{u}}{\partial\mathbf{x}}\right) \tag{4.62}$$

4.8 Physical interpretation revisited*

The length and angle changes for general deformations follow from the Lagrange tensor $\mathbf{L} = ½(\mathbf{C}-\mathbf{I})$ with **C** the right Cauchy-Green tensor and **I** the unit tensor. For the length change, using **r** and **x** as material and spatial co-ordinates, we have (Teodosiu, 1982)

$$\mathrm{d}\mathbf{x}^2 - \mathrm{d}\mathbf{r}^2 = 2\mathrm{d}\mathbf{r}\cdot\mathbf{L}\cdot\mathrm{d}\mathbf{r} \tag{4.63}$$

Let **m** again be the unit vector indicating the direction of the material vector d**r**, i.e. $\mathrm{d}\mathbf{r} = \mathrm{d}l_0\,\mathbf{m}$, where $\mathrm{d}l_0$ denotes the length of d**r**. Denoting the length of a material vector d**x** in the current configuration by dl, so that $\mathrm{d}\mathbf{x} = \mathrm{d}l\,\mathbf{n}$ with **n** the unit vector indicating the direction of d**x**, we have seen that the *stretch* $\lambda(\mathbf{m}) = \mathrm{d}l/\mathrm{d}l_0$ can be calculated from

$$\lambda^2(\mathbf{m}) = \mathbf{m}\cdot\mathbf{C}\cdot\mathbf{m} \tag{4.64}$$

leading to the *relative* or *unit extension* $\in(\mathbf{m}) = (\mathrm{d}l - \mathrm{d}l_0)/\mathrm{d}l_0 = \lambda(\mathbf{m}) - 1$. Particularly, if the material vector in the reference configuration is parallel to the unit vector $\mathbf{e}_1$, the stretch $\lambda(1)$ and unit extension $\in(1)$ are given by

$$\lambda^2(1) = C_{11} = 1 + 2L_{11} \qquad \text{and} \qquad \in(1) = \sqrt{C_{11}} - 1 = \sqrt{1 + 2L_{11}} - 1 \tag{4.65}$$

In the SDG approximation the unit extension reduces to

$$\in(1) = \sqrt{1 + 2L_{11}} - 1 \cong \sqrt{1 + 2\varepsilon_{11}} - 1 \cong \varepsilon_{11} \tag{4.66}$$

Consider now the angle change. Let $\mathrm{d}\mathbf{r}'$ and $\mathrm{d}\mathbf{r}''$ be two material vectors with unit vectors $\mathbf{m}'$ and $\mathbf{m}''$ having lengths $\mathrm{d}l_0'$ and $\mathrm{d}l_0''$, respectively, in the reference configuration. In the current configuration these vectors become $\mathrm{d}\mathbf{x}'$ and $\mathrm{d}\mathbf{x}''$ with unit vectors $\mathbf{n}'$ and $\mathbf{n}''$ having lengths $\mathrm{d}l'$ and $\mathrm{d}l''$, respectively. In that case we have

$$\cos(\mathbf{n}',\mathbf{n}'') = \mathbf{n}' \cdot \mathbf{n}'' = \frac{\mathrm{d}\mathbf{x}' \cdot \mathrm{d}\mathbf{x}''}{\mathrm{d}l'\,\mathrm{d}l''} = \frac{\mathrm{d}r_k{}' C_{km} \mathrm{d}r_m{}''}{\mathrm{d}l'\,\mathrm{d}l''} \tag{4.67}$$

Since $\mathrm{d}r_k' = m_k'\mathrm{d}l_0'$, $\mathrm{d}r_m'' = m_m''\mathrm{d}l_0''$, $\mathrm{d}l' = \lambda(\mathbf{m}')\mathrm{d}l_0'$ and $\mathrm{d}l'' = \lambda(\mathbf{m}'')\mathrm{d}l_0''$ we obtain

$$\cos(\mathbf{n}',\mathbf{n}'') = \frac{m_k{}' C_{km} m_m{}''}{\lambda(\mathbf{m}')\lambda(\mathbf{m}'')} \tag{4.68}$$

From $\cos(\mathbf{n}',\mathbf{n}'')$ in the reference configuration the angle change can be calculated by subtraction. For example, if the material vectors $\mathrm{d}\mathbf{r}'$ and $\mathrm{d}\mathbf{r}''$ are in the reference configuration parallel to $\mathbf{e}_1$ and $\mathbf{e}_2$, the angle change is $\pi/2-\theta_{12}$, where

$$\cos(\theta_{12}) = \frac{C_{12}}{\lambda(1)\lambda(2)} = \frac{C_{12}}{\sqrt{C_{11}C_{22}}} = \frac{2L_{12}}{\sqrt{(1+2L_{11})(1+2L_{22})}} \tag{4.69}$$

In the SDG approximation the above expression reduces to

$$\cos(\theta_{12}) \cong \frac{2\varepsilon_{12}}{\sqrt{(1+2\varepsilon_{11})(1+2\varepsilon_{22})}} \cong \frac{2\varepsilon_{12}}{(1+\varepsilon_{11})(1+\varepsilon_{22})} \cong 2\varepsilon_{12} \tag{4.70}$$

and since θ_{12} is small, via $\cos(\theta_{12}) = \sin[(\pi/2)-\theta_{12}] \cong (\pi/2)-\theta_{12}$, to $\theta_{12}/2 = \varepsilon_{12}$.

In conclusion, it will be clear that for describing general deformations both the right Cauchy-Green tensor $\mathbf{C}$ and the Lagrange strain tensor $\mathbf{L}$ can be used. In the SDG approximation $\mathbf{C} \cong \mathbf{I} + 2\boldsymbol{\varepsilon}$ and $\mathbf{L} \cong \boldsymbol{\varepsilon}$, leading to the unit extension $\in(\mathbf{n}) = \mathbf{n}\cdot\boldsymbol{\varepsilon}\cdot\mathbf{n}$ and the angle change between two orthogonal line elements $\theta(\mathbf{n},\mathbf{m})/2 = \mathbf{n}\cdot\boldsymbol{\varepsilon}\cdot\mathbf{m}$.

4.9 Bibliography

Kuiken, G.D.C. (1994), *Thermodynamics of irreversible processes*, Wiley, Chichester.

Lubliner, J. (1990), *Plasticity theory*, MacMillan, New York.

Maugin, G.A. (1992), *The thermomechanics of plasticity and fracture*, Cambridge University Press, Cambridge.

Teodosiu, C. (1982), *Elastic models of crystal defects*, Springer, Berlin.

Ziegler, H. (1983), *An introduction to thermomechanics*, 2nd ed., North-Holland, Amsterdam.

5

Kinetics

In this chapter the forces that act on a body and the way they influence the motion of the body, and henceforth its deformation, are discussed. To that purpose we briefly review Newton's three laws of motion for a collection of particles and for rigid bodies. The extension to deformable bodies is presented as well as an alternative presentation based on the principle of virtual work. The stress tensor is introduced, linear and angular momentum are discussed and the energy equation is interpreted. For other introductions to these topics, see the reference list at the end of this chapter.

5.1 Newton's laws of motion

From elementary mechanics we know the laws of motion, as presented by Newton. In these laws particles[a] are considered to be characterised by a mass m, a position vector $\mathbf{x}$ and a rate of change of $\mathbf{x}$, the velocity $\mathbf{v} = \mathrm{d}\mathbf{x}/\mathrm{d}t$. We consider a system of interacting particles. In this section a symbol with no further variable indicated denotes the quantity for the system while the variable (i) indicates a specific particle i. Hence e.g. $\mathbf{f}$ denotes the total force on the system and $m(1)$ the mass of particle 1. The notation $\mathbf{f}(12)$ indicates the force exerted by particle 2 on particle 1. *Newton's laws* read (Goldstein, 1950)

- If $\mathbf{f} = \mathbf{0}$, $\mathbf{v}$ = constant
- $\mathbf{f} = \mathrm{d}(m\mathbf{v})/\mathrm{d}t$
- $\mathbf{f}(12) + \mathbf{f}(21) = \mathbf{0}$

Two remarks must be made. First, in non-relativistic particle mechanics, m is constant and $\mathrm{d}(m\mathbf{v})/\mathrm{d}t = m\mathrm{d}\mathbf{v}/\mathrm{d}t = m\mathbf{a} = \mathbf{f}$. Obviously, in this case the first law is a consequence of the second law. Second, the third law implies that the forces are equal, oppositely directed and lie along a line joining the particles. Newton's laws lead to three immediate consequences, as indicated below.

The *linear momentum* $\mathbf{p}$ for a system with constant mass m and velocity $\mathbf{v}$ is defined by $\mathbf{p} = \Sigma_i\, \mathbf{p}(i) = \Sigma_i\, m(i)\mathbf{v}(i)$. The rate of change $\mathrm{d}\mathbf{p}/\mathrm{d}t$ is given by

$$\frac{\mathrm{d}\mathbf{p}}{\mathrm{d}t} = \frac{\mathrm{d}m\mathbf{v}}{\mathrm{d}t} = m\frac{\mathrm{d}\mathbf{v}}{\mathrm{d}t} = m\mathbf{a} = \mathbf{f} \tag{5.1}$$

Hence if $\mathbf{f} = \mathbf{0}$, $\mathrm{d}\mathbf{p}/\mathrm{d}t = \mathbf{0}$ or $\mathbf{p}$ = constant. Eq. (5.1) expresses the *conservation of linear momentum*.

The *angular momentum* $\mathbf{l}$ with respect to the origin for a system with particles at position $\mathbf{x}_i$ is defined by $\mathbf{l} = \Sigma_i\, \mathbf{l}(i) = \Sigma_i\, \mathbf{x}(i)\times\mathbf{p}(i) = \Sigma_i\, \mathbf{x}(i)\times[m(i)\,\mathrm{d}\mathbf{x}(i)/\mathrm{d}t]$ while the *moment of force* or *torque* is defined as $\mathbf{q} = \Sigma_i\, \mathbf{q}(i) = \Sigma_i\, \mathbf{x}(i)\times\mathbf{f}(i)$. The rate of change $\mathrm{d}\mathbf{l}/\mathrm{d}t$ is given by

$$\frac{\mathrm{d}\mathbf{l}}{\mathrm{d}t} = \frac{\mathrm{d}\mathbf{x}}{\mathrm{d}t}\times m\frac{\mathrm{d}\mathbf{x}}{\mathrm{d}t} + \mathbf{x}\times m\frac{\mathrm{d}^2\mathbf{x}}{\mathrm{d}t^2} = \mathbf{0} + (\mathbf{x}\times\mathbf{f}) = \mathbf{q} \tag{5.2}$$

[a] Here really meaning particles!

Hence if $\mathbf{q} = \mathbf{0}$, $\text{d}\mathbf{l}/\text{d}t = \mathbf{0}$. Like for linear motion $\mathbf{f} = \mathbf{0}$ leads to $\mathbf{p}$ = constant, for angular motion $\mathbf{q} = \mathbf{0}$ leads to $\mathbf{l}$ = constant. Eq. (5.2) expresses the *conservation of angular momentum*. In the literature the angular momentum is also referred to as *moment of momentum*. In case inertial forces can be neglected we have $\text{d}\mathbf{p}/\text{d}t = \mathbf{0}$ and $\text{d}\mathbf{l}/\text{d}t = \mathbf{0}$. The equations for conservation of linear and angular momentum then become $\mathbf{f} = \mathbf{0}$ and $\mathbf{x}\times\mathbf{f} = \mathbf{0}$ and we refer to them as force and moment (torque) equilibrium, respectively.

Consider now the motion of a system under the influence of a conservative force $\mathbf{f}$. A force $\mathbf{f}$ is conservative if the work done by that force, when moving a particle from one point to another, is independent of the path taken. Equivalently, the force $\mathbf{f}$ is conservative if it can be derived from a function $V(\mathbf{x})$ by $\mathbf{f} = -\nabla V(\mathbf{x})$. The function $V(\mathbf{x})$ is called the *potential energy*. We evaluate the integral I of $\mathbf{f}\cdot\text{d}\mathbf{x} = \mathbf{f}\cdot\mathbf{v}\,\text{d}t$. Here $\mathbf{f}\cdot\mathbf{v}$ is the power (work done by the force per unit time). On the one hand, we have for the integral

$$I = \int_{t'}^{t''} \mathbf{f}\cdot\text{d}\mathbf{x} = \int_{t'}^{t''}\left(m\frac{\text{d}^2\mathbf{x}}{\text{d}t^2}\cdot\frac{\text{d}\mathbf{x}}{\text{d}t}\right)\text{d}t = \left[\frac{m}{2}\left(\frac{\text{d}\mathbf{x}}{\text{d}t}\right)^2\right]_{t'}^{t''} = T(t'') - T(t') \tag{5.3}$$

where the *kinetic energy* $T = \frac{1}{2}m\mathbf{v}^2$ is introduced. On the other hand, we also have

$$I = \int_{t'}^{t''} \mathbf{f}\cdot\text{d}\mathbf{x} = -\int_{t'}^{t''}(\nabla V\cdot\text{d}\mathbf{x}) = -V(t'') + V(t') \tag{5.4}$$

For a conservative force the (total) *energy* $U = T(\dot{\mathbf{x}},\mathbf{x}) + V(\mathbf{x})$ is thus constant during the motion of the system and one speaks of *conservation of energy*. The quantities $\mathbf{p}$, $\mathbf{l}$ and U are called *constants of the motion*.

The above expressions are valid for a single particle as well as a collection of connected particles. For such a collection three types of forces can be distinguished. First, forces acting alike on all particles due to long-range external influences. Examples of this type of force are the gravity force or forces due to externally imposed electromagnetic fields. Anticipating a similar distinction in continuous matter we call them as *volume* (or *body*) *forces* and indicate them for a particle i by $\mathbf{f}_{\text{vol}}(i)$. Second, forces applied to a particle due to short-range external forces. Examples of this type are interactions with enclosures or a weight resting on a solid. Again anticipating a similar distinction in continuous matter we call them *surface* (or *contact*) *forces* and indicate them by $\mathbf{f}_{\text{sur}}(i)$. Volume and surface forces are collectively called *external forces*, $\mathbf{f}_{\text{ext}}(i)$, i.e. $\mathbf{f}_{\text{ext}}(i) = \mathbf{f}_{\text{vol}}(i) + \mathbf{f}_{\text{sur}}(i)$. Third, forces due to the presence of the other particles, e.g. internal loading. These *internal forces* are indicated by $\mathbf{f}_{\text{int}}(i)$. Let $\mathbf{f}_{\text{pp}}(ij)$ denote the force on particle i due to particle-particle interaction with particle j. Then according to Newton's third law we have

$$\mathbf{f}_{\text{pp}}(ij) = -\mathbf{f}_{\text{pp}}(ji) \tag{5.5}$$

The resultant internal force acting on particle i is then

$$\mathbf{f}_{\text{int}}(i) = \sum\nolimits_{i\neq j} \mathbf{f}_{\text{pp}}(ij) \tag{5.6}$$

and the total force on particle i is

$$\mathbf{f}(i) = \mathbf{f}_{\text{vol}}(i) + \mathbf{f}_{\text{sur}}(i) + \mathbf{f}_{\text{int}}(i) = \mathbf{f}_{\text{ext}}(i) + \sum\nolimits_{i\neq j} \mathbf{f}_{\text{pp}}(ij) \tag{5.7}$$

The system of particles is in equilibrium if the force $\mathbf{f}(i)$ on each particle i in the system is equal to its rate of change of linear momentum $\mathrm{d}\mathbf{p}(i)/\mathrm{d}t$, also known as the *inertial force*. Obviously in that case

$$\mathbf{f} = \Sigma_i\, \mathbf{f}(i) = \Sigma_i\, \mathrm{d}\mathbf{p}(i)/\mathrm{d}t = \mathrm{d}\mathbf{p}/\mathrm{d}t$$

holds as well. In quasi-static problems, where dynamic effects can be neglected and hence $\mathrm{d}\mathbf{p}/\mathrm{d}t = 0$, the equilibrium condition becomes

$$\mathbf{f} = \Sigma_i\, \mathbf{f}(i) = 0$$

Motions of the collection of all particles that leave the distances between particles unchanged are called *rigid body motions*. Obviously, according to Newton's third law, the work due to internal forces vanishes for a rigid body motion.

In Section 5.2 a direct application of momentum theorems is given. The resulting stress tensor is discussed in Section 5.3 while a frequently used graphical representation is discussed in Section 5.5. The energy aspects are treated in Section 5.6. Those interested in the description of the principle of virtual power and some of its consequences should also consult later sections.

Example 5.1: The harmonic oscillator

In a harmonic oscillator an external force f acts on a mass m and is linearly related to the extension x, i.e. $f = -kx$, where k is the spring constant. The force f can be obtained from the potential energy $V = \frac{1}{2}kx^2$ via $f = -\partial V/\partial x$. If the momentum and velocity are given by $p = mv$ and v, respectively, Newton's second law reads $f = \dot{p} = \dfrac{\mathrm{d}}{\mathrm{d}t}(mv) = m\ddot{x}$. Combining leads to $m\ddot{x} = -kx$. Defining the circular frequency $\omega = (k/m)^{1/2}$, the solution of this differential equation is $x = x_0 \exp(-\mathrm{i}\omega t)$ (or equivalently $x = x_0 \cos(\omega t + \varphi_0)$, where x_0 is the amplitude of the oscillator and φ_0 is the phase).

Isaac Newton (1642-1727)

Born in Woolsthorpe, England, a few months after the death of his father, he spent his childhood with his grandmother, a fact that some see as an important factor in the shaping of the suspicious and neurotic personality of the adult Newton. He was educated at Cambridge University but during 1665 and 1667 the pest ruled England and during that time Newton remained at his parental home where he discovered the basis of differential and integral calculus. In 1667 he returned to Cambridge to become fellow of Trinity College and two years later professor of mathematics. He studied the refraction of light, on the interpretation of which he had an intensive quarrel with Robert Hooke. He published his book *Opticks* only in 1704, a year after the death of Hooke. In 1672 he became a fellow of the Royal Society. In 1676 a long and bitter debate with Gottfried Wilhelm Leibniz (1646-1716) started which only ended with Leibniz's death. Newton would not believe that Leibniz had invented the differential calculus independently. In 1666 Newton had the idea about the universal gravity and described the motion of the moon around the earth but did not publish this. After a debate in 1684 in London between Christopher Wren, Robert Hooke and the astronomer Edmond Halley, the latter visited Newton and asked him what orbit a planet would follow if the gravity of the sun were inversely proportional to the square of the distance. Newton promptly replied: an ellipse. Upon the question how he knew this, Newton replied: I calculated it. However, he could not find his calculations but promised Halley to make a new calculation. The result was his well-known book *Philosophiae naturalis principia mathematica* (1687). Although at that time he had developed differential calculus quite a bit, the results in the book are largely presented from a geometrical point of view. In 1696 he joined the Mint to be the Warden and in 1699 became Master of the Mint. After the death of Hooke he became the chairman of the Royal Society and was knighted in 1705 by Queen Anne. He was also interested in religious matters and alchemy, the latter being not too strange if one recalls that for a Mint Master the transformation of any metal to gold would be very handy. This, together with his various quarrels, sketches the picture of a self-sufficient, secretive personality, mellowing somewhat in old age.

Problem 5.1

An object of mass m moves in a plane with speed v at a constant distance r to the centre of rotation. Let $\mathbf{x}$ be the position of the object in the plane with the origin as the centre of rotation. Show that

a) the angular speed ω and acceleration α is given by $\omega = v/r$ and $\alpha = \dot{v}/r$, respectively,

b) the angular momentum $l = mvr = I\omega$ with $I = mr^2$ the moment of inertia,

c) the torque $q = \dot{l} = mr^2\alpha = I\dot{\omega}$ and

d) the kinetic energy $T = \frac{1}{2}I\omega^2 = l^2/2I$.

5.2 Mechanical equilibrium

In this section we use the conservation of linear momentum and of angular momentum and apply these laws to an infinitesimal volume element of a continuous body. To do so we generalise the ideas about external and internal forces and introduce the stress vector and tensor (Ziegler, 1983).

Often we will refer to the continuous body as the system. To that purpose we consider a volume V of a continuum with a regular surface A (Fig. 5.1). A volume element $\mathrm{d}V$ contains a mass $\mathrm{d}m$ and $\mathrm{d}A$ is a surface element with exterior normal $\mathbf{n}$. The density ρ of the element is given by $\rho = \mathrm{d}m/\mathrm{d}V$. We have to distinguish again between external forces on V, those for which the reactions are acting outside V, and internal forces, those for which the reactions are acting inside V. We consider the

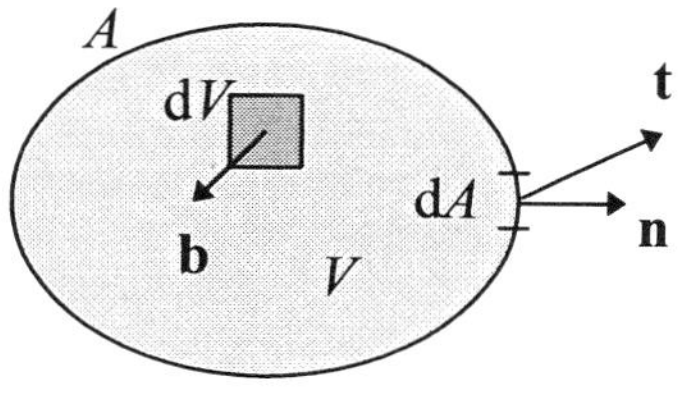

Fig. 5.1: External forces acting on a portion of a continuous body.

external forces first and return to the internal forces later.

Like for a collection of particles, there are two types of external forces for a continuous body. First, the *volume* (or *body*) *forces* acting on volume elements and distributed over the whole volume. The specific body force **b** is the body force per unit of mass. The most well-known body force is the specific weight, due to the acceleration of gravity. It is assumed that **b** is continuous and that couples acting on the volume element dV are excluded. Second, the *surface force* distributed over the surface A. Here it is convenient to refer to the force per unit area, which is usually called the *traction*. The traction depends on the position **x** and thus can be denoted by **t** = **t**(**x**). The volume and surface forces for a continuous body are the analogies of $\mathbf{f}_{\text{ext}}(i)$ for the case of a collection of particles.

We also need the *inertial force*, which is, like **b**, distributed over the volume of the entire body. The specific inertial force is again the force per unit mass and equal to the acceleration $\mathbf{a} = \dot{\mathbf{v}}$. This contribution is analogous to d**p**(i)/dt for the case of a collection of particles. Finally, the internal forces are the equivalents of the forces $\mathbf{f}_{\text{pp}}(ij)$ for a collection of particles. The precise expressions will be obtained later.

For an arbitrary body in equilibrium loaded by surface forces **t**(**x**) and body forces **b**(**x**), the total surface force and total body force are identically zero. If we cut the body into two pieces (Fig. 5.2), mechanical equilibrium can only be maintained if we apply at the same time extra forces at the cut, which compensate for the missing forces still acting on the other piece. The extra forces are equal in magnitude but of opposite direction since, obviously, their sum must be zero. The orientation of this cut is characterised by the normal vector **n**. If we now consider an infinitesimal surface element ΔA of the cut with a force $\Delta\mathbf{f}$ acting on this element, we can define the *stress vector* **t** by

$$\mathbf{t} = \mathbf{t}(\mathbf{x},\mathbf{n}) = \lim_{\Delta A \to 0} \frac{\Delta \mathbf{f}}{\Delta A} \tag{5.8}$$

We use again the symbol[b] **t** since this stress vector is an external surface force for the detached piece. The stress vector depends on the location (through **x**) and orientation (through **n**) of the surface element ΔA. This can be done for any point on the cut and since the cut itself is arbitrary, a stress vector can be defined for any surface element at any point in the body. Hence if we mention the stress at a point **x** in the body we mean the components of the stress vector on a certain plane.

Let us now apply the conservation of momentum to a small tetrahedron with volume ΔV whose three edges are parallel to the co-ordinate axes $\mathbf{e}_j$ (Fig. 5.3). If ΔA is

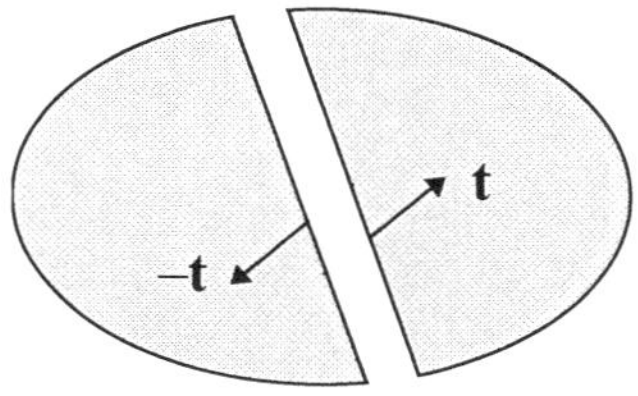

Fig. 5.2: Cut in a body.

[b] In the literature the terms stress vector and traction are used interchangeably. We use traction for an external force and stress vector for an internal force.

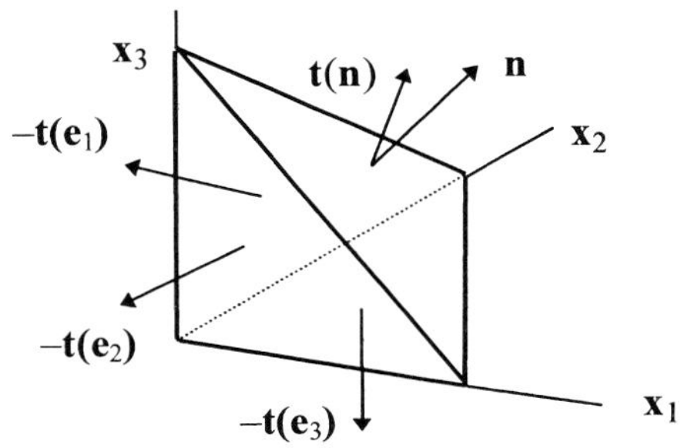

Fig. 5.3: Stresses acting on an infinitesimal tetrahedron.

the area of the oblique face with exterior normal **n**, the areas of the other three faces are $\Delta A_j = (\mathbf{n}\cdot\mathbf{e}_j)\Delta A$. The stress vector on ΔA is denoted by $\mathbf{t}(\mathbf{n})$ and the stress vectors on the other faces are denoted by $\mathbf{t}(-\mathbf{e}_j) = -\mathbf{t}(\mathbf{e}_j)$. The stress vector of a surface element is thus defined in such a way that a vector pointing in the positive x_j direction is $\mathbf{t}(\mathbf{e}_j)$. If the tetrahedron is sufficiently small it holds that

$$\int \rho\mathbf{b}\,\mathrm{d}V \cong \rho\mathbf{b}\,\Delta V\,, \tag{5.9}$$

$$\int \rho\mathbf{a}\,\mathrm{d}V \cong \rho\mathbf{a}\,\Delta V \quad \text{and} \tag{5.10}$$

$$\int \mathbf{t}(\mathbf{n})\,\mathrm{d}A \cong \mathbf{t}(\mathbf{n})\Delta A + \sum \mathbf{t}(-\mathbf{e}_j)\Delta A_j \tag{5.11}$$

Using $\mathbf{t}(\mathbf{e}_j) = -\mathbf{t}(-\mathbf{e}_j)$ and $\Delta A_j = (\mathbf{n}\cdot\mathbf{e}_j)\Delta A$ we thus approximately have from the conservation of momentum $\int \mathbf{t}(\mathbf{n})\,\mathrm{d}A + \int \rho\mathbf{b}\,\mathrm{d}V = \int \rho\mathbf{a}\,\mathrm{d}V$

$$\mathbf{t}(\mathbf{n}) = \mathbf{t}(\mathbf{e}_j)\mathbf{e}_j\cdot\mathbf{n} + \rho(\mathbf{b}-\mathbf{a})\frac{\Delta V}{\Delta A} \cong 0 \tag{5.12}$$

Taking the limit to an infinitesimal tetrahedron ($\Delta V/\Delta A \to 0$), we obtain

► $$\mathbf{t}(\mathbf{n}) = \mathbf{t}(\mathbf{e}_j)\mathbf{e}_j\cdot\mathbf{n} \equiv \boldsymbol{\sigma}\cdot\mathbf{n} = \sigma_{ij}n_j \tag{5.13}$$

where we have used the fact that $\mathbf{t}(\mathbf{e}_j)\mathbf{e}_j$ is a sum of three tensor products and therefore, a tensor $\boldsymbol{\sigma}$ itself. This tensor $\boldsymbol{\sigma}$ is called the *Cauchy stress tensor*[c]. Obviously **t** is a linear function of **n**. Using the definition of $\boldsymbol{\sigma}$ one finds that

$$\sigma_{ij} = \mathbf{e}_i\cdot\boldsymbol{\sigma}\cdot\mathbf{e}_j = \mathbf{e}_i\cdot\mathbf{t}(\mathbf{e}_j)$$

So, σ_{ij} is the component in the $\mathbf{e}_i$ direction of the stress vector $\mathbf{t}(\mathbf{e}_j)$ that acts on a surface element perpendicular to the $\mathbf{e}_j$ direction. A component is positive if it points in the positive direction for a plane with an outward normal also pointing in the positive direction (or in the negative direction for a plane with a normal also pointing in the negative direction). The components σ_{ij} ($i=j$) are the *normal stresses* while the components σ_{ij} ($i\neq j$) are the *shear stresses*.

---

[c] Note that this is another deviation from our 'tensor = uppercase bold' rule but, again, the notation is conventional.

Example 5.2

For the y-z plane, the normal vector **n** is represented by the matrix $\boldsymbol{n}^{\mathrm{T}} = (1,0,0)$ and the stress vector **t** for the y-z plane for an arbitrary stress tensor $\boldsymbol{\sigma}$ becomes

$$\begin{pmatrix} t_x \\ t_y \\ t_z \end{pmatrix} = \begin{pmatrix} \sigma_{xx} & \sigma_{xy} & \sigma_{xz} \\ \sigma_{yx} & \sigma_{yy} & \sigma_{yz} \\ \sigma_{zx} & \sigma_{zy} & \sigma_{zz} \end{pmatrix} \begin{pmatrix} 1 \\ 0 \\ 0 \end{pmatrix} = \begin{pmatrix} \sigma_{xx} \\ \sigma_{yx} \\ \sigma_{zx} \end{pmatrix}$$

If we choose the plane represented by $\boldsymbol{n}^{\mathrm{T}} = (1/\sqrt{2},1/\sqrt{2},0)$ the traction becomes

$$\begin{pmatrix} t_x \\ t_y \\ t_z \end{pmatrix} = \begin{pmatrix} \sigma_{xx} & \sigma_{xy} & \sigma_{xz} \\ \sigma_{yx} & \sigma_{yy} & \sigma_{yz} \\ \sigma_{zx} & \sigma_{zy} & \sigma_{zz} \end{pmatrix} \begin{pmatrix} 1/\sqrt{2} \\ 1/\sqrt{2} \\ 0 \end{pmatrix} = \frac{1}{\sqrt{2}} \begin{pmatrix} \sigma_{xx} + \sigma_{xy} \\ \sigma_{yx} + \sigma_{yy} \\ \sigma_{zx} + \sigma_{zy} \end{pmatrix}$$

Let us consider another volume element, in this case a simple cube with edges parallel to the co-ordinate axes of a local axes system with the origin at (x,y,z), and calculate the force and torque equilibrium. The edges of the cube are dx, dy and dz.

The force equilibrium in the x-direction, as shown in Fig. 5.4 and containing two normal forces and four shear forces, leads to the following equation (remember that $f_{,x}$ is an abbreviation of $\partial f/\partial x$ or $\partial_x f$)

$$\begin{aligned} &-\sigma_{xx}\,\mathrm{d}y\mathrm{d}z + (\sigma_{xx} + \sigma_{xx,x}\,\mathrm{d}x)\mathrm{d}y\mathrm{d}z \\ &-\sigma_{xy}\,\mathrm{d}z\mathrm{d}x + (\sigma_{xy} + \sigma_{xy,y}\,\mathrm{d}y)\mathrm{d}z\mathrm{d}x \\ &-\sigma_{xz}\,\mathrm{d}x\mathrm{d}y + (\sigma_{xz} + \sigma_{xz,z}\,\mathrm{d}z)\mathrm{d}x\mathrm{d}y \\ &+\rho b_x\,\mathrm{d}x\mathrm{d}y\mathrm{d}z = \rho a_x\,\mathrm{d}x\mathrm{d}y\mathrm{d}z \end{aligned}$$

where we included the volume force **b** and the acceleration **a**. Dividing by dxdydz and taking the limit dx, dy and dz → 0, implying an infinitesimal cube size so that the derivatives are to be evaluated at the point (x,y,z), leads to

$$\sigma_{xx,x} + \sigma_{xy,y} + \sigma_{xz,z} + \rho b_x = \rho a_x$$

Similarly for the y- and z-directions

$$\sigma_{yx,x} + \sigma_{yy,y} + \sigma_{yz,z} + \rho b_y = \rho a_y \quad \text{and} \quad \sigma_{zx,x} + \sigma_{zy,y} + \sigma_{zz,z} + \rho b_z = \rho a_z$$

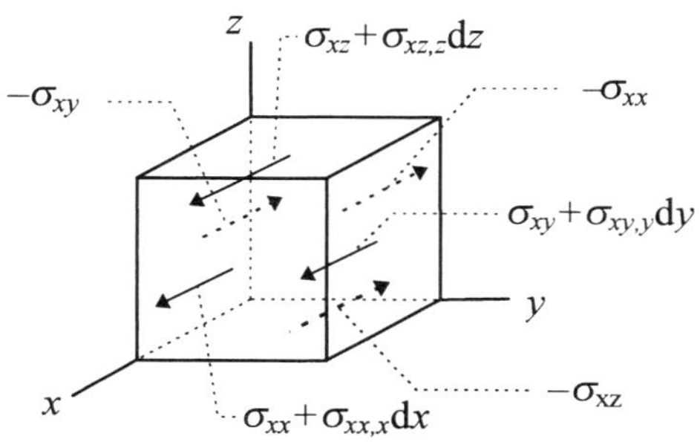

Fig. 5.4: Force equilibrium in the x-direction.

This can be further abbreviated to the so-called *equation of motion*[d]

► $\sigma_{ij,j} + \rho b_i = \rho a_i$ (5.14)

If changes in the structure are sufficiently slow so that the kinetic energy can safely be neglected, i.e. for quasi-static processes, we may neglect the inertia forces ρa_i. This leads to the so-called *equilibrium condition* $\sigma_{ij,j} + \rho b_i = 0$. Furthermore, for many quasi-static processes the body forces are relatively unimportant as compared to the external loading leading to a frequent use of the *reduced equilibrium equation* $\sigma_{ij,j} = 0$.

For the torque equilibrium we consider a rotation around the z-axis, as sketched in Fig. 5.5. Indicating in square brackets the forces and counting counterclockwise couples as positive, we obtain

$$[\sigma_{yx}\,\mathrm{d}y\mathrm{d}z]\cdot\tfrac{1}{2}\mathrm{d}x + [(\sigma_{yx}+\sigma_{yx,x}\,\mathrm{d}x)\mathrm{d}y\mathrm{d}z]\cdot\tfrac{1}{2}\mathrm{d}x$$

$$-[\sigma_{xy}\,\mathrm{d}z\mathrm{d}x]\cdot\tfrac{1}{2}\mathrm{d}y - [(\sigma_{xy}+\sigma_{xy,y}\,\mathrm{d}y)\mathrm{d}z\mathrm{d}x]\cdot\tfrac{1}{2}\mathrm{d}y = 0$$

Dividing by dxdydz and taking the limit d$y \to 0$ we obtain

$$\sigma_{yx} = \sigma_{xy}$$

Similarly for a rotation around the x- and y-axes we obtain

$$\sigma_{zx} = \sigma_{xz} \quad \text{and} \quad \sigma_{zy} = \sigma_{yz}$$

respectively. The Cauchy stress tensor $\boldsymbol{\sigma}$ is thus symmetric, or

► $\sigma_{ij} = \sigma_{ji}$ (5.15)

and contains only six independent components.

Summarising, in direct notation we have the (first Cauchy) equation of motion

$$\nabla\cdot\boldsymbol{\sigma} + \rho\mathbf{b} = \rho\mathbf{a}$$

or, if the body and inertia forces may be neglected, the (reduced) equilibrium condition

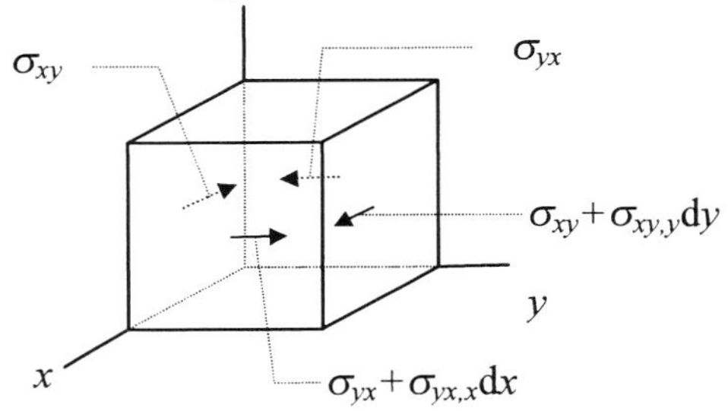

Fig. 5.5: Moment equilibrium around the z-direction.

[d] Note that this equation in index notation contains three individual equations, each with five terms. The derivation of the equation of motion using the Gauss theorem can be given by the one-liner: $\int\sigma_{ij}\,n_j\,\mathrm{d}A + \int\rho b_i\,\mathrm{d}V = \int\rho a_i\,\mathrm{d}V$ or $\int\sigma_{ij,j}\,\mathrm{d}V + \int\rho b_i\,\mathrm{dV} = \int\rho a_i\,\mathrm{d}V$ or $\sigma_{ij,j} + \rho b_i = \rho a_i$. In fact the arguments given in the main text are an outline of the proof of the Gauss theorem.

$$\nabla \cdot \boldsymbol{\sigma} = 0$$

and the conservation of angular momentum (or second Cauchy equation of motion)

$$\boldsymbol{\sigma} = \boldsymbol{\sigma}^{\mathrm{T}}$$

where $\boldsymbol{\sigma}^{\mathrm{T}}$ denotes the transpose of $\boldsymbol{\sigma}$. The equations describing mechanical equilibrium are partial differential equations and one needs *boundary conditions* if one actually wants to solve them. The most important types of conditions met in practice are the *displacement boundary conditions* where at a certain part A_u of the boundary A the displacements **u** of the particles are prescribed and the *traction boundary conditions* where at another part A_t of the boundary A the traction **t** should have a prescribed magnitude and direction.

Augustin-Louis Cauchy (1789-1857)
Born in Paris, France, a son of a father who had to leave Paris at the time of the French Revolution and took refuge in Arcueil, where at that time Laplace and Berthollet lived. Laplace's house became a meeting place where Lagrange noticed his unusual mathematical prowess. In 1805 he entered the École Polytechnique. After completion he was elected to enter the École des Ponts et Chaussées where he graduated in 1810. At the age of 21 he was doing important engineering work in Cherbourg. Being attracted more by mathematics, he returned to Paris where he became a member of the Academy in 1816. Cauchy started to teach at the École Polytechnique and at the Sorbonne, where he tried to present calculus in a more rigorous way. This attracted the interest of both his students and scientists from foreign countries, and his publication *Cours d'analyse de l'école polytechnique* in 1821 had an important effect on trends in mathematics. Cauchy then became interested in the theory of elasticity in which he introduced the concept of stress by his famous tetrahedron and the concept of strain. The terms principal directions, stress and strain are due to him. He subsequently derived the equilibrium equations, as we know them now, and applied them to isotropic bodies. One has to admit that Cauchy's narrow views, plus his many mean-spirited actions and disputations over priority with his mathematical contemporaries, do not make him an attractive personality, despite his extraordinary contributions.

5.3 The equilibrium conditions in cylindrical and spherical co-ordinates

As indicated in Section 4.4 in many applications Cartesian co-ordinates are not the most logical choice. Therefore we need to express the equilibrium conditions also in e.g. cylindrical and spherical co-ordinates (Lubliner, 1983).

For cylindrical co-ordinates we obtain, after some manipulation using the expressions of Section 3.10, for the equilibrium conditions

$$\frac{\partial \sigma_{rr}}{\partial r}+\frac{1}{r}\frac{\partial \sigma_{r\theta}}{\partial \theta}+\frac{\partial \sigma_{rz}}{\partial z}+\frac{\sigma_{rr}-\sigma_{\theta\theta}}{r}+\rho b_r=0$$
$$\frac{\partial \sigma_{r\theta}}{\partial r}+\frac{1}{r}\frac{\partial \sigma_{\theta\theta}}{\partial \theta}+\frac{\partial \sigma_{\theta z}}{\partial z}+2\frac{\sigma_{r\theta}}{r}+\rho b_\theta=0 \qquad (5.16)$$
$$\frac{\partial \sigma_{rz}}{\partial r}+\frac{1}{r}\frac{\partial \sigma_{\theta z}}{\partial \theta}+\frac{\partial \sigma_{zz}}{\partial z}+\frac{\sigma_{rz}}{r}+\rho b_z=0$$

In plane stress and strain the stresses σ_{rr}, $\sigma_{r\theta}$ and $\sigma_{\theta\theta}$ are constant with respect to z, σ_{zz} is either zero or $\nu(\sigma_{rr}+\sigma_{\theta\theta})$ and σ_{rz} and $\sigma_{\theta z}$ are zero. The above equations reduce to

$$\frac{\partial \sigma_{rr}}{\partial r}+\frac{1}{r}\frac{\partial \sigma_{r\theta}}{\partial \theta}+\frac{\sigma_{rr}-\sigma_{\theta\theta}}{r}+\rho b_r=0 \qquad \frac{\partial \sigma_{r\theta}}{\partial r}+\frac{1}{r}\frac{\partial \sigma_{\theta\theta}}{\partial \theta}+2\frac{\sigma_{r\theta}}{r}+\rho b_\theta=0$$

A similar exercise for spherical co-ordinates yields

$$\frac{\partial \sigma_{rr}}{\partial r}+\frac{1}{r}\frac{\partial \sigma_{r\theta}}{\partial \theta}+\frac{1}{r\sin\theta}\frac{\partial \sigma_{r\phi}}{\partial \phi}+\frac{2\sigma_{rr}-\sigma_{\phi\phi}-\sigma_{\theta\theta}+\sigma_{r\theta}\cot\theta}{r}+\rho b_r=0$$
$$\frac{\partial \sigma_{r\theta}}{\partial r}+\frac{1}{r}\frac{\partial \sigma_{\theta\theta}}{\partial \theta}+\frac{1}{r\sin\theta}\frac{\partial \sigma_{\theta\phi}}{\partial \phi}+\frac{\sigma_{\theta\theta}\cot\theta-\sigma_{\phi\phi}\cot\theta+3\sigma_{r\theta}}{r}+\rho b_\theta=0 \qquad (5.17)$$
$$\frac{\partial \sigma_{r\phi}}{\partial r}+\frac{1}{r}\frac{\partial \sigma_{\theta\phi}}{\partial \theta}+\frac{1}{r\sin\theta}\frac{\partial \sigma_{\phi\phi}}{\partial \phi}+\frac{3\sigma_{r\phi}+2\sigma_{\theta\phi}\cot\theta}{r}+\rho b_\phi=0$$

5.4 The stress tensor

As discussed in Section 5.2, the state of stress for a small volume element can be described by the stress tensor (Lubliner, 1983; Ziegler, 1983). Any volume element of a system can be detached from the system if we introduce at the same time the reaction forces for that element. These reaction forces then act as the external forces for that element. Consequently, the stress tensor is determined for the whole system and represents a field, i.e. the stress field. Once this field is given, the stress vector at any point acting on any surface element can be calculated. The stress tensor is a symmetric tensor so that all the properties of symmetric tensors, as discussed in Chapter 3, are applicable. In particular the tensor can brought at *principal axes* so that only diagonal elements remain. These elements are usually called *principal stresses*.

As discussed, the stress vector **t** acting on an infinitesimal plane with area dA, position vector **x** and outer normal **n** is given by

$$\mathbf{t}(\mathbf{x},\mathbf{n})=\sigma\cdot\mathbf{n} \qquad \text{or equivalently} \qquad t_i=\sigma_{ij}n_j \qquad (5.18)$$

This relation applies up to (and including) the boundary of a body. So, for a free boundary one has $\boldsymbol{\sigma}\cdot\mathbf{n}=\mathbf{0}$ and on those parts of a boundary where a traction **t** is prescribed one has $\boldsymbol{\sigma}\cdot\mathbf{n}=\mathbf{t}$. Often the normal stress, whose magnitude is indicated by $\sigma^{(n)}$ here, is required, i.e. the component of the stress vector normal to dA. The length of this stress component is given by the projection of the stress vector **t** on the normal vector **n** and thus equal to (Fig. 5.6)

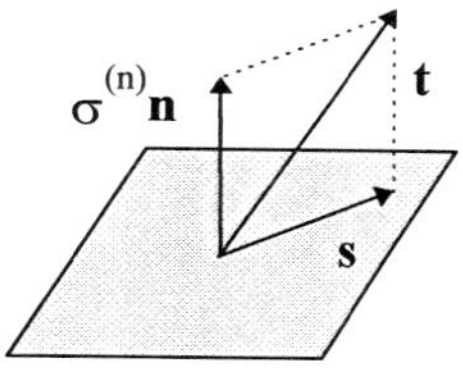

Fig. 5.6: Stress vector **t**, normal stress vector $\sigma^{(n)}\mathbf{n}$ and shear stress vector **s** at a plane with a normal vector **n**.

► $$\sigma^{(n)} = \mathbf{n}\cdot\mathbf{t} = \mathbf{n}\cdot\sigma\cdot\mathbf{n} \qquad \text{or equivalently} \qquad \sigma^{(n)} = n_i t_i = n_i\sigma_{ij}n_j \tag{5.19}$$

The normal stress vector is accordingly $\sigma^{(n)}\mathbf{n}$. The component of the stress vector in the plane dA is known as the shear component and, according to the Pythagorean theorem, its length $\sigma^{(s)}$ is equal to

$$\sigma^{(s)} = \sqrt{\mathbf{t}^2 - \left(\sigma^{(n)}\right)^2} \tag{5.20}$$

Using vector subtraction the shear stress vector[e] **s** can also be written as

► $$\mathbf{s} = \mathbf{t} - \sigma^{(n)}\mathbf{n} \qquad \text{or equivalently} \qquad \mathbf{s} = \mathbf{n}\times(\mathbf{t}\times\mathbf{n}) \tag{5.21}$$

Obviously it holds that

$$\sigma^{(s)} = \|\mathbf{s}\| \tag{5.22}$$

Similarly to the case of the strain tensor, the stress tensor can also be decomposed into an isotropic part $\sigma_{kk}\delta_{ij}/3$ and a deviator σ_{ij}' according to

$$\sigma_{ij} = \sigma_{ij}' + \sigma_{kk}\delta_{ij}/3 \tag{5.23}$$

The invariant σ_{kk} divided by 3, sometimes indicated by σ, is often called the *mean normal stress* and its negative $p = -\sigma_{kk}/3$ is the *pressure*. By the way, the isotropic and deviatoric parts of the stress tensor will be used in the discussion of the elastic energy (Chapter 9) and in the discussion of plasticity (Chapter 13).

Example 5.3

The stress tensor $\boldsymbol{\sigma}$ for a gas is $\boldsymbol{\sigma} = \begin{pmatrix} -p & 0 & 0 \\ 0 & -p & 0 \\ 0 & 0 & -p \end{pmatrix}$ where p is the classic pressure. The minus sign is used because in gases a compressive stress is reckoned positive whereas in mechanics a compressive stress is negative.

Example 5.4

Suppose we have a stress tensor $\boldsymbol{\sigma}$ and a normal to a plane **n** represented by

[e] Both the shear stress vector and the engineering stress (or first Piola-Kirchoff stress) are indicated by **s**. However, the meaning should be clear from the context and confusion is unlikely.

$$\sigma_{ij} = \begin{pmatrix} 2 & 1 & 1 \\ 1 & 2 & 1 \\ 1 & 1 & 3 \end{pmatrix} \quad \text{and} \quad n_k = \begin{pmatrix} 1/\sqrt{2} \\ 1/\sqrt{2} \\ 0 \end{pmatrix}$$

respectively. The stress vector **t** on this plane becomes

$$t_i = \sigma_{ij} n_j = \begin{pmatrix} 2 & 1 & 1 \\ 1 & 2 & 1 \\ 1 & 1 & 3 \end{pmatrix} \begin{pmatrix} 1/\sqrt{2} \\ 1/\sqrt{2} \\ 0 \end{pmatrix} = \begin{pmatrix} 3/\sqrt{2} \\ 3/\sqrt{2} \\ 2/\sqrt{2} \end{pmatrix}$$

The magnitude of the normal stress vector $\sigma^{(n)}$ is

$$\sigma^{(n)} = n_i \sigma_{ij} n_j = n_i t_i = \begin{pmatrix} 1/\sqrt{2} & 1/\sqrt{2} & 0 \end{pmatrix} \begin{pmatrix} 3/\sqrt{2} \\ 3/\sqrt{2} \\ 2/\sqrt{2} \end{pmatrix} = 3$$

while the shear stress vector **s** and its magnitude $\sigma^{(s)}$ are given by

$$s_i = t_i - \sigma^{(n)} n_i = \begin{pmatrix} 3/\sqrt{2} \\ 3/\sqrt{2} \\ 2/\sqrt{2} \end{pmatrix} - 3 \begin{pmatrix} 1/\sqrt{2} \\ 1/\sqrt{2} \\ 0 \end{pmatrix} = \begin{pmatrix} 0 \\ 0 \\ 2/\sqrt{2} \end{pmatrix} \quad \text{and} \quad \sigma^{(s)} = (s_i s_i)^{1/2} = \sqrt{2}$$

Alternatively, the length of the shear stress vector is given by

$$\sigma^{(s)} = \|\mathbf{s}\| = \left[t_i^2 - (\sigma^{(n)})^2\right]^{1/2} = \left[\left(\frac{9}{2} + \frac{9}{2} + \frac{4}{2}\right) - 3^2\right]^{1/2} = \sqrt{2}$$

Finally the isotropic part of σ_{ij} and the stress deviator σ_{ij}' are given by

$$\frac{\sigma_{kk}\delta_{ij}}{3} = \frac{(2+2+3)}{3} \begin{pmatrix} 1 & 0 & 0 \\ 0 & 1 & 0 \\ 0 & 0 & 1 \end{pmatrix} = \begin{pmatrix} 7/3 & 0 & 0 \\ 0 & 7/3 & 0 \\ 0 & 0 & 7/3 \end{pmatrix} \quad \text{and}$$

$$\sigma_{ij}' = \sigma_{ij} - \frac{\sigma_{kk}\delta_{ij}}{3} = \begin{pmatrix} 2 & 1 & 1 \\ 1 & 2 & 1 \\ 1 & 1 & 3 \end{pmatrix} - \begin{pmatrix} 7/3 & 0 & 0 \\ 0 & 7/3 & 0 \\ 0 & 0 & 7/3 \end{pmatrix} = \begin{pmatrix} -1/3 & 1 & 1 \\ 1 & -1/3 & 1 \\ 1 & 1 & 2/3 \end{pmatrix}$$

The mean normal stress $\sigma = \sigma_{kk}/3 = 7/3$ while the pressure $p = -\sigma_{kk}/3 = -7/3$.

Again similar to the strain case, a state of *plane stress* can arise. In that case we have, e.g. $\sigma_{i3} = 0$ ($i = 1,2,3$) and $\sigma_{ij,3} = 0$. This also implies that one of the principal stresses vanishes. If two of the principal stresses vanish, the stress state is *uniaxial*. If, on the other hand, the three principal stresses are equal, the stress distribution is *hydrostatic*.

From the principal axes transformation it is also clear that the principal stresses are also the extremes of the normal stresses as a function of orientation. One can ask: what are the extremes of the shear stresses and in which direction do they point? Maximising the length of the shear stress vector $\boldsymbol{\sigma}^{(s)}$ with respect to the components n_i, subject to the constraint $\|\mathbf{n}\| = 1$, yields the answer, whose the details are left as an exercise. If we indicate the shear stress by τ, it appears that

$$\tau^2 = (\sigma_{\rm I}-\sigma_{\rm II})^2 n_1^{\,2} n_2^{\,2} + (\sigma_{\rm II}-\sigma_{\rm III})^2 n_2^{\,2} n_3^{\,2} + (\sigma_{\rm III}-\sigma_{\rm I})^2 n_3^{\,2} n_1^{\,2}$$

The solutions as given in Table 5.1 arise. If the principal stresses obey the convention $\sigma_{\rm I} \geq \sigma_{\rm II} \geq \sigma_{\rm III}$, $\tau_{\rm II}$ is the maximum shear stress. The maximum shear stress planes bisect the principal axes planes and thus make an angle $\pi/4$ with them.

Table 5.1: Maximum shear directions.

n_1	n_2	n_3	τ
0	$\pm\frac{1}{2}\sqrt{2}$	$\pm\frac{1}{2}\sqrt{2}$	$\tau_{\rm I} = \frac{1}{2}(\sigma_{\rm II} - \sigma_{\rm III})$
$\pm\frac{1}{2}\sqrt{2}$	0	$\pm\frac{1}{2}\sqrt{2}$	$\tau_{\rm II} = \frac{1}{2}(\sigma_{\rm III} - \sigma_{\rm I})$
$\pm\frac{1}{2}\sqrt{2}$	$\pm\frac{1}{2}\sqrt{2}$	0	$\tau_{\rm III} = \frac{1}{2}(\sigma_{\rm I} - \sigma_{\rm II})$

Problem 5.2

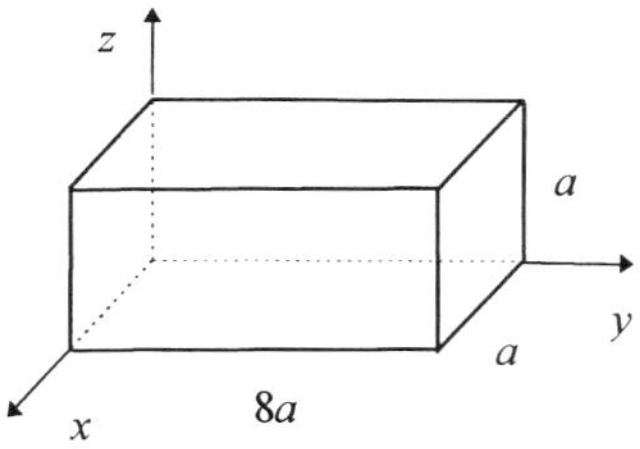

Consider a beam as shown in the accompanying figure. The stress field is given by $\boldsymbol{\sigma} = \begin{pmatrix} 0 & 0 & 0 \\ 0 & \alpha z + \beta & 0 \\ 0 & 0 & 0 \end{pmatrix}$ MPa, where α and β are parameters.

a) Determine the stress vector for a plane oriented perpendicular to the axis of the beam (normal in the positive y-direction) at an arbitrary point (x,y,z).
b) Determine the normal and shear stresses for that plane.
c) Make a sketch of the normal stress in a cross-section perpendicular to the axis of the beam.
d) Determine the normal and shear stresses for a plane parallel to the x-axis and at 45° with the y-axis and z-axis, through the point $(\frac{1}{2}a, 5a, \frac{3}{4}a)$.

Problem 5.3

The stress matrix for an infinitesimal cube-shaped volume element is given by

$$\boldsymbol{\sigma} = \begin{pmatrix} 4 & 2 & -1 \\ 2 & -2 & 3 \\ -1 & 3 & 1 \end{pmatrix} \text{MPa.}$$

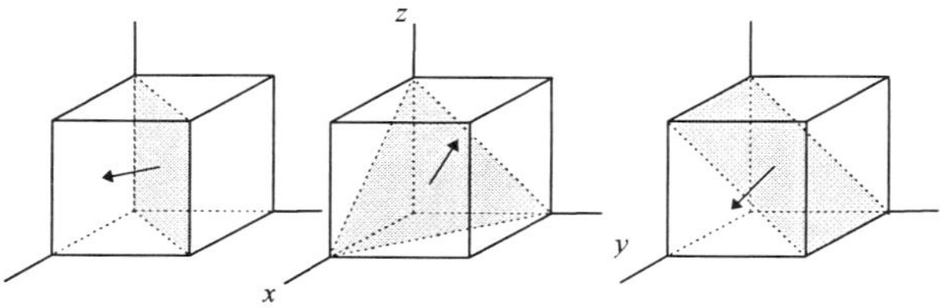

a) Determine the stress vector for the lattice planes as indicated in the figure.
b) Calculate the normal stress and shear stress vectors for these planes.

Problem 5.4

The stress matrix for an infinitesimal cube-shaped volume element is given by

$$\sigma = \begin{pmatrix} 30 & 60 & 45 \\ 60 & 0 & 0 \\ 45 & 0 & -15 \end{pmatrix} \text{MPa}$$

a) Determine the stress vector for a plane through a point P parallel to the plane $x+2y+2z = 2$.
b) Why are two solutions possible?
c) Calculate the normal and shear stresses and the normal and shear stress vectors for this plane.

Problem 5.5

Consider a plate as sketched in the accompanying figure, where p denotes the external pressure. The hatched area denotes a rigid connection.

a) Determine the boundary conditions for the edge AB in terms of the stress matrix.
b) Do the same for the edge BC, again in terms of the stress matrix.

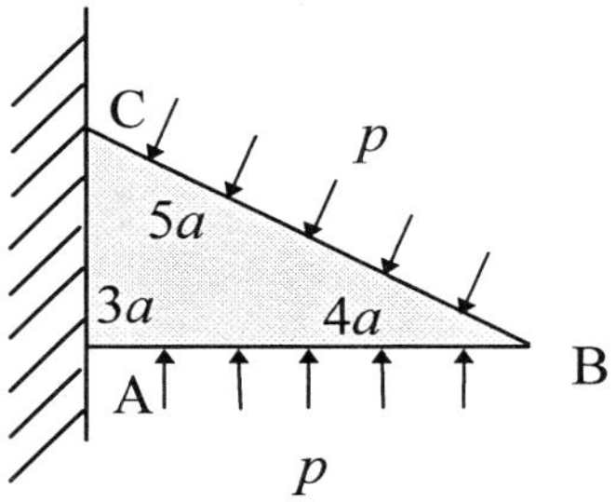

Problem 5.6

For a point P in a body the stress matrix is given by $\sigma = \begin{pmatrix} 4 & 3 & 2 \\ 3 & \alpha & 4 \\ 2 & 4 & 0 \end{pmatrix}$ MPa,

where α is an as yet undetermined parameter.

a) Determine the normal vector for the stress-free plane at point P.
b) Calculate the value of α.
c) Determine the principal stresses.
d) Determine the principal directions.

Problem 5.7

In the figure the stresses on an infinitesimal volume element are indicated, all equal to say τ.

a) Determine the stress matrix.
b) Determine the principal stresses.
c) Are the principal directions uniquely determined?

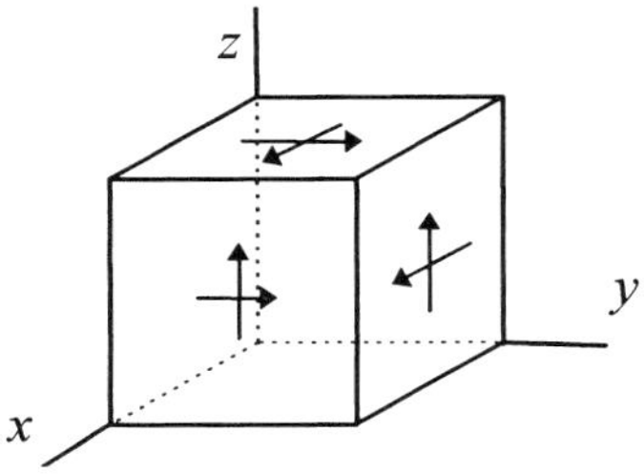

Problem 5.8

For a point in a body the stress matrix is given by $\sigma = 100\begin{pmatrix} \alpha & 0 & 0 \\ 0 & 1 & 2 \\ 0 & 2 & \beta \end{pmatrix}$ MPa. The principal stresses are $\sigma_{\text{I}} = 300$ MPa, $\sigma_{\text{II}} = 100$ MPa and $\sigma_{\text{III}} = -100$ MPa. Calculate α and β.

Problem 5.9*

a) Show that the traction expressed in principal values is given by $t_1 = \sigma_I n_1$, $t_2 = \sigma_{II} n_2$ and $t_3 = \sigma_{III} n_3$.
b) Show that the length of the shear stress vector $\boldsymbol{\sigma}^{(s)}$ is given by $\sigma^{(s)} = (t_1^2 - (\boldsymbol{\sigma}^{(n)})^2)^{1/2} = [(\sigma_I n_1)^2 + (\sigma_{II} n_2)^2 + (\sigma_{III} n_3)^2 - (\sigma_I n_1^2 + \sigma_{II} n_2^2 + \sigma_{III} n_3^2)^2]^{1/2}$ and that the normal stress is given by $\sigma^{(n)} = \sigma_I n_1^2 + \sigma_{II} n_2^2 + \sigma_{III} n_3^2$.
c) Using $1 - n_1^2 = n_2^2 + n_3^2$ and its permutations show that $\tau^2 = (\sigma_I - \sigma_{II})^2 n_1^2 n_2^2 + (\sigma_{II} - \sigma_{III})^2 n_2^2 n_3^2 + (\sigma_{III} - \sigma_I)^2 n_3^2 n_1^2$.
d) The extremes of τ^2 subject to $\|\mathbf{n}\| = 1$ can be obtained by constrained optimisation (Lagrange multipliers). Show that the solutions are as given in Table 5.1.

5.5 Mohr's circles

We have seen that the stress at any point in a material can be described by the stress tensor. If we know the stress tensor, the normal stress and shear stress at any plane can be calculated as has been discussed in Section 5.2. However, it is also possible to determine the normal stress and the shear stress by a convenient graphical representation of the principal axes known as the *Mohr circles*[f].

To that purpose we use the tensile test again and consider a plane with normal vector **n** making an angle α with the tensile axis. This implies that $\boldsymbol{n}^{\mathrm{T}} = (\cos\alpha, \sin\alpha, 0)$. The stress tensor is then

$$\sigma_{ij} = \begin{pmatrix} \sigma & 0 & 0 \\ 0 & 0 & 0 \\ 0 & 0 & 0 \end{pmatrix} \tag{5.24}$$

The stress vector **t** on the plane characterised by **n** is, as discussed, given by

$$t_i = \sigma_{ij} n_j = \begin{pmatrix} \sigma & 0 & 0 \\ 0 & 0 & 0 \\ 0 & 0 & 0 \end{pmatrix} \begin{pmatrix} \cos\alpha \\ \sin\alpha \\ 0 \end{pmatrix} = \begin{pmatrix} \sigma\cos\alpha \\ 0 \\ 0 \end{pmatrix} \tag{5.25}$$

The normal stress $\sigma^{(n)}$ is given by

$$\sigma^{(n)} = t_i n_i = \begin{pmatrix} \sigma\cos\alpha & 0 & 0 \end{pmatrix} \begin{pmatrix} \cos\alpha \\ \sin\alpha \\ 0 \end{pmatrix} = \sigma\cos^2\alpha \tag{5.26}$$

Hence the components of the normal stress vector $\sigma^{(n)}\mathbf{n}$ and shear stress vector **s** are

$$\sigma^{(n)} n_i = \sigma\cos^2\alpha \begin{pmatrix} \cos\alpha \\ \sin\alpha \\ 0 \end{pmatrix} \quad \text{and} \tag{5.27}$$

$$s_i = t_i - \sigma^{(n)} n_i = \begin{pmatrix} \sigma\cos\alpha \\ 0 \\ 0 \end{pmatrix} - \sigma\cos^2\alpha \begin{pmatrix} \cos\alpha \\ \sin\alpha \\ 0 \end{pmatrix} = \sigma\cos\alpha\sin\alpha \begin{pmatrix} \sin\alpha \\ -\cos\alpha \\ 0 \end{pmatrix} \tag{5.28}$$

---

[f] Otto Mohr (1835-1918). German engineer who introduced the circles in 1900.

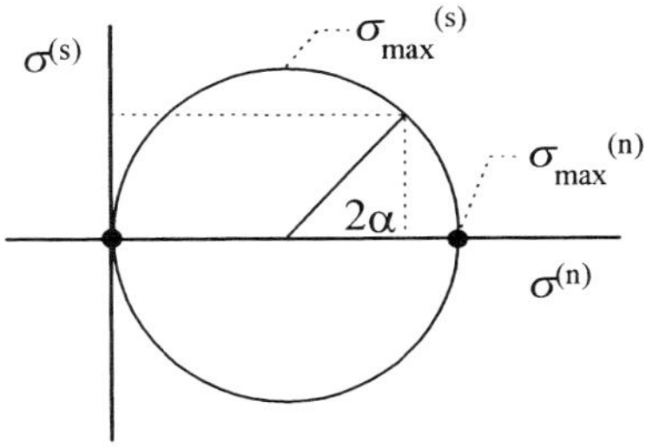

Fig. 5.7: Mohr circle for a tensile bar.

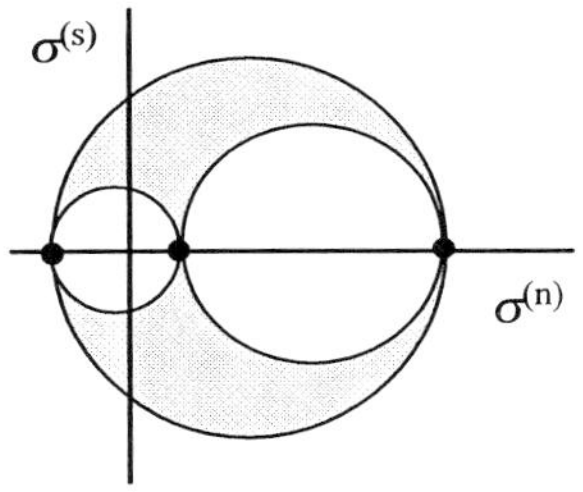

Fig. 5.8: Mohr circles for an arbitrary stress state.

respectively. The absolute value of the shear stress vector $\sigma^{(s)}$ is thus

$$\sigma^{(s)} = \|\mathbf{s}\| = \sigma\cos\alpha\sin\alpha\left[(\sin\alpha)^2 + (-\cos\alpha)^2\right]^{1/2} = \sigma\cos\alpha\sin\alpha \tag{5.29}$$

A parametric plot of the normal stress $\sigma^{(n)}$ and shear stress $\sigma^{(s)}$ as a function of α is thus given by

$$\begin{pmatrix} \sigma^{(n)} \\ \sigma^{(s)} \end{pmatrix} = \begin{pmatrix} \sigma\cos^2\alpha \\ \sigma\cos\alpha\,\sin\alpha \end{pmatrix} = \begin{pmatrix} \tfrac{1}{2}\sigma\cos 2\alpha + \tfrac{1}{2}\sigma \\ \tfrac{1}{2}\sigma\sin 2\alpha \end{pmatrix} \tag{5.30}$$

which represents the equation of a circle with centre ($\sigma/2$,0) and radius $\sigma/2$ in a $\sigma^{(n)}$-$\sigma^{(s)}$ plot (Fig. 5.7). The angle 2α is counted counterclockwise. This circle is known as the circle of Mohr. On the plane of which the normal coincides with the tensile stress, i.e. $\alpha = 0$, only a normal stress (and thus a principal stress) is acting. On a plane parallel to the tensile direction, i.e. $\alpha = \pi/2$, no stress acts at all. For the stress at a plane having a normal with an arbitrary angle α, we take an angle of 2α and read from the intersection with the circle the normal and shear stresses. This shows that the largest normal stress in the tensile bar is equal to the largest principal stress. The plane with the largest shear stress has an angle $\alpha = \pi/4$ and the shear stress is given by $\sigma_{max}^{(s)} = \tfrac{1}{2}\sigma_{max}^{(n)} = \tfrac{1}{2}\sigma$. The principal values are 0, 0 and σ along the normal stress axis.

The general case with three different principal values can be developed in principle in the same way[g]. The process is tedious though and we state only the final result in Fig. 5.8. Plot the three principal stresses along the $\sigma^{(n)}$-axis, as indicated by the dots, and draw three circles through these points as indicated. All possible normal stress and shear stress combinations for an arbitrary stress state are either on the circles or in the shaded area. Also in the general case from these circles a convenient representation of the stress state arises. Although this representation still has its educational value, the easy calculation of the principal values has largely superseded the graphical calculation via Mohr's circles.

Problem 5.10

Draw the Mohr circles for an isostatic compression, isostatic tension, biaxial compression and biaxial tension and for simple shear.

[g] Nadai, A. (1950), *The theory of flow and fracture in solids*, McGraw-Hill, New York.

Problem 5.11

Show that Mohr's expression for a two-dimensional stress state is given by

$$\sigma^{(\mathrm{n})} = \frac{\sigma_{xx} + \sigma_{yy}}{2} + \frac{\sigma_{xx} - \sigma_{yy}}{2}\cos 2\alpha + \sigma_{xy}\sin 2\alpha$$

$$\sigma^{(\mathrm{s})} = \frac{\sigma_{xx} - \sigma_{yy}}{2}\sin 2\alpha + \sigma_{xy}\cos 2\alpha$$

Draw the Mohr circle diagram.

5.6 Mechanical energy

So far we have used only the momentum theorems. The theorems of linear momentum and angular momentum are useful generalisations of the corresponding theorems in particle and rigid body mechanics. Similarly as before we can obtain the theorem of mechanical energy. It results in an expression for the power of the internal forces (Ziegler, 1983).

Multiplying both sides of the local form of the equilibrium equation $\rho \dot{v}_i = \rho b_i + \sigma_{ij,j}$ by v_i and integrating over the volume we obtain

$$\begin{aligned} \int \rho v_i \dot{v}_i \,\mathrm{d}V &= \int \rho b_i v_i \,\mathrm{d}V + \int \sigma_{ij,j} v_i \,\mathrm{d}V \\ &= \int \rho b_i v_i \,\mathrm{d}V + \int (\sigma_{ij} v_i)_{,j} \,\mathrm{d}V - \int \sigma_{ij} v_{i,j} \,\mathrm{d}V \end{aligned} \tag{5.31}$$

On account of the symmetry of σ_{ij} and the definition of the deformation rate $d_{ij} = v_{(i,j)}$, we can replace $\sigma_{ij}v_{i,j}$ by $\sigma_{ij}d_{ij}$. Moreover, we apply once more the Gauss theorem, in this case to the second term in the second line of the equation. Thus we obtain

► $$\int \rho v_i \dot{v}_i \,\mathrm{d}V = \int \rho b_i v_i \,\mathrm{d}V + \int \sigma_{ij} v_i n_j \,\mathrm{d}A - \int \sigma_{ij} d_{ij} \,\mathrm{d}V \tag{5.32}$$

We observe that in Eq. (5.32) the left-hand side represents the material derivative of the kinetic energy given by

$$T = \tfrac{1}{2} \int \rho v_i v_i \,\mathrm{d}V \tag{5.33}$$

and that the power of the inertial forces P_{acc} is given by

$$P_{\mathrm{acc}} = \dot{T} = \int \rho v_i \dot{v}_i \,\mathrm{d}V \tag{5.34}$$

The first two terms on the right-hand side of Eq. (5.32) are

$$P_{\mathrm{vol}} = \int \rho b_i v_i \,\mathrm{d}V \qquad \text{and} \qquad P_{\mathrm{sur}} = \int \sigma_{ij} v_i n_j \,\mathrm{d}A \tag{5.35}$$

and represent the power of the volume and surface forces, respectively. The third term

$$P_{\mathrm{int}} = -\int \sigma_{ij} d_{ij} \,\mathrm{d}V \tag{5.36}$$

can be interpreted as the power of the internal forces. In this way Eq. (5.32) can be interpreted as the global form of the theorem of *conservation of mechanical energy*[h]: the material derivative of the kinetic energy is equal to the power of the external forces $P_{ext} = P_{vol} + P_{sur}$ and internal forces P_{int}. Thus

► $$P_{acc} = P_{vol} + P_{sur} + P_{int} \tag{5.37}$$

Expressing the last integral in Eq. (5.32) as the sum of the powers of the external forces and the inertia forces,

$$\int \sigma_{ij} d_{ij} \, \mathrm{d}V = P_{vol} + P_{sur} - P_{acc} \tag{5.38}$$

we see that it represents that part of the power of the external forces that is not converted to kinetic energy associated with the global motion of the system. We call this remainder *internal energy* and it contains the potential and kinetic energies of the atoms or molecules constituting the solid. As long as dissipation is not involved, the conservation of energy interpretation in purely mechanical terms is correct. However, as soon as dissipation, or temperature for that matter, is involved, we need the first law of thermodynamics as an expression of the conservation of energy principle.

It may be useful to point out the physical reason for the minus sign in the definition of P_{int}. To that purpose consider two connected material particles (1) and (2) in equilibrium; particle (1) is the particle of interest. Neglecting external and acceleration forces we have $\mathbf{f} = \mathbf{f}_{int}(12) + \mathbf{f}_{int}(21) = \mathbf{0}$. Cutting the connection between the particles, for particle (1) a stress vector $\mathbf{t} = \boldsymbol{\sigma}\cdot\mathbf{n}$ has to be introduced equal to $\mathbf{f}_{int}(21)$. The stress $\boldsymbol{\sigma}$ thus corresponds to $\mathbf{f}_{int}(21)$, and since $\mathbf{f}_{int}(12) = -\mathbf{f}_{int}(21)$ is conventionally designated as internal force for particle (1), this leads to Eq. (5.36).
We finally remark that the work W_{int} of the internal forces can also be expressed as the integral of the power P_{int}. Since we have $P_{int} = \dot{W}_{int} = -\int \sigma_{ij} d_{ij} \, \mathrm{d}V \cong -\int \sigma_{ij} \dot{\varepsilon}_{ij} \, \mathrm{d}V$, we obtain $W_{int} = \int P_{int} \, \mathrm{d}t = -\int \int \sigma_{ij} \dot{\varepsilon}_{ij} \, \mathrm{d}t \mathrm{d}V = -\int \int \sigma_{ij} \mathrm{d}\varepsilon_{ij} \, \mathrm{d}V$. The mechanical work done on the system W_{mec} is given by the negative work of the internal forces, i.e. $W_{mec} = -W_{int}$.

5.7 Statically determined structures

Generally the equilibria of forces and momenta are insufficient to determine the stress distribution and we also need the stress-strain relation and strain compatibility. We deal with *statically indeterminate problems*. However, in a number of cases the stress distribution can be entirely determined using the laws of statics combined with the applied loads without considering deformation or the stress-strain relation. These situations are called *statically determined problems*. We explain the concepts by discussing two practically important examples.

Example 5.5: A simple truss

A simple but practically important problem example is provided by a *truss*, e.g. a pin-jointed frame made from straight bars linked by flexible joints which can only transmit uniaxial tensile or compressive forces. Fig. 5.9 shows a

[h] It is also frequently referred to as the *principle of virtual power* (PVP) although in this form the interpretation is more limited as in Section 5.8.

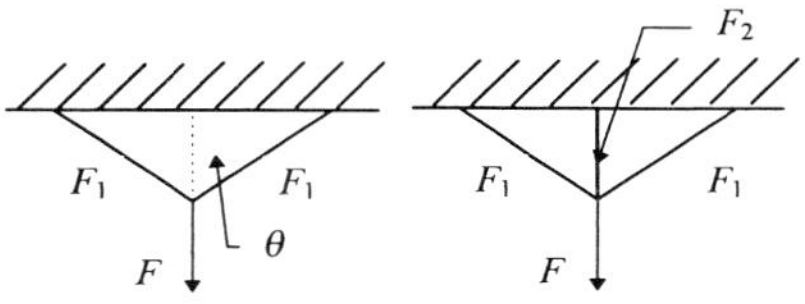

Fig. 5.9: A statically determined and undetermined pin-jointed frame.

simple example of a triangular truss with and without a central beam, attached to a rigid (hatched) plane. In the truss without a central beam the solution is $F = 2F_1 \cos\theta$, which is easily solved for F_1 because the load F and the angle θ are known. In the truss with the central beam we have $F = F_2 + 2F_1 \cos\theta$, which is an equation for two unknown forces F_1 and F_2. To solve for F_1 and F_2 we have to invoke the compatibility as well. The inclined bars elongate from l_1 to $l_1 + \delta l_1$ while the central bar elongates from l_2 to $l_2 + \delta l_2$. Since the vertical displacements must be equal (compatibility), we have $\delta l_1 = \delta l_2 \cos\theta$, where we tacitly assumed that the change in angle θ can be neglected. If all bars have the same cross-section A and Young's modulus E,

$$F_1 = AE\delta l_1/l_1 \quad \text{and} \quad F_2 = AE\delta l_2/l_2$$

which leads to

$$F_1 = F_2 \cos^2\theta$$

Solving these equations yields

$$F_1 = F\cos^2\theta/[1+2\cos^3\theta] \quad \text{and} \quad F_2 = F/[1+2\cos^3\theta]$$

The truss without the central beam is statically determinate and the elongations can be calculated after the forces are obtained. On the other hand, the truss with the central beam is statically indeterminate and the material behaviour is used to obtain a solution for the forces and elongations.

Example 5.6: A thin-walled pressure vessel

Consider a cylindrical vessel under a pressure p, which has a length l and a radius r. The wall thickness is t and we suppose that $l >> r >> t$. For this situation the principal axes are clear. The circumferential (or hoop) stress is denoted as σ_1 while σ_2 indicates the longitudinal stress. These stresses are

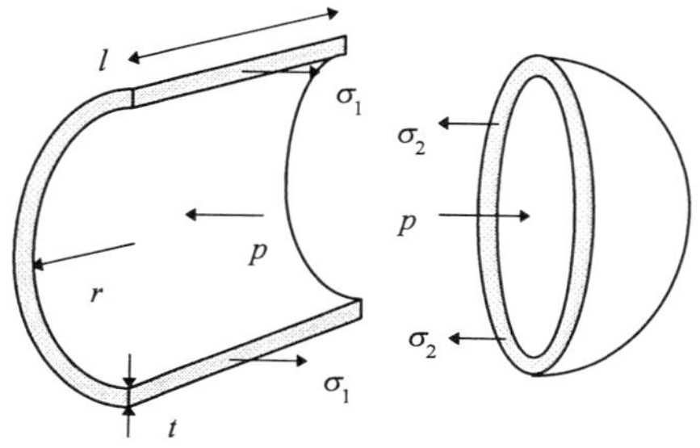

Fig. 5.10: A cylindrical pressure vessel.

supposed to be constant over the wall thickness. The third principal stress varies from $-p$ at the inner face to zero at the outer face. To calculate the stress σ_1 we consider a shell element as indicated in Fig. 5.10. The pressure provides an outward force $2rlp$ while the hoop stress provides $2tl\sigma_1$. Hence

$$\sigma_1 = pr/t \tag{5.39}$$

For the longitudinal stress we consider the end-cap. Here the outward force is $\pi r^2 p$ while the inward force is $2\pi rt\sigma_2$. Consequently

$$\sigma_2 = pr/2t \tag{5.40}$$

so that the longitudinal stress is half the hoop stress. The latter equation also provides the stress in a spherical vessel. So, the structures considered in this example are statically determined structures.

Although the above analysis is relatively simple, the results are nevertheless widely used. Statically determinate systems where a serial solution of stress and strain is possible are relatively rare and in general we have statically indeterminate systems for which a solution of stress and strain has to be obtained concomitantly. Finally we also note that in plasticity statically determined structures occur.

5.8 The principle of virtual power*

In this section an alternative way of summarising the (first and) second law of mechanics, usually called the principle of virtual work, is presented. Although entirely equivalent to Newton's laws, we nevertheless present this principle, the most important reasons being that it forms the basis of approximate analytical solutions and of modern numerical techniques. We start with a treatment for particles, as described in Section 5.1, discuss constraints and apply the results to a continuous body.

Discrete systems

A *virtual* (infinitesimal) displacement of a system refers to a change in the configuration of the system as a result of any arbitrary (infinitesimal) change of the independent co-ordinates $\delta\mathbf{x}$, consistent with the forces and constraints imposed on the system at any time t. The displacement is called virtual to distinguish it from an actual displacement occurring in a time interval $\mathrm{d}t$ during which the forces and constraints may change. For a system in equilibrium of N discrete particles and with no constraints, the total force on each particle $\mathbf{f}(i)$ is zero (Goldstein, 1950),

$$\mathbf{f}(i) = \mathbf{0}$$

while for a system in motion the total force on each particle equals the rate of change of the momentum

$$\mathbf{f}(i) = \mathrm{d}\mathbf{p}(i)/\mathrm{d}t$$

The latter equation reduces to the first for negligible velocities. Here again the variable (i) is used to indicate particle i. Clearly then the scalar product $\mathbf{f}(i)\cdot\delta\mathbf{x}(i)$, which is the virtual work of the total force $\mathbf{f}(i)$ on particle i in the displacements $\delta\mathbf{x}(i)$, equals $[\mathrm{d}\mathbf{p}(i)/\mathrm{d}t]\cdot\delta\mathbf{x}(i)$, the virtual work of the inertia forces. Similarly it holds that

$$\Sigma_i\,\mathbf{f}(i)\cdot\delta\mathbf{x}(i) = \Sigma_i\,[\mathrm{d}\mathbf{p}(i)/\mathrm{d}t]\cdot\delta\mathbf{x}(i)$$

The above equation is usually called the *principle of virtual work* (PVW). It states that the sum of the virtual work of the volume, surface and internal forces equals the virtual work of the inertia forces.

In many cases a rate formulation is advantageous, as will be shown at several occasions. We recall that the *power* or the *rate of work* is given by the scalar product of a force vector and the velocity at its point of application. Therefore using virtual velocities $\delta\mathbf{v}(i)$ rather than virtual displacements $\delta\mathbf{x}(i)$ leads in a similar way to the *principle of virtual power* (PVP)

► $$\Sigma_i\, \mathbf{f}(i)\cdot\delta\mathbf{v}(i) = \Sigma_i\, [\mathrm{d}\mathbf{p}(i)/\mathrm{d}t]\cdot\delta\mathbf{v}(i)$$

This expression is sometimes called *d'Alembert's principle*, although we will refer to it as the principle of virtual power.

In the form presented above the PVP is applicable to the analysis of dynamic systems. In case the system is quasi-static, one can neglect the inertial force obtaining

$$\Sigma_i\, \mathbf{f}(i)\cdot\delta\mathbf{v}(i) = 0$$

This is the form that will be used most often in this book since dynamic effects are generally not considered. The PVP can be 'derived' from Newton's laws as outlined above. However, it is also equally valid to consider the principle as the basic postulate and to derive from it Newton's laws and its consequences. In fact, the PVP is more general than the potential energy description since it includes forces, which cannot be derived from a potential energy, e.g. forces due to friction.

Justification 5.1

To show that the PVP leads to the conservation of linear momentum, angular momentum and energy, we note that in the equation

$$\Sigma_i\, [\mathbf{f}(i) - \mathrm{d}\mathbf{p}(i)/\mathrm{d}t]\cdot\delta\mathbf{v}(i) = 0$$

the variations $\delta\mathbf{v}(i)$ are independent. The equation can only be satisfied if

$$\Sigma_i\, [\mathbf{f}(i) - \mathrm{d}\mathbf{p}(i)/\mathrm{d}t] = \mathbf{f} - \mathrm{d}\mathbf{p}/\mathrm{d}t = 0$$

Hence if the total force $\mathbf{f} = 0$, the linear momentum $\mathbf{p}$ is constant. This represents the force equilibrium. To obtain the conservation of energy we take in the principle of virtual power

$$\delta\mathbf{v}(i) = \varepsilon\mathbf{v}(i)$$

where ε is a small but otherwise arbitrary scalar and $\mathbf{v}(i)$ is the actual speed of the particle. The PVP then leads to

$$\varepsilon\Sigma_i\, [\mathbf{v}(i)\cdot\mathbf{f}(i) - \mathbf{v}(i)\cdot\mathrm{d}\mathbf{p}(i)/\mathrm{d}t] = 0$$

If the forces are conservative, i.e. $\mathbf{f}(i) = -\partial V(i)/\partial\mathbf{r}$, we have

$$\mathbf{v}(i)\cdot\mathbf{f}(i) = -\,\mathrm{d}\mathbf{r}(i)/\mathrm{d}t\cdot\partial V(i)/\partial\mathbf{r}(i) = -\,\mathrm{d}V(i)/\mathrm{d}t$$

Recalling that the kinetic energy is given by $T = \mathbf{p}^2/2m$ we also have

$$\mathrm{d}T(i)/\mathrm{d}t = \mathbf{v}(i)\cdot\mathrm{d}\mathbf{p}(i)/\mathrm{d}t$$

and we therefore find, taking into account that ε is arbitrary, that

$$\mathrm{d}\left\{\sum_i [\,T(i)+V(i)]\right\}/\mathrm{d}t = 0 \qquad \text{or} \qquad T+V=\text{constant}$$

This represents the conservation of energy. To find the conservation of angular momentum we take in the PVP

$$\delta\mathbf{v}(i) = \boldsymbol{\varepsilon}\times\mathbf{v}(i)$$

where the vector $\boldsymbol{\varepsilon}$ is of arbitrary direction but of small magnitude. Invoking the PVP one obtains, after changing the order in the vector triple product,

$$\boldsymbol{\varepsilon}\cdot\left\{\sum_i [\mathbf{r}(i)\times\mathbf{f}(i)-\mathbf{r}(i)\times \mathrm{d}\mathbf{p}(i)/\mathrm{d}t]\right\} = 0$$

Since $\boldsymbol{\varepsilon}$ is arbitrary we obtain

$$\mathbf{q} - \mathrm{d}\mathbf{l}/\mathrm{d}t = 0.$$

Hence if the moment of the force $\mathbf{q} = 0$, the angular momentum $\mathbf{l}$ is constant. This represents the moment equilibrium.

The PVP can be expressed in matrix equations. To that purpose we note that $m(i)\mathbf{v}(i)$ can be written as $\boldsymbol{Mv}$, where $\boldsymbol{M}$ denotes a diagonal matrix with the various $m(i)$ as diagonal elements and $\boldsymbol{v}$ a column matrix containing the components of $\mathbf{v}(i)$. For the matrix $\boldsymbol{M}$ three consecutive elements are given by $m(i)$ since each component $v_j(i)$ of $\boldsymbol{v}$ has to correspond to the same mass $m(i)$, i.e.

$$\boldsymbol{M} = \begin{pmatrix} m(1) & 0 & 0 & 0 & .. & 0 \\ 0 & m(1) & 0 & 0 & .. & 0 \\ 0 & 0 & m(1) & 0 & .. & 0 \\ 0 & 0 & 0 & m(2) & .. & 0 \\ .. & .. & .. & .. & .. & 0 \\ 0 & 0 & 0 & 0 & 0 & m(N) \end{pmatrix} \tag{5.41}$$

Similarly, we can write $\boldsymbol{f}$ for the collection of $\mathbf{f}(i)$ and $\delta\boldsymbol{v}$ for the collection of $\delta\mathbf{v}(i)$, i.e.

$$\boldsymbol{f}^{\mathrm{T}} = [f_1(1), f_2(1), f_3(1), f_1(2), ...] \quad \text{and} \quad \delta\boldsymbol{v}^{\mathrm{T}} = [\delta v_1(1), \delta v_2(1), \delta v_3(1), \delta v_1(2), ...]$$

In this notation the PVP reads

$$\blacktriangleright \qquad \left(\boldsymbol{f} - \frac{\mathrm{d}}{\mathrm{d}t}\boldsymbol{Mv}\right)^{\mathrm{T}} \delta\boldsymbol{v} = 0 \tag{5.42}$$

Constraints

In the case of a system of N particles constraints are present, the variations $\delta\mathbf{v}(i)$ are not independent and the equations cannot be solved as indicated above. However, if the constraints can be expressed by $k = 3N\text{–}M$ equations (where $M < 3N$)

$$c_k(\delta\mathbf{r}(1), \ldots, \delta\mathbf{r}(i), \ldots, t) = 0 \qquad \text{or} \qquad c_k(\delta\mathbf{v}(1), \ldots, \delta\mathbf{v}(i), \ldots, t) = 0$$

a new set of M *generalised* (or *normal*) *co-ordinates* $\mathbf{q}(j)$ can be introduced. In general, the co-ordinates $\mathbf{r}(j)$ are non-linear functions of the generalised co-ordinates $\mathbf{q}(j)$ which can be expressed in matrix notation as indicated above by

$$\boldsymbol{r} = \boldsymbol{F}(\boldsymbol{q}) \tag{5.43}$$

Consequently,

$$\delta \boldsymbol{r} = \boldsymbol{A}\delta\boldsymbol{q} \qquad \text{and} \qquad \delta\boldsymbol{v} = \boldsymbol{A}\delta\dot{\boldsymbol{q}} \qquad \text{where} \qquad \boldsymbol{A} = \frac{\partial \boldsymbol{F}}{\partial \boldsymbol{q}} \tag{5.44}$$

Substitution in the PVP results in

$$\left(\boldsymbol{f} - \frac{\mathrm{d}}{\mathrm{d}t}\boldsymbol{M}\boldsymbol{v}\right)^{\mathrm{T}} \delta\boldsymbol{v} = \left(\boldsymbol{f} - \frac{\mathrm{d}}{\mathrm{d}t}\boldsymbol{M}\boldsymbol{v}\right)^{\mathrm{T}} \boldsymbol{A}\delta\dot{\boldsymbol{q}} = \left(\boldsymbol{A}^{\mathrm{T}}\boldsymbol{f} - \frac{\mathrm{d}}{\mathrm{d}t}\boldsymbol{A}^{\mathrm{T}}\boldsymbol{M}\boldsymbol{A}\dot{\boldsymbol{q}}\right)^{\mathrm{T}} \delta\dot{\boldsymbol{q}} = 0 \tag{5.45}$$

where use is made of $\boldsymbol{M} = \boldsymbol{M}^{\mathrm{T}}$. Defining $\boldsymbol{Q} = \boldsymbol{A}^{\mathrm{T}}\boldsymbol{f}$ and $\boldsymbol{M}' = \boldsymbol{A}^{\mathrm{T}}\boldsymbol{M}\boldsymbol{A}$ we obtain

$$(\boldsymbol{Q} - \boldsymbol{M}'\ddot{\boldsymbol{q}})^{\mathrm{T}}\, \delta\dot{\boldsymbol{q}} = 0 \tag{5.46}$$

which leads to

$$\blacktriangleright \qquad \boldsymbol{Q} = \boldsymbol{M}'\ddot{\boldsymbol{q}} \tag{5.47}$$

since the time derivatives of $\delta\boldsymbol{q}$ are independent and $\boldsymbol{M}' = \boldsymbol{M}'^{\mathrm{T}}$.

The elements of $\boldsymbol{Q}$ contain the *generalised forces* in terms of the generalised co-ordinates $\boldsymbol{q}$ and $\boldsymbol{M}'$ is the mass matrix associated with the generalised co-ordinates $\boldsymbol{q}$. Hence the equation $\boldsymbol{Q} = \boldsymbol{M}'\ddot{\boldsymbol{q}}$ is very similar to $\mathbf{f} = m\ddot{\mathbf{r}}$, which is Newton's second law in Cartesian co-ordinates.

Summarising, the original PVP states that the sum of the virtual power of the volume, surface and internal forces is equal to the virtual power of the inertia forces. In case constraints are present in which the constraint forces do not deliver work, the variations $\delta\boldsymbol{v}$ are not independent and we have to modify the PVP. Changing to generalised co-ordinates $\boldsymbol{q}$ and generalised forces $\boldsymbol{Q}$ restores the situation for which the PVP is again valid.

The PVP can be seen as an alternative formulation of Newton's (first and) second law. The third law is also frequently known as the *reaction principle* and leads to the statement that the power of the internal forces vanishes for rigid body motion. Together the PVP and the reaction principle constitute an equally valid basis for mechanical analysis as Newton's three laws. However, as a basis for numerical and approximate methods the PVP approach offers a great deal of advantage.

Continuous systems

In the following we shall evaluate the consequences of the principle of virtual power and the reaction principle for a continuous body instead of a collection of particles (Ziegler, 1983). Often we will refer to the continuous body as the system. To that purpose we consider again a volume V of a continuum with a regular surface A (Fig. 5.1). A volume element $\mathrm{d}V$ contains a mass $\mathrm{d}m$ and $\mathrm{d}A$ is a surface element with exterior normal $\mathbf{n}$. We have to distinguish again between external forces on V, those for which the reactions act outside V, and internal forces, those for which the reactions act inside V. These forces have been discussed in Section 5.2.

We recall that there are two types of external forces. First, the *volume* (or *body*) *forces* acting on volume elements and distributed over the whole volume. The specific body force $\mathbf{b}$ is the body force per unit of mass. Second, the *surface force* distributed over the surface A. Here it is convenient to refer to the force per unit area, which is usually called the *traction*. The traction not only depends on the position $\mathbf{x}$ but also on the orientation of the surface element $\mathrm{d}A$, indicated by the exterior normal $\mathbf{n}$, and thus can be denoted by $\mathbf{t} = \mathbf{t}(\mathbf{x},\mathbf{n})$. We also need the *inertial force* that is, like **b**, distributed over the volume of the entire body. The specific inertial force is again the force per unit mass and equal to the acceleration $\mathbf{a} = \dot{\mathbf{v}}$. Finally, we have the internal forces.

For the description of continuous matter we consider each volume element as a particle with as its label the position co-ordinate. Consequently, the summation over particles becomes an integral over the volume and surface of the system. In a continuum the forces thus appear in the form of a force field and the state of motion is described by a velocity field **v**.

Before we discuss the PVP for continuous systems we need a few more definitions. In the discussion of the PVP for discrete particles we stated that $\delta\mathbf{x}$ should be compatible with the constraints present. To be a bit more precise, in order to solve a boundary problem in solid mechanics, i.e. finding a displacement field **u** (velocity field **v**) and stress field $\boldsymbol{\sigma}$ in a volume V given the body force **b** in V, we also need the *boundary conditions* on the surface A. The latter implies that either the traction t_i or the displacement u_i (v_i) is prescribed at a certain point. The parts of A for which t_i or u_i (v_i) is prescribed are denoted by A_t and A_u (A_v), respectively. If we denote the prescribed values by $\bar{t}_i$ or $\bar{u}_i$ ($\bar{v}_i$), we thus have the following:

$$\text{traction boundary conditions: } t_i = \sigma_{ij} n_j = \bar{t}_i \text{ on } A_t \tag{5.48}$$

and

$$\text{displacement boundary conditions: } u_i = \bar{u}_i \text{ on } A_u \tag{5.49}$$

$$\text{(or velocity boundary conditions } v_i = \bar{v}_i \text{ on } A_v\text{)}$$

Since the conditions apply per component, the areas A_t and A_u (A_v) may actually overlap. The body forces b_i and tractions $\bar{t}_i$ are known as *loads* while the unknown tractions t_i at the point where the displacements are prescribed are called *reactions*. When displacements are prescribed one speaks of *external constraints*. In addition there may be *internal constraints*, e.g. incompressibility. A displacement field is *kinematically admissible* if it is mathematically well behaved (i.e. continuous and sufficiently differentiable) and obeys the external and internal constraints. For a static or quasi-static problem the equilibrium condition $\sigma_{ij,j} + \rho b_i = 0$ and the traction boundary conditions $\bar{t}_i = \sigma_{ij} n_j$ must be satisfied. A stress field that satisfies these conditions is called *statically admissible*[i].

A virtual displacement field $\delta\mathbf{u}$ is defined as the difference between two kinematically admissible displacement fields. It is assumed that $\delta\mathbf{u}$ is small and also that $|\delta u_{i,j}| << 1$. With the displacement field there is an associated velocity field $\delta\mathbf{v}$ and virtual strain field[j] $\delta\varepsilon_{ij} = \frac{1}{2}(\delta u_{i,j}+\delta u_{j,i})$. Finally we note that $\delta\mathbf{u} = \mathbf{0}$ on A_u.

The virtual power of the forces acting on the system is calculated from the actual forces at a certain moment t and the *virtual velocity field* $\delta\mathbf{v}$. As outlined above, the latter is independent of the actual state of motion and therefore generally not equal to the actual state of motion. The powers corresponding to the real velocity field **v** and the virtual velocity field $\delta\mathbf{v}$ are indicated by P and P^*, respectively. The kinetic behaviour is now derived from the *principle of virtual power* for the continuous body:

$$P_{\text{vol}}^* + P_{\text{sur}}^* + P_{\text{int}}^* = P_{\text{acc}}^*$$

---

[i] If the equation of motion $\sigma_{ij,j} + \rho b_i = \rho a_i$ is satisfied, the stress field is called *dynamically admissible*.

[j] The variation operator is a linear operator that commutes with (partial) differentiation (see Chapter 3).

At any time t, the total virtual power of the volume forces $P_{vol}*$, surface forces $P_{sur}*$ and internal forces $P_{int}*$ equals the virtual power of the inertia forces $P_{acc}*$ in any state of virtual motion (Maugin, 1992).

For the formulation of the reaction principle for a continuous body we note that it is always possible to detach a system from its surroundings if we introduce at the same time the external surface forces. The system may then be moved in particular as if it were rigid. The *reaction principle* for a continuous body can now again be stated as

$$P_{int}* = 0$$

or the virtual power of the internal forces vanishes for a rigid body motion.

Jean-le-Rond d'Alembert (1717-1783)

Born in Paris, France, he was the illegitimate child of the chevalier Destouches. Being abandoned by his mother, he was boarded out by the St. Jean-le-Rond parish with the wife of a glazier. His father paid for his going to a school where he obtained a fair mathematical education. An essay written by him in 1738 on the integral calculus and another in 1740 on ricochets, attracted attention, and in the same year he was elected a member of the French Academy, probably due to the influence of his father. It is to his credit that he absolutely refused to leave his adopted mother, with whom he continued to live until her death in 1757. It cannot be said that she sympathised with his success, for at the height of his fame she remonstrated with him for wasting his talents on such work, she said: *Vous ne serez jamais qu'un philosophe. Et qu'est-ce qu'un philosophe? c'est un fou que se tourmente pendant sa vie, pour qu'on parle de lui lorsqu'il n'y sera plus.* Nearly all his mathematical works were produced during the years 1743 to 1754. The most important of these was his *Traité de dynamique* (1743), in which he enunciates the principle known by his name, namely, that the 'internal forces of inertia' (that is, forces which resist acceleration) must be equal and opposite to the forces which produce the acceleration. The chief remaining contributions of D'Alembert to mathematics were on physical astronomy, especially on the precession of the equinoxes, and on variations in the obliquity of the ecliptic (*Système du monde*, 1754). During the last phase of his life he was mainly occupied with the French encyclopaedia. For this he wrote the introduction and numerous philosophical and mathematical articles. His style was brilliant, but not polished, and faithfully reflects his character, which was bold, honest and frank. He defended a severe criticism, which he had offered on some mediocre work by the remark *j'aime mieux être incivil qu'ennuyé*. With his dislike of sycophants and bores it is not surprising that he had more enemies than friends.

5.9 The momentum theorems and the energy function*

Consider an arbitrary rigid body motion of a volume element dV (Fig. 5.1) described by the virtual velocity field $\delta\mathbf{v}$. On account of the rigid body motion the virtual power of the internal forces P_{int}^* is zero. Hence there remains the virtual power of

- the body forces $P_{\text{vol}}^* = \int \rho b_i \delta v_i \text{d}V$,
- the surface forces $P_{\text{sur}}^* = \int t_i \delta v_i \text{d}A$ and
- the inertial force $P_{\text{acc}}^* = \int \rho \dot{v}_i \delta v_i \text{d}V = \int \rho a_i \delta v_i \text{d}V$

for which the principle of virtual power states that

$$P_{\text{acc}}^* = P_{\text{vol}}^* + P_{\text{sur}}^* \qquad \text{or} \qquad \int \rho \dot{v}_i \delta v_i \text{d}V = \int \rho b_i \delta v_i \text{d}V + \int t_i \delta v_i \text{d}A \tag{5.50}$$

Expression (5.50) holds for arbitrary virtual rigid motion.

Linear momentum

Applying Eq. (5.50) to translations, the velocity $\delta\mathbf{v}$ is constant throughout the body and can be dropped from the equation, leading to

$$\int \rho \dot{v}_i \text{ d}V = \int \rho b_i \text{ d}V + \int t_i \text{ d}A \tag{5.51}$$

The total linear momentum of the body is represented by

$$p_i = \int \rho v_i \text{ d}V \tag{5.52}$$

so that, since the mass density is conserved, the left-hand side of Eq. (5.51) represents the material derivative of the total linear momentum $\mathbf{p}$. Consequently, Eq. (5.51) represents the theorem of linear momentum: the material derivative of the linear momentum $\mathbf{p}$ is equal to the sum of the external forces $\mathbf{b}$ and $\mathbf{t}$. From this point on the analysis is identical to the one in Section 5.2, leading to $\mathbf{t}(\mathbf{x},\mathbf{n}) = \boldsymbol{\sigma}(\mathbf{x})\cdot\mathbf{n}$.

By means of the Gauss theorem we can transform the global expression for the linear momentum to an integral solely over the volume resulting in an expression for the conservation of linear momentum valid for an observer in the accompanying system:

$$\int \rho \dot{v}_i \text{ d}V = \int \rho b_i \text{ d}V + \int \sigma_{ij} n_j \text{ d}A = \int \left(\rho b_i + \sigma_{ij,j}\right) \text{d}V \tag{5.53}$$

If we note that this expression holds for any volume we obtain the local form or (first Cauchy) *equation of motion*

► $$\rho \dot{v}_i = \rho b_i + \sigma_{ij,j} \tag{5.54}$$

which reduces for an object at rest to the *equilibrium condition*

$$\rho b_i + \sigma_{ij,j} = 0 \tag{5.55}$$

Further reduction is obtained when the body forces are neglected and leads to

$$\sigma_{ij,j} = 0 \tag{5.56}$$

Recalling that

$$\dot{B}_{ij} = \int \rho \dot{b}_{ij} \text{d}V \quad \text{and} \quad \dot{A}_{ij} = \int \left[a_{ij,0} + (a_{ij} v_k)_{,k}\right] \text{d}V = \int \left[\dot{a}_{ij} + a_{ij} v_{k,k}\right] \text{d}V \tag{5.57}$$

we can write for Eq. (5.53), also using the Gauss theorem,

$$\int \rho v_{k,0}\,\mathrm{d}V = \int \rho b_k\,\mathrm{d}V + \int (\sigma_{kl} - \rho v_k v_l) n_l\,\mathrm{d}A \tag{5.58}$$

which is the conservation of the linear momentum valid for an observer at rest.

Angular momentum

Applying the PVP using a virtual rigid body rotation around the origin O leads to further insight. The virtual velocity for rigid body motion is given by

$$\delta v_k = e_{kij}\delta\omega_i x_j \tag{5.59}$$

where $\delta\omega_i$ denotes the virtual angular velocity and x_j the components of the position vector. Inserting the above equation in the PVP, Eq. (5.50), and noting that $\delta\omega_i$ can be dropped since it is independent of position, leads to

$$\int e_{ijk}\rho x_j \dot{v}_k\,\mathrm{d}V = \int e_{ijk}\rho x_j b_k\,\mathrm{d}V + \int e_{ijk} x_j t_k\,\mathrm{d}A \tag{5.60}$$

Noting that $e_{ijk}\dot{x}_j v_k = e_{ijk} v_j v_k = 0$ so that $x_j \dot{v}_k = (x_j v_k)^{\cdot}$ and using $t_i = \sigma_{ij} n_j$ yields

$$\int e_{ijk}\rho (x_j v_k)^{\bullet}\mathrm{d}V = \int e_{ijk}\rho x_j b_k \mathrm{d}V + \int e_{ijk} x_j \sigma_{kl} n_l \mathrm{d}A \tag{5.61}$$

The total angular momentum of the body is given by

$$l_i = \int e_{ijk}\rho x_j v_k\,\mathrm{d}V \tag{5.62}$$

so that, since the mass is conserved, the left-hand side of Eq. (5.61) is equal to the material derivative of the total angular momentum. Consequently, Eq. (5.61) represents the *theorem of angular momentum*: the material derivative of the angular momentum with respect to the origin O is equal to the sum of the moments of the external forces with respect to O.

If we apply the Gauss theorem to Eq. (5.61), analogously to the case of linear momentum, we can obtain a local form of the theorem of angular momentum from

$$\int e_{ijk}\rho x_j \dot{v}_k\,\mathrm{d}V = \int e_{ijk}\left[x_j \rho b_k + \left(x_j \sigma_{kl}\right)_{,l}\right]\mathrm{d}V \tag{5.63}$$

Using the equation of motion (5.54) and the identity $x_{j,l} = \delta_{jl}$ yields

$$\int e_{ijk} x_{j,l}\sigma_{kl}\,\mathrm{d}V = \int e_{ijk}\sigma_{kj}\,\mathrm{d}V = 0 \tag{5.64}$$

Recalling again that the volume is arbitrary yields the local form

$$e_{ijk}\sigma_{kj} = 0 \tag{5.65}$$

Considering the separate components we note that the local form of the angular momentum theorem states that the stress tensor is symmetric, e.g.

► $$\sigma_{ij} = \sigma_{ji} \tag{5.66}$$

(second Cauchy equation of motion). The theorem can be reformulated for an observer at rest, similar to the case of linear momentum. This leads to

$$\int e_{ijk}x_j(\rho v_k)_{,0}\,\mathrm{d}V = \int e_{ijk}x_j\rho f_k\,\mathrm{d}V + \int e_{ijk}x_j(\sigma_{kl} - \rho v_k v_l)n_l\,\mathrm{d}A \tag{5.67}$$

The derivation is left as an exercise.

Mechanical energy revisited

Similar to the derivation of the momentum theorems from the PVP, the conservation of energy can be derived from the PVP. Therefore consider again Eq. (5.32). We can interpret Eq. (5.32) as the principle of virtual power, applied to the real state of motion as a particular case of virtual motion, for which we can write (conform Section 5.8)

$$P_{\text{acc}} = P_{\text{vol}} + P_{\text{sur}} - \int \sigma_{ij}d_{ij}\,\mathrm{d}V$$

The volume V does not remain rigid and this accounts for the difference between Eq. (5.32) and Eq. (5.50) where $\delta\mathbf{v}$ has been replaced by $\mathbf{v}$. Since the integral on the left-hand side of Eq. (5.32) represents the power of the inertia forces P_{acc}, and the first two integrals on the right-hand side represent the power of the body forces P_{vol} and surface forces P_{sur}, respectively, the last term is the power of the internal forces P_{int}. Consequently, $P_{\text{int}} = -\int\sigma_{ij}d_{ij}\,\mathrm{d}V$. With this result the final interpretation of the PVP is

$$\begin{aligned} P^*_{\text{vol}} + P^*_{\text{sur}} + P^*_{\text{int}} &= \left(\int \rho b_i \delta v_i \mathrm{d}V + \int t_i \delta v_i \mathrm{d}A - \int \sigma_{ij}\delta d_{ij}\mathrm{d}V\right) \\ &= P^*_{\text{acc}} = \left(\int \rho \dot{v}_i \delta v_i \mathrm{d}V\right) \end{aligned} \tag{5.68}$$

stating that, at any time t, the virtual power of the external and internal forces is equal to the virtual power of the inertial forces for any virtual state of motion. The PVP can be seen as a disguised form of the conservation of (mechanical) energy theorem.

To show that this interpretation is consistent with our previous results, we note that the external virtual power is given by

$$P^*_{\text{ext}} = P^*_{\text{vol}} + P^*_{\text{sur}} = \int \rho b_i \delta v_i\,\mathrm{d}V + \int_{A_t} \bar{t}_i \delta v_i\,\mathrm{d}A \tag{5.69}$$

while the internal virtual power is defined by

$$P^*_{\text{int}} = -\int \sigma_{ij}\delta d_{ij}\,\mathrm{d}V \tag{5.70}$$

Since $\sigma_{ij}\delta d_{ij} = \sigma_{ij}\delta v_{i,j}$ and thus $\sigma_{ij}\delta d_{ij} = (\sigma_{ij}\delta v_i)_{,j} - \sigma_{ij,j}\delta v_i$ we get via the Gauss theorem

$$P^*_{\text{int}} = -\int \sigma_{ij} n_j \delta v_i\,\mathrm{d}A + \int \sigma_{ij,j}\delta v_i\,\mathrm{d}V \tag{5.71}$$

Because $\delta v_i = 0$ on A_v we restrict the integration to A_t and also write

$$\int \sigma_{ij} n_j \delta v_i\,\mathrm{d}A = \int_{A_t} \sigma_{ij} n_j \delta v_i\,\mathrm{d}A \tag{5.72}$$

Consequently for quasi-static loading

$$P^*_{\text{ext}} + P^*_{\text{int}} = 0 \qquad \text{or} \qquad \int (\sigma_{ij,j} + \rho b_i)\delta v_i\,\mathrm{d}V = \int_{A_t} (\sigma_{ij} n_j - \bar{t}_i)\delta v_i\,\mathrm{d}A \tag{5.73}$$

Hence, because the variations δv_i are arbitrary, the virtual power of external and internal forces vanishes only if

$$\sigma_{ij,j} + \rho b_i = 0 \text{ in } V \quad \sigma_{ij} n_j = \bar{t}_i \text{ on } A_t \quad \text{and} \quad v_i = \bar{v}_i \; (\delta v_i = 0) \text{ on } A_v \tag{5.74}$$

which are precisely the equilibrium condition and the boundary conditions, thus warranting our interpretation of the internal forces.

For completeness we mention that sometimes use is made of another principle called the *principle of complementary virtual power* (PCVP). The principle states that at any time the complementary virtual power of the external forces and internal forces equals the complementary virtual power of the inertia forces. In this case a virtual stress field, defined analogously to a virtual displacement field, is the difference between two statically admissable stress fields. A virtual stress thus satisfies

$$\delta\sigma_{ij,j} = 0 \text{ in } V \qquad \text{and} \qquad n_j \delta\sigma_{ij} = 0 \text{ on } A_t \qquad 5.75)$$

The internal and external complementary virtual powers are defined by

$$P_{\text{int}}^{*(\text{c})} = -\int d_{ij}\delta\sigma_{ij}\,\mathrm{d}V \qquad \text{and} \qquad P_{\text{ext}}^{*(\text{c})} = \int_{A_v} \bar{v}_i n_j \delta\sigma_{ij}\,\mathrm{d}A \tag{5.76}$$

An analysis for quasi-static conditions leads as before to

$$\begin{gathered} P_{\text{int}}^{*(\text{c})} + P_{\text{ext}}^{*(\text{c})} = 0 \qquad \text{or} \\ \int \left[d_{ij} - \tfrac{1}{2}\left(v_{i,j} + v_{j,i}\right)\right]\delta\sigma_{ij}\,\mathrm{d}V = \int_{A_v} \left(v_i - \bar{v}_i\right) n_j \delta\sigma_{ij}\,\mathrm{d}A \end{gathered} \tag{5.77}$$

Hence, because the variations $\delta\sigma_{ij}$ are arbitrary, the complementary virtual power of the internal and external forces vanishes only if $d_{ij} = \tfrac{1}{2}(v_{i,j}+v_{j,i})$ in V, $v_i = \bar{v}_i$ on A_v and $n_j \delta\sigma_{ij} = 0$ (or $t_i = \bar{t}_i$) on A_t. In this case we thus regain the compatibility for the rate of deformation and the boundary conditions.

5.10 Stress in the reference configuration*

The stress tensor as described above is expressed in spatial co-ordinates $\mathbf{x}$ and refers to the current configuration. Like for the strain sometimes it is convenient to refer to the stress in the reference configuration and have expressions in the material co-ordinates $\mathbf{r}$ (Teodosiu, 1982). We need, apart from the transformation of a line element as given by

$$\mathrm{d}\mathbf{x} = \mathbf{F} \cdot \mathrm{d}\mathbf{r} \tag{5.78}$$

where $\mathbf{F}$ denotes the deformation gradient, also the transformation for an area and volume element. We consider the change from material to spatial co-ordinates as a co-ordinate transformation and recall that, if $\mathrm{d}V$ and $\mathrm{d}V_0$ denote the volume element in the spatial and material co-ordinates respectively,

$$\mathrm{d}V = \det \mathbf{F}\,\mathrm{d}V_0 = J\,\mathrm{d}V_0 \tag{5.79}$$

where J = det $\mathbf{F}$ is the Jacobian determinant of the transformation. In terms of the derivatives of the spatial and material co-ordinates $\rho = V^{-1}$ and $\rho_0 = V_0^{-1}$, respectively, J is given by $J = \rho_0/\rho$. A similar result for the area element, known as *Nanson's formula* and given without proof, is less well known,

$$J\mathbf{m}\mathrm{d}A_0 = \mathbf{F}^{\mathrm{T}} \cdot \mathbf{n}\mathrm{d}A \tag{5.80}$$

Here dA and dA_0 denote the surface element in the spatial and material co-ordinates and **n** and **m** the unit vectors for these elements.

Whatever description we use, the actual force is the same. In the spatial description we have for a force

$$\mathbf{t}\,\mathrm{d}A = \boldsymbol{\sigma} \cdot \mathbf{n}\,\mathrm{d}A \tag{5.81}$$

We may write

$$\boldsymbol{\sigma} \cdot \mathbf{n}\,\mathrm{d}A = J\boldsymbol{\sigma}.\mathbf{F}^{-\mathrm{T}} \cdot \mathbf{m}\,\mathrm{d}A_0 \equiv \mathbf{s} \cdot \mathbf{m}\,\mathrm{d}A_0 \tag{5.82}$$

The Cauchy stress tensor is thus transformed to the (*first*) *Piola-Kirchhoff stress* $\mathbf{s} = J\boldsymbol{\sigma}\cdot\mathbf{F}^{-\mathrm{T}}$, which describes the stress in the reference configuration.

Example 5.7

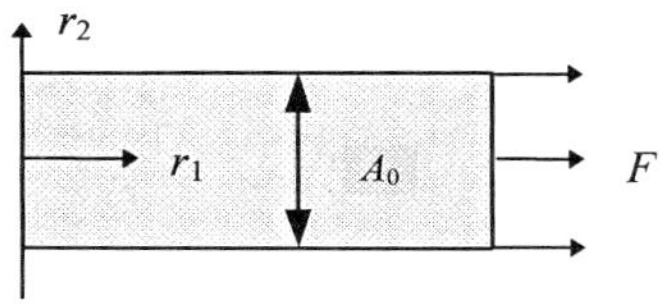

Consider a bar with cross-section area A_0 positioned along the r_1-axis, as indicated in the accompanying figure. If we load this bar axially with a force F, the cross-section area changes to A. The Cauchy stress $\sigma_{11} = F/A$ while the (first) Piola-Kirchhoff stress $s_{11} = F/A_0$. The former is frequently called the *true stress* while the latter is known as the *engineering stress*. The displacement field in this bar is

$$\begin{pmatrix} u_1(\mathbf{r}) \\ u_2(\mathbf{r}) \\ u_3(\mathbf{r}) \end{pmatrix} = \begin{pmatrix} \alpha r_1 \\ -\beta r_2 \\ -\beta r_3 \end{pmatrix}$$

where α denotes the relative change in length and β the relative change in lateral dimension. Hence

$$\begin{pmatrix} x_1 \\ x_2 \\ x_3 \end{pmatrix} = \begin{pmatrix} r_1 \\ r_2 \\ r_3 \end{pmatrix} + \begin{pmatrix} u_1(\mathbf{r}) \\ u_2(\mathbf{r}) \\ u_3(\mathbf{r}) \end{pmatrix} = \begin{pmatrix} (1+\alpha)r_1 \\ (1-\beta)r_2 \\ 1-\beta)r_3 \end{pmatrix} \quad \text{or} \quad \begin{pmatrix} x_1 \\ x_2 \\ x_3 \end{pmatrix} = \begin{pmatrix} 1+\alpha & 0 & 0 \\ 0 & 1-\beta & 0 \\ 0 & 0 & 1-\beta \end{pmatrix} \begin{pmatrix} r_1 \\ r_2 \\ r_3 \end{pmatrix}$$

This results in

$$\begin{pmatrix} r_1 \\ r_2 \\ r_3 \end{pmatrix} = \begin{pmatrix} (1+\alpha)^{-1} & 0 & 0 \\ 0 & (1-\beta)^{-1} & 0 \\ 0 & 0 & (1-\beta)^{-1} \end{pmatrix} \begin{pmatrix} x_1 \\ x_2 \\ x_3 \end{pmatrix}$$

Since $\mathbf{F} = \mathbf{1} + \partial\mathbf{u}/\partial\mathbf{r}$ and $\mathbf{x} = \mathbf{F}\cdot\mathbf{r}$, we obtain

$$\mathbf{F} = \begin{pmatrix} 1+\alpha & 0 & 0 \\ 0 & 1-\beta & 0 \\ 0 & 0 & 1-\beta \end{pmatrix}$$ and consequently $J = \det \mathbf{F} = (1+\alpha)(1-\beta)^2$.

Therefore the Cauchy stress, given by $\boldsymbol{\sigma} = J^{-1}\,\mathbf{s}\cdot\mathbf{F}^{\mathrm{T}}$, is

$$\boldsymbol{\sigma} = \frac{1}{(1+\alpha)(1-\beta)^2}\begin{pmatrix} F/A_0 & 0 & 0 \\ 0 & 0 & 0 \\ 0 & 0 & 0 \end{pmatrix}\begin{pmatrix} 1+\alpha & 0 & 0 \\ 0 & 1-\beta & 0 \\ 0 & 0 & 1-\beta \end{pmatrix} = \frac{1}{(1-\beta)^2}\begin{pmatrix} F/A_0 & 0 & 0 \\ 0 & 0 & 0 \\ 0 & 0 & 0 \end{pmatrix}$$

Hence $\sigma_{11} = (1-\beta)^{-2}F/A_0 = F/A$. Similarly, we obtain for the s_{11} component of the first Piola-Kirchoff stress **s**

$$s_{11} = (1+\alpha)\,(1-\beta)^2\,(F/A)(1+\alpha)^{-1} = F/[A/(1-\beta)^{-2}] = F/A_0.$$

The first Cauchy law of motion using the first Piola-Kirchoff stress reads

$$\nabla.\mathbf{s} + \rho_0\mathbf{b} = \rho_0\mathbf{a} \tag{5.83}$$

while the second Cauchy law of motion is given by

$$\mathbf{s}\cdot\mathbf{F}^{\mathrm{T}} = \mathbf{F}\cdot\mathbf{s}^{\mathrm{T}} \tag{5.84}$$

which shows that the first Piola-Kirchoff stress **s** is not a symmetric tensor. Therefore sometimes the (symmetric) *second Piola-Kirchoff stress* is introduced, defined by

$$\boldsymbol{\pi} = \mathbf{F}^{-1}\cdot\mathbf{s} = J\mathbf{F}^{-1}\cdot\boldsymbol{\sigma}\cdot\mathbf{F}^{-\mathrm{T}} \tag{5.85}$$

which reduces the second Cauchy law to a simpler form, namely

$$\boldsymbol{\pi} = \boldsymbol{\pi}^{\mathrm{T}} \tag{5.86}$$

In exchange the first Cauchy law becomes slightly more complicated reading

$$\nabla\cdot\mathbf{F}\cdot\boldsymbol{\pi} + \rho_0\mathbf{b} = \rho_0\mathbf{a} \tag{5.87}$$

Problem 5.12

Show that for the bar described in Example 5.7 the second Piola-Kirchoff stress component π_{11} is given by $\pi_{11} = (1+\alpha)^{-1}F/A_0$.

Problem 5.13

Derive Eqs. (5.83), (5.84), (5.86) and (5.87).

5.11 Work and power revisited*

In the previous sections we have discussed the deformation, resulting in various strain tensors, and the associated forces, resulting in various stress tensors. In principle many strain and stress tensors can be defined. They have to satisfy some conditions though. A strain should be *kinematically admissible*, i.e. satisfy the compatibility conditions. A stress should be *dynamically admissible*, i.e. satisfy the equations of motion. Although with these limitations still a great deal of strain and stress tensors can be defined, the only ones properly allowed are those for which the product of stress and strain increment yields the work done. This limits the various choices possible[k].

[k] McVean, D.B. (1968), Z. Angew. Math. Phys. **19**, 157-185.

Using spatial (or Euler) co-ordinates $\mathbf{x}$ the deformation is described by the symmetrised gradient of the velocity field, the so-called *rate of deformation* d_{ij}, given in index and direct notation by respectively

$$d_{ij} = \tfrac{1}{2}(v_{i,j} + v_{j,i}) \qquad \text{and} \qquad \mathbf{d} = \tfrac{1}{2}[\nabla\mathbf{v} + (\nabla\mathbf{v})^{\mathrm{T}}] \tag{5.88}$$

The velocity field v_j is given in index and direct notation by

$$v_j = v_j(x_k,t) = \frac{\mathrm{d}u_j(x_k,t)}{\mathrm{d}t} \qquad \text{and} \qquad \mathbf{v} = \mathbf{v}(\mathbf{x},t) = \frac{\mathrm{d}\mathbf{u}(x,t)}{\mathrm{d}t} = \dot{\mathbf{u}}(x,t), \tag{5.89}$$

respectively, where u_j (or $\mathbf{u}$) denotes the displacement, dependent on the spatial co-ordinates x_k (or $\mathbf{x}$) and time t. The associated stress measure is the *Cauchy stress* tensor σ_{ij} (or $\boldsymbol{\sigma}$). The *power* P, i.e. the work W done per unit time, per unit mass is represented by

$$P = \frac{\mathrm{d}W}{\mathrm{d}t} = \frac{1}{\rho}\sigma_{ij}d_{ij} \qquad \text{or} \qquad P = \frac{1}{\rho}\boldsymbol{\sigma}:\mathbf{d} \tag{5.90}$$

where $\rho = \rho(x_k,t)$ is the mass *density* distribution. This relation generalises the elementary definition of the power per unit volume $P = \sigma\mathrm{d}\dot{\varepsilon}$, given in Eq. (2.33).

Using material (or Lagrange) co-ordinates the deformation measure is the *Lagrange strain* L_{ij} (or $\mathbf{L}$) given by

$$L_{ij} = \tfrac{1}{2}(C_{ij} - \delta_{ij}) \qquad \text{or} \qquad \mathbf{L} = \tfrac{1}{2}(\mathbf{C} - \mathbf{I}) \tag{5.91}$$

where

$$C_{ij} = \frac{\partial x_i}{\partial r_k}\frac{\partial x_k}{\partial r_j} \qquad \text{or} \qquad \mathbf{C} = \mathbf{F}^{\mathrm{T}}\cdot\mathbf{F} \tag{5.92}$$

with the deformation gradient $F_{ij} = \partial x_i/\partial r_j$ (or $\mathbf{F} = \partial\mathbf{x}/\partial\mathbf{r}$). The (material) rate of change $\dot{\mathbf{L}}$ can be obtained as follows. Consider the derivative of $\mathbf{C}$

$$\begin{aligned}
\frac{\mathrm{d}}{\mathrm{d}t}\mathbf{F}^{\mathrm{T}}\cdot\mathbf{F} = \dot{\mathbf{F}}^{\mathrm{T}}\cdot\mathbf{F} + \mathbf{F}^{\mathrm{T}}\cdot\dot{\mathbf{F}} &= \frac{\mathrm{d}}{\mathrm{d}t}\left(\frac{\partial\mathbf{x}}{\partial\mathbf{r}}\right)^{\mathrm{T}}\cdot\frac{\partial\mathbf{x}}{\partial\mathbf{r}} + \left(\frac{\partial\mathbf{x}}{\partial\mathbf{r}}\right)^{\mathrm{T}}\cdot\frac{\mathrm{d}}{\mathrm{d}t}\frac{\partial\mathbf{x}}{\partial\mathbf{r}} \\
&= \left(\frac{\partial\mathbf{v}}{\partial\mathbf{r}}\right)^{\mathrm{T}}\cdot\frac{\partial\mathbf{x}}{\partial\mathbf{r}} + \left(\frac{\partial\mathbf{x}}{\partial\mathbf{r}}\right)^{\mathrm{T}}\cdot\frac{\partial\mathbf{v}}{\partial\mathbf{r}} = \left(\frac{\partial\mathbf{x}}{\partial\mathbf{r}}\right)^{\mathrm{T}}\cdot\left(\frac{\partial\mathbf{v}}{\partial\mathbf{x}}\right)^{\mathrm{T}}\cdot\frac{\partial\mathbf{x}}{\partial\mathbf{r}} + \left(\frac{\partial\mathbf{x}}{\partial\mathbf{r}}\right)^{\mathrm{T}}\cdot\frac{\partial\mathbf{v}}{\partial\mathbf{x}}\cdot\frac{\partial\mathbf{x}}{\partial\mathbf{r}} \\
&= \mathbf{F}^{\mathrm{T}}\cdot\mathbf{v}^{\mathrm{T}}\cdot\mathbf{F} + \mathbf{F}^{\mathrm{T}}\cdot\mathbf{v}\cdot\mathbf{F}
\end{aligned} \tag{5.93}$$

Hence we obtain

$$\dot{\mathbf{L}} = \tfrac{1}{2}(\mathbf{F}^{\mathrm{T}}\cdot\mathbf{v}^{\mathrm{T}}\cdot\mathbf{F} + \mathbf{F}^{\mathrm{T}}\cdot\mathbf{v}\cdot\mathbf{F}) = \mathbf{F}^{\mathrm{T}}\cdot\mathbf{d}\cdot\mathbf{F} \tag{5.94}$$

The associated stress measure is the *second Piola-Kirchhoff stress* tensor π_{ij} (or $\boldsymbol{\pi}$) and the power P per unit mass is

$$P = \frac{1}{\rho_0}\pi_{ij}\dot{L}_{ij} \qquad \text{or} \qquad P = \frac{1}{\rho_0}\boldsymbol{\pi}:\dot{\mathbf{L}} \tag{5.95}$$

with ρ_0 the mass density distribution expressed in material co-ordinates. This relation generalises the definition of the power per unit volume $P = s\mathrm{d}\dot{e}$, given in Eq. (2.33).

To show that this expression for the power is identical to the previous one, let us evaluate this expression. To this purpose we recall that

$$\boldsymbol{\pi} = \frac{\rho_0}{\rho}\mathbf{F}^{-1}\cdot\boldsymbol{\sigma}\cdot\mathbf{F}^{-\mathrm{T}}$$

Hence

$$P = \frac{1}{\rho_0}\boldsymbol{\pi}:\dot{\mathbf{L}} = \frac{1}{\rho_0}\frac{\rho_0}{\rho}\mathbf{F}^{-1}\cdot\boldsymbol{\sigma}\cdot\mathbf{F}^{-\mathrm{T}}:\mathbf{F}^{\mathrm{T}}\cdot\mathbf{d}\cdot\mathbf{F} \tag{5.96}$$

$$= \frac{1}{\rho}\mathbf{F}^{-1}\cdot\boldsymbol{\sigma}:\mathbf{d}\cdot\mathbf{F} = \frac{1}{\rho}\boldsymbol{\sigma}^{\mathrm{T}}\cdot\mathbf{F}^{-\mathrm{T}}:\mathbf{F}^{\mathrm{T}}\cdot\mathbf{d}^{\mathrm{T}} = \frac{1}{\rho}\boldsymbol{\sigma}:\mathbf{d}$$

where the last step can be made since both $\boldsymbol{\sigma}$ and $\mathbf{d}$ are symmetric. Hence the result is the same as obtained for the power expressed in spatial co-ordinates. Similarly, the power using the first Piola-Kirchoff stress tensor can be obtained. The final result is

$$P = \frac{1}{\rho_0}\mathbf{s}^{\mathrm{T}}:\dot{\mathbf{F}} \tag{5.97}$$

The development of continuum mechanics can be continued with either the spatial or the material description. We refer to the literature for details, e.g. see Fung (1965) or Malvern (1969). It has been shown that in the small displacement gradient approximation the stress and strain measures in the spatial and material co-ordinates coincide. We further denote them by σ_{ij} and ε_{ij}, respectively. Moreover, the density ρ can be considered as constant and the power reduces to

$$P = \frac{1}{\rho}\sigma_{ij}d_{ij} = \frac{1}{\rho}\sigma_{ij}\dot{\varepsilon}_{ij} \qquad \text{or} \qquad P = \frac{1}{\rho}\boldsymbol{\sigma}:\mathbf{d} = \frac{1}{\rho}\boldsymbol{\sigma}:\dot{\boldsymbol{\varepsilon}}$$

and the work increment dW to

► $$\mathrm{d}W = \frac{1}{\rho}\sigma_{ij}\mathrm{d}\varepsilon_{ij} \qquad \text{or} \qquad \mathrm{d}W = \frac{1}{\rho}\boldsymbol{\sigma}:\mathrm{d}\boldsymbol{\varepsilon}$$

We note that most of the literature dealing with continuum and/or thermomechanics is based on this approximation and for the remainder of this book we use it exclusively.

Problem 5.14

Derive Eq. (5.97).

5.12 Bibliography

Fung, Y.C. (1965), *Foundations of solid mechanics*, Prentice-Hall, Englewoods Cliffs, NJ.

Goldstein, H. (1950), *Classical mechanics*, Addison-Wesley, Reading, MA.

ter Haar, D. (1961), *Elements of Hamiltonian mechanics*, North-Holland, Amsterdam.

Lubliner, J. (1990), *Plasticity theory*, MacMillan, New York.

Malvern, L.E. (1969), *Introduction to the mechanics of a continuous medium*, Prentice-Hall, Englewoods Cliffs, NJ.

Maugin, G.A. (1992), *The thermomechanics of plasticity and fracture*, Cambridge University Press, Cambridge.

Teodosiu, C. (1982), *Elastic models of crystal defects*, Springer, Berlin.

Ziegler, H. (1983), *An introduction to thermomechanics*, 2nd ed., North-Holland, Amsterdam.

6

Thermodynamics

In many problems in materials science thermodynamics is required. This chapter briefly reviews some useful fundamental laws and relations. After the introduction of internal variables, a field formulation is presented. A brief introduction to irreversible processes concludes this chapter in which free use has been made of several books quoted in the bibliography.

6.1 Basic laws

Preliminary definitions

In thermodynamics the part of the physical world that is under consideration is for the sake of analysis considered to be separated from the rest. This separated part is known as the *system* while the remainder is called the *surroundings*. The system may have fixed or movable boundaries and may contain matter, radiation or both. A system is said to be *open* if matter can be exchanged with the surroundings. A *closed system* does not exchange matter with the surroundings but it may still be able to exchange energy with its surroundings. An *isolated system* has no interaction of any kind with its surroundings. The *thermodynamic state* of the system is assumed to be determined completely by a set of macroscopic, independent, external or *kinematical* co-ordinates a_i and one 'extra' other parameter[a] related to the thermal condition of the system. They constitute the set of *state variables* (also frequently known as *state parameters* or *generalised displacements*). Functions of the state variables are known as *state functions*. Sometimes the dependent and independent variables are collectively known as state functions. For the moment the properties of the system are taken to be the same throughout the system. In this case the system is called *homogeneous*. When a particular macroscopic co-ordinate is fixed in value by the conditions at the boundary of the system, there is said to be a *constraint* on the system. *Intensive* parameters, such as pressure and temperature, are independent of the size of the system while *extensive* parameters, such as volume and energy, are proportional to the extent of the system (amount of substance). It should be noted that there are, however, other properties, e.g. surface area, which are neither intensive nor extensive[b]. Moreover, the distinction between (global) intensive and extensive variables has no meaning in inhomogeneous systems. When the properties of a system do not change with time at an observable rate given certain constraints, the system is said to be in *equilibrium*. *Thermodynamics* is concerned with the equilibrium states available to systems, the transitions between them and the effect of external influences upon the systems and transitions (Ericksen, 1991; Callen, 1960).

The transition between two thermodynamic states is a (*thermodynamic*) *process*. The number of state variables required to describe a process is larger than the number required to describe the system at equilibrium. In the field of process thermodynamics

[a] For this parameter one has several choices, one of which is the most appropriate.

[b] This implies that, if not mentioned specifically otherwise, systems are taken sufficiently large so that surface effects can be neglected.

various formulations, not all consistent with each other, have been given. Here we use a rather generally accepted form based on the principle of *local state*. This principle implies that at any moment and at any point in the system a thermodynamic state can be defined. This in principle restricts the applicability of thermodynamics to sufficiently slow processes. In practice this is only a limited restriction.

For clarity, we mention that we adhere in the following to the convention that for the partial derivative of a function with respect to one of the variables, the other variables are kept constant without any further indication. Although it is customary in thermodynamics to indicate the variables that are kept constant by a subscript, once a choice of independent variables has been made this indication is redundant. Only in a few cases explicit indication is required.

With each independent kinematical parameter a_i a dependent parameter is associated, generally denoted as *force* A_i. A force is a quantity that if multiplied with a change in state variable yields the associated work. Work done on the system is counted positive and for an infinitesimal change per unit volume the work $\mathrm{d}W$ is given by

$$\mathrm{d}W = A_i \mathrm{d}a_i \tag{6.1}$$

where use has been made of the summation convention. The work $\mathrm{d}W$ is dependent on the path between the initial and the final state and is thus not a total differential, i.e. not a state function.

Example 6.1

We consider homogeneous situations only, i.e. the force A_i is constant throughout the volume. In Chapter 5 we showed that the total work W_{int} of the internal forces for a solid is given generally by $W_{\text{int}} = -\int\int \boldsymbol{\sigma}:\mathrm{d}\boldsymbol{\varepsilon}\, \mathrm{d}V = -\int\int \sigma_{ij}\mathrm{d}\varepsilon_{ij}\mathrm{d}V$, where σ_{ij} denotes the stress tensor and ε_{ij} the strain tensor. The mechanical work W_{mec} done on the system is thus $W_{\text{mec}} = -W_{\text{int}}$. For homogeneous loading the increment in work reduces to $\mathrm{d}W_{\text{mec}} = V\sigma_{ij}\mathrm{d}\varepsilon_{ij}$. For a fluid the stress is given by $\sigma_{ij} = -p\delta_{ij}$, where p is the external pressure, so that the increment in mechanical work is $\mathrm{d}W_{\text{mec}} = -Vp\delta_{ij}\, \mathrm{d}\varepsilon_{ij} = -p\, \mathrm{d}V$. Another example is the work necessary to create a new surface, given by $\mathrm{d}W_{\text{sur}} = \gamma\, \mathrm{d}A$, where γ is the surface tension and A is the area. The work associated with the transfer of charge $\mathrm{d}e$ across an electric potential ϕ in electrical systems is $\mathrm{d}W_{\text{ele}} = \phi\, \mathrm{d}e$. For dielectric systems[c] the work per unit volume is given by $\mathrm{d}w_{\text{die}} = -\mathbf{P}\cdot\mathrm{d}\mathbf{E} = -$

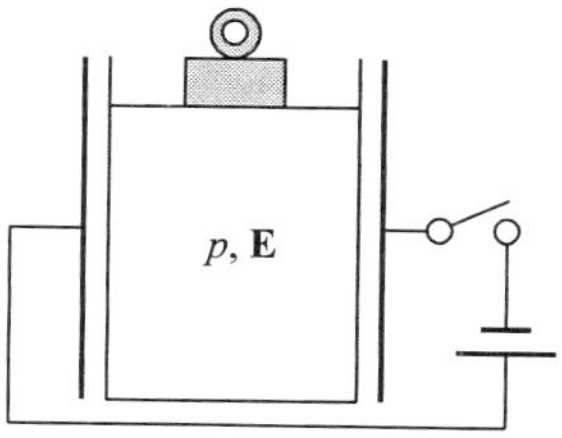

Fig. 6.1: A volume of gas in a cylinder with pressure p and electric field $\mathbf{E}$.

---

[c] The derivation of electrical and magnetic work is subtle and various authors use varying expressions. In many cases the work derived is actually $\mathrm{d}F$, see e.g. Guggenheim (1967) or Waldram (1984).

$P_i\mathrm{d}E_i$ (Fig. 6.1), where **P** denotes the polarisation and **E** the external electric field. Similarly for magnetic systems the work per unit volume is given by $\mathrm{d}w_{\mathrm{mag}} = -\mathbf{M}\cdot\mathrm{d}\mathbf{B} = -M_i\mathrm{d}B_i$, with **M** the magnetisation and **B** the external magnetic field.

To conclude this section, let us note that sufficient state variables should be included in the description of the system for proper characterisation, including specification of its chemical content, i.e. the amount of each substance contained. The exact choice of the kinematical variables depends upon the precision one wants to achieve in the description of the system and, of course, on the type of phenomena one wants to investigate. For example, if one wants to describe only the volume-temperature-pressure relationships for an amount of gas in a cylinder closed with a piston, the kinematical variable $a_i = V$, the force $A_i = -p$ and the temperature T suffice. The precision is determined by the equation of state used. However, if one is interested in the dielectric behaviour one must also consider the electric polarisation **P** due to the electric field **E** and apply a more elaborate equation of state. Obviously other combinations can arise. The selection of a particular set of independent state variables is important in each problem but the choice is arbitrary to a certain extent, as long as the variables are of the proper type.

Zeroth, first, second and third law

During a transition from one state to another also a certain amount of heat per unit volume $\mathrm{d}Q$ can enter the system. Heat entering the system is counted as positive and also depends on the path between the two states. If any two separate systems, each in thermal equilibrium, are brought in thermal contact through a thermally conducting wall, the two systems will gradually adjust themselves until they do reach mutual thermal equilibrium. The *zeroth law* states that if two systems are both in thermal equilibrium with a third system, they are also in thermal equilibrium with each other, i.e. they have the same (empirical) temperature T. If we consider two systems in thermal contact one of which is much smaller than the other, the state of the larger one will only change negligibly in comparison with the state of the smaller one if heat is transferred from one system to the other. If we are primarily interested in the small system the larger one is usually known as a *temperature bath* or *thermostat*. If, on the other hand, we are primarily interested in the large system, we regard the small system as a measuring device for registrating the temperature and refer to it as a *thermometer*. There is only one temperature scale which is independent of the properties of the small system used to measure the temperature and it is called the *thermodynamic* or *absolute temperature*. The associated units are *kelvins*, abbreviated as K. This scale is related to the conventional *Centigrade* or *Celsius* scale, using °C as a unit with the same size but with different origin, by

$$x\,°\mathrm{C} = (273.150 + x)\,\mathrm{K} \tag{6.2}$$

The origin of the absolute scale, 0 K = –273.150 °C, is referred to as *absolute zero*.

The *first law* states that there exists a (extensive) state function U, called *internal energy*, such that a change in the internal energy $\mathrm{d}U$ from one state to another is given by

► $$\mathrm{d}U = \mathrm{d}Q + \mathrm{d}W = \mathrm{d}Q + A_i\mathrm{d}a_i \tag{6.3}$$

The first law, Eq. (6.3), can be expressed as follows: for an isolated system the energy is constant. Each state is thus characterised by the set of state variables a_i and the internal energy U. The internal energy is thus the proper choice for the 'extra' state parameter. The energy of a composite system is additive over the constituent subsystems. The additivity property implies that the energy U is a homogeneous function of degree 1 of the extensive parameters. Although $\mathrm{d}Q$ and $\mathrm{d}W$ are dependent on the path between the initial and final states, $\mathrm{d}U$ is independent of the path and depends only on the initial and final states. Hence $\mathrm{d}U$ is a total differential. If for a process $\mathrm{d}Q = 0$, it is called *adiabatic*. If $\mathrm{d}W = 0$, we refer to it as pure *heating* or *cooling*. The first law is essentially due to Mayer[d], Joule[e] and Helmholtz.

Hermann von Helmholtz (1821-1894)

Helmholtz was educated in medicine as well as physics and physiology and made himself quite a reputation with the essay *Über die Erhaltung der Kraft*, published in 1847, and it is said by many that it contains the fundamental statement of the conservation of energy. Helmholtz was professor of physiology in Königsberg, Bonn and Heidelberg during the period from 1849 to 1871. He became professor of physics in Berlin in 1871 and in 1887 director of the new Physikalisch Technische Reichsanstalt in Charlottenburg, Berlin, the principal founder of which was the industrialist Werner Siemens. This institute was entirely devoted to research, contained a technical and a scientific section (the bill of the latter was about 1 million marks) and later became a model for independent institutes, including the National Physical Laboratory in Teddington, England (established in 1900) and the former National Bureau of Standards in Gaithersburg, United States (established in 1901), now the National Institute of Science and Technology. He worked extensively on electricity and his laboratory was a centre for many people amongst others, Albert A. Michelson (1852-1931). Heinrich Hertz was also one of his students.

The *second law* states that there exists another (extensive) state function S, called *entropy*, such that for a transition between two states

► $$T\mathrm{d}S \geq \mathrm{d}Q \tag{6.4}$$

where T is the temperature external to the element considered. The second law, Eq. (6.4), states that: for an isolated system the entropy can increase only or remain

---

[d] Julius Robert Mayer (1814-1874). German physician.

[e] James Prescott Joule (1818-1889). English scientist who established that units of energy could be measured. A devout Christian, he was one of the scientists who signed a declaration in 1864, following the publication of Charles Darwin's *On the origin of species*. The scientists stressed their confidence in the scientific integrity of the Bible.

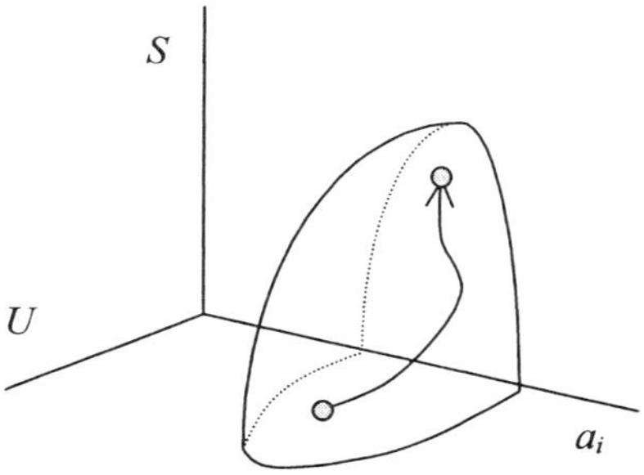

Fig. 6.2: The entropy S as a function of U and a_i and a quasi-static process from an initial to a final state.

constant at most. Since for any system the energy U is a characteristic of the state, the entropy S is a function of U (and the state variables a_i) and we write $S = S(U,a_i)$. The entropy of a composite system is additive over the constituent subsystems, continuously differentiable and monotonically increasing function of the energy. Again the additivity property implies that S is a homogeneous function of degree 1 of the extensive parameters. For a given U the equilibrium state is obtained when S is maximised or $\mathrm{d}S(U,a_i) = 0$ and $\mathrm{d}^2S(U,a_i) < 0$ (Fig. 6.2). If the equality in Eq. (6.4) holds, the process is *reversible*, otherwise it is *irreversible*. Rewriting $\mathrm{d}S$ to $\mathrm{d}S = \mathrm{d}S^{(r)} + \mathrm{d}S^{(i)}$, where $\mathrm{d}S^{(r)} = \mathrm{d}Q/T$ is the entropy supply from outside the system and $\mathrm{d}S^{(i)}$ is the entropy produced inside the system by irreversible processes, leads to $\mathrm{d}S^{(i)} \geq 0$. Unlike $\mathrm{d}S$, neither $\mathrm{d}S^{(r)}$ nor $\mathrm{d}S^{(i)}$ are total differentials. If for a process $\mathrm{d}S = 0$, it is called *isentropic*. If $\mathrm{d}T = 0$, we refer to it as *isothermal*. The second law is essentially due to Carnot, Clausius and Kelvin (see Section 18.2).

Nicolas-Léonard-Sadi Carnot (1796-1832)
French military engineer, born in the Palais du Petit-Luxembourg, educated at the École Polytechnique du Paris and died of cholera. He analysed the amount of work generated by engines that perform work at the expense of heat and realised that there is a limit to the amount of work that can be delivered, independent of the type of engine. He reported his work in only one publication. In 1824, 600 copies of his book *Réflexions sur la puissance motrice du feu, et sur les machines propres a développer cette puissance* were published at his own expense. The book did not attract much attention until Émile Clapeyron (1799-1864) came across the book in 1833 and made it known to the scientific community.

Rudolf Julius Emmanuel Clausius (1822-1888)
Born in Köslin, Prussia and educated at the Universities of Berlin and Halle, where he received his doctorate degree in 1847. In 1855 Clausius was appointed to the chair of mathematical physics at the Polytechnikum in Zürich and at the same time to the University of Zürich. In 1867 he became a professor at the University of Würzburg; in 1869 he accepted an offer of a chair at the University of Bonn, where he spent the remainder of his career. In 1884, he became rector of the University of Bonn. His most famous paper was *Uber die bewegende Kraft der Wärme* (1850), which contained a version of the Second Law of Thermodynamics. He was a theoretically oriented scientist investigating many topics but often with practical implications and made important contributions to thermodynamics (the introduction of the concept of entropy, which he named in 1865) and the kinetic theory of gases (the introduction of the mean free path). Clausius also coined the word 'virial' used in the virial equations. After 1875, he concentrated on electrodynamic theory. Clausius did not write very clearly and therefore the acceptance of entropy was slow. For example, Lord Kelvin never believed that the concept was of any help to the understanding of the second law. Clausius was an influential man with a fine character, received many honours and was exceptionally patient with those who did not accept his ideas. Clausius received the Iron Cross in 1871, for his leadership of an ambulance corps formed of Bonn students, during the Franco-German war, 1870-1871. He made the briefest summary of thermodynamics possible: 1) The energy of the universe is constant. 2) The entropy of the universe tends towards a maximum.

The entropy expressed as $S = S(U,a_i)$ is called a *fundamental equation*. Once it is known as a function of its *natural variables* U and a_i, all other properties can be calculated. The description $S = S(U,a_i)$ is called the *entropy representation*. Since S is a single valued continuously increasing function of U, the equation $S = S(U,a_i)$ can be inverted to $U = U(S,a_i)$ without ambiguity. For a given entropy S the equilibrium state is obtained when $\mathrm{d}U(S,a_i) = 0$. From stability considerations is follows that U is minimised or $\mathrm{d}^2U(S,a_i) > 0$. The description $U = U(S,a_i)$ is the *energy representation*. From $\mathrm{d}U = (\partial U/\partial S)\mathrm{d}S + (\partial U/\partial a_i)\mathrm{d}a_i$ we may define the (thermodynamic) *temperature* $T = \partial U(S,a_i)/\partial S$. It appears that $T^{-1} = \partial S(U,a_i)/\partial U$. One can show that this definition agrees with the intuitive concept of temperature we used before.

For completeness we mention the *third law*[f], stating that for any isothermal process involving only phases in internal equilibrium or, alternatively, if any phase is in frozen metastable equilibrium, provided the process does not disturb this frozen equilibrium,

$$\lim_{T\to 0} \Delta S = 0 \qquad (6.5)$$

[f] The status of this law is somewhat different from the others. It is also addressed as the Nernst postulate after its discoverer. For consistency we stick to the third law.

The discussion of the third law contains several subtleties for which we refer to Fowler and Guggenheim (1939). Its discovery is due to Nernst.

Hermann Walther Nernst (1864-1941)

Born in Briessen, Prussia and inventor of the Nernst lamp in 1898, the patent of which made him wealthy although the exploitation was a commercial failure. He studied in Zurich, Berlin and Graz before went to Würzburg were he graduated in 1887 with a thesis on electromotive forces produced by magnetism in heated metal plates. In 1896 the Institute for Physical Chemistry and Electrochemistry was founded in Göttingen and Nernst became its leader. There he worked on galvanic current production and attempted to provide a theoretical foundation for chemical empirical laws through physical chemistry. In 1905 he became professor of chemistry and later also of physics at the University of Berlin, where his earlier work on galvanic currents gave him the ideas for the third law, which he called 'my heat theorem', generally accepted only 10 years after its publication in 1906. He ended his 1937 honorary degree DSc Oxford by saying that the first law in thermodynamics had been discovered by three scientists (presumably Mayer, Joule and Helmholtz) and the second by two scientists (presumably Carnot or Kelvin and Clausius); for the third law, he said "Well, this I have done just by myself". Due to the same reasoning it is also said that he claimed that no fourth law could be found. His famous book *Theoretische Chemie*, first published in 1895 and the tenth edition in 1921, was rather influential. Nernst actually wanted to become a poet and used to love acting. He often played the role of somewhat ignorant and often astonished man, but he had a many-sided and sharp mind with a sarcastic sense of humour. He received the Nobel Prize in 1920 for his work on heat changes in chemical reactions.

Equation of state

Expressed in its natural variables S and a_i, $U(S,a_i)$ is also a fundamental equation. Again, once it is known, all other properties of the system can be calculated. If U is not expressed in its natural variables, but e.g. as $U = U(T,a_i)$ or the like, we have slightly less information. In fact it follows from $T = \partial U/\partial S$ that $U = U(T,a_i)$ represents a differential equation for $U(S,a_i)$. Generally functional relationships such as

$$T = T(S,a_i) \qquad \text{and} \qquad A_i = A_i\,(S,a_i)$$

expressing a certain (intensive) state variable as a single-valued function of the remaining (extensive) state variables are called *equations of state* and the variable so described is called a *state function*. Hence $U = U(T,a_i)$ is an equation of state while $U = U(S,a_i)$ is a fundamental equation. If all the equations of state are known, the fundamental equation can be inferred to within an arbitrary constant. To that purpose we note that $U(S,a_i)$ is homogeneous of degree 1 and thus from Euler's theorem (Section 3.3) that

$$U = \frac{\partial U}{\partial S}S + \frac{\partial U}{\partial a_i}a_i = TS + A_i a_i$$

Given $T = T(S,a_i)$ and $A_i = A_i(S,a_i)$, the energy U can be calculated. A similar argumentation for S leads to $S = (1/T)U - (A_i/T)a_i$. Only in some special cases, e.g. if $\mathrm{d}S(U,V) = f(U)\mathrm{d}U + g(V)\mathrm{d}V$, the differential can be integrated directly to obtain the fundamental equation.

Problem 6.1

An ideal gas is defined by $U = cNRT$ and $pV = NRT$, where c is a constant and the other symbols have their usual meaning. Show that the fundamental equation for the entropy S is given by $S = S_0 + NR \ln[(U/U_0)^c(V/V_0)]$.

Quasi-conservative and dissipative forces

Only for reversible systems we can associate $T\mathrm{d}S$ with $\mathrm{d}Q$ and in this case $\mathrm{d}W = A_i\mathrm{d}a_i$ is the reversible work. For irreversible systems $T\mathrm{d}S > \mathrm{d}Q$ and the work $\mathrm{d}W$ contains also an irreversible or dissipative part. To show this, note that the elementary work can be written as (Ziegler, 1983)

$$\mathrm{d}W = \mathrm{d}U - \mathrm{d}Q = \mathrm{d}U - T\mathrm{d}S^{(\mathrm{r})} = \mathrm{d}U - T\mathrm{d}S + T\mathrm{d}S^{(\mathrm{i})} \quad \textbf{(6.6)}$$

where $\mathrm{d}S = \mathrm{d}S^{(\mathrm{r})} + \mathrm{d}S^{(\mathrm{i})}$. Since $U = U(S,a_i)$ we have $\mathrm{d}U = (\partial U/\partial S)\mathrm{d}S + (\partial U/\partial a_k)\mathrm{d}a_k = T\mathrm{d}S + (\partial U/\partial a_k)\mathrm{d}a_k$. So we can also write Eq. (6.6) as

$$\mathrm{d}W = \frac{\partial U}{\partial a_k}\mathrm{d}a_k + T\mathrm{d}S^{(\mathrm{i})} \quad (6.7)$$

Therefore the expression $T\,\mathrm{d}S^{(\mathrm{i})}$ has also the form of elementary work and we can introduce forces $A_k^{(\mathrm{d})}$ by writing

$$T\mathrm{d}S^{(\mathrm{i})} = A_k^{(\mathrm{d})}\mathrm{d}a_k = \mathrm{d}W^{(\mathrm{d})} \quad \text{with} \quad A_k^{(\mathrm{d})} = A_k - \frac{\partial U}{\partial a_k} \quad (6.8)$$

The quantities $A_k^{(\mathrm{d})}$ are the *dissipative forces* and we refer to $\mathrm{d}W^{(\mathrm{d})} = A_k^{(\mathrm{d})}\,\mathrm{d}a_k$ as the *dissipative work*. Writing

$$A_k = A_k^{(\mathrm{d})} + A_k^{(\mathrm{q})} \quad (6.9)$$

we define the *quasi-conservative forces* $A_k^{(\mathrm{q})}$ by

$$A_k^{(\mathrm{q})} = \frac{\partial U}{\partial a_k} \quad (6.10)$$

The quasi-conservative force $A_k^{(\mathrm{q})}$ is also called as the variable *conjugated* to a_k. Summarising this leads to

$$\blacktriangleright \quad \begin{aligned} \mathrm{d}U &= \mathrm{d}W + \mathrm{d}Q = \left(A_k^{(\mathrm{q})}\mathrm{d}a_k + A_k^{(\mathrm{d})}\mathrm{d}a_k\right) + T\mathrm{d}S^{(\mathrm{r})} \\ &= A_k^{(\mathrm{q})}\mathrm{d}a_k + \left(T\mathrm{d}S^{(\mathrm{i})} + T\mathrm{d}S^{(\mathrm{r})}\right) = A_k^{(\mathrm{q})}\mathrm{d}a_k + T\mathrm{d}S \end{aligned} \quad (6.11)$$

known as the *Gibbs equation*. Introducing via a Legendre transform (see Section 3.5) of the internal energy U, by which the entropy S is eliminated in favour of the temperature T, the *Helmholtz energy* $F(T,a_i) = U - TS$, it is easily shown that

$$\blacktriangleright \quad S = -\frac{\partial F}{\partial T} \quad \text{and} \quad A_k^{(q)} = \frac{\partial F}{\partial a_k} \tag{6.12}$$

The Helmholtz energy $F(T,a_i)$ thus acts as a potential, similar to the potential energy in mechanics, and is called a *thermodynamic potential*. Its derivatives with respect to the kinematical parameters and temperature yield the quasi-conservative forces and the entropy, respectively. The adjective 'quasi-conservative' stems from the fact that F is still dependent on the temperature. The Helmholtz energy $F(T,a_i)$ is also a fundamental equation. Stable equilibrium state is reached when $\mathrm{d}F(T,a_i) = 0$ and when $\mathrm{d}^2F(T,a_i) > 0$.

Problem 6.2

Show that if one starts from the equation of state $U = U(T,a_i)$ the quasi-conservative force is given by

$$A_i^{(q)}(T,a_i) = \frac{\partial U(T,a_i)}{\partial a_i} - T\frac{\partial S(T,a_i)}{\partial a_i}$$

Rate formulation

For thermomechanics we need a rate formulation instead of the classic differential formulation. In a rate formulation we replace the increments $\mathrm{d}a_k$ by the *velocities* $\mathrm{d}a_k/\mathrm{d}t = \dot{a}_k$, to be interpreted as material derivatives. The work becomes the power $P_{\text{ext}} = A_k\dot{a}_k$ and introducing the heat supply per unit time $\mathrm{d}Q/\mathrm{d}t = \dot{Q}$, the first law (neglecting kinetic energy) reads (Ziegler, 1983)

$$\frac{\mathrm{d}U}{\mathrm{d}t} = A_i\frac{\mathrm{d}a_i}{\mathrm{d}t} + \frac{\mathrm{d}Q}{\mathrm{d}t} \quad \text{or} \quad \dot{U} = P_{\text{ext}} + \dot{Q} = A_k\dot{a}_k + \dot{Q} \tag{6.13}$$

Similarly the second law becomes

$$T\dot{S} \geq \dot{Q} \quad \text{with} \quad \dot{S} = \dot{S}^{(r)} + \dot{S}^{(i)} \quad \text{where} \tag{6.14}$$

$$\dot{S}^{(r)} = \frac{\dot{Q}}{T} \quad \text{and} \quad \dot{S}^{(i)} \geq 0 \tag{6.15}$$

Combining the above yields the *Gibbs equation* in rate form

$$\dot{U} = A_k^{(q)}\dot{a}_k + A_k^{(d)}\dot{a}_k + \dot{Q} = A_k^{(q)}\dot{a}_k + T\dot{S}^{(i)} + T\dot{S}^{(r)} = A_k^{(q)}\dot{a}_k + T\dot{S} \tag{6.16}$$

where the term $A_k^{(q)}\dot{a}_k$ denotes the power delivered by the quasi-conservative part of the external variables. As before, the term $\dot{Q} = T\dot{S}^{(r)}$ represents the heat input and the dissipation rate $T\dot{S}^{(i)}$ is governed by the *dissipation function* Φ, given by

$$\blacktriangleright \quad \Phi(\dot{a}_k,\dot{T},a_k,T) = T\dot{S}^{(i)} = A_k^{(d)}\dot{a}_k \geq 0 \tag{6.17}$$

The dissipation rate represents the *power of the dissipation.*

Specific quantities

For an infinitesimal element of solid the elements of the strain tensor ε_{ij}, together with the entropy S (or energy U), act as independent kinematical state variables for the energy U (or entropy S). Extensive quantities such as U when expressed per unit mass are conveniently referred to as specific quantities[g]. For example, from the internal energy per unit volume U (or energy density) we obtain the *specific energy* $u = U/\rho$, where ρ is the mass density. Similarly we obtain the *specific entropy* s, the *specific Helmholtz energy* $f = u - Ts$ and the *specific dissipiation function* φ. The specific energy u is thus a function of ε_{ij} and s, e.g. $u = u(\varepsilon_{ij},s)$. We have seen in Chapter 5 that the rate of work done by the deformation per unit volume (or power density) is given by $\sigma_{ij}d_{ij}$ so that the power per unit mass l (or specific power) is given by

$$l = \frac{1}{\rho}\sigma_{ij}d_{ij} \tag{6.18}$$

Since the total power P_{ext} is given by $P_{\text{ext}} = A_k\dot{a}_k$ we also have

$$l = \frac{P_{\text{ext}}}{\rho} = \frac{A_k\dot{a}_k}{\rho} = \frac{1}{\rho}\frac{\partial F}{\partial a_k}\dot{a}_k = \frac{\partial f}{\partial a_k}\dot{a}_k \tag{6.19}$$

and the forces per unit mass associated with the rate of deformation d_{ij} are the quotients σ_{ij}/ρ. If we decompose the stress into their quasi-conservative parts $\sigma_{ij}^{(q)}$ and dissipative part $\sigma_{ij}^{(d)}$, we obtain

► $$\sigma_{ij}^{(q)} = \rho\frac{\partial f}{\partial \varepsilon_{ij}} \quad \text{and} \quad s = -\frac{\partial f}{\partial T} \tag{6.20}$$

The specific power of dissipation is then

► $$l^{(\text{dl})} = \frac{1}{\rho}\sigma_{ij}^{(d)}d_{ij} \tag{6.21}$$

The general expression for the specific power of dissipation is, of course,

$$l^{(\text{dl})} = \frac{1}{\rho}A_i^{(d)}\dot{a}_i \tag{6.22}$$

6.2 Equilibrium

Next we have to consider under what conditions equilibrium is reached. Consider therefore an isolated system consisting of two subsystems only capable of exchanging heat ($da_i = 0$). We now ask ourselves under what conditions thermal equilibrium

---

[g] For completeness we mention that chemical thermodynamic quantities are often expressed per mole and indicated conventionally as *molar* quantities, e.g. molar energy. Guggenheim (1967) prefers the term *proper* on the argument that a generally accepted recommendation is absent and that the terminology 'molar energy' is as clumsy as if we speak of 'gramme energy' instead of 'specific energy'. If we indicate the specific quantity by q and the proper quantity by q', we have $q' = Mq$, where M is the molecular weight. Finally, the designation *density* is used for quantities expressed per unit volume, e.g. energy density.

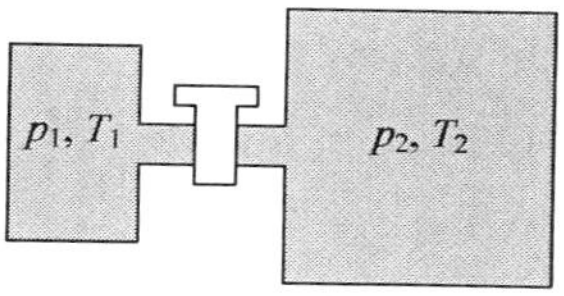

Fig. 6.3: Hydrostatic equilibrium.

occurs and to this purpose we consider arbitrary but small variations in energy in both subsystems. From the first law we have $\mathrm{d}U = \mathrm{d}U_1 + \mathrm{d}U_2 = 0$. If the system is in equilibrium we must also have $\mathrm{d}S = \mathrm{d}S_1 + \mathrm{d}S_2 = 0$. Because

$$\mathrm{d}S = \frac{\partial S_1}{\partial U_1}\mathrm{d}U_1 + \frac{\partial S_2}{\partial U_2}\mathrm{d}U_2 = \left(\frac{\partial S_1}{\partial U_1} - \frac{\partial S_2}{\partial U_2}\right)\mathrm{d}U_1 = \left(\frac{1}{T_1} - \frac{1}{T_2}\right)\mathrm{d}U_1 = 0 \tag{6.23}$$

and since $\mathrm{d}U_1$ is arbitrary, we obtain $T_1 = T_2$ in agreement with the zeroth law. Moreover, starting from a non-equilibrium state one can show that heat flows from high to low temperature, again in agreement with intuition (Callen, 1960).

The procedure outlined above is in fact the general method to deal with basic equilibrium thermodynamic problems: one considers an isolated system consisting of two subsystems of which one is the object of interest and the other represents the environment. Depending on the problem one assumes a specific type of membrane between the subsystems, i.e. a *diathermal* (contrary *adiabatic*) membrane allowing heat exchange, a *flexible* (contrary *rigid*) membrane allowing work exchange or a *permeable* (contrary *impermeable*) membrane allowing matter exchange. For an isolated system we have $\mathrm{d}U = 0$. Moreover in equilibrium we have $\mathrm{d}S = 0$. Together with the *closure relations*, i.e. the coupling relations between the displacements a_i of the system of interest and of the environment, a solution can be obtained.

Let us now consider hydrostatic equilibrium along these lines. Consider therefore again an isolated system consisting of two homogeneous subsystems but in this case capable of exchanging heat and work (Fig. 6.3). For this system to be in equilibrium we write again $\mathrm{d}S = \mathrm{d}S_1 + \mathrm{d}S_2 = 0$. However, the entropy S is now a function of U and the various a_i's. Since we are interested in hydrostatic equilibrium we take just the volume V for each subsystem. From Eq. (6.11) we obtain

$$\mathrm{d}S = \frac{1}{T}\mathrm{d}U + \frac{p}{T}\mathrm{d}V \tag{6.24}$$

and thus for the equilibrium configuration

$$\mathrm{d}S = \mathrm{d}S_1 + \mathrm{d}S_2 = \frac{1}{T_1}\mathrm{d}U_1 + \frac{p_1}{T_1}\mathrm{d}V_1 + \frac{1}{T_2}\mathrm{d}U_2 + \frac{p_2}{T_2}\mathrm{d}V_2 \tag{6.25}$$

In view of the fact that the total system is isolated we have $\mathrm{d}U = \mathrm{d}U_1 + \mathrm{d}U_2 = 0$ and the closure relation $\mathrm{d}V = \mathrm{d}V_1 + \mathrm{d}V_2 = 0$ and thus

$$\mathrm{d}S = \left(\frac{1}{T_1} - \frac{1}{T_2}\right)\mathrm{d}U_1 + \left(\frac{p_1}{T_1} - \frac{p_2}{T_2}\right)\mathrm{d}V_1 = 0 \tag{6.26}$$

Since $\mathrm{d}U_1$ and $\mathrm{d}V_1$ are arbitrary, we obtain

$$\frac{1}{T_1} = \frac{1}{T_2} \quad \text{or} \quad T_1 = T_2 \tag{6.27}$$

regaining thermal equilibrium and

$$\frac{p_1}{T_1}=\frac{p_2}{T_2} \qquad \text{or using } T_1=T_2, \qquad p_1=p_2 \tag{6.28}$$

The latter condition corresponds to hydrostatic equilibrium. Hence in equilibrium the temperature and the pressure of the subsystems are equal. One can show that for a non-homogeneous system where a full stress tensor should be used, equilibrium between internal volume elements requires the relation $\sigma_{ij,j} = 0$ to be fulfilled (in the absence of body and inertia forces) while at the surface the relation $\sigma_{ij}n_j = t_i$ has to be obeyed. This more general analysis of equilibrium dealing with stress, strain and thus with fields and the appropriate boundary conditions is given in Section 6.6.

6.3 Some further tools

The description given in the previous sections is complete but not convenient from an experimental point of view. Therefore auxiliary functions have to be introduced. Moreover, some special derivatives and their mutual relations are of importance.

Auxiliary functions

In this section we write $-p$ for $A_k^{(q)}$ and V for a_k. This simplifies somewhat some of the formulae given below while also the appearance becomes more familiar. The Gibbs equation then becomes

$$\mathrm{d}U = T\mathrm{d}S - p\mathrm{d}V \tag{6.29}$$

which gives the dependent variable U as a function of the independent, natural variables S and V. From consideration of the second law, the criterion for equilibrium for a closed system with fixed composition is that $\mathrm{d}S(U,V) = 0$ or S is a maximum at constant U. Equivalently, as stated earlier, $\mathrm{d}U(S,V) = 0$ or U is a minimum at constant S. Both criteria are, however, not very practical since it is difficult to keep the entropy constant and keeping the energy constant excludes interference from outside. Therefore auxiliary functions are introduced.

If we write the energy as $U(S,V)$, the *enthalpy* H is the Legendre transform[h] of U with respect to $p = -\partial U/\partial V$, which is obtained from

$$H = U + pV \tag{6.30}$$

After differentiation this yields

$$\mathrm{d}H = \mathrm{d}U + p\mathrm{d}V + V\mathrm{d}p \tag{6.31}$$

Combining with $\mathrm{d}U = T\mathrm{d}S - p\mathrm{d}V$ results in

$$\mathrm{d}H = T\mathrm{d}S + V\mathrm{d}p \tag{6.32}$$

The natural variables for the enthalpy H are thus S and p and the equilibrium condition becomes $\mathrm{d}H(S,p) = 0$.

Similarly writing $H(S,p)$, the *Gibbs energy* G is the Legendre transform of H with respect to $T = \partial H/\partial S$ and given by

$$G = H - TS \tag{6.33}$$

On differentiation and combination with $\mathrm{d}H = T\,\mathrm{d}S + V\,\mathrm{d}p$ this yields

---

[h] See Section 3.5. A particularly clear discussion of Legendre transforms is given by Callen (1960).

$$dG = dH - TdS - SdT = -SdT + Vdp \tag{6.34}$$

of which the natural variables are T and p. Consequently the equilibrium condition becomes $dG(T,p) = 0$. A third transform, already encountered, is the *Helmholtz energy* F, defined as $F = U - TS$, resulting in

$$dF = -SdT - pdV \tag{6.35}$$

with natural variables T and V and the corresponding equilibrium condition $dF(T,V) = 0$. The functions $U(S,V)$, $H(S,p)$, $F(T,V)$ and $G(T,p)$ are all thermodynamic potentials. Moreover, they are fundamental equations. For stable equilibrium these functions are all a minimum for a given set of their natural variables. Finally we remark that the advantage of using F and G is obvious: While it is possible to control either the set (V,T) or (p,T) experimentally, control of either the set (V,S) or (p,S) is virtually impossible.

Some derivatives and their relationship

Apart from the potentials defined above, we also need now and then some of their derivatives and the relationships between these derivatives. Consider the relations

$$TdS = dU + pdV \quad \text{and} \quad TdS = dH - Vdp$$

The *heat capacities* are defined by

$$C_X = \partial Q/\partial T = T(\partial S/\partial T)_X \quad (X = p \text{ or } V)$$

It thus follows that

$$C_V = T(\partial S/\partial T)_V = (\partial U/\partial T)_V \quad \text{and} \quad C_p = T(\partial S/\partial T)_p = (\partial H/\partial T)_p$$

Three other derivatives that occur frequently are the *compressibilities* β_X ($X = T$ or S) and the (linear) *thermal expansion coefficient* α.

$$\beta_T = -(1/V)(\partial V/\partial p)_T \qquad \beta_S = -(1/V)(\partial V/\partial p)_S \qquad 3\alpha = (1/V)(\partial V/\partial T)_p$$

Moreover, if $d\phi = Xdx + Ydy$ is a total differential, we can use the *Maxwell relations* $(\partial X/\partial y)_x = (\partial Y/\partial x)_y$. From the expressions for dU, dH, dF and dG we obtain

$$(\partial V/\partial S)_p = (\partial T/\partial p)_S \qquad (\partial S/\partial p)_T = -(\partial V/\partial T)_p$$

$$(\partial p/\partial T)_V = (\partial S/\partial V)_T \qquad (\partial T/\partial V)_S = -(\partial p/\partial S)_V$$

which can be used to reduce a set of thermodynamic quantities to one of measurable quantities. For example, only three of the five quantities just defined are independent because

$$\beta_T - \beta_S = 9TV\alpha^2/C_p \quad \text{and} \quad C_p - C_V = 9TV\alpha^2/\beta_T \tag{6.36}$$

Using the third law it can be shown that at $T = 0$, $C_p = C_V = 0$ and $\alpha = 0$.

Problem 6.3

Show for an ideal gas that $C_V = cNR$, $C_p = (c+1)NR$, $\beta_T = 1/p$ and $3\alpha = 1/T$.

Problem 6.4

Prove Eqs. (6.36).

6.4 Chemical aspects

For chemical aspects some specific information is sometimes required which will be addressed in this section.

Chemical content

The content of a system is defined by the amount of moles N_X of the various independent variable chemical species X in the system. For any extensive property Z we define the associated *partial* property $\overline{Z}_X$ as the partial derivative with respect to the number of moles N_X at constant temperature T and pressure p. For example, the partial volume $\overline{V}_X$ is defined by

$$\overline{V}_X = \left(\frac{\partial V}{\partial N_X}\right)_{T,p,N_Y} \qquad (Y \neq X) \qquad (6.37)$$

and similarly for the partial energy, partial entropy and so on. At constant T and p we thus have[i]

$$dZ = \frac{\partial Z}{\partial N_X} dN_X = \overline{Z}_X dN_X \qquad (p, T \text{ constant}) \qquad (6.38)$$

Since Z is homogeneous of the first degree in N_X, we have by the Euler theorem

$$Z = \frac{\partial Z}{\partial N_X} dN_X = \overline{Z}_X N_X \qquad (6.39)$$

Hence we may regard Z as the sum of the contributions $\overline{Z}_X$ for each of the species X. For chemical problems, i.e. where a change in chemical composition is involved, the fundamental equation thus becomes $U = U(S,a_i,N_X)$ or $S = S(U,a_i,N_X)$. This leads to

$$\begin{aligned} dU &= \frac{\partial U}{\partial S} dS + \frac{\partial U}{\partial a_i} da_i + \frac{\partial U}{\partial N_X} dN_X \\ &= T dS - p dV + \mu_X dN_X \end{aligned} \qquad (6.40)$$

where $a_i = V$ (and thus $A_i^{(q)} = \partial U/\partial a_i = -p$) for the last step is assumed. The partial derivative $\mu_X = \partial U/\partial N_X$ is called the *chemical potential* and is the conjugate intensive variable associated with the extensive variable N_X.

Applying a similar Legendre transformation as in the previous section to the second line of Eq. (6.40) and using the Gibbs energy $G = U + pV - TS + \mu_X N_X$ leads to

$$dG = \frac{\partial G}{\partial T} dT + \frac{\partial G}{\partial p} dp + \frac{\partial G}{\partial N_X} dN_X = -S dT + V dp + \mu_X dN_X \qquad (6.41)$$

The Gibbs energy is thus given by $G = G(T,p,N_X)$ and the equilibrium condition thus becomes $dG(T,p,N_X) = 0$. Similarly, for the Helmholtz energy F one obtains $dF(T,V,N_X) = -S dT - p dV + \mu_X dN_X = 0$. The chemical potential μ_X of the component X is thus equal to the partial Gibbs energy $\overline{G}_X$. Since G is homogeneous of the first degree in N_X, making use of the Euler theorem leads to $G = \overline{G}_X N_X = \mu_X N_X$. On the one hand we thus have

[i] Extending the summation convention to chemical species!

$$dG = \mu_X dN_X + N_X d\mu_X$$

while on the other hand we know that $G = G(T,p,N_X)$ and thus that

$$dG = \frac{\partial G}{\partial T}dT + \frac{\partial G}{\partial p}dp + \frac{\partial G}{\partial N_X}dN_X = -SdT + Vdp + \mu_X dN_X$$

Therefore we obtain by subtraction the so-called *Gibbs-Duhem relation*

$$SdT - Vdp + N_X d\mu_X = 0 \qquad (6.42)$$

implying a relation between the various differentials. It is particularly useful for constant p and constant T when it may be written as

$$N_X d\mu_X = 0 \qquad \text{(constant } p \text{ and } T)$$

Although the relation is here derived for T, p and N_X as the only independent variables, the extension to any number of variables is obvious. Finally we note that various quantities are used for amount of substance. So far we used the number of moles N_X. Frequently one is only interested in relative changes in composition. To that purpose one uses the *mole fraction* defined by $n_X = N_X/\Sigma N_X$.

Problem 6.5

Show that for the ideal mono-atomic gas (Problem 6.1) the fundamental equation $S = S(U,V,N)$ is given by

$$S = S_0 + RT \ln[(u/u_0)^c(v/v_0)(N_0/N)^{c+1}]$$

Use the Euler equation $S = (1/T)U + (P/T)V - (\mu/T)N$ and the Gibbs-Duhem equation for $d\mu$. Show also that the chemical potential μ is given by

$$\mu = \mu_0(T) + RT \ln[(u/u_0)^c(v/v_0)(N/N_0)^{c+1}]$$

where u, u_0, v and v_0 denote proper (or molar) quantities.

Pierre Maurice Marie Duhem (1861-1916)

Born in Paris, France, he studied at the École Normale Supérieure and submitted in 1884 his thesis in which he defined the criterion for chemical reactions in terms of free energy, thereby replacing the incorrect criterion, which Berthelot had put forward 20 years earlier. However, a scientist as influential as Berthelot was able to arrange for Duhem's thesis to be rejected. Nevertheless Duhem boldly published the rejected thesis in 1886, which did not help his relations with Berthelot. Duhem meanwhile worked on the second thesis on the mathematical theory of magnetism (accepted in 1888) but he suffered all his life because of Berthelot. He lectured in Lille from 1887 to 1893 on hydrodynamics, elasticity and acoustics, publishing these lectures in 1891 and briefly taught in Rennes during 1893-1894. He became professor of theoretical physics at the University of Bordeaux in 1894 but a move to Paris was blocked. Duhem was also at odds with Berthelot on religious issues and one biographer writes that *he was of a contentious and acrimonious disposition, with a talent for making personal enemies over scientific matters.* After becoming a corresponding member of the Académie des Sciences in 1900, he again requested a move from Bordeaux but again it was refused. His interests in science itself were mainly in the area of thermodynamics, hydrodynamics, elasticity, mathematical chemistry and mechanics. He considered that a generalised version of thermodynamics would provide a theory to explain all of physics and chemistry and this line of thinking was elaborated in *Traité d'énergétique générale* (1911). His scientific work led him towards the philosophy of science, then in turn to the history of science. His paper *L'évolution de la méchanique* (1902) is really an article on the philosophy of science, based heavily on historical examples. His most important work on philosophy of science *La Théorie physique, son objet et sa structure* (1906) depreciates pictorial models in favour of an axiomatic approach, according to which a physical theory is not an explanation, but a system of mathematical propositions that represents experimental laws. As a historian Duhem discovered important currents of medieval thought in physics, cosmology and astronomy. He disliked British science, in particular the work of Maxwell, and described it as broad and shallow while he said that French science was narrow and deep. German sciences he claimed were highly geometrical, which for Duhem was a criticism for he considered an approach using an analytical style of mathematics to be far superior to a geometrical one. Late in his career Duhem was offered a professorship in Paris as a historian of science. Duhem refused this chance, where he had always longed for, saying that he was a mathematical physicist and did not want to get to Paris through the back door.

Chemical equilibrium

We briefly discuss chemical equilibrium in this section. At constant p and constant T the Gibbs energy in a chemically reacting system then varies with composition as $\mathrm{d}G = \mu_X \mathrm{d}N_X$. For a reaction to occur spontaneously $\mathrm{d}G = \mu_X \mathrm{d}N_X < 0$ while at equilibrium[j] $\mathrm{d}G = \mu_X \mathrm{d}N_X = 0$. Let us consider a reaction given by

$$\alpha N_A + \beta N_B \leftrightarrow \gamma N_C + \delta N_D \tag{6.43}$$

where α, ..., δ denote the stoichiometric coefficients of components A, ..., D. It will be convenient to rearrange this notation to

$$0 = \gamma N_C + \delta N_D - \alpha N_A - \beta N_B$$

or even more compact, using the summation convention again for chemical species, to

$$0 = \nu_X N_X$$

with a positive value for the coefficient ν_X when X is a product (C, D) and a negative value when X is a reactant (A, B). We define a factor of proportionality $\mathrm{d}\xi(t)$ in such a way that $\mathrm{d}N_X = \nu_X \mathrm{d}\xi$. Starting at time zero with initially $N_X(0)$ moles of each species the changes of the number of moles of each species in the time interval $\mathrm{d}t$ are

$$\mathrm{d}N_X = \nu_X \frac{\mathrm{d}\xi}{\mathrm{d}t}\mathrm{d}t = \nu_X \dot{\xi}\,\mathrm{d}t \qquad \text{or} \qquad N_X = N_X(0) + \int \mathrm{d}N_X = N_X(0) + \nu_X \Delta\xi$$

where $\dot{\xi}$ is the *rate of reaction*. This leads to

$$\mathrm{d}G = \mu_X \mathrm{d}N_X = (\mu_X \nu_X)\dot{\xi}\,\mathrm{d}t = -D\dot{\xi}\,\mathrm{d}t \le 0$$

where we introduced the so-called *affinity* $D = -\nu_X \mu_X$. From $\mathrm{d}G \le 0$ we conclude that

---

[j] The condition of chemical equilibrium can also be derived similarly as for thermal and mechanical equilibrium.

$$\blacktriangleright \quad D\dot{\xi} \geq 0 \tag{6.44}$$

as the condition for a reaction to occur. So, D and $\dot{\xi}$ must have the same sign or be zero. At equilibrium $D = 0$. Since $dN_X = \nu_X d\xi$, the affinity D is related to the fundamental equations, the most important ones being

$$D = -\nu_X \mu_X = -\frac{\partial U(S,V,N_X)}{\partial \xi} = -\frac{\partial G(T,p,N_X)}{\partial \xi} = T\frac{\partial S(U,V,N_X)}{\partial \xi} \tag{6.45}$$

All N_X must be positive or zero and the reaction goes to completion if one of the components is exhausted. This implies a lower and upper value for $\Delta\xi$. Therefore the factor $\Delta\xi$ is sometimes normalised according to

$$\zeta = \frac{\Delta\xi - \Delta\xi_{min}}{\Delta\xi_{max} - \Delta\xi_{min}}$$

where ζ is referred to as the *degree of reaction*.

Example 6.2

A vessel contains a ½ mole of H_2S, ¾ mole of H_2O, 2 moles of H_2 and 1 mole of SO_2. The vessel is kept at constant temperature and pressure. The equilibrium condition is

$$-3\mu_{H_2} - \mu_{SO_2} + \mu_{H_2S} + 2\mu_{H_2O} = 0 \quad \text{and}$$

$$N_{H_2} = 2 - 3d\xi, \quad N_{SO_2} = 1 - d\xi, \quad N_{H_2S} = ½ + d\xi, \quad N_{H_2O} = ¾ + 2d\xi$$

If the chemical potentials are known as a function of T, p and the N_X's, the solution for $d\xi$ can be obtained. Suppose that the solution is $d\xi$ = ¼. If $d\xi$ = ⅔, $N_{H_2} = 0$ and therefore this is $d\xi_{max}$. If $d\xi$ = –⅜, $N_{H_2O} = 0$ and therefore this is $d\xi_{min}$. Therefore the degree of reaction ε = [¼–(–⅜)]/[⅔–(–⅜)] = 3/5.

The relation $D\dot{\xi} \geq 0$ is similar to the expression for the dissipation function $\Phi = A_k^{(d)}\dot{a}_k \geq 0$ derived before. We may thus interpret D as a force and $\dot{\xi}$ as a velocity. The above formulation leads immediately to the conventional chemical description. Let us introduce the *absolute activity* $\lambda_X = \exp(\mu_X/RT)$ and define, considering again the reaction given by Eq. (6.43), the *reaction product* by

$$\frac{\lambda_C^{\ \gamma}\lambda_D^{\ \delta}}{\lambda_A^{\ \alpha}\lambda_B^{\ \beta}} \tag{6.46}$$

where λ_A, ... denotes the activity of component A, ... Using the more compact notation introduced before we write more generally $\prod_X \lambda_X^{\nu_X}$. The equilibrium condition $D = -\nu_X \mu_X = 0$ can then be written as $\prod_X \lambda_X^{\nu_X} = 1$. We now distinguish between gases (X) and solids (Y). This allows us to write $\prod_X \lambda_X^{\nu_X} \prod_Y \lambda_Y^{\nu_Y} = 1$, where the first product contains all the terms relating to gaseous species and the second to the solid species. Now note that the chemical potential μ_X of a gaseous component X is given by $\mu_X = \mu_X{}^0 + RT\ln x_X/p^0$, where $\mu_X{}^0$ is the chemical potential in the standard state, R is the gas constant, $x_X = f_X n_X p$ (no sum) is the *activity* (or fugacity), f_X is the *activity* (or fugacity) *coefficient*, p is the total pressure and p^0 is the standard pressure. Hence for a gas $\lambda_X = \exp(\mu_X/RT) = \lambda_X{}^0\, x_X/p^0$, where $\lambda_X{}^0$ is the value of λ_X for $p = p^0$. For gases at low pressures (hence activity coefficient $f_X = 1$) the activity becomes the *partial pressure* p_X given by $p_X = n_X p$. For solids, on the other hand, we have $\lambda_Y \cong \lambda_Y{}^0$, only weakly dependent on the pressure. In total we have

$$\prod_X (\lambda_X^0)^{\nu_X} x_X^{\nu_X} (p^0)^{-\nu_X} \prod_Y (\lambda_Y^0)^{\nu_Y} = 1 \quad \text{or}$$

$$\prod_X x_X^{\nu_X} = K^* \quad \text{with} \quad K^* \equiv \prod_X (\lambda_X^0)^{-\nu_X} (p^0)^{\nu_X} \prod_Y (\lambda_Y^0)^{-\nu_Y}$$

The *equilibrium constant* K^* is related to the standard Gibbs energy of the reaction

$$\mu^* = \gamma\mu_C^0 + \delta\mu_D^0 - \alpha\mu_A^0 - \beta\mu_B^0 \quad \text{or, written more generally,} \quad \mu^* = \nu_X \mu_X^0 \qquad (6.47)$$

via $\mu^* = -RT\ln K^*$. Since μ^* is constant at constant temperature, the value of K^* is constant at constant temperature, which explains the name. The equilibrium constant allows one to calculate the activity of the dependent components given sufficient information on the others.

Problem 6.6

Derive the equilibrium conditions along the lines of Section 6.2.

Surface effects

In a few cases we need to be able to deal with some aspects of surfaces and in this section we provide some basics. Discussing a multi-component system we refer to component 1 as the reference component. We associate with the interface between two phases a geometrical surface positioned in such a way that the total amount of the reference compound is the same as if both bulk phases remained homogeneous up to the geometrical interface. The excess amount of component i over the amount when both phases remain homogeneous up to the interface is indicated by $n_i^{(1)}$ or per unit area as $\Gamma_i^{(1)} = n_i^{(1)}/A$, where A is the surface area. The Helmholtz energy of the surface F_{sur} is defined by $F = F^{(1)}+F^{(2)}+F_{sur}$, where F is the total Helmholtz energy while $F^{(1)}$ and $F^{(2)}$ denote the Helmholtz energies of the two phases calculated as if both phases remained homogeneous up to the geometrical interface. Thus we have

$$dF_{sur} = -S_{sur}dT + \gamma dA + \mu_i d(\Gamma_i^{(1)}A) \qquad (6.48)$$

where γ is the *surface tension* and S_{sur} is the surface entropy. Assuming additivity, Eq. (6.48) is homogeneous of the first degree in A and $n_i^{(1)} = \Gamma_i^{(1)}A$ (as well as F_{sur} and S_{sur}). Therefore from Euler's theorem (see Chapter 3) we have

$$F_{sur} = \gamma A + \mu_i n_i^{(1)} = \gamma A + \mu_i(\Gamma_i^{(1)}A) \qquad (6.49)$$

and by differentiation of Eq. (6.49) and subtraction from Eq. (6.48) the result is

$$A d\gamma = -S_{sur}dT - \Gamma_i^{(1)}A d\mu_i \qquad (6.50)$$

For a single component system Eq. (6.49) reduces to $F_{\text{sur}} = \gamma A$ and the surface tension is equal to the Helmholtz surface energy per unit area. Eq. (6.50) is the surface analogue of the Gibbs-Duhem equation, which at constant temperature leads to

$$\blacktriangleright \quad \mathrm{d}\gamma = -\Gamma_i^{(1)}\mathrm{d}\mu_i \tag{6.51}$$

known as the *Gibbs adsorption equation.*

6.5 Internal variables

Many problems in thermodynamics can be solved by the classical formulation as outlined in the previous sections. However, we need two extra ingredients for thermomechanics. The first is the concept of internal variables and the second is a field formulation. In this section the concept of internal variables is described, while in the next section a field formulation of the theory is given (Ziegler, 1983).

In the classical theory the state variables can be varied at will, at least in principle and in many cases also in practice. If such a state variable is varied, the system reacts with a change in associated force and the work performed in this way enters the first law. The classical example is given by a cylinder closed with a piston and containing a gas. If we decrease the volume of the gas by moving the piston, the pressure increases. Similarly, if we deform a solid, a stress occurs.

Deformation of an elastic solid can be thought of as prescribing the strain. However, the state of many systems is also determined by variables that cannot be prescribed at will and we call them *internal variables*[k]. In fact we have already encountered such a variable, the degree of reaction ζ, during the discussion of chemical equilibrium. The degree of reaction in a closed system cannot be influenced directly but only indirectly by changing T and p. Other examples are provided by the variables describing the microstructure of a ceramic and the (magnetic or ferroelectric) domain structure in a material. In a later stage we will encounter the structure and density of a dislocation network and the extent and configuration of cracks. Although internal variables cannot be influenced directly this does not mean that they cannot be influenced at all. Consider as an example the following 1D model.

Example 6.3: The Maxwell model

Fig. 6.4 shows a (black) box containing a spring in series with a dashpot. This system is frequently referred to as a *Maxwell element*. The elongation in the dashpot is denoted by α while the total elongation of the element is represented by ε. For the spring there remains an elongation of $\varepsilon-\alpha$. Hence ε and α are the independent variables for the element. However, the only variable that can be influenced directly is the total elongation ε. The force on the element σ is the same for the spring and the dashpot, is determined by ε and α and in fact decreases slowly at constant elongation ε. While the state

[k] First introduced by P.W. Bridgman (1941), *The nature of thermodynamics*, Harvard University Press and Harper and Brothers (1961), although in particular some French authors claim that in work of P. Duhem the idea is already implicated. Bridgman wrote: "I believe that in general the analysis of such systems will be furthered by the recognition of a new type of large-scale thermodynamic parameter of state, namely the parameter of state which can be measured but not controlled… These parameters are measurable, but they are not controllable, which means that they are coupled to no external forces variable which might provide the means of control. And not being coupled to a force variable, they cannot take part in mechanical work". See also Phys. Rev. **22** (1950), 56.

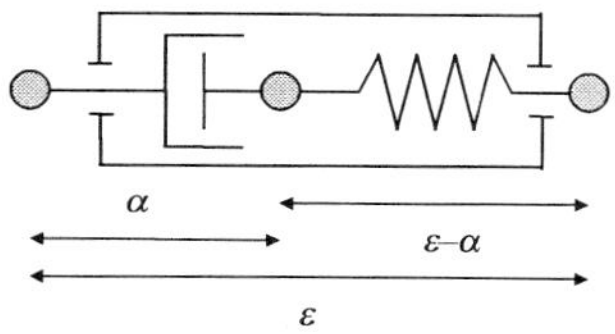

Fig. 6.4: The Maxwell model with a spring and dashpot in series.

functions like u and s depend on ε, α and T, the force depends not only on ε, α and T but also on $\dot{\varepsilon}$ and $\dot{\alpha}$. The total elongation is a classic, external variable and the elongation of the dashpot is an internal variable. The Maxwell element is one of the simplest analogous models, as used in modelling visco-elastic behaviour of e.g. polymers.

Although simple models like the one discussed above cannot describe material behaviour in sufficient generality, the concept of internal variable is nevertheless extremely useful. Just like the classic external variables, the internal variables have thermodynamic forces associated with them but the internal variables in conjunction with these forces do not contribute to the mechanical work in the first law. We will denote these variables by b_i, or if we anticipate their tensorial character for solids by b_{ij}, similar to the strain tensor ε_{ij}. We will denote general external variables still by a_i and identify their meaning in particular cases. The energy and entropy are now of the form $u(a_i,b_i,s)$ and $s(a_i,b_i,u)$. While the definition of the specific Helmholtz energy f remains unaltered, it now reads $f(a_i,b_i,T)$. The forces associated with the internal variables are denoted by B_i and are referred to as *internal forces*. Their quasi-conservative parts are defined by

$$B_i^{(q)} = \rho \frac{\partial f}{\partial b_i} \tag{6.52}$$

They are state functions and, like u and s, depend on a_i, b_i and T. The dissipative parts $B_i^{(d)}$ also depend on $\dot{a}_i$ and $\dot{b}_i$. The specific power of dissipation is given by

$$l^{(d2)} = \frac{1}{\rho} B_i^{(d)} \dot{b}_i \tag{6.53}$$

which should be added to Eq. (6.22). The internal forces do not appear in the expression for the total specific power and for an arbitrary process we conclude that

$$\frac{1}{\rho} B_i \dot{b}_i = \frac{1}{\rho}\left(B_i^{(q)} + B_i^{(d)}\right)\dot{b}_i = 0 \qquad \text{or} \qquad B_i^{(q)} = -B_i^{(d)} \tag{6.54}$$

if the various internal variables $\dot{b}_i$ are independent.

While the rates of the external variables $\dot{a}_i$ can be controlled by the experimenter, the rates of the internal variables $\dot{b}_i$ are controlled by the system itself and only indirectly by the rates of the external variables $\dot{a}_i$. Therefore to describe the evolution in time of the internal variables one needs an extra equation, typically an *evolution equation* like

$$\dot{b}_i = x(a_k, b_l) + y(a_k, b_l)\dot{a}_m$$

where x and y are functions of the variables a_k and b_l. Often it is possible to suppose that we have been able to select the internal variables b_l in such a way that $y(a_k,b_l) = 0$ and in this case the instantaneous change in a_m does not cause an instantaneous change in b_i. For the mechanical case this corresponds to the fact that instantaneous strains are either elastic or zero. For example, for the Maxwell element (Example 6.3) the evolution equation describes the ratio of the damper force σ and the rate of change of the damper length $\dot{\alpha}$ as the damper constant η. In fact this equation describes the viscous behaviour of the damper and reads explicitly, since $\sigma = \sigma(\varepsilon,\alpha)$, $\dot{\alpha} = \sigma(\varepsilon,\alpha)/\eta$. A method of deriving the evolution equation is discussed in Section 6.7.

Finally, we elaborate a bit on the character of the internal variables. We recall that a particle (in the sense of continuum mechanics) is actually a representative volume element or meso-cell (Chapter 1) for which the local internal energy u can be seen as the global energy of the meso-cell. Similarly, the strain ε_{ij} may be viewed as the global geometrical description of the meso-cell. With such a meso-cell a characteristic time scale τ is associated, given by $\tau_{\rm m} = \varepsilon / \dot{\varepsilon}$. Furthermore, there may be a bound $\sigma_{\rm m}$ for the stress acting on the meso-cell. Now there are two types of internal variables. First, the *relaxation type* that are characterised by a relaxation time $\tau_\alpha = (\alpha - \alpha^{(\rm equ)}) / \dot{\alpha}$. If $\tau_\alpha / \tau_{\rm m} << 1$, the corresponding internal variables take their equilibrium values $\alpha^{(\rm equ)}$ (for u, ε and the other α's constant), while for $\tau_\alpha / \tau_{\rm m} >> 1$ the corresponding internal variables are said to be *frozen in* (on the time scale of strain evolution) and can be neglected. For $\tau_\alpha / \tau_{\rm m} \cong 1$, the internal variables do change on the time scale of the strain evolution. Second, internal variables of the *rate-independent type*, which do not involve a characteristic time but have a limiting stress, often known as yield stress $\sigma_{\rm y}$. If $\sigma_{\rm y}/\sigma_{\rm m} \leq 1$, the corresponding internal variables must be kept or, if $\sigma_{\rm y}/\sigma_{\rm m} >> 1$, the corresponding variables may be considered as frozen in.

Problem 6.7

Assuming $\sigma = \eta\dot{\alpha}$ for the dissipative element and $\sigma = E(\varepsilon - \alpha)$ for the elastic element, write down the expression for the Helmholtz energy F and dissipation function Φ for the Maxwell element. Derive the time dependence.

Problem 6.8

In the spirit of Example 6.3 consider a box with a spring and damper connected in parallel, generally referred to as a *Voigt element*.

a) Does this element contain an internal variable?

b) Is the element dissipative?

Problem 6.9

Again assuming $\sigma = \eta\dot{\alpha}$ for the dissipative element and $\sigma = E(\varepsilon - \alpha)$ for the elastic element, write down the expression for the Helmholtz energy F and dissipation function Φ for the Voigt element. Derive the time dependence.

Problem 6.10*

Write down the expression for the Helmholtz energy F and dissipation function Φ for a *standard linear solid* in which a spring and Maxwell element are connected in parallel using the same assumptions as in Problem 6.7 and Problem 6.9. Derive the time dependence.

---

*The local accompanying state**

For the moment we consider the meso-cell as our system. So far we have two sets of equations that describe the behaviour of that system. The first is the equation-of-state set, which describes the quasi-conservative forces $A_i^{(q)}$ by

$$A_i^{(q)} = \rho\frac{\partial u(s,a_i,b_i)}{\partial a_i} = \rho\frac{\partial f(T,a_i,b_i)}{\partial a_i} = \rho T\frac{\partial s(u,a_i,b_i)}{\partial a_i}$$

with u the specific internal energy, f the specific Helmholtz energy and s the specific entropy. As before a_i and b_i denote the external (controllable) and internal (non-controllable) variables, respectively, and T is the absolute temperature. The second is the evolution equation set describing the rate of change of the internal variables b_i by

$$\dot{b}_i = x(a_k,b_l) + y(a_k,b_l)\dot{a}_m$$

Elimination of the internal variables b_i, e.g. via time integration of the evolution equation and substitution in the equation of state, yields a functional in time t for A_i with respect to a_i. This functional could be used to describe the material behaviour (although the link with experiment might be cumbersome). However, in internal variable thermodynamics one keeps the b_j's explicitly, since one hopes to describe the system with a limited set for which all the relevant thermodynamics equations derived can be used. Moreover, the use of an evolution equation, which is reasonable anyway, leads to a simpler mathematical implementation.

To achieve a description of an irreversible process we use again the division between system and surroundings (or environment). The surroundings we take, as usual, sufficiently large so that it is always in equilibrium and therefore the temperature and forces of the environment are equilibrium quantities. The equilibrium state is described by the set E = $\{u,a_i\}$. As stated before, a description of a thermodynamic process in the system requires a larger number of variables than an equilibrium state. The first was the introduction of the internal variables b_j so that we obtain the constrained equilibrium state space described by the set C = $\{u,a_i,b_j\}$. However, one more parameter is needed, which describes the temperature for a non-equilibrium system. This is the so-called *contact temperature*[1] Θ, which is independent of u for non-equilibrium states but becomes identical to the thermostatic temperature in equilibrium. This temperature is the dynamical analogue of the dynamical (non-equilibrium) kinematical variables. For example, the pressure $p^{(\text{equ})}$ of a gas in an adiabatic cylinder closed by a piston exerts on the surface is measured by a device gauged in equilibrium, e.g. the length of a spring installed in the piston rod (Fig. 6.5). The pressure p, one of the several forces A_i, now can be defined by the zero of the time derivative of the volume V of the system

$$\left(p - p^{(\text{equ})}\right)\dot{V} \geq 0 \qquad Q = 0 \;\text{ and }\; A_i \neq p = 0$$

where Q is the heat exchange rate. Similarly we may write

$$\left(\Theta^{-1} - T^{-1}\right)Q \geq 0 \qquad \text{power} = 0$$

---

[1] Muschik, W. (1979), J. Non-Equilib. Thermodyn. **4**, 277 and 377.

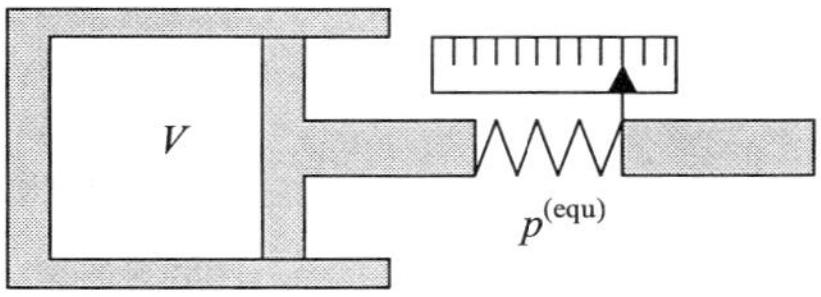

Fig. 6.5: The dynamical pressure as illustrated by a gas in a cylinder.

Therefore we may envisage a *constrained equilibrium state space*, described by the set C = $\{u,a_i,b_j\}$ and an *extended state space* described by the set N = $\{u,a_i,b_j,\Theta;T^*,A_k^*\}$ which depends parametrically on the temperature T^* and force A_k^* of the environment of the system. We will further omit the dependence on T^* and A_k^*. A reversible process can be represented as a path in the equilibrium state space while an irreversible process requires a path in the extended state space (Fig. 6.6). Consider the transition from equilibrium state A to B and let these two states be connected by an irreversible process. Now with each point of the path in extended state space N we can associate a point in the equilibrium state space C. This can be done essentially in two ways.

The first way is to isolate the system by what might be called *adiabatic closure*, i.e. we isolate the system in a 'Gedankenexperiment' so that no heat or work is exchanged with the environment. Therefore for this process du = 0. After isolation the temperature of the system will change from the contact temperature Θ to the equilibrium temperature $T^{(U)}$ as calculated from $T^{-1} = \partial s(u,a_i,b_j)/\partial u$. Due to the nature of the adiabatic closure, the temperature and forces of the surroundings no longer play a role. The process can be seen as a projection P^U of the path in space N on space C for which the energy of the system is kept constant so that we may write

$$\mathrm{P}^U\{u,a_i,b_j,\Theta\} = \{u,a_i,b_j\}$$

In the associated constrained equilibrium state u, a_i and b_j have the same value as in the non-equilibrium state. The temperature changed, however, from Θ to $T^{(U)}$.

The second way is by means of *thermal equilibration*; again we consider a Gedankenexperiment in which the system is purely heated or cooled so that no work is exchanged with environment but this time we keep Θ constant, so that after equilibration the internal energy changes from its initial value u to $u^{(\Theta)}$, corresponding to the temperature Θ. This process can also be seen as a projection P^Θ of the path in space N on space C for which we write

$$\mathrm{P}^\Theta\{u,a_i,b_j,\Theta\} = \{u^{(\Theta)},a_i,b_j\}$$

In this case the associated constrained equilibrium state has the same values of Θ, a_i and b_j as the non-equilibrium state but a different energy $u^{(\Theta)}$. These two projections are shown in Fig. 6.6.

Now a (non-equilibrium) entropy in the extended state space can be defined, albeit a non-unique one, by analogy of the Gibbs equation for the equilibrium case. Using $s^{(\mathrm{equ})} = s^{(\mathrm{equ})}(u^{(\mathrm{equ})},a_i,b_j)$ and remembering that $b_j^{(\mathrm{equ})} = b_j(u,a_i)$, we have at equilibrium

$$T^{(\mathrm{equ})}\dot{s}^{(\mathrm{equ})} = \dot{u}^{(\mathrm{equ})} - A_i^{(\mathrm{equ},\mathrm{q})}\dot{a}_i \tag{6.55}$$

We may take for the P^Θ associated equilibrium state, using $s^{(\Theta)} = s^{(\Theta)}(u^{(\Theta)},a_i,b_j)$,

$$\Theta\dot{s}^{(\Theta)} = \dot{u}^{(\Theta)} - A_i^{(\Theta,q)}\dot{a}_i$$

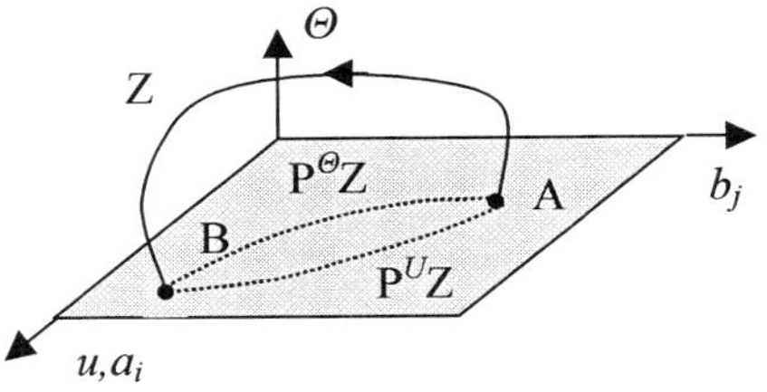

Fig. 6.6: Schematic of the constrained equilibrium state space C = $\{u,a_i,b_j\} \equiv \{(u,a_i),b_j\}$ and the extended state space N = $\{u,a_i,b_j,\Theta;T^*,A_k^*\} \equiv \{(u,a_i),b_j,\Theta\}$ and of the role of internal variables in thermomechanics depicting the transition from state A to state B via the process Z. While the process Z is described in extended state space, characterised by the energy u, the set of external and internal variables a_i and b_j as well as Θ, T^* and A_k^*, the accompanying process $\mathrm{P}^U\mathrm{Z}$ (or $\mathrm{P}^\Theta\mathrm{Z}$) is described in constrained equilibrium state space, characterised by the energy u (or u^Θ) and the set of external and internal variables a_i and b_j.

while for the P^U associated equilibrium state we have, using $s^{(U)} = s^{(U)}(u,a_i,b_j)$,

$$T^{(U)}\dot{s}^{(U)} = \dot{u} - A_i^{(U,q)}\dot{a}_i$$

where the corresponding forces are introduced. We now write generally for the non-equilibrium state

$$\dot{s}(u,a_i,b_j;\varphi_1,\varphi_2,\varphi_3) = \frac{\dot{u}}{\varphi_1} - \frac{A_i^{(q)}}{\varphi_2}\dot{a}_i - \frac{B_i^{(q)}}{\varphi_3}\dot{b}_i \tag{6.56}$$

where the φ_i's denote different temperatures and, as usual, u is the internal energy. For the difference between the non-equilibrium state and the equilibrium state we subtract Eq. (6.55) from Eq. (6.56), meanwhile using $\dot{u}^{(\mathrm{equ})} = \dot{q}^{(\mathrm{equ})} + A_i^{(\mathrm{equ,q})}\dot{a}_i$, $\dot{u} = \dot{q} + A_i\,\dot{a}_i$ and $B_i^{(\mathrm{d})} = -B_i^{(\mathrm{q})}$, to obtain

$$\Delta\dot{s} = \dot{s} - \dot{s}^{(\mathrm{equ})} = \frac{\dot{q}}{\varphi_1} - \frac{\dot{q}^{(\mathrm{equ})}}{T^{(\mathrm{equ})}} - A_i^{(\mathrm{q})}\left(\frac{1}{\varphi_2} - \frac{1}{\varphi_1}\right)\dot{a}_i + \sigma \qquad \sigma = \frac{A_i^{(\mathrm{d})}\dot{a}_i}{\varphi_2} + \frac{B_j^{(\mathrm{d})}\dot{b}_j}{\varphi_3}$$

with σ the entropy production. To be consistent with the equilibrium entropy definition we must require that for the process from state A to state B the time integral over the entropy rate equals the entropy difference between state A and B so that

$$\int_{\mathrm{A}}^{\mathrm{B}} \dot{s}\,\mathrm{d}t = s^{(\mathrm{B})} - s^{(\mathrm{A})}$$

Using for s the difference between non-equilibrium and equilibrium entropy Δs, the integral should vanish and we have

$$0 = \int_{\mathrm{A}}^{\mathrm{B}} (\dot{s} - \dot{s}^{(\mathrm{equ})})\,\mathrm{d}t = \int_{\mathrm{A}}^{\mathrm{B}} \left(\frac{\dot{q}}{\varphi_1} - \frac{\dot{q}^{(\mathrm{equ})}}{T^{(\mathrm{equ})}}\right)\mathrm{d}t - \int_{\mathrm{A}}^{\mathrm{B}} A_i^{(\mathrm{q})}\left(\frac{1}{\varphi_2} - \frac{1}{\varphi_1}\right)\dot{a}_i\,\mathrm{d}t + \int_{\mathrm{A}}^{\mathrm{B}} \sigma\,\mathrm{d}t$$

If we make the special choice $\varphi_1 = \varphi_2 = \varphi_3 = \varphi$, this equation reduces to

$$0 = \int_{\mathrm{A}}^{\mathrm{B}} \left(\frac{\dot{q}}{\varphi} - \frac{\dot{q}^{(\mathrm{equ})}}{T^{(\mathrm{equ})}}\right)\mathrm{d}t - \int_{\mathrm{A}}^{\mathrm{B}} \sigma\,\mathrm{d}t \equiv \oint \left(\frac{\dot{q}}{\varphi}\right)\mathrm{d}t + \int_{\mathrm{A}}^{\mathrm{B}} \sigma\,\mathrm{d}t \tag{6.57}$$

with $\sigma = \varphi^{-1}\left(A_i^{(q)}\dot{a}_i + B_j^{(q)}\dot{b}_j\right)$. To make further progress we realise that in equilibrium the properties of the system are described by (conventional) state functions, i.e. we cannot judge how the system has reached its equilibrium state. Hence for any state function $Z = Z(u,a_i,b_j)$ in equilibrium state space, which in extended state space is given by $Z = Z(u,a_i,b_j,\Theta)$, we have

$$\oint^{*} Z(t)\,\mathrm{d}t = 0$$

for any loop (cyclic process) in extended state space which contains at least one equilibrium state. The asterisk added to the contour integral sign is to denote this last condition. If the theorem does not hold, two different values of Z will apply to a single equilibrium states. Therefore we can write

$$\int_{\mathrm{A}}^{\mathrm{B}} \left(\frac{\dot{q}}{\varphi} - \frac{\dot{q}^{(\mathrm{equ})}}{T^{(\mathrm{equ})}}\right)\mathrm{d}t \equiv \oint^{*} \left(\frac{\dot{q}}{\varphi}\right)\mathrm{d}t$$

The next step is to use the Clausius inequality or its extension as given by Muschik[m]. They read, respectively,

$$\oint \frac{\dot{q}}{T^{*}}\,\mathrm{d}t \le 0 \qquad \text{and} \qquad \oint \frac{\dot{q}}{\Theta}\,\mathrm{d}t \le 0$$

where, as before, T^* is the equilibrium temperature of the environment and Θ is the contact temperature. Therefore if we take $\varphi = T^*$ or $\varphi = \Theta$, we obtain from Eq. (6.57)

$$\int_{\mathrm{A}}^{\mathrm{B}} \sigma\,\mathrm{d}t \ge 0$$

which states that the entropy production is positive. Although derived here for a system exchanging only work and heat, a similar derivation can be made for systems also exchanging matter. Muschik gives a more elaborate discussion, also dealing with the various subtleties concerning the existence of the various fields involved. This derivation, however, indicates the existence of a non-equilibrium entropy but also that it is non-unique. In fact the latter observation would have been also clear from $s = s(u,a_i,b_j)$ since the Gibbs equation

$$\dot{s} = \frac{\partial s}{\partial u}\dot{u} + \frac{\partial s}{\partial a_i}\dot{a}_i + \frac{\partial s}{\partial b_j}\dot{b}_j = T^{-1}\left(\dot{u} - \rho^{-1}A_i^{(q)}\dot{a}_i - \rho^{-1}B_i^{(q)}\dot{b}_i\right)$$

yields the equations of state

$$T^{-1} = \frac{\partial s}{\partial u} \qquad A_i^{(q)} = -\rho T\frac{\partial s}{\partial a_i} \qquad \text{and} \qquad B_i^{(q)} = -\rho T\frac{\partial s}{\partial b_j}$$

which are evidently dependent on the choice of the internal parameters.

The approximation of the *local* (*accompanying*) *state* proclaims that it is sufficient for the description of a non-equilibrium process to use the entropy and temperature of the associated states in constrained equilibrium state space. Since either the temperature or the energy can be projected identically from extended state space to constrained equilibrium state space but not both, the different projections define a different local state. Nevertheless often within internal variable thermodynamics the

[m] Muschik, W. and Riemann, H. (1979), J. Non-Equilib. Thermodyn. **4**, 17.

projection P^U is used, so that $u = u^{(equ)}$, but in conjunction with $\Theta = T^*$, which is only true for an irreversible process sufficiently close to equilibrium. Finally, it will be clear that the entropy of a non-equilibrium state is higher value than $s^{(U)}$, since the change from Θ to $T^{(U)}$ occurs adiabatically.

6.6 Field formulation*

Up to now we have described the behaviour of a representative volume element that is using a homogeneous distribution of the properties throughout the element. However, for continuum thermomechanics we want to use gradients of parameters through the system, in particular the temperature gradient. Hence we consider the situation anew (Maugin, 1992; Ziegler, 1983; Woods, 1975).

The first law

The energy of the system U is obtained by integration of the specific energy u times the mass density over the volume of the system yielding

$$U = \int \rho u \, \mathrm{d}V \tag{6.58}$$

The power of the external forces is given according to Section 5.6 by

$$P_{\text{ext}} = \int \rho b_i v_i \, \mathrm{d}V + \int \rho \sigma_{ij} v_i n_j \, \mathrm{d}A = P_{\text{vol}} + P_{\text{sur}} \tag{6.59}$$

Further we need the heat supply. Neglecting any volume mechanism, e.g. by radiation, the heat supply $\dot{Q}$ is due to conduction across the surface[n] A. If q_k denotes the heat flow vector per unit time across a surface element dA with unit outward normal n_k, the total heat supply per unit time is given by

$$\dot{Q} = -\int q_k n_k \, \mathrm{d}A \tag{6.60}$$

Finally we need the kinetic energy[o] K or, actually, its material derivative given by

$$\dot{K} = \int \rho v_i \dot{v}_i \, \mathrm{d}V \tag{6.61}$$

The first law in global form is now written as

$$\dot{K} + \dot{U} = P_{\text{ext}} + \dot{Q} \tag{6.62}$$

or equivalently

$$\int \rho(v_k \dot{v}_k + \dot{u}) \mathrm{d}V = \int \rho b_k v_k \, \mathrm{d}V + \int (\sigma_{kl} v_k - q_l) n_l \, \mathrm{d}A \tag{6.63}$$

It states that the rate of increase of the kinetic and internal energy equals the power of the external forces plus the heat supply per unit time. Applying the divergence theorem to the first term of the last integral yields

---

[n] In the literature often an extra term r is given representing the generation of heat within the volume. Since the term drops out of the Clausius-Duhem inequality (see the next section) we omit it here. For a critical discussion, see Lavenda (1978).

[o] Here we denote the kinetic energy by K to avoid confusion with the usual notation T, which is also used for the temperature.

$$\int \rho(v_k \dot{v}_k + \dot{u})\mathrm{d}V = \int (\rho b_k v_k + \sigma_{kl} v_{k,l} + \sigma_{kl,l} v_k)\mathrm{d}V - \int q_k n_k \,\mathrm{d}A \tag{6.64}$$

which reduces on account of the equation of motion $\rho\dot{v}_k = \rho b_k + \sigma_{kl,l}$ to[p]

$$\int \rho\dot{u}\,\mathrm{d}V = \int \sigma_{kl} d_{kl}\,\mathrm{d}V - \int q_k n_k \,\mathrm{d}A \qquad \text{or in direct notation} \tag{6.65}$$

$$\int \rho\dot{u}\,\mathrm{d}V = \int \boldsymbol{\sigma}\mathbf{:d}\,\mathrm{d}V - \int \mathbf{q}\cdot\mathbf{n}\,\mathrm{d}A \tag{6.66}$$

where use has been made of the definition of the rate of deformation d_{ij} and the symmetry of σ_{ij}. Using the divergence theorem once more on the last term we obtain

$$\int \rho\dot{u}\mathrm{d}V = \int \sigma_{kl} d_{kl}\mathrm{d}V - \int q_{k,k}\mathrm{d}V \tag{6.67}$$

Observing that this result is valid for any volume we obtain the first law in local form

► $$\rho\dot{u} = \sigma_{kl} d_{kl} - q_{k,k} \qquad (\text{or more generally } \rho\dot{u} = A_i \dot{a}_i - q_{k,k}) \tag{6.68}$$

It states that the material derivative of the internal energy is equal to the power of the stress tensor plus the rate of heat supply.

The second law

In a similar way as for the energy the entropy S of the system is obtained from

$$S = \int \rho s \,\mathrm{d}V \tag{6.69}$$

where $S = S^{(\mathrm{r})} + S^{(\mathrm{i})}$. One can define a local entropy flow vector by q_k/T, where T is the local temperature. The entropy supply per unit time is then given by

$$\dot{S}^{(\mathrm{r})} = -\int \frac{q_k n_k}{T}\,\mathrm{d}A \tag{6.70}$$

To obtain the local form of the second law a similar operation can be done as for the first law. Again we start from the global form

$$\dot{S} \geq \dot{S}^{(\mathrm{r})} \tag{6.71}$$

Inserting Eq. (6.70) and using the divergence theorem we obtain

$$\int \rho\dot{s}\,\mathrm{d}V \geq -\int \frac{q_k}{T} n_k \mathrm{d}A = -\int \left(\frac{q_k}{T}\right)_{,k} \mathrm{d}V = \int \left(\frac{q_k}{T^2} T_{,k} - \frac{q_{k,k}}{T}\right)\mathrm{d}V \tag{6.72}$$

Since this inequality holds for any volume, we find the local form of the second law

$$\rho\dot{s} \geq -\left(\frac{q_k}{T}\right)_{,k} = \frac{q_k}{T^2} T_{,k} - \frac{q_{k,k}}{T} \tag{6.73}$$

This inequality states that the rate of entropy increase per unit volume is never less than rate of entropy supply. The first term on the right-hand side is due to the temperature gradient while the last term is the classical entropy increase due to heat supply.

---

[p] Remember that the equation of motion is supposed to hold at all points.

For more clarity in the final dissipation expression we need again the specific Helmholtz energy $f(a_i,b_i,T)$. Its material derivative

$$\rho\dot{f} = \rho\left(\dot{u} - T\dot{s} - s\dot{T}\right) = \rho\frac{\partial f}{\partial a_i}\dot{a}_i + \rho\frac{\partial f}{\partial b_i}\dot{b}_i + \rho\frac{\partial f}{\partial T}\dot{T} \tag{6.74}$$

leads us to

$$\rho\dot{f} = A_i^{(q)}\dot{a}_i + B_i^{(q)}\dot{b}_i - \rho s\dot{T} \tag{6.75}$$

or

$$\rho\left(\dot{f} + s\dot{T}\right) = \rho\left(\dot{u} - T\dot{s}\right) = A_i^{(q)}\dot{a}_i + B_i^{(q)}\dot{b}_i = A_i\dot{a}_i - A_i^{(d)}\dot{a}_i - B_i^{(d)}\dot{b}_i \tag{6.76}$$

Solving for $\rho\dot{s}$ and using the first law, Eq. (6.68), yields

$$\rho\dot{s} = \frac{1}{T}\left(A_i^{(d)}\dot{a}_i + B_i^{(d)}\dot{b}_i - q_{i,i}\right) = \frac{1}{T}\left(A_i^{(d)}\dot{a}_i + B_i^{(d)}\dot{b}_i\right) - \frac{T_{,i}}{T^2}q_i - \left(\frac{q_i}{T}\right)_{,i} \tag{6.77}$$

Remembering that $\dot{s} = \dot{s}^{(r)} + \dot{s}^{(i)}$ where $\dot{s}^{(r)} = -(q_i/T)_{,i}$, we obtain the final local form of the second law

► $$T\dot{s}^{(i)} = \frac{1}{\rho}\left(A_i^{(d)}\dot{a}_i + B_i^{(d)}\dot{b}_i - \frac{T_{,i}}{T}q_i\right) = l^{(d1)} + l^{(d2)} + l^{(d3)} \geq 0 \tag{6.78}$$

often referred to as the *Clausius-Duhem inequality*. If we write down the Clausius-Duhem expression for the specific case of a strained solid we obtain

$$T\dot{s}^{(i)} = \frac{1}{\rho}\left(\sigma_{ij}^{(d)}d_{ij} + B_{ij}^{(d)}\dot{b}_{ij} - \frac{T_{,i}}{T}q_i\right) = l^{(d1)} + l^{(d2)} + l^{(d3)} \geq 0 \tag{6.79}$$

where the internal variables have been chosen as tensors. In this case the first term $l^{(d1)}$ and second term $l^{(d2)}$ are the rates of work of the external and internal forces, as obtained before. The third term $l^{(d3)}$ reflects an additional entropy production due to the heat flow across the element under consideration and does not arise for a homogeneous situation. For the first two terms the interpretation in forces and velocities is evident. The quantities $\sigma_{ij}^{(d)}$ and $B_{ij}^{(d)}$ are the forces while the quantities d_{ij} and $\dot{b}_{ij}$ are the corresponding velocities. In the third term the expression $-T_{,i}/\rho T = -(\ln T)_{,i}/\rho$ can be interpreted as the force and q_i as the corresponding velocity. Although the second law requires that the left-hand side of the Clausius-Duhem inequality is non-negative as a whole and there is no experimental evidence that in general the inequality should hold separately for the sum of the first two terms and the third term, it is nevertheless often assumed that this is the case. We will do likewise and restrict us to materials for which $l^{(d1)} + l^{(d2)} \geq 0$ and $l^{(d3)} \geq 0$ apply separately.

Equilibrium revisited

Finally we have to consider general mechanical equilibrium. Consider, to that purpose, again an isolated system consisting of two subsystems. We will use the Helmholtz formulation. Equilibrium is reached when the Helmholtz energy of the system $F = F_1+F_2$ is minimal or $\mathrm{d}F = 0$. Since in a general mechanical system the total Helmholtz energy is given by the integral $F = \int\rho f\,\mathrm{d}V$, we have to consider variations in strain and temperature, i.e. $\delta\varepsilon$ and δT, instead of the differentials $\mathrm{d}\varepsilon$ and

$\mathrm{d}T$. For convenience we take isothermal conditions and, considering system 1 indicated by a superscript as the system of interest, we have

$$\delta F^{(1)} = \frac{\partial F^{(1)}}{\partial \varepsilon_{ij}}\delta\varepsilon_{ij} = \int \rho \frac{\partial f}{\partial \varepsilon_{ij}}\delta\varepsilon_{ij}\,\mathrm{d}V = \int \sigma_{ij}\delta\varepsilon_{ij}\,\mathrm{d}V$$

Let us assume that the loading system, i.e. system 2, provides a loading of body forces and tractions independent of the deformation of system 1, neither in direction, nor in magnitude, and also that on part A_u of system 1 the displacements are prescribed. We write for the (prescribed) body force b_i and for the (prescribed) traction[q] $\bar{t}_i$ on part A_t of system 1. The variation for the loading system thus becomes[r]

$$\delta F^{(2)} = -\int \rho b_i \delta u_i\,\mathrm{d}V - \int_{A_t} \bar{t}_i \delta u_i\,\mathrm{d}A$$

Because $\delta u_i = 0$ on A_u we may replace the integration over A_t by the one over the complete surface $A = A_u + A_t$ and write

$$\delta F^{(2)} = -\int \rho b_i \delta u_i\,\mathrm{d}V - \int \bar{t}_i \delta u_i\,\mathrm{d}A$$

At equilibrium the total variation $\delta F = 0$ and since

$$\sigma_{ij}\delta\varepsilon_{ij} = ½\sigma_{ij}\delta(u_{i,j} + u_{j,i}) = \sigma_{ij}\delta u_{i,j} = (\sigma_{ij}\delta u_i)_{,j} - \sigma_{ij,j}\delta u_i$$

we obtain, also making use of the Gauss theorem for the term $(\sigma_{ij}\delta u_i)_{,j}$,

$$\delta F = \delta F^{(1)} + \delta F^{(2)} = \int \left(-\sigma_{ij,j} - \rho b_i\right)\delta u_i\,\mathrm{d}V + \int \left(\sigma_{ij} n_j - \bar{t}_i\right)\delta u_i\,\mathrm{d}A = 0 \qquad (6.80)$$

Because the variations δu_i are arbitrary, this can only be true if

$$\sigma_{ij,j} + \rho b_i = 0 \ \text{ in } V \qquad \sigma_{ij} n_j = \bar{t}_i \ \text{ on } A_t \qquad \text{and} \qquad \delta u_i = 0 \ \text{ on } A_u$$

thus regaining again the (mechanical) equilibrium condition and the boundary conditions. In a homogeneous system without body forces, as is usually considered in thermodynamics, the equilibrium condition is obviously fulfilled. If we write for the stress vector on the boundary of system 1, $t_i^{(1)} = \sigma_{ij}^{(1)} n_j^{(1)}$, and remember that both the traction of the loading system 2 and the stress vector of the system of interest 1 are defined along their own outer normal, so that at any point of the interface between systems 1 and 2 we have $\mathbf{n}^{(1)} = -\mathbf{n}^{(2)}$, we obtain the traction boundary condition $\mathbf{t}^{(1)} = -\mathbf{t}^{(2)}$. This is the familiar Newton's third law. If the loading system can also be considered as a continuous material, the boundary condition becomes

$$\sigma_{ij}^{(1)} n_j^{(1)} = \sigma_{ij}^{(2)} n_j^{(2)}$$

This equation represents the continuity of force over an interface.

6.7 Non-equilibrium processes*

For irreversible processes the second law was given as

---

[q] We write $\bar{t}_i$ for prescribed tractions and use t_i for the reactions at the prescribed displacement $\bar{u}_i$.

[r] The work is defined as positive if delivered *on* system 2. Here the work is delivered *by* system 2, hence the minus sign.

$$T\dot{s}^{(i)} = \frac{1}{\rho}\left(\sigma_{ij}^{(d)} d_{ij} + B_{ij}^{(d)} \dot{b}_{ij} - \frac{T_{,i}}{T} q_i\right) = l^{(d1)} + l^{(d2)} + l^{(d3)} \geq 0 \qquad (6.81)$$

The three contributions to the dissipation function are thus associated with the external variables d_{ij}, the internal variables b_{ij} and the heat flux q_i. As we have seen the dissipative force associated with the heat flux vector q_i is the logarithmic temperature gradient $-(\ln T)_{,i}/\rho$. Together with the external forces $\sigma_{ij}^{(d)}$ and the internal forces $B_{ij}^{(d)}$ they provide the driving force for a deviation from equilibrium. For the moment we denote the forces collectively with $A_i^{(d)}$ and the associated velocities (also denoted as *fluxes* or *flows*) with $\dot{a}_i$. A generic expression for the dissipation function ϕ is thus

$$T\dot{s}^{(i)} = \phi(\dot{a}_i) = A_i^{(d)} \dot{a}_i \qquad (6.82)$$

At equilibrium state the entropy is maximum and the irreversible part is zero. Therefore we will use the equilibrium state a reference and calculated the state variables a_k from this reference. The entropy production rate $\dot{s}^{(i)}$ can be developed in a Taylor series (de Groot, 1961)

$$\dot{s}^{(i)} = \dot{s}_0^{(i)} + \frac{\partial \dot{s}^{(i)}}{\partial \dot{a}_i} \mathrm{d}\dot{a}_i + \tfrac{1}{2} \frac{\partial^2 \dot{s}^{(i)}}{\partial \dot{a}_i \partial \dot{a}_j} \mathrm{d}\dot{a}_i \mathrm{d}\dot{a}_j + \cdots \qquad (6.83)$$

At equilibrium the rate of entropy production is zero and therefore $\dot{s}_0^{(i)} = 0$. Moreover, for a stable equilibrium state, the first derivatives $\partial \dot{s}^{(i)} / \partial \dot{a}_i$ are zero. It also holds that

$$\mathrm{d}\dot{a}_i = \dot{a}_i - \dot{a}_i^{(\mathrm{equ})} = \dot{a}_i \qquad (6.84)$$

since at equilibrium the velocity $\dot{a}_i^{(\mathrm{equ})} = 0$. This means that one can write for a non-equilibrium state, not too far removed from equilibrium, for the dissipation function $T\dot{s}^{(i)}$, a quadratic expression in the state velocities $\dot{a}_k$

$$T\dot{s}^{(i)} = \tfrac{1}{2} T \frac{\partial^2 \dot{s}^{(i)}}{\partial \dot{a}_i \partial \dot{a}_j} \dot{a}_i \dot{a}_j = \dot{a}_i L_{ij}^{-1} \dot{a}_j \qquad (6.85)$$

where L_{ij}^{-1} denotes a matrix of constants (the reason for taking the reciprocal is to be consistent with literature). Comparing with Eq. (6.82) results in

$$A_i^{(d)} = L_{ij}^{-1} \dot{a}_j \qquad (6.86)$$

and their reciprocal equivalents, the *phenomenological equations*

$$\dot{a}_i = L_{ij} A_j^{(d)} \qquad (6.87)$$

The matrix L_{ij}^{-1} in our approach is a matrix of second derivatives and for that reason symmetric. So, its inverse L_{ij} is also symmetric. Historically, researchers started with the matrix equation (6.87) and in this case the symmetry of L_{ij} is less obvious. The symmetry was shown by Onsager and for that reason the symmetry relations $L_{ij} = L_{ji}$ are in literature known as *Onsager's reciprocal relations*[s] and they apply provided the

[s] Lars Onsager (1903-1976). Norwegian chemist who worked most of his life in the USA and who received the Noble Prize in chemistry in 1968 for the discovery of the reciprocal relations bearing his name, which are fundamental for the thermodynamics of irreversible processes.

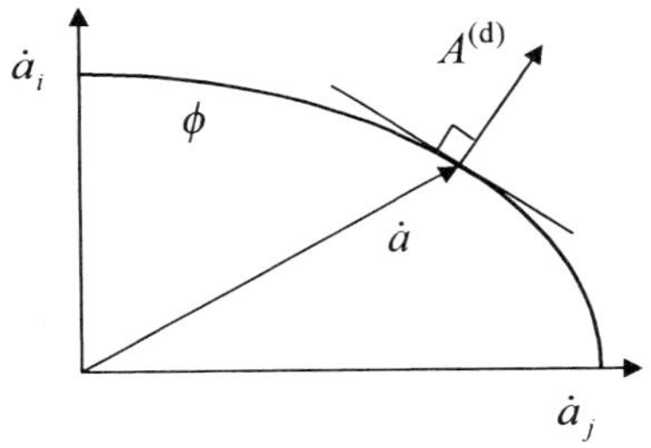

Fig. 6.7: Graphical illustration of the orthogonality principle.

flux is chosen as the time derivative of a state variable and the force is chosen as the appropriate conjugate variable.

Usually the entropy production rate $\dot{s}^{(i)}$ is more easily calculated than the force $A_i^{(d)}$. Typically the expression for $\dot{s}^{(i)}$ has a simple form so that it is particularly easy to choose the proper velocities $\dot{a}_i$ and forces $A_i^{(d)}$, once $\dot{s}^{(i)}$ is found. There is a certain freedom in the choice of velocities and forces because $\dot{s}^{(i)}$ can be split in several ways into a sum of conjugated $\dot{a}_i$ and A_i. As indicated before, for all proper choices the symmetry relation $L_{ij} = L_{ji}$ remains valid. Using Eq. (6.85) the force can be written as

$$\blacktriangleright \quad A_i^{(d)} = \xi \frac{\partial \phi(\dot{a}_j)}{\partial \dot{a}_i} \tag{6.88}$$

where ξ is chosen such that $A_i^{(d)}\dot{a}_i = \phi$, resulting in $\xi = ½$. The latter formulation for the force, in particular advocated by Ziegler (1983), is entirely equivalent to the Onsager formulation. While the Onsager formulation is only valid for small deviations from equilibrium, Ziegler proposed that his alternative formulation is also valid for larger deviations from equilibrium, e.g. for more elaborate, i.e. non-quadratic, dissipation functions. This postulate is known as the *orthogonality principle*. If we consider the function ϕ as a surface in the space where the velocities $\dot{a}_i$ act as co-ordinate axes, the forces $A_i^{(d)}$ are perpendicular to surfaces of constant values for the dissipation function ϕ (Fig. 6.7). To be consistent in this more general case the constant $\lambda = ½$ in front of Eq. (6.88) has to be replaced by

$$\blacktriangleright \quad \xi = \phi \bigg/ \left(\frac{\partial \phi}{\partial \dot{a}_i} \dot{a}_i \right) \tag{6.89}$$

For homogeneous dissipation functions of degree r, it follows that $\xi = 1/r$. Hence for a dissipation function homogeneous of degree 2, like $\phi = \dot{a}_i L_{ij}^{-1} \dot{a}_j$, $\xi = ½$ as before.

Example 6.4: Newtonian liquids

A Newtonian liquid is an isotropic liquid for which the deviatoric stress is purely dissipative while the spherical stress is purely conservative. Hence from the Helmholtz energy f for isotropic materials

$$\rho f = ½\lambda\varepsilon_{ii}\varepsilon_{jj} + \mu\varepsilon_{ij}\varepsilon_{ij} \qquad \text{(where } \lambda \text{ and } \mu \text{ are Lamé's constants)}$$

we have, using hydrostatic deformation with $\varepsilon_{ij} = \varepsilon_{\mathrm{m}}\delta_{ij} = ⅓\varepsilon_{kk}\delta_{ij}$,

$$\rho f = \frac{1}{2}\lambda\varepsilon_{ii}\varepsilon_{jj} + \mu\left(\frac{1}{3}\varepsilon_{kk}\delta_{ij}\right)\left(\frac{1}{3}\varepsilon_{ll}\delta_{ij}\right) = \left(\frac{1}{2}\lambda + \frac{1}{3}\mu\right)\varepsilon_{kk}\varepsilon_{ll}$$

and the quasi-conservative stress becomes

$$\sigma_{kk}^{(\mathrm{q})} = \frac{\partial\rho f}{\partial\varepsilon_{kk}} = \left(\lambda + \frac{2}{3}\mu\right)\varepsilon_{kk} = K\varepsilon_{kk}$$

where K is the bulk modulus. Similarly we have for the dissipation function ϕ for isotropic materials

$$\phi = \eta\dot{\varepsilon}_{ii}'\dot{\varepsilon}_{jj}' \qquad \text{(with } \eta \text{ the shear viscosity)}$$

Since this function is homogeneous of degree $r = 2$, we have $\xi = ½$ (where ξ is the multiplier for the dissipative stress). The dissipative stress $\sigma_{ij}^{(\mathrm{d})}$ is therefore

$$\sigma_{ij}^{(\mathrm{d})} = \xi\frac{\partial\phi}{\partial\dot{\varepsilon}_{ij}} = \eta\dot{\varepsilon}_{ij}'$$

which is *Newton's law of viscosity*. The energy dissipated is $\phi = \sigma_{ij}^{(\mathrm{d})}\dot{\varepsilon}_{ij}'$.

Example 6.5: Thermal conductivity

The dissipation function per unit volume for heat flow q_k can be written as

$$\rho\phi = \frac{1}{\kappa T}q_k q_k \qquad \text{(with } \kappa \text{ the thermal conductivity)}$$

This function is homogeneous of degree $r = 2$ and therefore we have $\xi = ½$. The dissipative stress $Q_k^{(\mathrm{d})}$ is thus

$$Q_k^{(\mathrm{d})} = \xi\rho\frac{\partial\phi}{\partial q_k} = ½\frac{2}{\kappa T}q_k = \frac{1}{\kappa T}q_k$$

Since we know from Eq. (6.79) that the force $Q_k^{(\mathrm{d})} = -T_{,k}/T$, we obtain

$$q_k = -\kappa T_{,k}$$

which is *Fourier's law of heat conductivity*. The dissipation is thus $\phi = Q_k^{(\mathrm{d})}q_k$.

It can be shown that the orthogonality principle in velocity space can be inverted to the one in force space with a dissipation function $\phi' = \phi'(A_i^{(\mathrm{d})})$ resulting in

$$\dot{a}_i = \xi'\frac{\partial\phi'}{\partial A_i^{(\mathrm{d})}} \qquad \text{with} \qquad \xi' = \phi'\bigg/\left(\frac{\partial\phi'}{\partial A_i^{(\mathrm{d})}}A_i^{(\mathrm{d})}\right) \tag{6.90}$$

provided the dissipation function ϕ is a function of the (internal and external) displacement rates alone, i.e. $\phi = \phi(\dot{a}_i)$. In case ϕ is also a function of the variables themselves, i.e. $\phi = \phi(a_i, \dot{a}_j, T)$, the orthogonality principle in force space does not hold and one has to use the original formulation.

A few remarks are in place:

- The dissipation function ϕ can be shown to be *convex*. This guarantees that the dissipation function is positive semi-definite ($\phi \geq 0$), conform the second law.
- A process can be dependent on, say, two sets of state variables a_i and b_i. In this case there may be a coupling between the two sets implying that terms like a_ib_j are present in ϕ . However, these coupling terms may also be absent. The former process is referred as *complex* while the latter is addressed as *compound*. In the first case the orthogonality principle is valid for the complex process. In the second case the orthogonality principle has to be applied to the separate processes constituting the compound process.
- The orthogonality principle is only valid for *purely dissipative processes*, i.e. processes where gyroscopic forces are absent. Gyroscopic forces are forces for which the power is always zero, such as Lorentz and Coriolis forces. In fact, this restriction is always tacitly assumed in thermodynamics and does not constitute a serious restriction.

A clear and elaborate discussion of the orthogonality principle has been given by Ziegler (1983) to which we refer for further details. It should be stated, though, that the extension of the orthogonality principle to more general dissipation functions is not generally accepted: the theory for larger deviations from equilibrium has not been settled. Fortunately, the 'small-deviation-from-equilibrium' theory, usually known as *linear irreversible thermodynamics*, is frequently sufficiently accurate. In this case the dissipation function acts as a convenient device in the description of thermomechanical problems once a proper choice of ϕ for a set of independent state variables a_i and their rates $\dot{a}_i$ has been made.

Problem 6.11

a) Show that for a general dissipation function Eq. (6.89) holds.
b) Show that for a dissipation function homogeneous of degree r, the relation $\xi = 1/r$ holds.

6.8 Type of materials*

Although the constitutive behaviour as such is not part of thermodynamics, the framework presented makes a systematic discussion of the material behaviour feasible. To that purpose we have to perform three tasks (Malvern, 1969).

The first task is to survey the basic principles that govern the formulation of the constitutive behaviour. We mention here just three. The first two principles are obvious enough but the third requires more discussion. The first principle is the *principle of determinism*, which states that although materials can have a memory, they cannot have foresight. Hence the stress in a body is determined by the history of that body so that in principle the behaviour has to be described by a functional of the external variables. It can be shown that an alternative description using a function of internal variables is equally possible. As example we mention visco-elasticity, which can be treated using external variable functionals or using internal variable functions. If the response of the material is momentary, i.e. only dependent on the present time, the behaviour can always be described by a function. This is e.g. the case for elasticity. The second principle is the *principle of local action*, which states that in determining the stress at a given material point the motion outside an arbitrary neighbourhood of that point may be discarded. If only the co-ordinates of one point are involved, the material is called *simple*. In this form the behaviour may still depend, not only on, the co-ordinates of that point but also on the derivatives at that point, giving rise to *gradient theories*. If the material behaviour depends on the co-ordinates of two (or more) points it is addressed as *non-local*.

The third principle is the *principle of objectivity*. This principle expresses the belief that the response of a material to a given (history of) stimulus is independent of any motion of the observer. This is essentially the same as requiring that the response is indifferent to rigid body translations and rotations of the body. To formalise this principle we need to describe a *change of frame of reference* (or observer), by which we mean a time-dependent, spatially homogeneous transformation of space and time. An event $(\mathbf{x}, t)$, consisting of a point in space $\mathbf{x}$ and a time t, is transformed to the event $(\mathbf{x}^*, t^*)$ according to

$$\mathbf{x}^* = \mathbf{Q}(t)\cdot\mathbf{x} + \mathbf{c}(t) \quad \text{and} \quad t^* = t - a \tag{6.91}$$

where $\mathbf{c}(t)$ is an arbitrary vector function of time t, $\mathbf{Q}(t)$ represents an arbitrary time-dependent orthogonal transformation and a is an arbitrary constant. Quantities which are invariant under a change of frame of reference, as expressed by Eq. (6.91), are said to be *frame-indifferent* or *objective*. For a scalar s we have $s^* = s$. A vector $\mathbf{v}$ may be thought to represent a directed line from point $\mathbf{x}$ to point $\mathbf{y}$ given by the point-difference relation $\mathbf{v} = \mathbf{y}-\mathbf{x}$ (and $\mathbf{v}^* = \mathbf{y}^*-\mathbf{x}^*$) and therefore we immediately find $\mathbf{v}^* = \mathbf{Q}(t)\cdot\mathbf{v}$. For a tensor $\mathbf{T}$ in the relation $\mathbf{u} = \mathbf{T}\cdot\mathbf{v}$ we find

$$\mathbf{u}^* = \mathbf{Q}(t)\cdot\mathbf{u} = \mathbf{Q}(t)\cdot\mathbf{T}\cdot\mathbf{v} = \mathbf{Q}(t)\cdot\mathbf{T}\cdot\mathbf{Q}^{\mathrm{T}}(t)\cdot\mathbf{v}^* = \mathbf{T}^*\cdot\mathbf{v}^*$$

Summarising we have

$$s^* = s$$

$$\blacktriangleright \quad \mathbf{v}^* = \mathbf{Q}(t)\cdot\mathbf{v} \quad \text{or} \quad v_i^* = Q_{ij}(t)v_j \tag{6.92}$$

$$\mathbf{T}^* = \mathbf{Q}(t)\cdot\mathbf{T}\cdot\mathbf{Q}^{\mathrm{T}}(t) \quad \text{or} \quad T_{ij}^* = Q_{im}(t)T_{mn}Q_{nj}(t)$$

Not all quantities representing physical properties are objective. We discuss only the important example of the deformation gradient $\mathbf{F}$. A change of frame for the Lagrange co-ordinate $\mathbf{r}$ is given by $\mathbf{r}^* = \mathbf{Q}(t)\cdot\mathbf{r} + \mathbf{c}(t)$ while for the Euler co-ordinate $\mathbf{x}$ we obtain $\mathbf{x}^*(t^*) = \mathbf{Q}(t^*)\cdot\mathbf{x}(t^*) + \mathbf{c}(t^*)$. From the definition of $\mathbf{F}^*(t^*) = \partial\mathbf{x}^*(t^*)/\partial\mathbf{r}^*$ we obtain

$$\mathbf{F}^*(t^*) = \frac{\partial\mathbf{x}^*(t^*)}{\partial\mathbf{r}^*} = \frac{\partial\mathbf{x}^*(t^*)}{\partial\mathbf{x}(t^*)}\cdot\frac{\partial\mathbf{x}(t^*)}{\partial\mathbf{r}}\cdot\frac{\partial\mathbf{r}}{\partial\mathbf{r}^*} = \mathbf{Q}(t^*)\cdot\mathbf{F}(t^*)\cdot\mathbf{Q}^{\mathrm{T}}(t)$$

where $\mathbf{F}(t^*) = \partial\mathbf{x}(t^*)/\partial\mathbf{r}$ is used. At the reference time t the two Lagrange frames coincide, i.e. $\mathrm{d}\mathbf{r}^* = \mathrm{d}\mathbf{r}$, and therefore the result is

$$\mathbf{F}^*(t^*) = \mathbf{Q}(t^*)\cdot\mathbf{F}(t^*)$$

so that the deformation gradient transforms like a vector.

Let us now apply these ideas to constitutive behaviour. Restricting us to simple materials a general constitutive relation $\boldsymbol{\sigma}$ is dependent on the deformation gradient $\mathbf{F}$. The relation can be written as $\boldsymbol{\sigma} = \mathbf{g}(\mathbf{F})$, where the response function $\mathbf{g}$ is a tensor function. Now we recall that a second-order tensor $\boldsymbol{\sigma}$ transforms as $\boldsymbol{\sigma}^* = \mathbf{Q}\cdot\boldsymbol{\sigma}\cdot\mathbf{Q}^{\mathrm{T}}$ and the deformation gradient $\mathbf{F}$ as $\mathbf{F}^* = \mathbf{Q}\cdot\mathbf{F}$ and we thus have, if objectivity has to be true, $\mathbf{g}(\mathbf{Q}\cdot\mathbf{F}) = \mathbf{Q}\cdot\mathbf{g}(\mathbf{F})\cdot\mathbf{Q}^{\mathrm{T}}$ for an arbitrary orthogonal transformation $\mathbf{Q}$. Further reduction can be obtained if we take for $\mathbf{Q}^{\mathrm{T}}$ the orthogonal tensor $\mathbf{R}$ implied by the polar decomposition theorem of $\mathbf{F}$ given by $\mathbf{F} = \mathbf{R}\cdot\mathbf{U}$, where $\mathbf{U}$ is the right stretch tensor. Multiplying $\mathbf{F}$ from the left by $\mathbf{R}^T$ we find $\mathbf{R}^T\cdot\mathbf{F} = \mathbf{R}^T\cdot\mathbf{R}\cdot\mathbf{U} = \mathbf{U}$, since $\mathbf{R}^T\cdot\mathbf{R} = \mathbf{R}^{-1}\cdot\mathbf{R} = \mathbf{I}$. Therefore we have

$$\mathbf{g}(\mathbf{R}^T\cdot\mathbf{F}) = \mathbf{R}^T\cdot\mathbf{g}(\mathbf{F})\cdot\mathbf{R} \quad \text{or} \quad \mathbf{R}\cdot\mathbf{g}(\mathbf{U})\cdot\mathbf{R}^T = \mathbf{R}\cdot\mathbf{R}^T\cdot\mathbf{g}(\mathbf{F})\cdot\mathbf{R}\cdot\mathbf{R}^T = \mathbf{g}(\mathbf{F})$$

and the response becomes

$$\boldsymbol{\sigma} = \mathbf{R}\cdot\mathbf{g}(\mathbf{U})\cdot\mathbf{R}^T = \mathbf{R}\cdot\mathbf{f}(\mathbf{C})\cdot\mathbf{R}^T = \mathbf{R}\cdot\mathbf{h}(\mathbf{L})\cdot\mathbf{R}^T$$

where $\mathbf{f}(\mathbf{C}) = \mathbf{g}(\mathbf{C}^{1/2}) \equiv \mathbf{g}(\mathbf{U})$ and $\mathbf{h}(\mathbf{L}) = \mathbf{f}(\mathbf{I}+2\mathbf{L})$. As before, $\mathbf{C} = \mathbf{F}^T\cdot\mathbf{F} = \mathbf{U}^2 = \mathbf{I} + 2\mathbf{L}$ represents the right Cauchy-Green tensor and $\mathbf{L}$ the Lagrange strain tensor. In the small deformation gradient approximation to which we will adhere, $\mathbf{R} \cong \mathbf{I}$ and $\mathbf{L} \cong \boldsymbol{\varepsilon}$ so that objectivity is always fulfilled. The principle of objectivity has a confusing history, see e.g. Lavenda (1978),and realistic systems can be conceived for which it is not true. Malvern (1969) provides a readable, more elaborate introduction to all this.

Now we turn to the second task, which is to choose the independent, external variables and survey the related energetics. In the small deformation gradient theory it is logical for a solid to take the strain ε_{ij} and temperature T. For the moment we consider materials for which internal variables are absent. The behaviour is then governed by the Helmholtz energy f and the dissipation function ϕ

$$f = f(\varepsilon_{ij}, T) \qquad \text{and} \qquad \phi = \phi(\varepsilon_{ij}, \dot{\varepsilon}_{ij}, T) \tag{6.93}$$

An *elastic material* now may be defined as a material for which the dissipative stresses $\sigma_{ij}^{(d)}$ are identically zero. In this case the stress is entirely quasi-conservative and the dissipation function is zero. The absence of hysteresis is implied.

The stress tensor can always be decomposed into an isotropic tensor $\sigma_{kk}\delta_{ij}/3$ and a deviator σ_{ij}'. A similar decomposition can be made for the quasi-conservative stress $\sigma_{ij}^{(q)}$ and the dissipative stress $\sigma_{ij}^{(d)}$. All known materials are capable of sustaining an isotropic quasi-conservative stress $\sigma_{kk}^{(q)}\delta_{ij}/3$. If the material is also capable of sustaining deviatoric stresses, it is a *solid*. The (thermo-)elastic solid, either isotropic or anisotropic, is the most important example. On the other hand, if the material is incapable of sustaining quasi-conservative deviatoric stresses, it is a *liquid* and the Helmholtz function is only a function of the temperature and the trace ε_{ii} or equivalently the density. A material for which the quasi-conservative stresses $\sigma_{ij}^{(q)}$ are identically zero is a *purely viscous material*. In this case the Helmholtz function is a function of temperature only. The viscosity is due to internal friction in the liquid. We will see that rigid-ideally plastic materials provide another example.

Generally neither $\phi = 0$ nor $f = f(T)$ will hold. The stress tensor then contains a quasi-conservative and a dissipative part. These materials are called *visco-elastic*. The simplest example is a compressible liquid with internal friction. Here the isotropic stress contains only an elastic part while the stress deviator contains only dissipative components. We have also the *Kelvin material* for which both the isotropic stress and the stress deviator contain a quasi-conservative and a dissipative part. An even more general visco-elastic material is obtained by admitting internal variables. The simplest is the *Maxwell body* for which the strain instantaneously drops to a lower value upon unloading and remains constant after that. While the drop is associated with an elastic strain $\varepsilon_{ij}^{(e)}$, the level is associated with the internal variables and known as the dissipative strain $\varepsilon_{ij}^{(d)}$. For the external parameters we take the total strains $\varepsilon_{ij} = \varepsilon_{ij}^{(e)} + \varepsilon_{ij}^{(d)}$ while for the internal parameters either the dissipative strains $\varepsilon_{ij}^{(d)}$ or the elastic strains $\varepsilon_{ij}^{(e)}$ can be chosen. It is conventional and appropriate to take the former.

geophysics, theory of heat and sound, optics and electricity. He became a full professor in 1829. He was the first to introduce a seminar in theoretical physics and mathematics, where each student had to prepare a paper in which some advanced topic was discussed. This new way of training proved to be very successful and was much copied. His research was mainly on the effect of symmetry in the theory of elasticity. Among his students were Clebsch, Kirchoff and Voigt.

6.9 Bibliography

Bhagavantam, S. (1966), *Crystal symmetry and physical properties*, Academic, London.

Callen, H.B. (1960), *Thermodynamics*, Wiley, New York.

Ericksen, J.L. (1991), *Introduction to the thermodynamics of solids*, Chapman and Hall, London.

Fowler, R.H. and Guggenheim, E.A. (1939), *Statistical thermodynamics*, Cambridge University Press, London.

de Groot, S.R. (1961), *Thermodynamics of irreversible processes*, North-Holland, Amsterdam.

Guggenheim, E.A. (1967), *Thermodynamics*, North-Holland, Amsterdam.

Lavenda, B.H. (1978), *Thermodynamics of irreversible processes*, MacMillan, London (see also Dover, 1997).

Malvern, L.E. (1969), *Introduction to the mechanics of a continuous medium*, Prentice-Hall, Englewood Cliffs, NJ.

Maugin, G.A. (1992), *The thermomechanics of plasticity and fracture*, Cambridge University Press, Cambridge.

Nye, J.F. (1957), *Physical properties of crystals*, Oxford University Press, London.

Woods, L.C. (1975), *The thermodynamics of fluid systems*, Clarendon, Oxford.

Wooster, W.A. (1973), *Tensors and group theory for the physical properties of crystals*, Clarendon, Oxford

Ziegler, H. (1983), *An introduction to thermomechanics*, 2nd ed., North-Holland, Amsterdam.

7

C, Q and S mechanics

In the introduction to this book it was indicated that its main topic is the thermomechanical behaviour of solids as exhibited on the meso-scale (microstructure/morphology), applied on the macro-scale and explained via the micro-scale (atomic/molecular aspects). In particular for the description of atomic level aspects we need some physics and chemistry. It is therefore useful to review also briefly a number of basic concepts from physics and chemistry. We start with a number of concepts from classical mechanics, continue with quantum mechanics and conclude with a brief treatment of statistical mechanics. In the next chapter bonding, structural and microstructural (or morphological) aspects are dealt with. The treatment is necessarily brief and more condensed than the rest of this book. More details may be found in the books referred to in the bibliography of which free use has been made.

7.1 Classical mechanics

As a preliminary to the rest of this chapter and some topics of this book we briefly review classical mechanics as formulated by Lagrange and Hamilton and used as a basis for both quantum and statistical mechanics. In Chapter 5 we have indicated that the principle of virtual power can be regarded as a basis for mechanics. Although this formulation is quite sufficient for continuum mechanics, the description of a set of particles benefits from some concepts to be discussed below (Goldstein, 1981).

Generalised co-ordinates

Within an abstract framework it is convenient to describe the motion of systems with the generalised co-ordinates q_i and velocities $\dot{q}_i$. The generalised co-ordinates q_i, similar to the Cartesian co-ordinates r_i, are used to describe the instantaneous configuration of the system of interest while the generalised velocities $\dot{q}_i$ specify the instantaneous motion of the system. It is supposed that given a set of initial co-ordinates and velocities the development of the system in time can be calculated. The number of co-ordinates is equal to $3N$ where N is the number of particles. The description of the configuration of a diatomic molecule may serve as an example. The orientation in space of the molecule as a whole can be described by the three co-ordinates of its centre of gravity and the two orientation angles of the line joining both atoms while the distance between the atoms provides the last co-ordinate. In this example five co-ordinates deal with the translation and rotation of the molecule as a whole and there is only one internal, vibration co-ordinate. In general, for non-linear molecules the number of internal co-ordinates is $3N–6$ while for linear molecules this number is $3N–5$. In Chapter 5 it has been shown that constraints of a certain type, the so-called *holonomic constraints*,

$$C(q_i) = c$$

where $C(q_i)$ is a function of the co-ordinates q_i and c is a constant, can be taken into account relatively easily. If M constraints are present the number of degrees of freedom reduces from $3N$ to $3N–M$. If taken care of these constraints, the system itself is referred to as *holonomic*. We refrain from explicit considerations of constraints and refer to the literature for further discussion. We will use in this section the matrix notation as introduced in Chapter 5 in which the masses of the individual particles are collected in a diagonal matrix $\boldsymbol{M}$ given by

$$\boldsymbol{M} = \begin{pmatrix} m(1) & 0 & 0 & \ldots & 0 \\ 0 & m(1) & 0 & \ldots & 0 \\ 0 & 0 & m(1) & \ldots & 0 \\ \ldots & \ldots & \ldots & \ldots & 0 \\ 0 & 0 & 0 & 0 & m(N) \end{pmatrix}$$

and the forces $\boldsymbol{f}$ and position co-ordinates $\boldsymbol{r}$ are given by

$$\boldsymbol{f}^{\mathrm{T}} = [f_1(1), f_2(1), f_3(1), f_1(2), \ldots, f_3(N)] \qquad \text{and}$$
$$\boldsymbol{r}^{\mathrm{T}} = [r_1(1), r_2(1), r_3(1), r_1(2), \ldots, r_3(N)]$$

respectively, and where, as before, the N particles are indicated by the label (i) and the components by a subscript 1, 2 or 3. In this notation the relation between the (stationary) generalised co-ordinates $\boldsymbol{q}$ and the Cartesian co-ordinates $\boldsymbol{r}$ is given by[a]

$$\boldsymbol{r} = \boldsymbol{r}(\boldsymbol{q}) \qquad \text{or} \qquad \mathrm{d}\boldsymbol{r} = \boldsymbol{A}\,\mathrm{d}\boldsymbol{q} \tag{7.1}$$

where the matrix $\boldsymbol{A}$ is given by

$$\boldsymbol{A} = \frac{\partial \boldsymbol{r}}{\partial \boldsymbol{q}} = \begin{pmatrix} \dfrac{\partial r_1(1)}{\partial q_1} & \dfrac{\partial r_1(1)}{\partial q_2} & \ldots & \dfrac{\partial r_1(1)}{\partial q_{3N}} \\ \dfrac{\partial r_2(1)}{\partial q_1} & \dfrac{\partial r_2(1)}{\partial q_2} & \ldots & \ldots \\ \ldots & \ldots & \ldots & \ldots \\ \dfrac{\partial r_3(N)}{\partial q_1} & \ldots & \ldots & \dfrac{\partial r_3(N)}{\partial q_{3N}} \end{pmatrix}$$

For the velocity we obtain

$$\boldsymbol{v} = \dot{\boldsymbol{r}} = \frac{\mathrm{d}\boldsymbol{r}}{\mathrm{d}t} = \frac{\partial \boldsymbol{r}}{\partial \boldsymbol{q}}\frac{\mathrm{d}\boldsymbol{q}}{\mathrm{d}t} = \boldsymbol{A}\dot{\boldsymbol{q}} \tag{7.2}$$

From Eq. (7.1) we also obtain the variation of $\boldsymbol{r}$ and from Eq. (7.2) a further useful relation, namely

$$\delta\boldsymbol{r} = \frac{\partial \boldsymbol{r}}{\partial \boldsymbol{q}}\delta\boldsymbol{q} = \boldsymbol{A}\,\delta\boldsymbol{q} \qquad \text{and} \qquad \frac{\partial \dot{\boldsymbol{r}}}{\partial \dot{\boldsymbol{q}}} = \frac{\partial \boldsymbol{r}}{\partial \boldsymbol{q}} = \boldsymbol{A} \tag{7.3}$$

In generalised co-ordinates the kinetic energy T can be expressed as

$$T = \tfrac{1}{2}\boldsymbol{v}^{\mathrm{T}}\boldsymbol{M}\boldsymbol{v} = \tfrac{1}{2}\dot{\boldsymbol{q}}^{\mathrm{T}}\boldsymbol{A}^{\mathrm{T}}\boldsymbol{M}\boldsymbol{A}\dot{\boldsymbol{q}} \equiv \tfrac{1}{2}\dot{\boldsymbol{q}}^{\mathrm{T}}\boldsymbol{M}^*(\boldsymbol{q})\dot{\boldsymbol{q}}$$

with $\boldsymbol{M}^* = \boldsymbol{A}^{\mathrm{T}}\boldsymbol{M}\boldsymbol{A}$ the generalised mass matrix. In general the kinetic energy T can thus be considered as a function of $\boldsymbol{q}$ and $\dot{\boldsymbol{q}}$, i.e. $T = T(\boldsymbol{q},\dot{\boldsymbol{q}})$. The force $\boldsymbol{f}$ becomes $\boldsymbol{Q} = \boldsymbol{A}^{\mathrm{T}}\boldsymbol{f}$. If the force $\boldsymbol{f}$ is conservative, i.e. if $\boldsymbol{f} = -(\partial V/\partial \boldsymbol{r})^{\mathrm{T}}$ where we used the potential energy V, the generalised force $\boldsymbol{Q}$ is conservative as well and given by

[a] For the general transformations $\boldsymbol{r} = \boldsymbol{r}(\boldsymbol{q},t)$ we refer to the literature.

$$Q = A^{\mathrm{T}} f = -A^{\mathrm{T}}\left(\frac{\partial V}{\partial r}\right)^{\mathrm{T}} = -\left(\frac{\partial r}{\partial q}\right)^{\mathrm{T}}\left(\frac{\partial V}{\partial r}\right)^{\mathrm{T}} = -\left(\frac{\partial V}{\partial q}\right)^{\mathrm{T}}$$

Example 7.1

The water molecule is non-linear with an O–H bond length of 0.0958 nm and an included angle of about 105°. The internal co-ordinates can be taken as the bond length r and the bond angle ϕ. If we take small deviations from the equilibrium configuration we may approximate the vibration energy by a harmonic model and write the potential energy V as $V = \frac{1}{2}k_r\Delta r^2 + \frac{1}{2}k_r\Delta r^2 + \frac{1}{2}k_\phi\Delta\phi^2$, where k_r and k_ϕ denote the force constants for the change in bond length Δr and bond angle $\Delta\phi$, respectively. Of the other co-ordinates, three refer to translation and three to the rotation of the molecule as a whole.

Hamilton's principle

The description of thermomechanics was based on the principle of virtual power (PVP). In classical (particle) mechanics the usual basis for further discussion is Hamilton's principle. To obtain this principle, we recall that the PVP is given by

$$\left(f - \frac{\mathrm{d}p}{\mathrm{d}t}\right)^{\mathrm{T}} \delta r = 0 \tag{7.4}$$

where f, $p = Mv$ and r denote the force, momentum and position co-ordinate column matrices, respectively. If we consider the motion of the system from a certain fixed position at time t_1 to another fixed position at t_2, we have to integrate this equation between these times and we obtain

$$\int_{t_1}^{t_2}\left(f - \frac{\mathrm{d}p}{\mathrm{d}t}\right)^{\mathrm{T}} \delta r \,\mathrm{d}t = 0 \tag{7.5}$$

Integrating by parts the second term we obtain

$$-\int_{t_1}^{t_2}\left(\frac{\mathrm{d}p}{\mathrm{d}t}\right)^{\mathrm{T}} \delta r \,\mathrm{d}t = -\int_{t_1}^{t_2}\frac{\mathrm{d}}{\mathrm{d}t}\left(p^{\mathrm{T}}\,\delta r\right)\mathrm{d}t + \int_{t_1}^{t_2} p^{\mathrm{T}}\,\frac{\mathrm{d}}{\mathrm{d}t}(\delta r)\,\mathrm{d}t$$

$$= p^{\mathrm{T}}\,\delta r\Big|_{t_1}^{t_2} + \int_{t_1}^{t_2} v^{\mathrm{T}} M\,\delta v\,\mathrm{d}t = \int_{t_1}^{t_2}\delta\left(\tfrac{1}{2}\,v^{\mathrm{T}} M v\right)\mathrm{d}t$$

$$= \int_{t_1}^{t_2}\delta T\,\mathrm{d}t = \delta\int_{t_1}^{t_2} T\,\mathrm{d}t$$

where we introduced the kinetic energy $T = \frac{1}{2}v^{\mathrm{T}}Mv$. Since we consider variation between fixed positions the boundary term $p^{\mathrm{T}}\,\delta r\Big|_{t_1}^{t_2}$ is identically zero. Moreover, for fixed times t_1 and t_2 the variation of the integrand is identical to the variation of the integral. The most general form of Hamilton's principle is then

$$\int_{t_1}^{t_2}\left(\delta T + f^{\mathrm{T}}\,\delta r\right)\mathrm{d}t = 0$$

If the system is conservative we may write for the first term of Eq. (7.5)

$$\int_{t_1}^{t_2} \boldsymbol{f}^{\mathrm{T}}\, \delta \boldsymbol{r}\, \mathrm{d}t = -\int_{t_1}^{t_2} \delta V\, \mathrm{d}t = -\delta \int_{t_1}^{t_2} V\, \mathrm{d}t \tag{7.6}$$

Defining the *Lagrange function* $L = T - V$ the total result, conventionally known as *Hamilton's principle*, is

$$\delta \int_{t_1}^{t_2} L\, \mathrm{d}t = \delta \int_{t_1}^{t_2} (T - V)\, \mathrm{d}t = 0 \tag{7.7}$$

It states that the motion of an arbitrary conservative system occurs in such a way that the variation integral of the Lagrange function, the so-called *action*, vanishes provided that the initial and final states are prescribed. Since both T and V are scalars, it is immaterial in what co-ordinate system they are expressed. In particular they may be expressed in the generalised co-ordinates resulting in

► $$\delta \int_{t_1}^{t_2} L\, \mathrm{d}t = \delta \int_{t_1}^{t_2} L(\boldsymbol{q}, \dot{\boldsymbol{q}}, t)\, \mathrm{d}t = 0 \tag{7.8}$$

From this variational equation classical mechanics is developed to great heights (Lanczos, 1970) and we only briefly mention a limited number of aspects of it (Jeffreys and Jeffreys, 1956). We note that, while the PVP describes the motion with $3N$ second-order equations, Hamilton's principle uses only one scalar equation.

William Rowan Hamilton (1805-1865)

Born in Dublin, Ireland he was educated at Trinity College, Dublin. He was appointed Astronomer Royal of Ireland and professor of astronomy at the age of 21 based on his work on caustics, even before he graduated in 1827. This optical theory had a large impact on the design of telescopes. Largely because of that work he was knighted in 1835. On 16 October 1843 he invented, while walking with Lady Hamilton to the Irish Academy of Sciences, quaternions[b], which are an extension of imaginary numbers, obeying a non-commutative algebra. During that walk he carved with his penknife the rules that govern quaternions in Broughham Bridge. He spent a great deal of his life on the theory of quaternions, unfortunately in a dogmatic way. Vectors as we know them now eventually replaced them, to a large extent due to the work of Gibbs. The last 22 years of his life were not happy ones, some say due to amorous problems (his wife was semi-invalid) and/or alcoholism, others say due to the inherent contradictions in quaternion theory. Three months before his death the newly formed American National Society of Sciences named him as the top living scientist in his role as the foreign associate member, recognising therewith his important contributions. At

[b] For the interesting story of quaternions I strongly recommend *Icons and symmetries* (1992) by S.L. Altmann, Clarendon, Oxford.

present better known contributions of Hamilton are the Hamilton function, the Hamilton-Jacobi differential equations of dynamics and the Cayley-Hamilton theorem in matrix theory.

Lagrange's equations

From Hamilton's principle we can derive a set of equations of motion in terms of the generalised co-ordinates and velocities. Hamilton's principle in index notation reads

$$\int_{t_1}^{t_2}\left(\delta T(q_i,\dot{q}_i)+Q_i\delta q_i\right)\mathrm{d}t=0 \tag{7.9}$$

Evaluating the kinetic energy term results in

$$\int_{t_1}^{t_2}\delta T(q_i,\dot{q}_i)\,\mathrm{d}t=\int_{t_1}^{t_2}\left(\frac{\partial T}{\partial q_i}\delta q_i+\frac{\partial T}{\partial \dot{q}_i}\delta \dot{q}_i\right)\mathrm{d}t=\int_{t_1}^{t_2}\left(\frac{\partial T}{\partial q_i}\delta q_i+\frac{\partial T}{\partial \dot{q}_i}\frac{\mathrm{d}}{\mathrm{d}t}\delta q_i\right)\mathrm{d}t$$

Integrating the second term of the previous equation by parts

$$\int_{t_1}^{t_2}\frac{\partial T}{\partial \dot{q}_i}\frac{\mathrm{d}}{\mathrm{d}t}\delta q_i\,\mathrm{d}t=\frac{\partial T}{\partial \dot{q}_i}\delta q_i\bigg|_{t_1}^{t_2}-\int_{t_1}^{t_2}\frac{\mathrm{d}}{\mathrm{d}t}\frac{\partial T}{\partial \dot{q}_i}\delta q_i\,\mathrm{d}t$$

Again the boundary term vanishes since δq_i vanish at the boundary. Combining results in

$$\int_{t_1}^{t_2}\left(\frac{\partial T}{\partial q_i}-\frac{\mathrm{d}}{\mathrm{d}t}\frac{\partial T}{\partial \dot{q}_i}+Q_i\right)\delta q_i\,\mathrm{d}t=0$$

and, since the variations δq_i are arbitrary, we obtain

$$\frac{\partial T}{\partial q_i}-\frac{\mathrm{d}}{\mathrm{d}t}\frac{\partial T}{\partial \dot{q}_i}+Q_i=0$$

These equations describe the behaviour of a holonomic system and are sometimes addressed as Lagrange's equations, although usually this name is reserved for those equations dealing with a conservative system. To obtain the latter we use the property of a conservative system, $\boldsymbol{f}=-(\partial V/\partial \boldsymbol{r})^{\mathrm{T}}$, so that we may write the generalised force as $\boldsymbol{Q}=\boldsymbol{A}^{\mathrm{T}}\boldsymbol{f}=-(\partial V/\partial \boldsymbol{q})^{\mathrm{T}}$. Using the *Lagrange function* $L=T-V$ and realising that $\partial V/\partial\dot{\boldsymbol{q}}=0$, since V is independent of the generalised velocities, we may write

$$\blacktriangleright\quad \frac{\mathrm{d}}{\mathrm{d}t}\left(\frac{\partial L}{\partial \dot{\boldsymbol{q}}}\right)-\frac{\partial L}{\partial \boldsymbol{q}}=0 \quad\text{or equivalently}\quad \frac{\mathrm{d}}{\mathrm{d}t}\left(\frac{\partial L}{\partial \dot{q}_j}\right)-\frac{\partial L}{\partial q_j}=0 \tag{7.10}$$

which are the *Lagrange equations of motion* for a conservative system, containing again $3N$ second-order equations.

Finally we note for completeness that for more complex cases with potentials dependent on $\dot{q}_j$, although we do not treat them, the force Q_j can be derived from

$$Q_j=-\frac{\partial V'}{\partial q_j}+\frac{\mathrm{d}}{\mathrm{d}t}\frac{\partial V'}{\partial \dot{q}_j}$$

where the function V' is addressed as the *generalised* or *velocity-dependent potential*.

Example 7.2: The harmonic oscillator

For a harmonic oscillator the kinetic energy T is given by $T = ½mv^2$ while the potential energy V is described by $V = ½kx^2$. Here m is the mass, $v = \mathrm{d}x/\mathrm{d}t$ is the velocity, k is the force constant and x is the position co-ordinate of the particle. For the generalised co-ordinates we thus take the Cartesian co-ordinates. From $\delta \int L\,\mathrm{d}t = \delta \int (T-V)\,\mathrm{d}t = 0$ we obtain

$$\delta \int_{t_1}^{t_2} L\,\mathrm{d}t = \delta \int_{t_1}^{t_2} (T-V)\,\mathrm{d}t = \int_{t_1}^{t_2} \delta(½mv^2 - ½kx^2)\,\mathrm{d}t$$

$$= \int_{t_1}^{t_2} [mv\delta v - kx\delta x]\,\mathrm{d}t = \int_{t_1}^{t_2} \left[mv\frac{\mathrm{d}}{\mathrm{d}t}\delta x - kx\delta x \right]\mathrm{d}t$$

Now we integrate the first term by parts and obtain

$$\int_{t_1}^{t_2} mv\frac{\mathrm{d}}{\mathrm{d}t}\delta x\,\mathrm{d}t = mv\delta x\Big|_{t_1}^{t_2} - \int_{t_1}^{t_2} \frac{\mathrm{d}}{\mathrm{d}t} mv\delta x\,\mathrm{d}t = -\int_{t_1}^{t_2} m\ddot{x}\delta x\,\mathrm{d}t$$

Combination yields

$$\int_{t_1}^{t_2} (m\ddot{x} + kx)\delta x\,\mathrm{d}t = 0$$

and thus, since δx is arbitrary, $m\ddot{x} = -kx$. Solving this differential equation, which we recognise as Newton's second law, leads to $x = x_0 \exp(\mathrm{i}\omega t)$ with $\omega = (k/m)^{1/2}$ and x_0 the amplitude. Of course, the same result is obtained directly from the Lagrange equation

$$\frac{\mathrm{d}}{\mathrm{d}t}\left(\frac{\partial L}{\partial \dot{x}}\right) - \frac{\partial L}{\partial x} = 0$$

Hamilton's equations

The above formalism may be put in more convenient form by introducing the *generalised momentum* p_i defined by

$$p_i = \frac{\partial L}{\partial \dot{q}_i} \tag{7.11}$$

Defining the *Hamilton function H* by the Legendre transform

$$H = p_i\dot{q}_i - L(q_k, \dot{q}_l, t)$$

we obtain for its differential

$$\mathrm{d}H = p_i\,\mathrm{d}\dot{q}_i + \dot{q}_i\,\mathrm{d}p_i - \frac{\partial L}{\partial q_i}\mathrm{d}q_i - \frac{\partial L}{\partial \dot{q}_i}\mathrm{d}\dot{q}_i - \frac{\partial L}{\partial t}\mathrm{d}t$$

$$= \left(p_i - \frac{\partial L}{\partial \dot{q}_i}\right)\mathrm{d}\dot{q}_i + \dot{q}_i\,\mathrm{d}p_i - \frac{\partial L}{\partial q_i}\mathrm{d}q_i - \frac{\partial L}{\partial t}\mathrm{d}t = \dot{q}_i\,\mathrm{d}p_i - \frac{\partial L}{\partial q_i}\mathrm{d}q_i - \frac{\partial L}{\partial t}\mathrm{d}t$$

where the last step is made by virtue of Eq. (7.11). We calculate

$$\dot{q}_i = \frac{\partial H}{\partial p_i} \qquad \frac{\partial H}{\partial q_i} = -\frac{\partial L}{\partial q_i} \qquad \text{and} \qquad \frac{\partial H}{\partial t} = -\frac{\partial L}{\partial t}$$

Finally using Eq. (7.11) in the Lagrange equation (7.10) we also have

$$\dot{p}_i = -\frac{\partial H}{\partial q_i}$$

For the equations of motion we thus obtain

$$\blacktriangleright \qquad \dot{p}_i = -\frac{\partial H}{\partial q_i} \qquad \text{and} \qquad \dot{q}_i = \frac{\partial H}{\partial p_i} \tag{7.12}$$

known as *Hamilton's* (or the *canonical*) equations of motion. In this way the system is described by $6N$ first-order equations instead of the $3N$ second-order equations.

Example 7.3: The harmonic oscillator again

We treat again the harmonic oscillator with kinetic energy $T = ½mv^2 = p^2/2m$ (with $p = mv$) and potential energy $V = ½kx^2$. From Hamilton's equation $\dot{p} = -\frac{\partial H}{\partial x}$ we get $m\dot{v} = m\ddot{x} = -kx$, resulting in the solution of the previous example. From Hamilton's $\dot{x} = \frac{\partial H}{\partial p}$ the identity $\dot{x} = \frac{p}{m} = \frac{m\dot{x}}{m} = \dot{x}$ is obtained.

Change with time

For the change with time of any property not explicitly dependent on time, say $X = X(q_i,p_i)$, we have to consider its differential and get

$$\frac{\mathrm{d}X}{\mathrm{d}t} = \frac{\partial X}{\partial q_i}\frac{\mathrm{d}q_i}{\mathrm{d}t} + \frac{\partial X}{\partial p_i}\frac{\mathrm{d}p_i}{\mathrm{d}t} = \frac{\partial X}{\partial q_i}\frac{\partial H}{\partial p_i} - \frac{\partial X}{\partial p_i}\frac{\mathrm{d}H}{\partial q_i} = \{X, H\}$$

where we used for the second step the Hamilton equations. The last step uses the so-called *Poisson brackets* defined for any pair of properties X and Y by

$$\{X, Y\} \equiv \frac{\partial X}{\partial q_i}\frac{\partial Y}{\partial p_i} - \frac{\partial X}{\partial p_i}\frac{\mathrm{d}Y}{\partial q_i}$$

If we apply the general expression for the Poisson brackets to the Hamilton function, explicitly independent of time, we obtain

$$\frac{\mathrm{d}H}{\mathrm{d}t} = \frac{\partial H}{\partial q_i}\frac{\partial q_i}{\partial t} + \frac{\partial H}{\partial p_i}\frac{\mathrm{d}p_i}{\partial t} = \frac{\partial H}{\partial q_i}\frac{\partial H}{\partial p_i} - \frac{\partial H}{\partial p_i}\frac{\mathrm{d}H}{\partial q_i} = 0$$

The Hamilton function is thus constant, $H(q_i,p_i) = E$, where E is the constant. To interpret this constant for simple mechanical systems described by a Lagrange function $L = T - V$, we consider the Hamilton function H and obtain

$$\blacktriangleright \qquad H = p_i\dot{q}_i - L = \dot{q}_i\frac{\partial L}{\partial \dot{q}_i} - T + V = \dot{q}_i\frac{\partial T}{\partial \dot{q}_i} - T + V = 2T - T + V = T + V \tag{7.13}$$

The final result shows that the constant E is equal to the energy[c] expressed in terms of position and momentum as independent co-ordinates. For a conservative system with a Hamilton function explicitly independent of time the energy is thus constant. In terms of the Poisson brackets the equation of motion for an arbitrary, explicitly time-dependent function $X = X(t,q_i,p_i)$ obviously becomes

$$\frac{\mathrm{d}X}{\mathrm{d}t} = \frac{\partial X}{\partial t} + \{X,H\}$$

This concludes our brief overview of classical mechanics. Apart from the use of classical mechanics for its own sake, in particular for theoretical derivations, its formulation has been proven crucial in the proper development of quantum mechanics, the basics of which are discussed in the next section.

Problem 7.1

Show that for a conservative system $\partial S/\partial t = -H$, where S is the *action* $S = \int L(\mathbf{q},\dot{\mathbf{q}},t)\,\mathrm{d}t$.

7.2 Quantum mechanics

Atomic and molecular phenomena need to be described by quantum mechanics. In this section we introduce the principles, deal briefly with a few exact single particle problems and discuss some approximate solution methods for many-particle systems.

Principles

Since the discovery of quantum mechanics we know that in the Schrödinger[d] picture the state of a system[e] is given by the *wavefunction*, where **x** stands for the full set of co-ordinates and t for time. This function and its gradient must be single valued, finite and continuous for all values of its arguments. A function satisfying these conditions is said to be *well-behaved*. It also must have a finite quadratic integral if integrated over the complete range of co-ordinates. Generally, we require that $\widetilde{\Psi}(\mathbf{x},t)$ is *normalised* (Merzbacher, 1970; Schiff, 1955), i.e.

$$\int \widetilde{\Psi}^*(\mathbf{x},t)\widetilde{\Psi}(\mathbf{x},t)\,\mathrm{d}\mathbf{x} = \int \left|\widetilde{\Psi}(\mathbf{x},t)\right|^2 \mathrm{d}\mathbf{x} = 1$$

where the asterisk denotes the complex conjugate. Every physical observable is represented by a linear Hermitian operator[f] acting on the wavefunction. To construct this operator we proceed as follows. The classical variable is expressed as a symmetrised function of the (conjugated) spatial co-ordinates **x** and momenta **p**, as

---

[c] In this chapter E indicates the energy, as usual, in the disciplines discussed in this chapter.

[d] Erwin Schrödinger (1887-1961). Austrian physicist who received the Nobel Prize for Physics in 1933 for the discovery of new fertile forms of the quantum theory. Before embarking on quantum physics he made several contributions to the theory of solid state. In 1938 he fled for the nazi regime to Ireland, together with his wife and mistress. In later years he also became interested in the nature of life and his thoughts were reflected in the little book *What is life?*

[e] In this section we denote atoms, molecules or the solid at hand by a system.

[f] For a Hermitian operator A it holds that $\int f^*Ag\,\mathrm{d}\tau = \int g(Af)^*\,\mathrm{d}\tau$ where the integration is over all of the domain of the functions f and g. Linearity implies $A(c_1f+c_2g) = c_1Af+c_2Ag$ with c_1 and c_2 constants.

obtained from the Hamilton formulation of classical mechanics. In the Schrödinger picture the co-ordinates **x** and momenta **p** become operators. Moreover the energy E is conjugate to time t and also becomes an operator. In total we have the prescription

$$\mathbf{x} \to \mathbf{x}, \quad \mathbf{p} \to -\mathrm{i}\hbar\nabla \qquad \text{and} \qquad t \to t, \quad E \to \mathrm{i}\hbar\frac{\partial}{\partial t}$$

where $\hbar = h/2\pi$ and h denotes *Planck's constant*[g]. In general, an operator, denoted by A, operating on $\widetilde{\Psi}(\mathbf{x},t)$ yields another function, say $\widetilde{\Psi}'(\mathbf{x},t)$, i.e. $A\widetilde{\Psi}(\mathbf{x},t) = \widetilde{\Psi}'(\mathbf{x},t)$. If the operator operating on $\widetilde{\Psi}(\mathbf{x},t)$ yields a multiple of $\widetilde{\Psi}(\mathbf{x},t)$, say $a\widetilde{\Psi}(\mathbf{x},t)$, the resulting equation $A\widetilde{\Psi}(\mathbf{x},t) = a\widetilde{\Psi}(\mathbf{x},t)$ is an *eigenvalue equation*. The function $\widetilde{\Psi}(\mathbf{x},t)$ is the *eigenfunction* and a the *eigenvalue*. The only possible values, which a measurement of the observable with operator A can yield, are the eigenvalues of the equation $A\widetilde{\Psi}(\mathbf{x},t) = a\widetilde{\Psi}(\mathbf{x},t)$. For the co-ordinate operator $\mathbf{x}(i)$ the eigenvalues ξ of a single particle i are the values for which the equation

$$\mathbf{x}\widetilde{\Psi}(\mathbf{x},t) = \xi\widetilde{\Psi}(\mathbf{x},t)$$

which is an ordinary algebraic equation, possesses solutions. Rewriting this as

$$(\mathbf{x} - \xi)\widetilde{\Psi}(\mathbf{x},t) = 0$$

it is evident that $\mathbf{x} = \xi$ or $\widetilde{\Psi}(\mathbf{x},t) = 0$. This corresponds exactly to the definition of the Dirac δ-function, which is thus the eigenfunction $\delta(\mathbf{x} - \xi)$ associated with the operator **x** for the eigenvalue ξ.

The time development of $\widetilde{\Psi}(\mathbf{x},t)$ is given by the Schrödinger equation. We recall that the Hamilton function H given in terms of the co-ordinates **x** and conjugated momenta **p** for conservative systems equals the energy, i.e. $H(\mathbf{p},\mathbf{x}) = E$. The Hamilton operator so formed also operates on the wavefunction and this yields, in view of the prescription $E \to \mathrm{i}\hbar\,\partial/\partial t$, the *time-dependent Schrödinger equation*

$$► \quad H(\mathbf{p},\mathbf{x},t)\widetilde{\Psi}(\mathbf{x},t) = \mathrm{i}\hbar\frac{\partial}{\partial t}\widetilde{\Psi}(\mathbf{x},t) \qquad (7.14)$$

For stationary solutions, i.e. for a time-independent H, we may write

$$\widetilde{\Psi}(\mathbf{x},t) = \Psi(\mathbf{x})f(t)$$

which upon substitution in Eq. (7.14) leads to

$$\frac{H\Psi(\mathbf{x})}{\Psi(\mathbf{x})} = \frac{\mathrm{i}\hbar}{f(t)}\frac{\partial f(t)}{\partial t}$$

Since both sides are independent of each other, they both must be equal to a (so-called separation) constant, say E. Solving the time-dependent part leads to

$$f(t) = C\exp(-\mathrm{i}Et/\hbar) \qquad C = \text{constant}$$

[g] Max Planck (1858-1947). German theoretical physicist who received the Nobel Prize for Physics in 1918 for the discovery of the quantum of action h, now known as Planck's constant. After World War II he was accidentally found in a long trail of moving citizens by an American officer who happened to be a physicist and remembered his face. His son Erwin was executed by the Gestapo early 1945, suspected to be involved in the plot to assassinate Hitler on 20 July, 1944.

while the space-part yields the *time-independent Schrödinger equation*

► $$H(\mathbf{p},\mathbf{x})\Psi(\mathbf{x}) = E\Psi(\mathbf{x}) \quad (7.15)$$

The latter is an eigenvalue equation that shows that the eigenvalue E represents the energy. Since E is time independent, the eigenvalue E is a constant of the motion, like in classical mechanics. The constant C may be chosen to normalise $\Psi(\mathbf{x})$, if necessary.

The various wavefunctions $\Psi_k(\mathbf{x})$ form a complete, orthonormal set, i.e.

$$\int \Psi_k^* \Psi_l \, \mathrm{d}\mathbf{x} = \langle \Psi_k | \Psi_l \rangle = \delta_{kl}$$

where the Dirac[h] *bra-ket notation*[i] is introduced. In this connection the designation complete set implies that an arbitrary, well-behaved function Φ that depends on the same variables in the same domain and obeys the same boundary conditions can be expressed as a sum in terms of these functions, i.e. as $\Phi = \sum_i c_i \Psi_i$ where c_i's are constants whose value depends on the function being expressed.

When a system is in a state Φ the expected mean of a sequence of measurements of the observable with operator A is given by

► $$\langle A \rangle = \int \Phi^* A \Phi \, \mathrm{d}\mathbf{x} = \langle \Phi | A | \Phi \rangle \quad (7.16)$$

where $\langle A \rangle$ is referred to as the *expectation value* and is to be interpreted as $\langle A \rangle = \sum_i \rho_i a_i$, where a_i denotes the measured value and ρ_i its frequency. If we expand Φ in terms of Ψ_k's, we have $\Phi = \sum_i c_k \Psi_k$ and the expectation value becomes

$$\langle A \rangle = \langle \sum_k c_k \Psi_k | A | \sum_l c_l \Psi_l \rangle = \sum_{k,l} c_k c_l \langle \Psi_k | A | \Psi_l \rangle = \sum_k c_k^2 a_k$$

Hence when the system is in a state Φ a measurement of the observable A yields the value a_k with a probability c_k^2, where c_k is the expansion coefficient (or *probability amplitude*) in the expansion of Φ in terms of Ψ_k's, or

$$\rho_k = c_k^2 \quad (7.17)$$

The probability amplitude may be expressed in terms of Φ and Ψ_k as

$$\langle \Psi_k | \Phi \rangle = \langle \Psi_k | \sum_l c_l \Psi_l \rangle = c_k \quad (7.18)$$

If Φ is one of the eigenfunctions Ψ_l of A we obviously have $\rho_k = \langle \Psi_k | \Psi_l \rangle = \delta_{kl}$. In this case the expectation value $\langle A \rangle$ is thus equal to the eigenvalue a_l. The above considerations lead directly to the interpretation of the wavefunction. Consider the probability that a measurement of the position of a single particle i will give the value ξ. The eigenfunction corresponding to the co-ordinate operator $\mathbf{x}(i)$ for the eigenvalue ξ has shown to be $\Psi_\xi = \delta(\mathbf{x} - \xi)$. From Eqs. (7.17) and (7.18) we obtain

$$\rho_\xi = |\langle \delta(\mathbf{x} - \xi) | \Phi(\mathbf{x}) \rangle|^2 = |\Phi(\xi)|^2$$

---

[h] Paul Adrien Maurice Dirac (1902-1984). English theoretical physicist who received the Nobel Prize for physics in 1933 for the discovery of new fertile forms of the quantum theory. Apart from being a physicist renowned for his elegant theory, he was a man of few words, whose vocabulary was usually limited to 'yes', 'no' and 'I don't know'. Margit Dirac, sister of Eugene Wigner, was once introduced to a visitor with the words 'Have you met Wigner's sister?' Wolfgang Pauli characterised him with 'There is no God and Dirac is his prophet'.

[i] The association with brackets is obvious but the meaning of the bra $\langle \Psi |$ and ket $| \Psi \rangle$ is much deeper, see Dirac, P.A.M. (1957), *The principles of quantum mechanics*, 4th ed., Clarendon, Oxford.

Hence the probability of finding the particle i at $\mathbf{x}(i) = \xi$ is given by the square of its wavefunction evaluated at ξ. Generalisation to many particles is straightforward.

The equations of motion of a quantum system can be concisely described as follows. Consider an arbitrary operator X acting on a state function Φ_j. Multiplying with Φ_i^* and integrating over all co-ordinates we obtain the *matrix element* $\langle \Phi_i|X|\Phi_j\rangle$ for the operator X and together they form the matrix $\boldsymbol{X}$. In classical mechanics the equations of motions are summarised by $\mathrm{d}X/\mathrm{d}t = \partial X/\partial t + \{X,H\}$ where $\{X,H\}$ are the Poisson brackets. The analogue of the Poisson brackets in quantum mechanics is referred to as the *commutator*. For two operators A and B it is defined by

$$[A,B] \equiv (AB - BA)$$

If we now take the time derivative of any operator X we find after some manipulation

$$\frac{\mathrm{d}\boldsymbol{X}}{\mathrm{d}t} = \frac{\partial \boldsymbol{X}}{\partial t} + \frac{1}{\mathrm{i}\hbar}(\boldsymbol{XH} - \boldsymbol{HX}) = \frac{\partial \boldsymbol{X}}{\partial t} + \frac{1}{\mathrm{i}\hbar}[\boldsymbol{X},\boldsymbol{H}] \tag{7.19}$$

These equations are the equations of motion in the matrix representation of quantum mechanics and are commonly known as the Heisenberg[j] *equations of motion*.

An important next item is that conjugated variables cannot be determined precisely at the same time or, in other words, if one of the conjugated variables is exactly known, the other is fully undetermined. For the energy this is already clear: in a stationary state at *any* time t the energy is *exactly* E. In general any process that shortens the lifetime, broadens the energy level. Since transitions to other states are required for a change in time, there is a limited residence time Δt and this leads to a small uncertainty in energy ΔE given by $\Delta E \Delta t = \frac{1}{2}\hbar$. More generally, defining the variance $\Delta A^2 = \langle (A - \langle A\rangle)^2\rangle$ we have for any pair of conjugated variables $\mathbf{q}$ and $\mathbf{p}$

$$\Delta\mathbf{q}\Delta\mathbf{p} \geq \hbar/2$$

which are known as the Heisenberg *uncertainty relations*. An important example is given by the pair co-ordinate $\mathbf{x}$ and momentum $\mathbf{p}$. A precisely localised particle thus has an indeterminate momentum and vice versa.

Another important point is that particles also possess *spin angular momentum*, for which no classical analogue is available, but which obeys the same quantisation rules as orbital angular momentum. For an electron the eigenvalue for the spin operator is either $\frac{1}{2}\hbar$ or $-\frac{1}{2}\hbar$ with the eigenfunctions generally denoted by α and β, respectively. Finally, we focus our attention on systems with a Hamilton operator with additive properties. For many-particle systems the Hamilton operator H is a function of the co-ordinates and the momenta of all the particles $i,j,k, \cdots$, i.e. $H = H(i,j,k,\cdots)$. In a number of cases, however, the Hamilton operator is the sum (or can be approximated as the sum) of operators $h(i)$ for particle i only, e.g. $H = \sum_i h(i)$. These single-particle operators satisfy the (single particle) Schrödinger equation

$$h(i)\phi_k(i) = \varepsilon_k\phi_k(i)$$

where $\phi_k(i)$ and ε_k denote the single-particle wavefunction and eigenvalue (particle energy), respectively. The total wavefunction for the N-particle system can be taken

[j] Werner Heisenberg (1901-1976). German physicist who received the Nobel Prize for physics in 1932 for the creation of quantum mechanics. The judicium for his Ph.D. degree from Wilhelm Wien (1864-1928, Nobel prize 1911) for experimental physics and Arnold Sommerfeld (1869-1951) for theoretical physics was 'bottomless ignorance' and 'unique genius', respectively. He received a three, the narrowest of passes between a one (the highest possible) and five (the most abject).

as the product of the individual particle wavefunctions $\Psi = \prod_k \phi_k(i)$ so that the N-particle Schrödinger equation reads

$$H\Psi = \sum_i h(i) \prod_j \phi_k(j) = \sum_k \varepsilon_k \prod_j \phi_k(j) = E\Psi$$

and the total energy is given by $E = \sum_k \varepsilon_k$. However, we recall that the individual particles in a system are indistinguishable. This implies that the wavefunction must be either symmetrical or anti-symmetrical with respect to exchange of particle co-ordinates including spin. Electrons are particles with a half-integer spin, the so-called *fermions* (or Fermi[k]-Dirac particles), for which the wavefunction must be anti-symmetric in the co-ordinates of all electrons. A direct consequence, usually referred to as *Pauli's principle*[l], is that each particle state can be occupied with only one electron. For an electron with spin ½ this implies either spin up ($\frac{1}{2}\hbar$) or with spin down ($-\frac{1}{2}\hbar$). Making allowance for Pauli's principle, a many-electron wavefunction can be expanded in anti-symmetrised products of one-electron wavefunctions (or *spin orbitals*) ϕ_j, each ϕ_j consisting of a spatial part (or *orbital*) and a spin function ($\sigma_j = \alpha$ or β). A convenient form for such an anti-symmetrised product is the *Slater determinant*, in shorthand written as

$$|\phi_a(1)\,\phi_b(2)\ldots\phi_n(N)| \equiv \begin{vmatrix} \phi_a(1) & \phi_a(2) & \ldots & \phi_a(N) \\ \phi_b(1) & \phi_b(2) & \ldots & \phi_b(N) \\ \ldots & \ldots & \ldots & \ldots \\ \phi_n(1) & \phi_n(2) & \ldots & \phi_n(N) \end{vmatrix} \tag{7.20}$$

where (i) denotes the co-ordinates of an electron including spin and N the number of electrons. Similarly, for integer spin particles, the so-called *bosons* (or Bose-Einstein particles), the wavefunction must be symmetric in all the co-ordinates of the particles. This can be realised by taking the *permanent* of individual particle wavefunctions indicated by

$$\|\phi_a(1)\,\phi_b(2)\ldots\phi_n(N)\| \tag{7.21}$$

The permanent is constructed similarly as the determinant but upon expanding one takes all signs as positive. In order to obtain a normalised wavefunction Ψ, if all ϕ's are orthonormal, one has to include the factor $(N!)^{-1/2}$, i.e.

$$\Psi = \frac{1}{\sqrt{N!}}|\phi_a(1)\,\phi_b(2)\ldots\phi_n(N)| \quad \text{or} \quad \Psi = \frac{1}{\sqrt{N!}}\|\phi_a(1)\,\phi_b(2)\ldots\phi_n(N)\|$$

for fermions and bosons, respectively. The factor $(N!)^{-1/2}$ is often incorporated in the definition of the anti-symmetrised product function so that we write for the wavefunction $\Psi = |\phi_a(1)\ \phi_b(2)\ \ldots\ \phi_n(N)|$ or even, assuming a fixed order of co-ordinates $\Psi = |\phi_a\ \phi_b\ \ldots\ \phi_n|$. The same convention is used for the symmetrised product

---

[k] Enrico Fermi (1901-1954). Italian physicist who received the Nobel Prize for physics in 1938 for the discovery of making artificial radioactive elements from neutron irradiation.

[l] Wolfgang Pauli (1900-1958). Austrian physicist who received the Nobel Prize for physics in 1945 for the discovery of the quantum exclusion principle and also known for his indiscriminate rudeness. The second Pauli principle states that his approach spelled destruction to any scientific apparatus, evidenced by an explosion at the University of Berne coinciding with passage through town of a train bearing Pauli to Zürich.

function. A straightforward calculation shows that for the Hamilton operator with additive properties $H = \sum_i h(i)$ the N-particle system eigenvalue is still

$$E = \sum\nolimits_k \varepsilon_k$$

When $H = \sum_i h(i)$ is no longer true, the total energy E is no longer the sum of the particle energies. Moreover, the product function and determinant (or permanent) no longer yield the same total energy.

To conclude this section we mention that the number of exact solutions for realistic systems is very limited in quantum mechanics[m]. Therefore, as has been emphasised many times before, the Schrödinger equation needs drastic approximations for almost all cases, even to obtain approximate solutions. We now deal first with a few single-particle problems that can be solved exactly and after that with approximation methods for many-particle systems. We note that in many cases so-called atomic units[n] are used, connected with the names of Bohr[o], Hartree[p] and Rydberg[q], and we will do so in a number of cases.

Problem 7.2

Show that the eigenvalues of a Hermitian operator are real.

Problem 7.3

Prove the orthogonality of the eigenfunctions of a Hermitian operator.

Problem 7.4

Derive the equation of motion (7.19).

Problem 7.5

Show that the squared deviation of the energy from its expectation value for a stationary state is zero, i.e. $\Delta H^2 = \langle (H - \langle H \rangle)^2 \rangle = 0$.

Problem 7.6

Show that the use of the product function $\Psi(\mathbf{x})f(t)$ results in $f(t) = \exp(-\mathrm{i}Et/\hbar)$.

---

[m] Virtually the only realistic one is the hydrogen atom, at least if a non-relativistic Hamilton operator is used. Fortunately several model systems can be solved exactly.

[n] In the *atomic unit system* the length unit is the *Bohr radius*, 1 $a_0 = (4\pi\varepsilon_0)\hbar^2/me^2 = 0.529\times10^{-10}$ m and the energy unit is the *Rydberg*, 1 Ry $= me^4/2(4\pi\varepsilon_0\hbar)^2 = 2.18\times10^{-18}$ J = 13.61 eV. Here $\hbar = h/2\pi$ with h Planck's constant, m and e the mass and the charge of an electron, respectively, and ε_0 the permittivity of the vacuum. In this system $e^2/(4\pi\varepsilon_0) = 2$ and $\hbar^2/2m = 1$. Unfortunately also another convention for the energy atomic unit exists, i.e. the *Hartree*, 1 Ha = 2 Ry. In the Rydberg convention the kinetic energy reads $-\nabla^2$ and the electrostatic potential $2/r$, while in the Hartree convention they read $-\frac{1}{2}\nabla^2$ and $1/r$, respectively. Chemists seem to prefer Hartrees and physicists Rydbergs.

[o] Niels Bohr (1885-1963). Danish physicist who received the Nobel Prize for physics in 1922 for the study of structure and radiation of atoms, founder of a famous institute in Copenhagen, stimulator of many physicists and well known for his debate with Einstein on the interpretation of quantum mechanics.

[p] Douglas Rayner Hartree (1897-1958). English theoretical physicist who developed powerful methods in numerical analysis. He wrote a number of important books including *Numerical analysis* (1952).

[q] Johannes Robert Rydberg (1854-1919). Swedish mathematical physicist whose most important work is on spectroscopy where he found a relatively simple expression relating the various lines in the spectra of the elements in 1890.

Single-particle problems

We briefly discuss two single-particle problems that are useful throughout materials science: the particle-in-a-box and the harmonic oscillator. We also illustrate the solution for the H atom providing the concept of orbitals, necessary in Chapter 8.

Example 7.4: The particle-in-a-box

For a particle in a one-dimensional box the potential energy is given by

$$V(x)=0 \qquad \text{for} \quad 0<x<w \qquad \text{and} \qquad V(x)=\infty \quad \text{otherwise} \tag{7.22}$$

The kinetic energy operator is

$$T(x)=-\frac{\hbar^2}{2m}\frac{\mathrm{d}^2}{\mathrm{d}x^2} \tag{7.23}$$

where m is the mass of the particle, so that the Schrödinger equation reads

$$-\frac{\hbar^2}{2m}\frac{\mathrm{d}^2}{\mathrm{d}x^2}\Psi=E\Psi \tag{7.24}$$

The solutions are

$$\Psi_n=\sqrt{\frac{2}{w}}\sin\left(\frac{2\pi nx}{w}\right) \tag{7.25}$$

where n = 1,2,... is an integer, the so-called *quantum number*, arising since only for discrete wavelengths the solution obeys the boundary conditions. These wavelengths are given by $\lambda = 2w/1,\ 2w/2,\ \ldots,\ 2w/n$ with allowed energies

$$E_n=\frac{h^2}{8m}\left(\frac{n^2}{w^2}\right) \tag{7.26}$$

In the ground state with $n = 1$ the energy is still not zero. For a three-dimensional box with potential energy $V(\mathbf{x}) = 0$ for $\mathbf{0}<\mathbf{x}<\mathbf{w}$ and $V(\mathbf{x}) = \infty$ otherwise, the energy levels are

$$E_{\mathbf{n}}=\frac{h^2}{8m}\left(\frac{n_1^2}{w_1^2}+\frac{n_2^2}{w_2^2}+\frac{n_3^2}{w_3^2}\right) \tag{7.27}$$

If one of the dimensions of the box is equal to another, the energy levels become *degenerate*, i.e. there are two (or more) levels with the same energy. For example, if $w_1 = w_2 = w_3$ the energies for $\mathbf{n}$ = (1,2,2), $\mathbf{n}$ = (2,1,2) and $\mathbf{n}$ = (2,2,1) are the same and the system is said to be three-fold degenerate.

Example 7.5: The harmonic oscillator

For a one-dimensional harmonic oscillator the potential energy and kinetic energy operator are given by

$$V(x)=\tfrac{1}{2}k(x-x_0)^2 \qquad \text{and} \qquad T(x)=-\frac{\hbar^2}{2m}\frac{\mathrm{d}^2}{\mathrm{d}x^2} \tag{7.28}$$

respectively, so that the Schrödinger equation reads

$$\left[-\frac{\hbar^2}{2m}\frac{\mathrm{d}^2}{\mathrm{d}x^2}+\tfrac{1}{2}k(x-x_0)^2\right]\Psi=E\Psi \tag{7.29}$$

The solutions for the wavefunctions are

$$\Psi_n(\xi)=\left(\frac{\sqrt{\alpha/\pi}}{2^n n!}\right)^{1/2}H_n(\xi)\exp(-\tfrac{1}{2}\xi^2) \tag{7.30}$$

with $\quad \xi=\sqrt{\alpha}\,x \qquad \alpha=\dfrac{\sqrt{mk}}{\hbar} \qquad n=0,1,2,\ldots$

and where $H_n(x)$ are known as Hermite functions defined by

$$H_n(x)=(-1)^n\exp(x^2)\frac{\mathrm{d}^n}{\mathrm{d}x^n}\exp(-x^2)$$

Explicitly the first few Hermite functions are

$$H_0(x)=1 \qquad H_1(x)=2x \qquad H_2(x)=4x^2-2 \qquad H_3(x)=8x^3-12x$$
$$H_4(x)=16x^4-48x^2+12 \qquad H_5(x)=32x^5-160x^3+120x,\ldots$$

The energy levels are given by

$$E_n=\hbar\omega(n+\tfrac{1}{2}) \qquad \text{with circular frequency } \omega=\sqrt{k/m} \tag{7.31}$$

For a three-dimensional oscillator with potential energy

$$V(x,y,z)=\tfrac{1}{2}k_x(x-x_0)^2+\tfrac{1}{2}k_y(y-y_0)^2+\tfrac{1}{2}k_z(z-z_0)^2 \tag{7.32}$$

the energy levels are given by

$$E_{\mathbf{n}}=\hbar\omega_x(n_x+\tfrac{1}{2})+\hbar\omega_y(n_y+\tfrac{1}{2})+\hbar\omega_z(n_z+\tfrac{1}{2}) \tag{7.33}$$

Again, if one of the frequencies is equal to another, the system is degenerate. For example, if $\omega_x=\omega_y=\omega_z$, for $\mathbf{n}=(1,1,2)$, $\mathbf{n}=(1,2,1)$ and $\mathbf{n}=(2,1,1)$ the energy levels are the same and given by

$$E_{\mathbf{n}}=\hbar\omega(n_x+n_y+n_z+3/2) \tag{7.34}$$

and the system is three-fold degenerate. Finally, we note that the presence of the zero-point energy is in accord with the uncertainty relations. If the oscillator had no zero-point energy, it would have zero momentum and be located exactly at the minimum of $V(x)$. The necessary uncertainties in position and momentum thus give rise to the zero-point energy.

Example 7.6: The hydrogen atom

For the hydrogen atom the Schrödinger equation reads

$$\left[-\frac{\hbar^2}{2m}\nabla^2+\frac{e^2}{4\pi\varepsilon_0 r}\right]\Psi=E\Psi$$

Although an exact solution can be obtained, the detailed procedure as given by Pauling[r] is elaborate. It appears that the wavefunction Ψ is separable due to the

[r] Linus Carl Pauling (1901-1994). American scientist who received the Nobel Prize for Chemistry (1954) for his research into the nature of the chemical bond and its application to the elucidation of

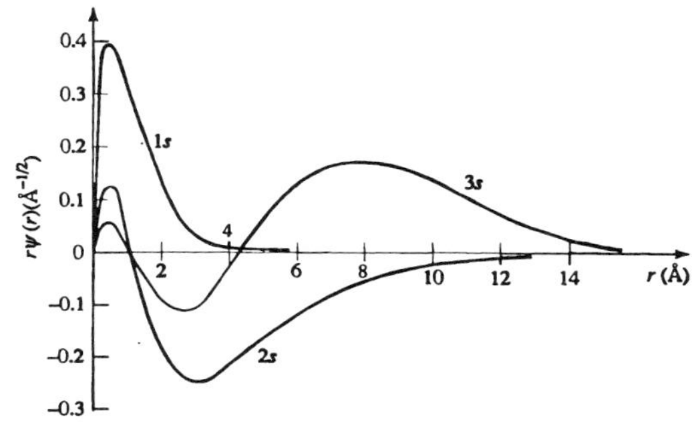

Fig. 7.1: The nodes in the 1s, 2s and 3s wavefunctions.

spherical symmetry in a radial part $R_{nl}(r)$ and a *spherical harmonic* $Y_{lm}(\theta,\varphi)$, describing the angular part. The solutions[s] are labelled by four quantum numbers:

- The principal quantum number n = 1,2,3,... . The energy of a state is proportional to $1/n^2$ and independent of l and m. This is no longer true for many-electron atoms.
- The angular quantum number $l = 0,1,2,\ldots,n$. States with l = 0, 1 or 2 are designated as s, p or d, respectively. The overall angular momentum is $[l(l+1)]^{1/2}\hbar$.
- The magnetic quantum number $m = -l,-l+1,\ldots,l-1,l$. The angular momentum in a specified direction, conventionally taken as x, is $m\hbar$.
- Finally, there is the spin quantum number.

The wavefunctions $\Psi_{nlm} = R_{nl}(r)Y_{lm}(\theta,\varphi)$ are orthonormal, i.e.

$$\int \Psi^*_{nlm}\Psi_{n'l'm'}\,\mathrm{d}\mathbf{r} = \delta_{n,n'}\delta_{l,l'}\delta_{m,m'}$$

Since the spherical harmonics themselves are orthonormal, orthonormality for Ψ_{nlm} is guaranteed if $l \neq l'$ and/or $m \neq m'$. However, if $l = l'$ and $m = m'$ the radial components have to satisfy $\int R_{nl}R_{n'l'}r^2\mathrm{d}r = 0$. This leads to nodes in the radial functions, as illustrated in Fig. 7.1. In practice it is convenient and conventional to take linear combinations of spherical harmonics. If done so, the first few wavefunctions are

$$\Psi_{1s} = \pi^{-1/2}a_0^{-3/2}\exp(-r/a_0)$$
$$\Psi_{2s} = (2\pi)^{-1/2}a_0^{-3/2}(2-r/a_0)\exp(-r/2a_0)$$
$$\Psi_{2p_x} = \tfrac{1}{4}(2\pi)^{-1/2}a_0^{-3/2}(r/a_0)\exp(-r/2a_0)\sin\theta\cos\varphi$$
$$\Psi_{2p_y} = \tfrac{1}{4}(2\pi)^{-1/2}a_0^{-3/2}(r/a_0)\exp(-r/2a_0)\sin\theta\sin\varphi$$
$$\Psi_{2p_z} = \tfrac{1}{4}(2\pi)^{-1/2}a_0^{-3/2}(r/a_0)\exp(-r/2a_0)\cos\theta$$

They are often addressed as *orbitals* and their shape is shown in Fig. 7.2. Finally, the energy is given by

---

the structure of complex substances and the Nobel Peace Prize (1962). His book *The nature of the chemical bond* was rather influential. In later years he advocated the abundant use of vitamin C to prevent illness.

[s] Pauling, L. and Wilson, E.B. (1935), *Introduction to quantum mechanics*, McGraw-Hill, New York.

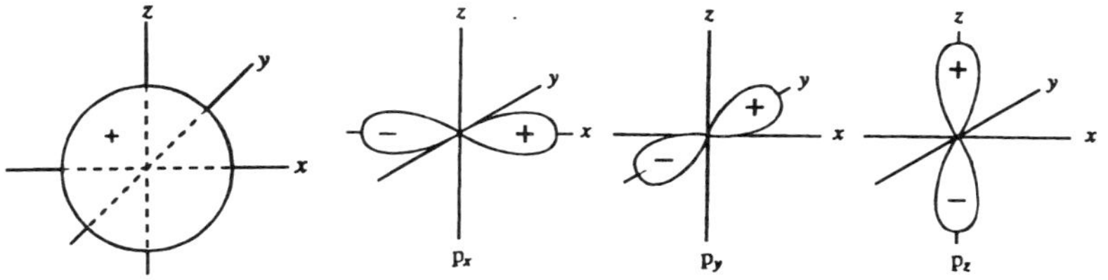

Fig. 7.2: The shape of s and p orbitals for the H atom.

$$E_n = -\frac{me^4}{(4\pi\varepsilon_0)^2 2\hbar^2}\frac{1}{n^2} = -\frac{1}{n^2}\text{Ry} \cong -\frac{13.61}{n^2}\text{ eV}$$

For the ground state, $n = 1$, we have $E_1 = -13.61$ eV or $E_1 = -1$ Ry.

The Born-Oppenheimer approximation

In a molecule or solid both electrons and nuclei are considered as particles. In this section, we denote the co-ordinates of the electrons by **x** and the co-ordinates of the nuclei by **X**. Neglecting relativistic terms, spin interactions and the possibility of explicit time dependence, the Hamilton operator $H(\mathbf{x},\mathbf{X})$ consists of three terms representing the kinetic energy of the nuclei $T^{(n)}(\mathbf{X})$ and electrons $T^{(e)}(\mathbf{x})$ and the potential energy of interaction $V(\mathbf{x},\mathbf{X})$:

$$H(\mathbf{x},\mathbf{X}) = T^{(n)}(\mathbf{X}) + \left(T^{(e)}(\mathbf{x}) + V(\mathbf{x},\mathbf{X})\right) = T^{(n)}(\mathbf{X}) + H^{(e)}(\mathbf{x},\mathbf{X}) \tag{7.35}$$

where the labels (n) and (e) indicate the nuclei and electrons, respectively. A further step can be made by adopting the *adiabatic* (or *Born-Oppenheimer*) *approximation*

$$\Psi(\mathbf{x},\mathbf{X}) = \Psi^{(e)}(\mathbf{x};\mathbf{X})\Psi^{(n)}(\mathbf{X}) \tag{7.36}$$

where the separate factors are obtained from

$$H^{(e)}(\mathbf{x};\mathbf{X})\Psi^{(e)}(\mathbf{x};\mathbf{X}) = E(\mathbf{X})\Psi^{(e)}(\mathbf{x};\mathbf{X}) \quad \text{and} \tag{7.37}$$

$$\left(T^{(n)}(\mathbf{X}) + E(\mathbf{X})\right)\Psi^{(n)}(\mathbf{X}) = E_{\text{tot}}\Psi^{(n)}(\mathbf{X}) \tag{7.38}$$

Here E and E_{tot} denote the electronic and total energy, respectively. Eq. (7.37) is usually solved for a fixed configuration of nuclei so that the co-ordinates enter the electronic wavefunction parametrically, hence $\Psi^{(e)}(\mathbf{x};\mathbf{X})$. From Eq. (7.37) we see that in this approximation $E(\mathbf{X})$ can be considered as the *effective potential energy* for the motion of the nuclei[t]. The approximation thus allows us to discuss the electronic structure and nuclear motion separately.

The variation principle

Let us now focus on the electronic Schrödinger equation (McWeeny and Sutcliffe, 1976; Pilar, 1968). Dropping the label (e) and the parameter **X**, Eq. (7.37) reads

$$H(\mathbf{x})\Psi_j(\mathbf{x}) = E_j\Psi_j(\mathbf{x}) \tag{7.39}$$

---

[t] However, see appendix VIII of Born, M. and Huang, K. (1954), *Dynamical theory of crystal lattices*, Oxford University Press, Oxford.

where the index j refers to a specific electronic wavefunction Ψ_j with associated energy E_j. Since this equation can be solved exactly only in a few cases; in particular, for single particle problems, one invokes approximation methods. The most well-known methods are perturbation theory and use of the variation principle. The latter is discussed in this section and we deal with perturbation theory in the next section.

Indicating from now on the co-ordinates of electron i with $\mathbf{x}_i$, in general a many-electron wavefunction $\Psi(\mathbf{x}_1,\ldots,\mathbf{x}_N)$ can be expanded as

$$\Psi(\mathbf{x}_1,\ldots,\mathbf{x}_N) = \sum c_k \Phi_k(\mathbf{x}_1,\ldots,\mathbf{x}_N) \tag{7.40}$$

where c_k's are the coefficients of the expansion and Φ_k's are the members of a proper complete set. For example, for many-electron systems Φ_k may be a proper anti-symmetrised product function of single-electron functions or Slater determinant

$$\Phi_k(\mathbf{x}_1,\ldots,\mathbf{x}_N) = \left|\phi_i(\mathbf{x}_1)\phi_j(\mathbf{x}_2)\ldots\phi_p(\mathbf{x}_N)\right|$$

where k is a collective index representing $i, \ldots, p$. The functions Φ_k form the *basis*. If we collect the basis functions Φ_k and the coefficients c_k in a column matrix we may write in matrix notation $\boldsymbol{\Psi} = \boldsymbol{\Phi}^{\mathrm{T}}\boldsymbol{c}$. Now the Hamilton operator acting on Ψ yields a new function Ψ', i.e. $H\Psi = \Psi'$. Expanding Ψ' similarly as Ψ we have

$$H\Psi = \Psi' \quad \text{or} \quad H\boldsymbol{\Phi}^{\mathrm{T}}\boldsymbol{c} = \boldsymbol{\Phi}^{\mathrm{T}}\boldsymbol{c}'$$

Multiplying the last equation by $\boldsymbol{\Phi}^*$ and integrating over all co-ordinates results in

$$\left(\int \boldsymbol{\Phi}^* H \boldsymbol{\Phi}^{\mathrm{T}}\,\mathrm{d}\mathbf{x}\right)\boldsymbol{c} = \left(\int \boldsymbol{\Phi}^* \boldsymbol{\Phi}^{\mathrm{T}}\,\mathrm{d}\mathbf{x}\right)\boldsymbol{c}' \quad \text{or} \quad \left(\int \Phi_i^* H \Phi_j\,\mathrm{d}\mathbf{x}\right)c_j = \left(\int \Phi_i^* \Phi_j\,\mathrm{d}\mathbf{x}\right)c_j' \quad \text{or}$$

$$H_{ij}c_j = S_{ij}c_j' \quad \text{or} \quad \boldsymbol{Hc} = \boldsymbol{Sc}'$$

where H_{ij} and S_{ij} denote $\langle\Phi_i|H|\Phi_j\rangle$ and $\langle\Phi_i|\Phi_j\rangle$, respectively, and are referred to as *Hamilton* and *overlap matrix elements*. For an orthonormal basis $\boldsymbol{S} = \boldsymbol{1}$ and $\boldsymbol{Hc} = \boldsymbol{c}'$.

Consider now the Schrödinger equation (7.39). Applying the same procedure as before, i.e. multiplying by $\boldsymbol{\Phi}^*$ and integrating over all co-ordinates, we obtain

$$\boldsymbol{Hc} = E\boldsymbol{Sc} \quad (\text{or if } \boldsymbol{S} = \boldsymbol{1},\ \boldsymbol{Hc} = E\boldsymbol{c}).$$

In practice we use a truncated set of n basis functions. Let us write for this case $\boldsymbol{H}^{(n)}\boldsymbol{c}^{(n)} = E^{(n)}\boldsymbol{S}^{(n)}\boldsymbol{c}^{(n)}$. This equation will have eigenvalues $E_1^{(n)}$, $E_2^{(n)}$, etc. It can be shown that if we use a truncated set with n+1 basis functions $E_1^{(n)} > E_1^{(n+1)}$ or in general

$$E_i^{(n)} > E_i^{(n+1)} > \cdots > E_i^{(\infty)} = E_0$$

where $i = 1, \ldots, n$ and E_0 the exact solution. This result implies that by admitting one more function to an n-truncated complete set, an improved estimate is made for the eigenvalues. This is true for the ground state as well as all excited states. The conventional *variation theorem* arises by taking $n = 1$ so that $\boldsymbol{c}$ drops out and we have $H_{11} - E_1^{(1)}S_{11} = 0$ or (omitting the superscript and subscripts)

► $$E = \frac{\langle\Phi\,|\,H\,|\,\Phi\rangle}{\langle\Phi\,|\,\Phi\rangle} \geq E_0 \tag{7.41}$$

where E_0 is the energy of the first state of the corresponding symmetry, usually the ground state. According to this theorem an upper bound E can be obtained from an approximate wavefunction. Only in cases where Φ is an exact solution the equality sign holds. In all other cases the quotient in the middle results in a larger value than the exact value E_0. This implies that one can introduce one or more parameters α in the electronic wavefunction so that it reads $\Phi(\mathbf{x};\alpha)$ and minimise the left-hand side of Eq. (7.41) with respect to the parameters α. In this way the best approximate wavefunction, given a certain functional form, is obtained. The variation principle is used for nearly all approximate solutions.

Example 7.7: The helium atom

As an example of the variation theorem we discuss briefly the helium atom having a nuclear charge $Z = 2$ and 2 electrons. The Hamilton operator for this atom in atomic units is thus

$$H = -(\nabla_1^2+\nabla_2^2) - Z(2/r_1 + 2/r_2) + 2/r_{12}$$

where the first term indicates the kinetic energy of electrons 1 and 2, the second term the nuclear attraction for electrons 1 and 2 and the third term the electron-electron repulsion. As a normalised trial wavefunction we use a product of two simple exponential functions, namely $\Psi = \phi(1)\phi(2) = (\eta^3/\pi)\exp[-\eta(r_1+r_2)]$, where the parameter η is introduced. The expectation value of the Hamilton operator is $\langle\Psi|H|\Psi\rangle$. For the kinetic energy integral we obtain

$$K = -<\phi(1)|\nabla_1^2|\phi(1)> = -\frac{\eta^3}{\pi}<\exp(-\eta r)|\nabla^2|\exp(-\eta r)> =$$

$$-\frac{\eta^3}{\pi}\int_0^\infty\int_0^{2r}\int_0^\pi \exp(-\eta r)\left(\frac{1}{r^2}\frac{\partial}{\partial r}r^2\frac{\partial}{\partial r}\right)\exp(-\eta r)r^2\sin\theta\,\mathrm{d}\theta\,\mathrm{d}\phi\,\mathrm{d}r =$$

$$-4\eta^3\int_0^\infty\left(\eta^2r^2 - 2\eta r\right)\exp(-\eta r)\,\mathrm{d}r = \eta^2$$

Here the standard integral $\int_0^\infty x^n\exp(-ax)\,\mathrm{d}x = n!/a^{n+1}$, valid for $n = 0,1,2,\ldots$ and $a > 0$, is used. For the nuclear attraction integral we obtain

$$N = -2Z<\phi(1)|\frac{1}{r_1}|\phi(1)> = -\frac{2Z\eta^3}{\pi}<\exp(-\eta r)|\frac{1}{r}|\exp(-\eta r)>$$

$$= -\frac{2Z\eta^3}{\pi}\int_0^\infty\int_0^{2\pi}\int_0^\pi r\exp(-2\eta r)\sin\theta\,\mathrm{d}\theta\,\mathrm{d}\phi\,\mathrm{d}r$$

$$= -8Z\eta^3\int_0^\infty r\exp(-2\eta r)\,\mathrm{d}r = -2Z\eta$$

The electron-electron repulsion integral is complex (see, e.g., Pilar, 1968) and we quote the result without calculation:

$$I = 2\langle\Psi|\frac{1}{r_{12}}|\Psi\rangle = \frac{5\eta}{4}$$

The total energy is thus $\langle\Psi|H|\Psi\rangle = 2(K+N)+I = 2\eta^2-4Z\eta+5\eta/4$. This expression is at minimum when

$$\partial\langle\Psi|H|\Psi\rangle/\partial\eta = 0 = 4\eta - 4Z + 5/4 \quad \text{or when} \quad \eta = (4Z-5/4)/4 = 1.688.$$

The corresponding energy is –5.70 Ry, to be compared with the experimental value of –5.81 Ry.

We now return to the original linear expansion

$$\Psi = \sum\nolimits_k c_k \Phi_k$$

Multiplying $H\Psi = E\Psi$ by Ψ^*, integrating over all electron co-ordinates and using the expansion above results in

$$\sum\nolimits_{i,j} c_i c_j H_{ij} = E\sum\nolimits_{i,j} c_i c_j S_{ij} \qquad \text{or} \qquad \sum\nolimits_{i,j} c_i c_j \left(H_{ij} - ES_{ij}\right) = 0$$

where H_{ij} and S_{ij} denote the Hamilton matrix elements $\langle\Phi_i|H|\Phi_j\rangle$ and overlap matrix elements $\langle\Phi_i|\Phi_j\rangle$, respectively. In matrix notation we have

$$E = \frac{\boldsymbol{c}^\dagger \boldsymbol{H}\boldsymbol{c}}{\boldsymbol{c}^\dagger \boldsymbol{S}\boldsymbol{c}}$$

where † denotes the Hermitian transpose. Minimising with respect to c_i yields

$$\frac{\partial}{\partial c_i}\sum\nolimits_{i,j} c_i^* c_j \left(H_{ij} - ES_{ij}\right) = \sum\nolimits_j c_j \left(H_{ij} - ES_{ij}\right) + \text{complex conjugate} = 0$$

Because c_j's are independent, the determinant of $(H_{ij} - ES_{ij})$ must vanish. In matrix notation we thus have $\det(\boldsymbol{H} - E\boldsymbol{S}) = 0$, corresponding to the generalised eigenvalue equation $\boldsymbol{Hc} - E\boldsymbol{Sc} = \boldsymbol{0}$. This equation can be transformed to the conventional eigenvalue equation by applying a unitary transformation to the basis functions. Since $\det(\boldsymbol{S}) \neq 0$, we can take

$$\Phi_k' = S_{kl}^{-1/2}\Phi_l \quad \text{or} \quad \boldsymbol{\Phi}' = \boldsymbol{S}^{-1/2}\boldsymbol{\Phi}$$

and we find

$$\langle\boldsymbol{\Phi}'|\boldsymbol{\Phi}'\rangle = 1 \quad \text{and} \quad \boldsymbol{H}' = \langle\boldsymbol{\Phi}'|H|\boldsymbol{\Phi}'\rangle = \boldsymbol{S}^{-1/2}\boldsymbol{H}\boldsymbol{S}^{-1/2}$$

In this way we are led to the conventional eigenvalue problem $\boldsymbol{H}'\boldsymbol{c} - E\boldsymbol{c} = \boldsymbol{0}$, which can be solved in the usual way to find the eigenvalues. The lowest one approximates the energy of the ground state. Substituting the eigenvalues back in the secular equation results in the coefficients $\boldsymbol{c}$ from which the approximate wavefunction can be constructed. In case the functions Φ are already orthonormal, $\boldsymbol{S} = \boldsymbol{1}$ and thus $\boldsymbol{S}^{-1/2} = \boldsymbol{1}$ and we arrive directly at the standard eigenvalue problem.

Since the eigenfunctions are all orthonormal or can be made to be so, the approximate wavefunctions are, like the exact ones, orthonormal. Moreover, they provide stationary solutions. As usual with eigenvalue problems we can collect the various eigenvalues E_k in a diagonal matrix $\boldsymbol{E}$ where each eigenvalue takes a diagonal position. The eigenvectors can be collected in a matrix $\boldsymbol{C} = (c_1|c_2|\ldots)$. In this case we can summarise the problem by writing

$$\blacktriangleright \quad \boldsymbol{HC} = \boldsymbol{ESC} \quad \text{or equivalently} \quad \boldsymbol{E} = \boldsymbol{C}^\dagger\boldsymbol{HC} \quad \text{and} \quad \boldsymbol{C}^\dagger\boldsymbol{SC} = \boldsymbol{1} \qquad (7.42)$$

which shows that the matrix $\boldsymbol{C}$ brings $\boldsymbol{H}$ and $\boldsymbol{S}$ simultaneously to a diagonal form, $\boldsymbol{E}$ and $\boldsymbol{1}$, respectively.

This approach is used nowadays almost exclusively for the calculation of molecular wavefunctions, either in *ab-initio* (rigorously evaluating all integrals) or semi-empirically (neglecting and/or approximating integrals by experimental data or empirical formulae). The usual method is the Hartree-Fock method of which a brief account is given in the next chapter.

Perturbation theory

Another way to approximate wavefunctions is by perturbation theory. If we use the same matrix notation again we may partition the equation (Pilar, 1968)

$$\boldsymbol{Hc} = E\,\boldsymbol{c} \qquad \text{to} \qquad \begin{pmatrix} \boldsymbol{H}_{\text{AA}} & \boldsymbol{H}_{\text{AB}} \\ \boldsymbol{H}_{\text{BA}} & \boldsymbol{H}_{\text{BB}} \end{pmatrix}\begin{pmatrix} \boldsymbol{a} \\ \boldsymbol{b} \end{pmatrix} = E\begin{pmatrix} \boldsymbol{a} \\ \boldsymbol{b} \end{pmatrix}$$

Solving for $\boldsymbol{b}$ from $\boldsymbol{H}_{\text{BA}}\boldsymbol{a}+\boldsymbol{H}_{\text{BB}}\boldsymbol{b} = E\boldsymbol{b}$ we obtain

$$\boldsymbol{b} = (E\boldsymbol{1}_{\text{BB}}-\boldsymbol{H}_{\text{BB}})^{-1}\boldsymbol{H}_{\text{BA}}\boldsymbol{a}$$

which upon substitution in $\boldsymbol{H}_{\text{AA}}\boldsymbol{a}+\boldsymbol{H}_{\text{AB}}\boldsymbol{b} = E\boldsymbol{a}$ yields

$$\boldsymbol{H}_{\text{eff}}\boldsymbol{a} = E\boldsymbol{a} \quad \text{with} \quad \boldsymbol{H}_{\text{eff}} = \boldsymbol{H}_{\text{AA}}+\boldsymbol{H}_{\text{AB}}(E\boldsymbol{1}_{\text{BB}}-\boldsymbol{H}_{\text{BB}})^{-1}\boldsymbol{H}_{\text{BA}}.$$

Let us take the number of elements in $\boldsymbol{a}$ equal to 1. In this case the coefficient drops out and we have $E = \boldsymbol{H}_{\text{eff}}(E)$. Since $\boldsymbol{H}_{\text{eff}}$ depends on E we have to iterate to obtain a solution and we do so by inserting as a first approximation $E = \boldsymbol{H}_{\text{AA}} = H_{11}$ in $\boldsymbol{H}_{\text{eff}}$. By expanding the inverse matrix to second order in off-diagonal elements we obtain

► $$E = H_{11} + \sum\nolimits_{m\neq 1} \frac{H_{1m}H_{m1}}{H_{11} - H_{mm}} \tag{7.43}$$

For completeness, we mention that if we allow for non-orthogonal basis the general result becomes (McWeeny and Sutcliffe, 1976)

$$E = H_{11} + \sum\nolimits_{m\neq 1} \frac{(H_{1m} - H_{11}S_{1m})(H_{m1} - H_{11}S_{m1})}{H_{11} - H_{mm}}$$

This is a general form of perturbation analysis since the basis is entirely arbitrarily; in particular, it is not necessary to assume a complete set, and so far it has not been necessary to divide the Hamilton operator into a perturbed and unperturbed part.

The conventional *Rayleigh-Schrödinger perturbation* equations result if we do divide the Hamilton operator into an unperturbed part H_0 and perturbed part H' or

$$H = H_0 + \lambda H'$$

where λ is the order parameter to be used for classifying orders, e.g. a term in λ^n being of nth order. The perturbation may be regarded as switched on by changing λ from 0 to 1, assuming that the energy levels and wavefunctions are continuous functions of λ. Furthermore, we assume that we have the exact solutions of

$$H_0\Phi_k = E_k^{(0)}\,\Phi_k$$

and we use these solutions Φ_k in Eq. (7.43). The matrix elements become

$$H_{11} = \langle\Phi_1|\,H_0 + \lambda H'|\Phi_1\rangle = E_1^{(0)} + \lambda\langle\Phi_1|\,\lambda H'|\Phi_1\rangle$$

$$H_{1m} = \langle\Phi_1|\,H_0 + \lambda H'|\Phi_m\rangle = \lambda\langle\Phi_1|\,H'|\Phi_m\rangle$$

Since the label 1 is arbitrary we replace it by k to obtain the final result

$$\blacktriangleright \quad E_k = E_k^{(0)} + \langle \Phi_k | H' | \Phi_k \rangle + \sum_{m \neq k} \frac{\langle \Phi_k | H' | \Phi_m \rangle \langle \Phi_m | H' | \Phi_k \rangle}{E_k^{(0)} - E_m^{(0)}} \tag{7.44}$$

where λ has been suppressed. This expression is used amongst others for the calculation of van der Waals forces between molecules.

Example 7.8: The helium atom revisited

As an example of perturbation theory we discuss briefly the helium atom again. The Hamilton operator in atomic units is $H = -(\nabla_1^2+\nabla_2^2) - Z(2/r_1 + 2/r_2) + 2/r_{12}$, where the first term indicates the kinetic energy of electrons 1 and 2, the second term the nuclear attraction for electrons 1 and 2 and the third term the electron-electron repulsion. We consider the electron-electron repulsion as the perturbation and as zeroth-order wavefunction we use a product of two simple exponential functions, namely $\Psi = \phi(1)\phi(2) = (Z^3/\pi)\exp[-Z(r_1+r_2)]$, where the nuclear charge Z instead of the variation parameter η, is introduced. For the kinetic energy integral K we obtained in Example 7.7 $K = Z^2$ and for the nuclear attraction integral N we derived $N = -2Z^2$. The zeroth-order energy ε_0 is $\varepsilon_0 = 2(K+N) = 2Z^2-4Z^2 = -2Z^2$. For the electron-electron repulsion integral I, which in this case represents the first-order energy ε_1, we obtained $I = 5Z/4$. The total energy ε is thus

$$\varepsilon = \varepsilon_0 + \varepsilon_1 = -2Z^2+5Z/4$$

The corresponding energy is –5.50 Ry, to be compared with the experimental value of –5.81 Ry and the variation result of –5.70 Ry.

Time-dependent perturbation theory

A slightly different line of approach is to start with the time-dependent Schrödinger equation right away. This is particularly useful for time-dependent perturbations. In this case we divide the Hamilton operator directly into $H = H_0 + \lambda H'$ and suppose that the solutions of $H_0\Phi_n = E_n\Phi_n$ are available. Using the same matrix notation as before and allowing the coefficients c_k to be time dependent, we obtain

$$\Psi = \sum_n c_n(t)\Phi_n(\mathbf{x})\exp(-\mathrm{i}E_n t/\hbar) \quad \text{or} \quad \Psi = \tilde{\boldsymbol{\Phi}}^{\mathrm{T}}(\mathbf{x},t)\boldsymbol{c}(t)$$

where $\tilde{\boldsymbol{\Phi}}^{\mathrm{T}}(\mathbf{x},t) = \Phi_k(\mathbf{x})\ \exp(-\mathrm{i}E_k t/\hbar)$. Insertion in the Schrödinger equation leads to

$$H\tilde{\boldsymbol{\Phi}}^{\mathrm{T}}(\mathbf{x},t)\boldsymbol{c}(t) = \mathrm{i}\hbar\frac{\partial}{\partial t}\tilde{\boldsymbol{\Phi}}^{\mathrm{T}}(\mathbf{x},t)\boldsymbol{c}(t)$$ or omitting the arguments for $\tilde{\boldsymbol{\Phi}}$ and $\boldsymbol{c}$,

$$H\tilde{\boldsymbol{\Phi}}^{\mathrm{T}}\boldsymbol{c} = \mathrm{i}\hbar\tilde{\boldsymbol{\Phi}}^{\mathrm{T}}\frac{\mathrm{d}}{\mathrm{d}t}\boldsymbol{c} + \mathrm{i}\hbar\left(\frac{\partial}{\partial t}\tilde{\boldsymbol{\Phi}}^{\mathrm{T}}\right)\boldsymbol{c} = \mathrm{i}\hbar\tilde{\boldsymbol{\Phi}}^{\mathrm{T}}\dot{\boldsymbol{c}} + \mathrm{i}\hbar\left(-\frac{\mathrm{i}}{\hbar}\boldsymbol{E}\tilde{\boldsymbol{\Phi}}^{\mathrm{T}}\boldsymbol{c}\right)$$

with $\boldsymbol{E}$ a diagonal matrix containing all eigenvalues E_k. Evaluating using

$$H\tilde{\boldsymbol{\Phi}}^{\mathrm{T}}\boldsymbol{c} = (H_0 + \lambda H')\tilde{\boldsymbol{\Phi}}^{\mathrm{T}}\boldsymbol{c} = \boldsymbol{E}\tilde{\boldsymbol{\Phi}}^{\mathrm{T}}\boldsymbol{c} + \lambda H'\tilde{\boldsymbol{\Phi}}^{\mathrm{T}}\boldsymbol{c}$$

leads to

$$\lambda H'\tilde{\boldsymbol{\Phi}}^{\mathrm{T}}\boldsymbol{c} = \mathrm{i}\hbar\tilde{\boldsymbol{\Phi}}^{\mathrm{T}}\dot{\boldsymbol{c}}$$

Multiplying by $\tilde{\boldsymbol{\Phi}}^*$ and integrating over spatial co-ordinates yields

$$\lambda\tilde{\boldsymbol{H}}'\boldsymbol{c}(t) = \mathrm{i}\hbar\frac{\mathrm{d}}{\mathrm{d}t}\boldsymbol{c}(t) = \mathrm{i}\hbar\dot{\boldsymbol{c}}(t) \quad \text{with} \quad \tilde{H}'_{ij} = \exp\left[\frac{\mathrm{i}(E_i - E_j)t}{\hbar}\right]\langle\Phi_i\,|\,H'\,|\,\Phi_j\rangle$$

Finally, integrating over time, we obtain the result

$$\boldsymbol{c}(t) = (\mathrm{i}\hbar)^{-1}\lambda\int_0^t \tilde{\boldsymbol{H}}'\boldsymbol{c}(t')\,\mathrm{d}t' \tag{7.45}$$

which is an equation that can be solved formally by iteration. Further particular solutions depend on the nature of the Hamilton operator and the boundary conditions.

We leave the perturbation H' undetermined but assume that at $t = 0$ the time-dependent effect is switched on, i.e. $\lambda = 1$, and that at that moment the system is in a definite state, say q. In that case $c_q(0) = 1$ and all others $c_p(0) = 0$ or equivalently $c_p = \delta_{pq}$. We thus write to first order

$$c_p(t) = (\mathrm{i}\hbar)^{-1}\int_0^t \sum\nolimits_q \tilde{H}_{pq}{}'c_q(t')\,\mathrm{d}t' \quad\Rightarrow\quad (\mathrm{i}\hbar)^{-1}\int_0^t H_{pq}{}'\exp\left[\frac{\mathrm{i}(E_p - E_q)t'}{\hbar}\right]\mathrm{d}t'$$

Using the abbreviation $(E_p{-}E_q)/\hbar = \omega_{pq}$ and evaluating leads to

$$c_p(t) = -H'_{pq}\frac{\exp(\mathrm{i}\omega_{pq}t) - 1}{\hbar\omega_{pq}}$$

if we may assume that $H_{pq}{}'$ varies but weakly (or not at all) with t. The probability that the system is in state p at the time t is given by $|c_p(t)|^2$ and this evaluates to

$$|c_p(t)|^2 = 4|H_{pq}{}'|^2\frac{\sin^2 \frac{1}{2}\omega_{pq}t}{(\hbar\omega_{pq})^2}$$

Introducing the final density of states $g_p = \mathrm{d}n_p/\mathrm{d}E$ of states p with nearly the same energy as the initial state q, the transition probability w_p to state p is given by

$$w_p = t^{-1}\int |c_p(t)|^2 g_p\,\mathrm{d}E_p = t^{-1}4\frac{|H_{pq}{}'|^2}{\hbar}g_p\int_{-\infty}^{+\infty}\frac{\sin^2 \frac{1}{2}\omega_{pq}t}{\omega_{pq}^2}\,\mathrm{d}\omega_{pq}$$

where the last step can be made if g_p varies slowly with energy. Since the last integral evaluates to $\frac{1}{2}\pi t$, the final equation becomes

$$\blacktriangleright \qquad w_p = \frac{2\pi}{\hbar}|H_{pq}{}'|^2 g_p \tag{7.46}$$

In fact, the energy available is slightly uncertain due to the Heisenberg uncertainty principle. Therefore the state p is actually one out of a group having nearly the same energy and labelled A. Similarly q is one out of a set B with nearly the same energy and Eq. (7.46) should be interpreted accordingly. We will use Eq. (7.46), known as *Fermi's golden rule*, for the formulation of statistical mechanics.

Here our brief overview on quantum mechanics ends. We have only chosen a few examples that are directly useful for materials science in general and in particular for statistical mechanics, the basis of which is discussed in the next section.

7.3 Statistical mechanics

Quantum physics yields the energy levels and associated wavefunctions of individual quantum systems. Alternatively one may try to describe a system by classical mechanics. If the system is relatively small, say a single molecule or particle, one may proceed as follows. In classical mechanics the system is described by *Hamilton's equations*. Each system can be characterised by a set of n co-ordinates (degrees of freedom) and the associated momenta. A state of a system thus can be depicted as a point in a $2n$-dimensional (Cartesian) space whose axes are labelled by the allowed momenta and co-ordinates of the particles of the system. This space is called the (molecule) *phase space* or μ-space. An example is provided in Example 7.1. If we have a system containing many particles, such as molecules in a volume of gas, a collective of points describes the gas, each point describing a molecule. Similar considerations hold in quantum mechanics. However, in many cases in materials science we need the time average behaviour of a macroscopic system. There are four problems. The first is the size of the macroscopic system, which contains a large number of particles. The large number of degrees of freedom present renders a general solution highly unlikely to be found. The second is that the initial conditions of such systems are unknown so that, even if a general solution was possible in principle, a particular solution cannot be obtained. Third, even if the initial conditions were known, they have a limited accuracy. Since it appears that the relevant equations are extremely sensitive to small changes in initial conditions, this rapidly leads to chaotic behaviour. Fourth, it is difficult to incorporate interaction between the molecules and the interaction with environment. To avoid these problems a statistical approach is followed, sometimes referred to as *statistical physics* but more often as *statistical mechanics*. To avoid unnecessary complications we limit our attention initially to systems with a single kind of particles only.

In 1902 Gibbs made a major step. To overcome the aforementioned problems he introduced the *ensemble*: a large collection of identical systems with the same Hamilton function but different initial conditions. By taking a $2nN$-dimensional space, where N is the total number of particles in the system, we can make a similar representation as before. The axes are labelled with the allowed momenta and co-ordinates of all the particles. This enlarged space is called the (gas) *phase space* or Γ-space. To each system in the ensemble there corresponds a *representative point* in the $2nN$-dimensional Γ-space. Since the system evolves in time the representative point will describe a path as a function of time in Γ-space and this path is called a *trajectory*. From general considerations of differential equations it follows that through each point in phase space (or phase point) can go one and only one trajectory so that trajectories do not cross. The ensemble thus can be depicted as a swirl of points in the extended phase space and the macroscopic state of a system is described by the average behaviour of this swirl. Since the macroscopic system considered has a large number of degrees of freedom, the density of phase points can be considered as continuous in many cases and this density is denoted by $\rho(\mathbf{p},\mathbf{q})$, where $\mathbf{p}$ and $\mathbf{q}$ denote the collective of momenta and co-ordinates, respectively. Obviously $\rho(\mathbf{p},\mathbf{q}) \geq 0$ and we take it normalised, i.e. $\int\rho(\mathbf{p},\mathbf{q})\,\mathrm{d}\mathbf{p}\mathrm{d}\mathbf{q} = 1$. A function $F(\mathbf{p},\mathbf{q})$ in phase space representing a certain property is called a *phase function*. The Hamilton function provides an important example. The time average of a phase function is given by

$$\hat{F} = \lim_{t\to\infty}\frac{1}{t}\int_0^t F[\mathbf{p}(t'),\mathbf{q}(t')]\,\mathrm{d}t' \tag{7.47}$$

However, as stated before, this average cannot be calculated in general since neither the solution $\mathbf{p} = \mathbf{p}(t)$ and $\mathbf{q} = \mathbf{q}(t)$ nor the initial conditions $\mathbf{p} = \mathbf{p}(0)$ and $\mathbf{q} = \mathbf{q}(0)$ are known. To obtain nevertheless estimates of properties the phase-average

$$\overline{F} = \int_{\Gamma} F(\mathbf{p},\mathbf{q})\rho(\mathbf{p},\mathbf{q})\,\mathrm{d}\mathbf{p}\mathrm{d}\mathbf{q} \tag{7.48}$$

is introduced. The assumption, known as the *ergodic theorem* and originally introduced by Boltzmann in 1887, is now that

$$\overline{F} = \hat{F} \tag{7.49}$$

and essentially implies that each trajectory visits each infinitesimal volume element of phase space. This assumption has been proved false but can be replaced by the *quasi-ergodic theorem*, proved by Birkhoff in 1931 for metrically transitive systems[u]. It states that trajectories approach all phase points as closely as desired, given sufficient time to the system. In practice, this means that we accept the equivalence of the time- and phase-average. For details (including the conditions for which the theorem is not valid) we refer to the literature.

Since the remainder of this section deals primarily with the quantum description, we anticipate and remark that maximum correspondence between classical and quantum statistical mechanics can be obtained if we associate in the classical approach, with each degree of freedom a volume of size h, where h denotes Planck's constant. This implies in μ-space a volume of h^n and for the Γ-space a volume of size h^{nN}. The terminology of classical statistical mechanics is frequently used though in the literature even when dealing with quantum systems.

Ludwig Boltzmann (1844-1906)
Born in Vienna, Austria and educated at the University of Vienna where he received his doctorate degree in 1867. His work was mainly on the kinetic and statistical theory but he also made contributions to other area of mathematics, chemistry and physics. Extension of Maxwell's kinetic theory of gases led to the famous Boltzmann distribution with constant k, now known as Boltzmann's constant. May be the most well-known contribution is the relationship $S = k \log W$, which is engraved on his tombstone in Vienna. In his days the abbreviation 'log' denoted the natural logarithm, nowadays usually indicated by 'ln'. Here S is the entropy and W is the number of possible configurations corresponding to a given state of a system. Although nowadays considered as a great scientist, in his time there was a great

[u] See e.g. Khinchin (1949). A mechanical system is metrically transitive if the energy surface cannot be divided into two finite regions such that orbits starting from points in one region always remain in that region. Many people accept this for systems in physics without proof.

controversy about some of his ideas, in particular the statistical theories, which tried to explain the second law in purely mechanical terms. The famous H-theorem uses time reversal for the atoms but nevertheless can explain macroscopic irreversibility. In a reply to comments that this was impossible since if we could reverse the motion of atoms, we would have reversibility, he allegedly seems to have stated: make them to move in a reversed way! Although he defended his position, he took the criticisms so severely that, while on holiday in Duino at the Adriatic Sea near Trieste with his wife and daughter, he committed suicide.

Example 7.9: The harmonic oscillator

A single harmonic oscillator provides a good demonstration for the factor h. In quantum mechanics the energy for an oscillator with spring constant k and mass m is given by $E_n = \hbar\omega(n+½)$ with $\omega = \sqrt{k/m}$ so that the energy difference ΔE between two successive states is $\Delta E = \hbar\omega$. In classical mechanics the total energy can be written as $E(p,q) = p^2/2m + ½m\omega^2q^2$, where p is the momentum and q is the coordinate. We may also write $p^2/\alpha^2+q^2/\beta^2 = 1$ with $\alpha = (2mE)^{1/2}$ and $\beta = (2E/\,m\omega^2)^{1/2}$. If we plot constant energy curves in μ-space, which is in this case a two-dimensional space, we obtain ellipses. The area enclosed by such an ellipse is given by the integral $I = \oint p\,\mathrm{d}q$, where the integration is over one period, or $I = \pi\alpha\beta = 2\pi E/\omega$. Alternatively, we have $q = q_0\sin(\omega t)$, $p = m\dot{q} = m\omega q_0\cos(\omega t)$ and $I = \int_0^{2\pi/\omega} p\dot{q}\,\mathrm{d}t = \pi m\omega q_0^{\;2} = 2\pi E/\omega$. Let us draw ellipses with energies corresponding to n–1, n and n+1. The area between two successive ellipses is the area in classical phase space associated with one quantum state and this area corresponds to $2\pi\Delta E/\omega = 2\pi\hbar\omega/\omega = h$. For large n the classical energy is almost constant in the phase region between n–½ and n+½, so that the approximation $\iint\exp[-E(p,q)/kT]\,\mathrm{d}p\mathrm{d}q \cong h\,\Sigma_n \exp(-E_n/kT)$ can be made. For $T \to \infty$ and $n \to \infty$ the argument becomes exact.

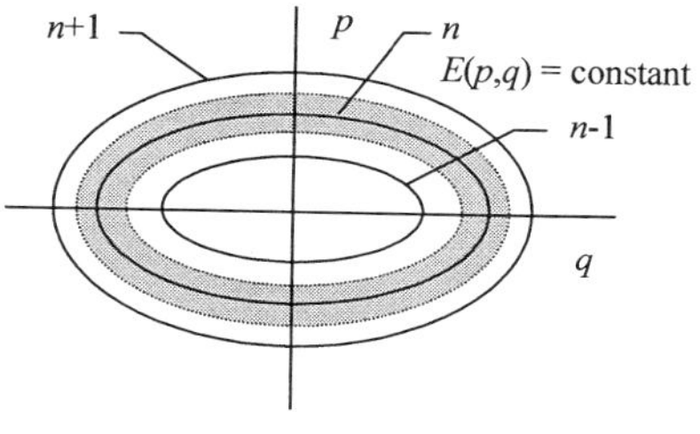

The Boltzmann distribution

In quantum mechanics even a macroscopic system is described by the wavefunction, obtained from the *Schrödinger equation* and obeying the *exchange principle*. The main question is how to characterise a macroscopic state. It might be useful to recall that a macroscopic system is considered to have a *thermodynamic* or *macro-state*, characterised by a limited set of macroscopic parameters. However, such a system contains many particles, e.g. atoms, molecules, electrons and photons. As said, in this connection a macroscopic system is also a quantum system, i.e. an object that contains energy and particles and is identified by a set of distinct and particular states. Identical particles are objects that have access to the same (*quantum*) or *micro-states* of the system. Hence *each macro-state contains many micro-states*. We now describe the macroscopic state of the system by the fraction $\tilde{p}_i$ of each possible quantum state Φ_i in which the macroscopic system remains. Obviously we have[v]

$$\sum_i \tilde{p}_i = 1 \qquad (7.50)$$

For a sufficiently large system the energy levels of the system will form a quasi-continuum. This continuum can be described by the *density of states* g indicating the number of energy states $\mathrm{d}n$ for a certain energy range $\mathrm{d}E$, i.e.

$$g(E) = \mathrm{d}n/\mathrm{d}E$$

For a macroscopic system g is astronomically large and increases extremely rapidly with the size of the system.

Example 7.10

Consider a particle of mass m in a 3D box with edge a and energy $\varepsilon_\mathbf{n} = \dfrac{h^2}{8ma^2}\mathbf{n}^2 = \dfrac{h^2}{8ma^2}\left(n_1^2+n_2^2+n_3^2\right)$. The degeneracy of energy levels is then given by the number of ways the integer $8ma^2\varepsilon_\mathbf{n}/h^2$ can be written as a sum of squares of three positive integers. To estimate the degeneracy, consider a three-dimensional space with the axes labelled with n_1, n_2 and n_3. The number of lattice points with positive integer label, hence in one octant of a sphere of radius $\left(n_1^2+n_2^2+n_3^2\right) = \dfrac{8ma^2\varepsilon_\mathbf{n}}{h^2} \equiv R^2$, gives the number of energy levels. For large values of $\mathbf{n}$, the volume G of the octant of the sphere is given by

$$G(\varepsilon_\mathbf{n}) = \frac{1}{8}\frac{4\pi R^3}{3} \cong \frac{\pi}{6}\left(\frac{8ma^2\varepsilon_\mathbf{n}}{h^2}\right)^{3/2}$$

The density of states is then, omitting the label $\mathbf{n}$,

$$g(\varepsilon) = \frac{\partial G(\varepsilon)}{\partial \varepsilon} = \frac{\pi}{4}\left(\frac{8ma^2}{h^2}\right)^{3/2}\sqrt{\varepsilon}$$

Using $\varepsilon = 3kT/2$, $T = 300$ K, $m = 10^{-22}$ g, $a = 10^{-2}$ m, we obtain $g \cong 10^{30}$, a large number. A similar estimate can be made for N indistinguishable particles. In that case, $E = \dfrac{h^2}{8ma^2}\sum_i^{3N} n_i^2$. Now we need the volume V_N of an N-dimensional sphere of radius x, which is given by $V_N = \pi^{N/2}x^N/\Gamma\left(\tfrac{1}{2}N+1\right)$, where Γ denotes the gamma function (see Appendix D). Hence the number of indistinguishable states G in one 'octant' is given by

---

[v] It is a lucky situation that the probability of each system being in state i can also be interpreted in classical terms as the fraction of systems in the ensemble in state i. In that case i refers, of course, to a volume in Γ-space of size h^{nN}, labelling a particular set $(\mathbf{p},\mathbf{q})$.

$$G(E)=\frac{1}{\Gamma(N+1)}\frac{1}{2^{3N}}\frac{\pi^{3N/2}}{\Gamma\left(\frac{3N}{2}+1\right)}x^{3N}=\frac{1}{\Gamma(N+1)}\frac{1}{\Gamma\left(\frac{3N}{2}+1\right)}\left(\frac{2\pi ma^2E}{h^2}\right)^{3N/2}$$

The factor $1/\Gamma(N+1) = 1/N!$ stems from the indistinguishability. We thus have

$$g(E)=\frac{\partial G(E)}{\partial E}\sim E^{\frac{3N}{2}-1}$$

Making a similar estimate as before again using $T = 300$ K, $m = 10^{-22}$ g, $a = 10^{-2}$ m but now with $N = 6\times10^{23}$ and $E = 3NkT/2$, we obtain $g \cong 10^N$, which is an extremely large number. Moreover, the expression for g shows that it increases extremely rapidly with the size of the system.

The occupation probability $\tilde{p}_i$ of each of the states i may change in time. If a system in state i within a set of states A makes a transition to any state within a set of states B, the *one-to-many jump rate* w_B is given by Fermi's golden rule

$$w_B = 2\pi|H_{BA}'|^2 g_B/\hbar \tag{7.51}$$

where H_{BA}' is the matrix element of the perturbation between states A and B. Since we are interested in the *one-to-one jump rate* ν_{ji} from any state i in A to every state j in B we make the *accessibility assumption*, i.e. all states within a (small) energy range δE, the *accessibility range*, are equally accessible. It can be shown that the exact size of δE is unimportant for macroscopic systems[w]. We thus obtain

$$\nu_{ji}=\frac{w_B}{g_B\delta E}=\frac{2\pi|H_{BA}'|^2}{\hbar\delta E} \tag{7.52}$$

From quantum mechanics we know that $H_{AB}' = (H_{BA}')^*$ so that we obtain *jump rate symmetry* $\nu_{ji} = \nu_{ij}$.

If we consider now the rate of change $d\tilde{p}_i/dt$ for a certain state i, there are two contributions. The first is associated with jumps *from* state i given by $-\sum \tilde{p}_i\nu_{ji}$. The second is associated with jumps *to* state i given by $\sum \tilde{p}_j\nu_{ij}$. With $\nu_{ji} = \nu_{ij}$ we obtain

$$\blacktriangleright\quad \frac{d\tilde{p}_i}{dt}=\sum\nolimits_j \nu_{ij}\left(\tilde{p}_j-\tilde{p}_i\right) \tag{7.53}$$

known as *Fermi's master equation*. The sum runs over all states within the accessibility range. Since we made essential use of the constancy of energy, the master equation applies to thermally isolated systems only. In case thermal equilibrium is reached the probabilities $\tilde{p}_i$ all have become equal to the equilibrium

---

[w] The basic reason is that for macroscopic systems ln δE is completely negligible as compared with ln g, so that in the following ln($g\delta E$) may be replaced by ln g whatever (reasonable) units for E are used. Apart from being unimportant, one may also wonder why a certain energy interval is required at all. In fact, as stated before, precisely defined energies are not allowed in quantum mechanics due to the Heisenberg uncertainty principle and have uncertainty ΔE. Moreover, exactly defined energies are not realisable experimentally and cover a range δE. Typically it holds that $\delta E >> \Delta E$.

probabilities p_i. For an isolated system these equilibrium probabilities for all accessible states are equal, generally referred to as the *principle of equal equilibrium probability*. If for an isolated system an internal constraint is relaxed, implying that two subsystems with energies E_1 and E_2 get coupled to one combined system with energy $E = E_1 + E_2$, a larger set of states becomes available. During the process of equilibration the new set of accessible states becomes increasingly populated and in the new equilibrium situation all the states are again equally populated. It appears that equilibration maximises the new density of states of the combined system $g(E)$ or

$$\mathrm{d}g(E) \geq 0 \quad \text{subject to } E = E_1 + E_2$$

Now consider a system X with energy E_X weakly coupled to a reservoir R (or *heat bath*) with energy E_R, both sufficiently large and isolated as a whole so that the total energy $E = E_X + E_R$. Weak coupling implies that the overall density of states can be written as the product $g_X g_R$. In that case equilibration maximises $g_X g_R$ or $\mathrm{d}(g_X g_R) \geq 0$. Equivalently one may write $\mathrm{d}\ln g_X + \mathrm{d}\ln g_R \geq 0$. When exchange of a small amount of energy takes place from X to R or vice versa, we obviously have $\mathrm{d}E_X = -\mathrm{d}E_R$ and thus

$$\frac{\partial \ln g_X}{\partial E_X}\mathrm{d}E_X + \frac{\partial \ln g_R}{\partial E_R}\mathrm{d}E_R = \left(\frac{\partial \ln g_X}{\partial E_X} - \frac{\partial \ln g_R}{\partial E_R}\right)\mathrm{d}E_X \geq 0 \tag{7.54}$$

Suppose that heat flows from R to X, so that X is colder than R. In that case we have $\mathrm{d}E_X > 0$ and thus $\partial \ln g_X/\partial E_X > \partial \ln g_R/\partial E_R$. Similarly for R colder than X, $\mathrm{d}E_X < 0$ and we have $\partial \ln g_X/\partial E_X < \partial \ln g_R/\partial E_R$. The derivative $\partial \ln g/\partial E$ is thus a measure of the 'coldness' for which we usually write $1/kT$ where T is the (statistical) temperature and k is Boltzmann's constant, a constant chosen to match the statistical temperature with the thermodynamic temperature. Equilibrium thus entails

$$\left(\frac{1}{kT_X} - \frac{1}{kT_R}\right)\mathrm{d}E_X \geq 0$$

or in equilibrium equal temperature for X and R, in agreement with the zeroth law. Now once we have established equilibrium, the various probabilities of the joint system X+R are still proportional to $g_X g_R$. If we are interested in the probability of the system of interest X to be in a particular state i with probability p_i and energy E_i so that g_X is fixed, we thus have (remarkably enough) $p_i \sim g_R$ since the reservoir is assumed to be large and therefore the temperature of the reservoir remains constant. Because $g_R \sim \exp(E_R/kT)$, combination leads to

$$p_i \sim g_R \sim \exp(E_R/kT) \sim \exp(-E_i/kT) \tag{7.55}$$

where the last step can be made since the total energy $E_i + E_R$ = constant. The proportionality factor can be obtained from $\sum_i p_i = 1$ leading to

► $$p_i = \exp(-E_i/kT)/Z \tag{7.56}$$

where $Z = \sum_i \exp(-E_i/kT)$ is the *partition function*. Eq. (7.56) is known as the *Boltzmann distribution* and describes the equilibrium probability distribution for states of systems at constant temperature T. It can be shown that $g_X(E_X)g_R(E_R)$ is extremely strongly peaked at the energy E_X. This is the basic reason that fluctuations for physical properties are typically of order N^{-1} or $N^{-1/2}$, where N is the total number of particles, and thus entirely negligible in most cases. Moreover, the difference between mode (maximum) and average is negligible.

Example 7.11

Consider as system X a particle in a 3D box with edge a for which the density of states is given by $g_X(\varepsilon_X) \sim \varepsilon_X^{1/2}$. Take as reservoir R also such a system, so that the total energy becomes $\varepsilon = \varepsilon_X+\varepsilon_R$. The total density of states $g = g_X(\varepsilon_X)g_R(\varepsilon_R) = g_X(\varepsilon_X)g_R(\varepsilon-\varepsilon_X) \sim \varepsilon_X^{1/2}(\varepsilon-\varepsilon_X)^{1/2}$. Maximising g using $\partial g/\partial\varepsilon_X = 0$ yields $\varepsilon_X = \frac{1}{2}\varepsilon$. The total energy is thus equally divided over the system and reservoir, as expected. The function g is not sharply peaked. However, consider now a many-particle system with N_X particles with density of states $g_X \sim E_X^{\alpha}$ with $\alpha = (3N_X/2)-1$. Couple this system to a many-particle reservoir with N_R particles, so that $g_X \sim E_R^{\beta}$ with $\beta = (3N_R/2)-1$. In this case $g = g_X(E_X)g_R(E_R) \sim E_X^{\alpha}(E-E_X)^{\beta}$. For large N_X and N_R this function is extremely sharp peaked at $E_X = \alpha E/(\alpha+\beta)$, where $E = E_X+E_R$.

For a statistical representation of the entropy S various expressions can be given. Since the macroscopic system is characterised by $\tilde{p}_i$'s, we will have $S = S(\tilde{p}_i)$. Probably the most direct way for a system having n states is to require the following:

- $S(\tilde{p}_i) \le S(1/n)$. This statement implies that equilibrium and thus equi-probability for each state yields the maximum entropy.
- $S(\tilde{p}_i,0) = S(\tilde{p}_i)$. This statement requires that if a state cannot be occupied it will not contribute to the entropy.
- $S = S_A + S_B$ for non-interacting systems A and B. This condition delivers the usual additivity of entropy. For interacting systems the condition changes to $S = S_A + S_B^{(A)}$, where $S_B^{(A)}$ denotes the entropy of system B given that of system A.

These requirements have a unique solution (Landsberg, 1978) apart from a multiplicative constant. This constant is chosen in such a way that the entropy thus defined corresponds to the conventional thermodynamical entropy and this constant appears to be Boltzmann's constant k. The canonical entropy S is then defined by

► $$S = -k\sum\nolimits_i \tilde{p}_i \ln\tilde{p}_i \tag{7.57}$$

It can be shown that for an isolated system $\mathrm{d}S/\mathrm{d}t \ge 0$ and that for the total entropy S of two independent systems A and B indeed the equality $S = S_A + S_B$ holds. To show that the canonical entropy just defined (for all situations) equals the thermodynamic entropy (defined for equilibrium situations only), consider the total energy of the system of interest $E = \sum \tilde{p}_i E_i$ and its derivative $\mathrm{d}E = \sum(\tilde{p}_i\,\mathrm{d}E_i + E_i\,\mathrm{d}\tilde{p}_i)$. From thermodynamics we know that the internal energy U is given by the first law of thermodynamics $\mathrm{d}U = \mathrm{d}W + \mathrm{d}Q$, where $\mathrm{d}W$ and $\mathrm{d}Q$ denote the increment in work and heat, respectively. If we identify U with E, we can associate $\sum \tilde{p}_i\,\mathrm{d}E_i$ with $\mathrm{d}W$ and $\sum E_i\,\mathrm{d}\tilde{p}_i$ with $\mathrm{d}Q$. For a large system in contact with a heat bath or small systems in contact with each other we may assume that we are always near equilibrium and thus that $\tilde{p}_i \cong p_i$. If we then calculate $\mathrm{d}S$ we obtain

$$\mathrm{d}S = -k\sum\nolimits_i \mathrm{d}(p_i \ln p_i) = -k\sum\nolimits_i \mathrm{d}p_i\left(-\frac{E_i}{kT} - \ln Z\right) = \sum\nolimits_i \frac{E_i\,\mathrm{d}p_i}{T} \tag{7.58}$$

where we have used the definition of the Boltzmann distribution and twice the fact that $\Sigma_i \, \mathrm{d}p_i = 0$. Hence for reversible heat flow we have $\mathrm{d}Q = T\,\mathrm{d}S$. From Eq. (7.58) one can easily show that $\partial S/\partial E = 1/T$.

Calculating the entropy in this approximation results in

$$S = -k\sum_i p_i \ln p_i = -k\sum_i p_i\left(-\frac{E_i}{kT} - \ln Z\right) = \frac{E}{T} + k\ln Z \tag{7.59}$$

Using the definition of the Helmholtz energy $F = E - TS$, we easily obtain

► $$F = -kT\ln Z \quad \text{or} \quad Z = \sum_i \exp(-E_i/kT) = \exp(-F/kT) \tag{7.60}$$

From the expression for F the expressions for G, H, c_V, etc. can be obtained in the usual (thermodynamic) way. For large systems in equilibrium we may assume that $p_i \sim (g\,\delta E)^{-1}$ where again g is the density of states of the system and δE is the width of the accessible energy distribution. Hence the entropy S becomes

$$S = -k\sum_i p_i \ln p_i \cong -k(\ln p_i)_{\text{ave}} = k\ln(g\delta E) \cong k\ln g \tag{7.61}$$

where the last step can be made since $\ln \delta E$ is small as compared to $\ln g$ anyway. In equilibrium maximum g thus corresponds to maximum S, coherent with $\mathrm{d}S/\mathrm{d}t \geq 0$ always. If, as occurs in many cases, $g = g_0 W$ with W the number of possible arrangements with the same g_0, the expression reduces to the Boltzmann relation

$$S = k\ln W \tag{7.62}$$

Hence we have three options to calculate the behaviour of a closed quantum system:

1) Use Boltzmann's equation for systems directly. For the average value of a property we obtain $X = \sum_n p_n X_n$. For the energy, entropy and pressure[x] we find

$$E = \sum_i p_i E_i \qquad S = \sum_i p_i \ln p_i \qquad \text{and} \qquad P = \sum_i p_i \frac{\partial E_i}{\partial V}$$

respectively. This approach should be considered before using the other methods.

2) Use the expression for the Helmholtz energy of the system. Using $F = -kT\ln Z$ we obtain for the energy, entropy and pressure

$$E = -\left(\frac{\partial \beta F}{\partial \beta}\right)_V = -\left(\frac{\partial \ln Z}{\partial \beta}\right)_V \quad \text{where} \quad \beta = 1/kT,$$

$$S = -\left(\frac{\partial F}{\partial T}\right)_V = \left(\frac{\partial kT\ln Z}{\partial T}\right)_V \quad \text{and} \quad P = \left(\frac{\partial F}{\partial V}\right)_T = -\left(\frac{\partial kT\ln Z}{\partial V}\right)_T, \text{ respectively.}$$

This method is often the best to use when more than one thermodynamic parameter is required for a relatively small system at constant temperature.

3) Use the entropy expression for large systems. From $S = k\ln g$ or $S = k\ln W$ the entropy can be calculated as a function of energy. This way is especially convenient for highly degenerate systems. Using the conventional thermodynamic relations the other quantities can be obtained relatively easily.

The statistical method outlined above is used at several places for the calculation of thermodynamic quantities. Below we provide as an example of the first method the results for the harmonic oscillator. The second method is quite generally used, also

[x] We use P instead of the earlier used p to avoid confusion with p_i's.

using a macroscopic expression for F. The third method is used for the calculation of defect statistics in the next chapter.

Example 7.12: Statistical mechanics of the harmonic oscillator

For the harmonic oscillator we obtain from quantum physics

$$\varepsilon_n = \hbar\omega(n+½) \quad \text{with} \quad n = 0,1,2,\ldots .$$

Hence for the average energy E we find

$$E = Z^{-1}\sum\nolimits_n \exp\left[-\frac{\hbar\omega}{kT}(n+½)\right]\hbar\omega(n+½) \text{ with } Z = \textstyle\sum_n \exp[-\hbar\omega(n+½)/kT]$$

If we write $\beta = (kT)^{-1}$, the partition function Z can be written as

$$Z = \textstyle\sum_n \exp[-\hbar\omega\beta(n+½)]$$

and, since this is a geometric series, we can evaluate the sum as

$$Z = \exp(-½\hbar\omega\beta)(1-\exp(-\hbar\omega\beta))^{-1}.$$

Since $\sum_n \exp[-\hbar\omega\beta(n+½)]\ \hbar\omega\ (n+½) = -\partial Z/\partial\beta$ we can write

$$E = -\frac{\partial \ln Z}{\partial\beta} = -\frac{\partial Z/\partial\beta}{Z} = ½\hbar\omega + \frac{\hbar\omega}{\exp(\hbar\omega/kT)-1} \tag{7.63}$$

The Helmholtz energy F is given by

$$F = -kT\ln Z = ½\hbar\omega + kT\ln(1-\exp(-\hbar\omega/kT))$$

so that the entropy S is given by

$$S = -\frac{\partial F}{\partial T} = -k\ln[1-\exp(-\hbar\omega/kT)] + \frac{k(\hbar\omega/kT)\exp(-\hbar\omega/kT)}{1-\exp(-\hbar\omega/kT)}$$

The energy can also be calculated from $E = F-TS$. The specific heat c_V reads

$$c_V = \frac{\partial E}{\partial T} = \frac{k(\hbar\omega/kT)^2\exp(\hbar\omega/kT)}{[\exp(\hbar\omega/kT)-1]^2}$$

Finally the average occupation number is given via $E = \langle n\rangle\,\hbar\omega$ and leads to

$$\langle n\rangle = \frac{1}{\exp(\hbar\omega/kT)-1}$$

Problem 7.7

Consider a system of N harmonic oscillators with Q quanta with energy $\hbar\omega$. Show that the density of states can be written as

$$g = W/\hbar\omega \quad \text{with} \quad W = (Q+N-1)!/Q!(N-1)!$$

where W is the number of possible arrangements of quanta over oscillators. For N approaching macroscopic numbers, say about $N \sim 10^{23}$ oscillators, and also $Q \sim 10^{23}$ quanta, show that an estimate for W becomes $\sim 4^{10^{23}}$.

Problem 7.8

Show that
a) the canonical entropy always increases with time and
b) the entropy of two independent systems equals the sum of the entropies of the individual systems.

Problem 7.9

Show using Eq. (7.58) that $\partial S/\partial E = 1/T$.

The Gibbs distribution

So far we have considered systems with a constant number of particles. We now allow for a variable number of particles N_X for the system of interest X and consider again the system of interest plus reservoir as an isolated system. In that case not only the total energy $E = E_X + E_R$ = constant but the total number of particles $N = N_X + N_R$ = constant as well. A similar reasoning as before leads to

$$\left(\frac{\partial \ln g}{\partial E}\right)_N = \frac{1}{kT} \quad \text{and} \quad \left(\frac{\partial \ln g}{\partial N}\right)_E = \frac{\mu}{kT} \tag{7.64}$$

where the new parameter μ is known as the *chemical potential*. The equilibrium condition becomes

$$\mathrm{d}S = \left(\frac{1}{kT_X} - \frac{1}{kT_R}\right)\mathrm{d}E_X + \left(\frac{\mu_X}{kT_X} - \frac{\mu_R}{kT_R}\right)\mathrm{d}N_X \geq 0$$

In equilibrium thus $\mu_X = \mu_R$, in addition to $T_X = T_R$. Because the reservoir is large we consider T and μ as constant for small changes in E_R and N_R. This leads to

$$g_R \approx \exp[(E_R - \mu N_R)/kT] \tag{7.65}$$

for the ranges of E_R and N_R of interest. Since E_R+E = constant, N_R+N = constant and $p_i \sim g_R$ we obtain the *Gibbs distribution*

$$\blacktriangleright \quad p_i = \exp[(\mu N_i - E_i)/kT]/\Xi \quad \text{with} \quad \Xi = \sum\nolimits_i \exp[(\mu N_i - E_i)/kT] \tag{7.66}$$

where the normalisation factor Ξ is the *grand partition function*. The entropy is recalculated similarly and the probability is spread over approximately an accessibility range of $g\delta E\delta N$ states. Similar to $\ln \delta E$, $\ln \delta N$ is completely negligible as compared to $\ln g$. The grand canonical entropy is thus numerically very nearly equal to the canonical entropy.

However, $kT \ln\Xi$ is significantly different from $kT \ln Z$ because of the extra factor $\exp(\mu N_i/kT)$. The grand canonical entropy yields

$$TS = -kT \textstyle\sum_i p_i \ln p_i = -kT \sum_i p_i[(\mu N_i - E_i)/kT - \ln\Xi] = -\mu \overline{N} + \overline{E} + kT \ln\Xi \tag{7.67}$$

where we now write $\overline{N}$ and $\overline{E}$ for the averages. Defining the *grand potential* Φ by the Legendre transform

$$\blacktriangleright \quad \Phi = F - \mu N = U - TS - \mu N \tag{7.68}$$

and identifying U with $\overline{E}$ and $\overline{N}$ with the macroscopic number of particles N, we have

$$\blacktriangleright \quad \Phi = -kT\ln\Xi \quad \text{or} \quad \Xi = \sum\nolimits_i \exp[(\mu N_i - E_i)/kT] = \exp(-\Phi/kT) \qquad (7.69)$$

comparable to the expression for F and Z.

Josiah Willard Gibbs (1839-1903)

Born in New Haven, Connecticut, USA and graduated as engineer at the University of Yale in 1859. He received the first engineering doctorate of the USA in 1863. During 1866-69 he studied in Europe. After returning to Yale he became professor of mathematical physics in 1871, rather surprisingly before he published any work. Most of the present day thermodynamics and the basis of statistical mechanics are due to Gibbs. In 1876 (part I) and 1878 (part II) he introduced the chemical potential in a long paper entitled *On the equilibrium of heterogeneous substances* in the *Transactions of the Connecticut Academy of Sciences*, a journal which was not widely read. In 1881 he published privately *The elements of vector analysis*, which contains the essence of modern vector calculus (and was also the most influential work in replacing quaternions with vectors). His book *The elementary principles in statistical mechanics* was published in 1902. Gibbs never married and living in sister's household he performed his part in the household chores. He was approachable and kind, if unintelligible, to students. From his studies in Europe he returned more a European than an American scientist in spirit – one of the reasons why general recognition in his native country came so slowly. The other might be that, in spite of his immense importance, "his exposition is not noted for its clarity", as Herbert Callen puts it.

Another approach

Another approach leading to the same results but via a much shorter route is as follows. We assume the system to be in contact with a thermal bath characterised by the temperature T, a work bath characterised by the pressure p and a particle bath characterised by the chemical potential μ. The probabilities of the system are given, as before, by p_i and they are normalised, i.e. $\Sigma_i\, p_i = 1$. In addition we require that the average energy is constant, i.e. $\Sigma_i\, p_iE_i = \overline{E}$ where $\overline{E}$ is the average energy, and that the average number of particles is constant, i.e. $\Sigma_i\, p_iN_i = \overline{N}$, where $\overline{N}$ is the average number of particles. We use the expression for the entropy considered before

$$S = -k\sum\nolimits_i p_i \ln p_i \qquad (7.70)$$

where k denotes Boltzmann's constant. For this expression we seek the average given the above-mentioned constraints. Since the distribution is extremely narrow, we can use the maximum as representing the average. This maximum can be obtained by using the Lagrange method of undetermined multipliers (see Chapter 3). We take these multipliers as $-k\alpha$, $-k\beta_E$ and $-k\beta_N$. The maximum is now obtained from

$$\frac{\partial}{\partial p_i}\left[S(p_i)-k\alpha\left(\sum_i p_i-1\right)-k\beta_E\left(\sum_i p_iE_i-\overline{E}\right)-k\beta_N\left(\sum_i p_iN_i-\overline{N}\right)\right]=0 \quad (7.71)$$

and after some calculation using the Stirling approximation[y] this expression leads to

$$p_i=\exp[-(1+\alpha+\beta_E E_i+\beta_N N_i)]$$

The normalisation condition yields

$$\sum_i p_i=1=\exp[-(1+\alpha]\sum_i\exp[-(\beta_E E_i+\beta_N N_i)]$$

and we have

$$p_i=\exp[-(\beta_E E_i+\beta_N N_i)]/\sum_i\exp[-(\beta_E E_i+\beta_N N_i)]$$

Defining the (*grand*) *partition function* (or *sum-over-states*) by

$$\Xi=\sum_i\exp[-(\beta_E E_i+\beta_N N_i)]=\sum_i\exp(-\beta_E E_i)\sum_j\exp(-\beta_N N_j)$$

and using the normalisation condition $\Sigma_i\, p_i = 1$, we find for the entropy

$$S=k\beta_E\overline{E}+k\beta_N\overline{N}+kT\ln\Xi$$

If we compare this expression with the thermodynamic expression for the entropy

$$S=(1/T)(U-\mu N+pV)$$

with U the internal energy, N the number of particles, p the pressure and V the volume, and identify the average microscopic energy $\overline{E}$ with the internal energy U, the average number of particles $\overline{N}$ with the macroscopic N, we obtain

$$\beta_E = 1/kT \qquad \beta_N = -\mu/kT \qquad \text{and} \qquad \ln\Xi = pV/kT$$

so that we finally have

► $$p_i=\exp[(\mu N_i-E_i)/kT]/\Xi \quad (7.72)$$

If the number of particles is fixed the constraint characterised by β_N is removed and one can take $\beta_N = 0$. Therefore in that case

$$p_i=\exp(-E_i/kT)/Z \qquad \text{with} \qquad Z=\sum_i\exp(-E_i/kT)=\exp(-F/kT)$$

where the last step can be made using $S = (1/T)(U - F)$ with F the Helmholtz energy. Here Z denotes the (conventional) *partition function.* Finally, if also the energy is fixed, the constraint characterised by β_E is also removed and one can take $\beta_E = 0$ as well. In this case the probability p_i reduces to

$$p_i = 1/W$$

---

[y] The Stirling approximation for factorials reads $\ln x! = x\ln x - x + ½\ln(2\pi x)$. This approximation is excellent even for $x = 3$, the difference with the exact value being about 2%. Often the term $½\ln(2\pi x)$ is neglected. Although the latter approximation is considerably less accurate, for $x = 50$ it deviates only about 2% from the exact value. Since typically much larger numbers are used, the approximation $\ln x! = x\ln x - x$ is usually quite sufficient.

where W is the number of accessible states for the macroscopic system. In this case the entropy $S(p_i) = -k\,\Sigma_i\,p_i\,\ln p_i$ becomes again the Boltzmann relation

$$S = k \ln W$$

To conclude this section we remark that in fact even this line-of-thought is approximate since the mode has been taken as representing the average behaviour and Stirling's approximation, although quite accurate, is used. A completely rigorous derivation is via the Darwin-Fowler method (see e.g. Schrödinger, 1952).

The Bose-Einstein and Fermi-Dirac distribution

The descriptions given above allows us to conclude that the system as a whole obeys either the Boltzmann or Gibbs distribution but we can say nothing about the occupation of the particle states within a system. For strongly interacting particles there is no escape from the above-described route but for weakly interacting particles in a system we can make further progress. In this respect weakly interacting particles imply that there is some interaction between the particles so that equilibrium can be obtained but that the total energy can be approximated as the sum of the particle energies. Moreover, we recall that identical, weakly interacting (or formally non-interacting) particles in a system are indistinguishable and therefore the wavefunction for the system must be either symmetric or anti-symmetric with respect to exchange in particle co-ordinates. We cannot speak of the probability of any particular particle in some particular particle state but we can speak of the mean occupation number of that state. If the system can exchange energy and particles with the reservoir we may consider the *individual particle states* ϕ_j approximately as 'systems'. Each of these particle states k associated with the system state i has a set of occupation numbers n_{ij} indicating the number of particles with energy ε_j. The total energy and total number of particles of system state i are thus

$$E_i = \sum\nolimits_j n_{ij}\varepsilon_j \quad \text{and} \quad N_i = \sum\nolimits_j n_{ij} \tag{7.73}$$

respectively. The index i in p_i for state i actually indicates the combination of the energy E_i and number of particles N_i in that state, i.e. $p(E_i,N_i)$. Since Eq. (7.73) holds, the set n_{ij} also accounts for all states. We replace the double index ij with a single one, say j, so that we have $p(n_j)$. Returning to index notation we write p_{n_j} and obtain

$$p_{n_j} = \exp[\sum\nolimits_j n_j(\mu - \varepsilon_j)/kT]\big/\Xi = \prod\nolimits_j t_j^{n_j}\big/\Xi \tag{7.74}$$

where the abbrevation $t_j = \exp[(\mu-\varepsilon_j)/kT]$ is used. The grand partition function Ξ is obtained by summing over all states

$$\sum\nolimits_{n_j} p_{n_j} = \sum\nolimits_{n_j}\prod\nolimits_j t_j^{n_j}\big/\Xi = 1 \tag{7.75}$$

The sum we evaluate as

$$\sum\nolimits_{n_j}\prod\nolimits_j t_j^{n_j} = \sum\nolimits_{n_1}\sum\nolimits_{n_2}..\left(t_1^{n_1}t_2^{n_2}..\right) = \sum\nolimits_{n_1} t_1^{n_1}\sum\nolimits_{n_2} t_2^{n_2}.. = \prod\nolimits_j\sum\nolimits_{n_j} t_j^{n_j} = \prod\nolimits_j \Xi_j$$

introducing the partial grand partition function Ξ_j given by

$$\Xi_j = \sum\nolimits_{n_j} t_j^{n_j}$$

and the grand partition function is thus

$$\Xi = \prod_j \Xi_j$$

For *bosons* (also called Bose-Einstein particles) the wavefunction must be symmetric and this leads to occupation numbers that can be any integer. The occupation number n_j of the system state j is given by

$$n_j = 0,1,2,\ldots,\infty \tag{7.76}$$

The partial grand partition function Ξ_j can be evaluated as

$$\Xi_j = \sum_{n_j} t_j^{n_j} = \frac{1}{1-t_j} \qquad \text{or in full} \qquad \Xi_j = \frac{1}{1-\exp[(\mu-\varepsilon_j)/kT]} \tag{7.77}$$

For *fermions* (also called Fermi-Dirac particles) on the other hand, the wavefunction must be anti-symmetric and this implies that the occupation numbers can be only 0 or 1 with corresponding energies 0 and ε_j. Therefore we obtain for Ξ_j directly

$$\Xi_j = \sum_{n_j} t_j^{n_j} = 1+t_j \qquad \text{or in full} \qquad \Xi_j = 1+\exp[(\mu-\varepsilon_j)/kT] \tag{7.78}$$

These results can be combined as

$$\Xi = \prod_j \Xi_j = \prod_j (1 \pm t_j)^{\pm 1} \tag{7.79}$$

where the upper sign (+) is for Fermi-Dirac and the lower sign (–) for Bose-Einstein particles. From Ξ the average occupation number $\langle n_j \rangle$ for state j can be calculated as

$$\langle n_j \rangle = -\frac{\partial \ln \Xi}{\partial \eta_j} \qquad \text{where} \quad \eta_j = \varepsilon_j / kT \tag{7.80}$$

Inserting Eq. (7.79) we obtain

$$\langle n_j \rangle = \mp \frac{\partial}{\partial \eta_j} \sum_j \ln(1 \pm t_j) = \frac{t_j}{1 \pm t_j} = \frac{1}{t_j^{-1} \pm 1} \qquad \text{or in full}$$

► $$\langle n_j \rangle_{\mathrm{BE}} = \frac{1}{\exp[(\varepsilon_j-\mu)/kT]-1} \qquad \text{and} \qquad \langle n_j \rangle_{\mathrm{FD}} = \frac{1}{\exp[(\varepsilon_j-\mu)/kT]+1} \tag{7.81}$$

The above-given distributions apply for a system of weakly interacting particles if the mean state occupation numbers are substantial, i.e. $\langle n_j \rangle \cong 1$. In this case we denote the system as *condensed*. At high temperature both for fermions and bosons the chemical potential μ becomes large and negative and, neglecting the term 1, the above distributions reduce to a Boltzmann distribution over particle states

$$\langle n_j \rangle_{\mathrm{B}} = \exp[(\mu-\varepsilon_j)/kT] \tag{7.82}$$

This is the result for *uncondensed* systems. From this expression we can calculate the thermodynamic properties as follows. Consider first the partial grand partition function Ξ_j, which is written as $\Xi_j = \sum_{n_j} t_j^{n_j}$ and where the summation is over all possible sets of occupation numbers n_j. Since the system is uncondensed we may neglect the small terms with $n_j > 1$ for bosons and write for bosons and fermions

$$\Xi_j \cong 1+t_j = 1+\exp[(\mu-\varepsilon_j)/kT]$$

The corresponding partial grand potential Φ_j is given by

$$\Phi_j = -kT\ln\Xi_j = -kT\ln\{1+\exp[(\mu-\varepsilon_j)/kT]\} \cong -kT\exp[(\mu-\varepsilon_j)/kT]$$

where the last step can be made since the exponential term is small in an uncondensed system. Summing over all Φ_j we have for the system as a whole

$$\begin{aligned}\Phi &= \sum\nolimits_j \Phi_j = -kT\sum\nolimits_j \exp[(\mu-\varepsilon_j)/kT] \\ &= -kT\exp(\mu/kT)\sum\nolimits_j \exp[-\varepsilon_j/kT] \equiv -kT\exp(\mu/kT)\,z\end{aligned}$$

where we introduced the *one-particle partition function* z, the partition function for a single particle alone in the system. From Eq. (7.68) we have

$$N = -\left(\frac{\partial\Phi}{\partial\mu}\right)_{T,V} = \exp(\mu/kT)\,z$$

and therefore we obtain

► $$\mu = kT\ln(N/z) \quad \text{or} \quad \mu = \mu_0 + kT\ln N \quad \text{with} \quad \mu_0 = -kT\ln z \qquad (7.83)$$

which is the familiar result from thermodynamics for ideal systems.

Example 7.13: The translation partition function

For a non-interacting particle with mass m in a volume $V = l^3$ the energy is given by $\varepsilon = p^2/2m = \hbar^2q^2/2m$ with $\mathbf{p} = \hbar\mathbf{q}$ the momentum and $\mathbf{q}$ the wave vector (deviating from the normal $\mathbf{k}$ to distinguish $k = |\mathbf{k}|$ from Boltzmann's constant k). Since the spacing of the energy levels is close, the summation may be replaced by an integral resulting in

$$z = \int\exp(-p^2/2mkT)\,\mathrm{d}\mathbf{p}\mathrm{d}\mathbf{x} = \int_V \mathrm{d}\mathbf{x}\int_0^\infty \exp(-\hbar^2q^2/2mkT)\frac{q^2}{2\pi^2}\mathrm{d}q = \frac{V}{h^3}(2\pi mkT)^{3/2}$$

For a single degree of freedom the partition function is thus $z = \dfrac{l}{h}(2\pi mkT)^{1/2}$.

Let us calculate the partition function Z for N independent particles. From $\mu = kT\ln(N/z)$, $\Phi = -kT\ln(\mu/kT)z$ and $F = \Phi + \mu N$, we obtain

$$\begin{aligned}F &= \Phi + \mu N = -kT\exp(\mu/kT)\,z + kT\ln(N/z)\,N = -kT\,N + kT\ln(N/z)\,N \\ &= -kT(N - N\ln N + N\ln z) \cong -kT(-\ln N! + \ln z^N) = -kT\ln\frac{z^N}{N!}\end{aligned}$$

Hence, comparing with $F = -kT\ln Z$, the result is

► $$Z = z^N/N! \qquad (7.84)$$

and the contributions of the various particles to the Helmholtz energy are thus additive, provided a weak coupling prevails[z].

---

[z] In the literature extensive discussions about indistinguishability in classical statistics have appeared rationalising the appearance of the factor $N!$. Confusion about the introduction of $N!$ arises only if one starts directly in a classical way. This derivation indicates that it is due to the exchange principle.

Finally, let us assume that for the particles the energy can be written as the sum of the energies of the contributing mechanisms. Let us thus assume that $\varepsilon_j = \varepsilon_\alpha + \varepsilon_\beta + \cdots$ where each of the indices α,β, ... indicates a different mechanism. In that case

$$z = \sum_j \exp(-\varepsilon_j/kT) = \sum_{\alpha,\beta,...} \exp[-(\varepsilon_\alpha + \varepsilon_\beta + \cdots)/kT)$$
$$= \sum_\alpha \exp(-\varepsilon_\alpha/kT) \sum_\beta \exp(-\varepsilon_\beta/kT) \cdots = z_\alpha z_\beta \cdots$$

The one-particle partition function can thus be written as the product of the partition functions for each of the (independent) mechanisms. Also in this case the contributions of the various mechanisms to the Helmholtz energy are thus additive since $F = -kT \ln[(z_\alpha z_\beta \cdots)^N/N!]$.

Summarising, the distribution over states in weakly interacting particle systems is given by the Bose-Einstein or Fermi-Dirac distribution, depending on whether the particles are bosons or fermions, and reduces to a Boltzmann (Gibbs) distribution over particle states at high temperature. Note that the Boltzmann distribution applies exactly for system states but only approximately for particle states. For independent particles or independent contributions to the energy of a single-particle the partition function factorises in the product of the individual partition functions of the particles or contributing mechanisms.

This concludes our overview of statistical mechanics. The framework outlined has been used to elucidate the behaviour of gases, solids and to a lesser extent fluids. For the application to various systems and a further background in the principles, in particular the applicability of the master equation, which is easily grasped but has further deeper considerations, we refer to the literature, e.g. Waldram (1985) or van Kampen[aa]. In the following chapters we occasionally make use of the results presented above for the atomic or molecular explanation of mesoscopic observations. In the next sections we elaborate briefly to obtain the transition state equations, used throughout materials science and to make the connection to irreversible thermodynamics. For further study we refer to the bibliography. In particular the classical monograph by Tolman (1938) is still very useful. Other classics are Fowler and Guggenheim (1936) and Slater (1939). Callen (1985) and Reif (1965) provide highly readable, more modern presentations. More condensed are Landsberg (1978) and Schrödinger (1952).

Problem 7.10

Derive Eq. (7.80).

Problem 7.11

Evaluate, using Eq. (7.84) and Example 7.13, the pressure for an ideal gas.

7.4 Transition state theory

Transition state theory, also known as activated complex theory, is an important follow-up from statistical mechanics, used throughout in materials science, which has been developed mainly by Eyring and co-workers. We will illustrate this theory for chemical reactions but with proper changes it can be used in plasticity and fracture.

[aa] Van Kampen, N.G. (1993), Physica A **194**, 542.

Henry Eyring (1901-1981)
Born in Colonia Juarez, Mexico and educated at the Universities of Arizona (in mining engineering) and California, Berkeley where he received his degree in physical chemistry in 1927. In 1935 he was naturalised as an American. After occupying positions in Berlin and Princeton he spent from 1946 to the end of his life his academic career at the University of Utah. Most of his work was on chemical and physical kinetics such as diffusion and the structure of liquids. His research was almost entirely theoretical. Probably his most important contribution is the introduction of transition state theory and its applications to many chemical and physical processes. Eyring was a friendly man, always full of ideas many of which were wrong but, as Laidler puts it, always welcoming criticism and willing to discuss them. He was an active member of the Mormon Church and, to the surprise of many, never received the Nobel Prize.

The equilibrium constant

The first result we need is the equilibrium condition. Associating as before U with $\overline{E}$, N with $\overline{N}$ and recalling that $\ln \Xi = pV/kT$ we conclude from Eq. (7.67) that $\partial S/\partial N = -\mu/T$. The equilibrium condition at constant U and V for a system of several components thus becomes $\mathrm{d}S = \sum_j (\partial S_j/\partial N_j)\,\mathrm{d}N_j = -(\mu_j/T)\,\mathrm{d}N_j = 0$, where the index j denotes the various reactants and products. If we have a reaction

$$\alpha N_A + \beta N_B \leftrightarrow \gamma N_C + \delta N_D \qquad \text{or equivalently} \qquad \textstyle\sum_\mathrm{R} \nu_\mathrm{R} N_\mathrm{R} \leftrightarrow \sum_\mathrm{P} \nu_\mathrm{P} N_\mathrm{P}$$

where R and P denote reactants and products, respectively, we obtain

$$\frac{1}{T}\sum\nolimits_\mathrm{R} \nu_\mathrm{R}\mu_\mathrm{R} = \frac{1}{T}\sum\nolimits_\mathrm{P} \nu_\mathrm{P}\mu_\mathrm{P}$$

Using the relation $\mu = kT\ln(N/z)$ the result is

$$\sum\nolimits_\mathrm{R} \nu_\mathrm{R}\ln(N_\mathrm{R}/z_\mathrm{R}) = \sum\nolimits_\mathrm{P} \nu_\mathrm{P}\ln(N_\mathrm{P}/z_\mathrm{P}) \qquad \text{or}$$

► $$\frac{\prod_\mathrm{P} N_\mathrm{P}^{\nu_\mathrm{P}}}{\prod_\mathrm{R} N_\mathrm{R}^{\nu_\mathrm{R}}} = K(T,V) \qquad \text{with} \qquad K(T,V) = \frac{\prod_\mathrm{P} z_\mathrm{P}^{\nu_\mathrm{P}}}{\prod_\mathrm{R} z_\mathrm{R}^{\nu_\mathrm{R}}} \tag{7.85}$$

where $K(T,V)$ is the *equilibrium constant*. This relationship is known as the *law of mass action*. The reference level of energy for each factor in Eq. (7.85) is the same. However, for the evaluation of the various partition functions it is more convenient to use the ground state of each species as a reference. Let us take the gas phase chemical reaction AB+C $\leftrightarrow$ A+BC as a simple example. In this case $K(T,V) = z_\mathrm{A}z_\mathrm{BC}/z_\mathrm{AB}z_\mathrm{C}$ where z_X denotes the partition function of species X. Shifting the reference level of all species to an arbitrary level and denoting the partition function with respect to the ground state for each species by z', we obtain

$$K(T,V) = \frac{z'_{\mathrm{A}} e^{-E_{\mathrm{A}}/kT} z'_{\mathrm{BC}} e^{-E_{\mathrm{BC}}/kT}}{z'_{\mathrm{AB}} e^{-E_{\mathrm{AB}}/kT} z'_{\mathrm{C}} e^{-E_{\mathrm{C}}/kT}} = \frac{z'_{\mathrm{A}} z'_{\mathrm{BC}}}{z'_{\mathrm{AB}} z'_{\mathrm{C}}} \exp[-(E_{\mathrm{A}} + E_{\mathrm{BC}} - E_{\mathrm{AB}} - E_{\mathrm{C}})/kT]$$

$$= \frac{z_{\mathrm{A}} z_{\mathrm{BC}}}{z_{\mathrm{AB}} z_{\mathrm{C}}} \exp[-\Delta E/kT] \quad (7.86)$$

where in the last step the primes have been removed (using, from now on, as reference level for each of the species the ground state) and $\Delta E = (E_{\mathrm{A}}+E_{\mathrm{BC}}-E_{\mathrm{AB}}-E_{\mathrm{C}})$ represents the difference in ground state energies of the reactants and products.

Potential energy surfaces

The second concept we need is the potential energy surface. In Section 7.2 we noticed that the potential energy of a system containing atoms or molecules could be written as a function of special combinations of the nuclear spatial coordinates, usually referred to as *generalised* or *normal co-ordinates*. Generally the pictorial representation of the potential energy hyper-surface is difficult. The idea can be grasped from a simple example for which we take the collinear reaction between three atoms A, B and C. In Fig. 7.3 a map is given for this reaction with as axes the distances AB and BC, respectively. It shows two valleys separated by a col[bb]. In general, thermal fluctuation creates continuous attempts to pass from one valley to another. In order to calculate the rate constant for the chemical reaction AB+C ↔ A+BC we have to calculate all trajectories on the potential energy surface. With this in mind we need the concept of a *dividing surface*, defined as a surface, which cannot be passed without passing a barrier. In Fig. 7.3 e.g. one of the dividing surfaces is given by $R_{\mathrm{AB}} = R_{\mathrm{BC}}$. For the calculation of the rate constant we have to take into

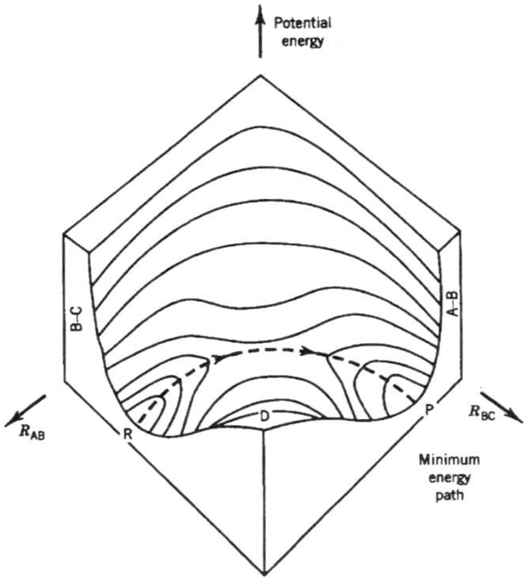

Fig. 7.3: The potential energy surface for the reaction AB+C ↔ A+BC showing in a col between two valleys.

[bb] Instead of a col, also a deeper basin or two cols with a small basin in between may separate the valleys. We limit further discussion to the case mentioned since the others can be derived from it.

account only those trajectories that do pass the dividing surface. An upper limit to the reaction rate constant $R(T)$ is then given by

$$R(T) \leq F(T) \tag{7.87}$$

where $F(T)$ represents the total flux of reactant systems crossing a dividing surface per unit volume and per unit time at a temperature T. Eq. (7.87) is known as the *Wigner variational theorem*. All statistical methods for computing reaction rates are based on this principle. The calculation of all allowed trajectories to estimate the flux F provides an enormous task if done in a rigorous way. Therefore approximate methods are usually introduced. Comparable to the variation theorem in quantum mechanics, this theorem provides the opportunity to make an optimum choice for the dividing surface, i.e. to pick that surface that provides the lowest flux F. This may be done by using a surface depending on several parameters and choosing these parameters in such a way that a minimum F is obtained.

The activated complex

The simplest approach along the above lines is to take into account only the trajectory following the minimum energy path from one valley to the other valley. In the case of the chemical reaction AB+C ↔ A+BC this means the path from the configuration represented by AB+C (point P in Fig. 7.3) to the configuration represented by A+BC (point R in Fig. 7.3). The coordinate along this path is called the *reaction coordinate*. The top of the col between the two valleys, actually a saddle point of the potential energy surface, is known as the *transition state*. A configuration in the neighbourhood of the transition state is addressed as an *activated complex*. Transition state theory is based on a number of assumptions. The first assumption is that the optimum dividing surface passes through the transition state and is perpendicular to the reaction coordinate. The second is that activated complexes are at all times in equilibrium with both the reactants and the products. For the forward reaction A+BC ↔ (ABC)*, where (ABC)* denotes the activated complex, this implies that the equilibrium constant K_{for} is given by

$$K_{for} = [(ABC)^*]/[A][BC] \tag{7.88}$$

where, as usual, [X] denotes the concentration of species X. The equilibrium constant K_{rev} for the reverse reaction AB+C↔ (ABC)* is

$$K_{rev} = [(ABC)^*]/[AB][C]$$

The combination of these two conditions together is called the *quasi-equilibrium assumption*. Finally, it is assumed that once the transition state is reached the reaction completes, i.e. the reactants do not return to the non-reacted state. We consider the forward reaction in some detail. Eq. (7.86) shows that the equilibrium constant for the example mentioned can be written as

$$K_{for} = \frac{z_{(ABC)^*}}{z_A z_{BC}} \exp(-\Delta E / kT) \tag{7.89}$$

The partition function of each species X can be (approximately) factorised as

$$z = z_{ele} z_{tra} z_{vib} z_{rot}$$

where z_{ele}, z_{tra}, z_{vib} and z_{rot} represent the electronic, translation, vibration and rotation partition functions, respectively. The contribution of z_{ele} reduces to the degeneracy number of the ground state, usually 1, since we assume the reaction to proceed on a potential energy surface, which exists only by virtue of the adiabatic assumption. Therefore no electronic transitions are allowed.

If we have sufficient information on the transition state, i.e. know the structure and relevant force constants, we can construct its partition function. The translation and rotation partition function do not provide a problem in principle, as they can be constructed as for normal molecules. The vibration partition function requires some care though. The usual normal coordinate analysis of an N-atom molecule can be done and from that we obtain $3N–6$ normal coordinates for a non-linear molecule or $3N–5$ normal coordinates for a linear molecule. Of these vibration coordinates all but one has a positive coefficient in the second-order terms in the potential energy expansion. The last, negative coefficient corresponds to an imaginary frequency for this vibration coordinate and represents the reaction co-ordinate. This implies that in the transition state a small fluctuation in the reaction coordinate leads to an unstable configuration with respect to this coordinate. It is customary to treat this coordinate as a translation coordinate over a small length δ at the top of the potential energy barrier with a partition function $z = (\delta/h)(2\pi m^*kT)^{1/2}$. However, since the concentration of activated complexes is actually due to both the forward and reverse reaction we need only half of this for the forward reaction. The complete partition function for the activated complex is then

$$z_{(\mathrm{ABC})^*} = (\delta/2h)(2\pi m^*kT)^{1/2}z^\# \tag{7.90}$$

where $z^\#$ represents the partition function for the remaining co-ordinates, i.e. the true vibration coordinates, the translation and rotation co-ordinates. Further m^* denotes the mass of the activated complex and k, T and h have their usual meaning.

The concentration of the activated complexes due to the forward reaction is given by Eq. (7.88) where the forward equilibrium constant is given by Eq. (7.89). Since there are [(ABC)*] complexes per unit volume which populate the length δ and which are moving forward, the forward reaction rate R_{for} is $R_{\mathrm{for}} = [(\mathrm{ABC})^*]/\tau$, where τ is the average time to traverse the length δ, given by $\tau = \delta/v_{\mathrm{ave}}$. Thus we need the average velocity v_{ave} over the length δ. Using the same approximation of a free translatory motion again, we borrow from kinetic gas theory

$$v_{\mathrm{ave}} = (2kT/\pi m^*)^{1/2} \tag{7.91}$$

Combining Eqs. (7.88), (7.89), (7.90) and (7.91) we obtain

$$R_{\mathrm{for}} = f[\mathrm{A}][\mathrm{BC}] \qquad \text{with} \qquad f \equiv \frac{kT}{h}\frac{z^\#}{z_{\mathrm{A}}z_{\mathrm{BC}}}\exp(-\Delta E/kT) \equiv \frac{kT}{h}K^\# \tag{7.92}$$

A similar expression is obtained for the reverse reaction and it is easily verified that the forward and reverse reactions are in equilibrium. The forward rate constant f thus can be calculated given the relevant information. However, for a first-principles calculation many pieces of information are required: the energy barrier for the reaction, the structure and the force constants associated with the dynamics of the reactants, products and activated complex. Typically this information is incompletely available. This is also true for other mechanisms although, of course, for other mechanisms than gas reactions different arguments apply for details. The exponential dependence on the barrier energy is generally valid, however, and rationalises the generally observed *Arrhenius-type behaviour*.

It remains to be discussed how to connect the results for chemical reactions to experimental data. Experimentally, it is often observed that, if the logarithm of the rate constant is plotted against the reciprocal temperature, a straight line is obtained. The gradient of this line is used to define the *activation energy* E_{act} by

$$\frac{\mathrm{d}\ln f}{\mathrm{d}(1/T)} \equiv -\frac{E_{act}}{R} \quad \text{or} \quad \frac{\mathrm{d}\ln f}{\mathrm{d}T} \equiv \frac{E_{act}}{RT^2} \tag{7.93}$$

To connect to the theory discussed above, the forward equilibrium constant K_{for} is expressed as $K_{for} = (\delta/2h)(2\pi m^* kT)^{1/2}K^{\#}$, so that f becomes

$$f = \frac{kT}{h}\frac{K_{for}}{(\delta/2h)(2\pi m^* kT)^{1/2}}$$

Taking logarithms and differentiating with respect to T yields

$$\frac{\mathrm{d}\ln f}{\mathrm{d}T} = \frac{\mathrm{d}\ln K_{for}}{\mathrm{d}T} + \frac{1}{2T}$$

Using the standard thermodynamic relation $\mathrm{d}\ln K_{for}/\mathrm{d}T = \Delta U/RT^2$, where ΔU is the change in internal energy, and Eq. (7.93) we obtain

$$\frac{E_{act}}{RT^2} = \frac{\Delta U}{RT^2} + \frac{1}{2T} \quad \text{or} \quad E_{act} = \Delta U + \frac{RT}{2}$$

This equation thus provides a link between the experimentally observed E_{act} and the internal energy change barrier ΔU. It should be noted that neither E_{act} nor ΔU is identical to ΔE, which appears in Eq. (7.89) because of the temperature dependence of the various partition functions. The differences are probably not very large.

Here our brief overview on transition state theory ends. We have chosen to discuss some of the basics and an example that is directly useful for any process that can be depicted as a chemical reaction. Diffusion, plastic deformation and fracture are important examples of these processes in materials science. It will be clear that for a detailed calculation a considerable amount of information is required.

Svante August Arrhenius (1859-1927)

Born in Vik, Sweden and educated at the Universities of Uppsala and Stockholm, where he received his doctoral degree in 1884. His thesis was poorly received and did not contain explicitly the suggestion of dissociation of weak electrolytes in solution. This idea came only after discussions with others, in particular Ostwald, and was published in 1887. After holding some teaching positions he was appointed professor of physics at Stockholm University in 1895. He mainly worked on electrolytic solutions but he could never accept that strong electrolytes are fully dissociated in solution and influenced by interionic forces, in spite of his own suggestion. He received the Nobel Prize for chemistry in 1903 in recognition of the extraordinary services he has rendered to the advancement of chemistry by his electrolytic theory of dissociation.

7.5 The transition to irreversible thermodynamics

The discussion of statistical mechanics so far has been limited entirely to reversible processes. For the treatment of irreversible phenomena, as abundantly present in thermomechanics, we need an extension of classical statistical mechanics as discussed in Section 7.3. Although there are several approaches possible, we limited ourselves to the simple approach by Ziegler[cc].

Let us reiterate what has been established so far in classical terms and elaborate a bit. A micro-system is described by a Hamilton function $H = H(\mathbf{p},\mathbf{q},\mathbf{a},\dot{\mathbf{a}})$ where $\mathbf{p}$ denotes the collective of momenta associated with the generalised co-ordinates $\mathbf{q}$. To account for external influences the deformation co-ordinates $\mathbf{a}$ and their rates $\dot{\mathbf{a}}$ have been introduced. Generally, we consider only quasi-static processes so that the rates $\dot{\mathbf{a}}$ can be neglected and then the expression for H reduces to $H = H(\mathbf{p},\mathbf{q},\mathbf{a})$. Examples of deformation co-ordinates are the volume V of a gas, the strains ε_{ij} for a solid or the electric field $\mathbf{E}$. The differential of H reads

$$\mathrm{d}H = \left(\frac{\partial H}{\partial \mathbf{p}}\cdot\dot{\mathbf{p}} + \frac{\partial H}{\partial \mathbf{q}}\cdot\dot{\mathbf{q}}\right)\mathrm{d}t + \frac{\partial H}{\partial \mathbf{a}}\cdot\mathrm{d}\mathbf{a}$$

Since the momenta and co-ordinates obey the Hamilton relations

$$\dot{\mathbf{p}} = -\frac{\partial H}{\partial \mathbf{q}} \quad \text{and} \quad \dot{\mathbf{q}} = \frac{\partial H}{\partial \mathbf{p}} \tag{7.94}$$

$\mathrm{d}H$ reduces to

$$\mathrm{d}H = \frac{\partial H}{\partial \mathbf{a}}\cdot\mathrm{d}\mathbf{a} \tag{7.95}$$

Hence the total energy of a micro-system depends only on the deformation variables $\mathbf{a}$ and is constant as long as these variables remain fixed.

The macro-system is described by an ensemble of micro-systems. Each of the micro-systems in the ensemble is represented by a representative point $\mathbf{p}$,$\mathbf{q}$ in Γ-space. Since the number of systems in the ensemble is assumed to be large the density can be considered as continuous and the behaviour of the macro-system is given by the density of representative points $\rho(\mathbf{p},\mathbf{q})$. The number of systems in a volume element $\mathrm{d}\mathbf{p}\mathrm{d}\mathbf{q}$ is given by $\rho(\mathbf{p},\mathbf{q})\mathrm{d}\mathbf{p}\mathrm{d}\mathbf{q}$. The behaviour of the macro-system is given by the motion of the representative points and can be considered as a flow in Γ-space of a fictitious fluid, sometimes known as phase fluid. The velocity of the flow, the phase velocity, is a vector in Γ-space with components $\dot{\mathbf{p}},\dot{\mathbf{q}}$. Conventionally it is assumed that the representative points are conserved, although the flow is not the one of a real fluid. This leads to the *continuity equation* for the phase fluid

[cc] Ziegler, H. (1963), page 91 in *Progress in Solid Mechanics* IV, I.N. Sneddon and R. Hill, eds., North-Holland, Amsterdam. See also Ziegler, H. (1970), Z. Angew. Math. Phys. **21**, 853.

$$\frac{\mathrm{d}\rho}{\mathrm{d}t}+\rho\left(\frac{\partial \dot{\mathbf{p}}}{\partial \mathbf{p}}+\frac{\partial \dot{\mathbf{q}}}{\partial \mathbf{q}}\right)=\frac{\partial \rho}{\partial t}+\frac{\partial \rho\dot{\mathbf{p}}}{\partial \mathbf{p}}+\frac{\partial \rho\dot{\mathbf{q}}}{\partial \mathbf{q}}=0 \tag{7.96}$$

completely analogous to the continuity equation of a real fluid (see Eqs. (4.35) and (4.36)). As usual, $\mathrm{d}\rho/\mathrm{d}t$ is the material rate of change obtained by an observer moving with the fluid and $\partial\rho/\partial t$ is the local rate of change obtained by a stationary observer. The velocity of the representative points is given by Eq. (7.94). It follows that

$$\frac{\partial \dot{\mathbf{p}}}{\partial \mathbf{p}}+\frac{\partial \dot{\mathbf{q}}}{\partial \mathbf{q}}=0 \tag{7.97}$$

i.e. the divergence of the phase velocity is zero, a result known as *Liouville's theorem*. From Eq. (7.96) we see that the phase fluid should be considered as incompressible.

Since Eq. (7.95) implies that the energy of a micro-system cannot change unless the deformation co-ordinates change, the macro-system can only exchange work with the surroundings and is thus an adiabatic system. However, also heat must be able to enter or leave the macro-system and for this the energy of the micro-system must change without changing the deformation co-ordinates. The usual way to solve this is to enlarge the micro-system so that it includes the surroundings. The micro-system plus surroundings then is again considered as a closed micro-system. Another way to realise heat exchange is by admitting the creation and annihilation of representative points due to change in $\mathbf{p}$ and $\mathbf{q}$ but still with a constant total number of micro-systems. These changes are indicated by the vectors $\boldsymbol{\alpha}$ and $\boldsymbol{\beta}$, respectively. In this case we replace the continuity equation by the *transport equation*

$$\frac{\mathrm{d}\rho}{\mathrm{d}t}+\rho\left(\frac{\partial \dot{\mathbf{p}}}{\partial \mathbf{p}}+\frac{\partial \dot{\mathbf{q}}}{\partial \mathbf{q}}\right)+\frac{\partial \boldsymbol{\alpha}}{\partial \mathbf{p}}+\frac{\partial \boldsymbol{\beta}}{\partial \mathbf{q}}=\frac{\partial \rho}{\partial t}+\frac{\partial \rho\dot{\mathbf{p}}}{\partial \mathbf{p}}+\frac{\partial \rho\dot{\mathbf{q}}}{\partial \mathbf{q}}+\frac{\partial \boldsymbol{\alpha}}{\partial \mathbf{p}}+\frac{\partial \boldsymbol{\beta}}{\partial \mathbf{q}}=0$$

According to this equation there are two types of transport in phase space. First, the flow (or convection) with velocity $\dot{\mathbf{p}},\dot{\mathbf{q}}$ and, second, the flux (or conduction) described by the vectors $\boldsymbol{\alpha}$, $\boldsymbol{\beta}$. Using the Liouville theorem we easily obtain

$$\frac{\mathrm{d}\rho}{\mathrm{d}t}+\frac{\partial \boldsymbol{\alpha}}{\partial \mathbf{p}}+\frac{\partial \boldsymbol{\beta}}{\partial \mathbf{q}}=\frac{\partial \rho}{\partial t}+\frac{\partial \rho}{\partial \mathbf{p}}\dot{\mathbf{p}}+\frac{\partial \rho}{\partial \mathbf{q}}\dot{\mathbf{q}}+\frac{\partial \boldsymbol{\alpha}}{\partial \mathbf{p}}+\frac{\partial \boldsymbol{\beta}}{\partial \mathbf{q}}=0$$

So far the development sketched was treated in Section 7.3 albeit in slightly different terms. We now turn to the extension to deal with processes.

In the way indicated above the micro-systems in the ensemble can change their energy by exchange work and heat but it is assumed that the density of representative points remains always close to the equilibrium one. For the calculation of a macroscopic value the phase average is used. For example, for the macroscopic (internal) energy U we have

$$U=\overline{H}=\int H\rho\,\mathrm{d}\mathbf{p}\mathrm{d}\mathbf{q} \tag{7.98}$$

Since the density ρ is a function of the energy of the micro-system we have to indicate the dependence on energy. In this connection it is convenient to introduce the so-called *index of probability* η given by $\eta = k\ln\rho$ where k denotes Boltzmann's constant. For a system with a constant number of particles the canonical ensemble is used with an index of probability given by

$$\eta=(F-H)/T$$

where F and T are parameters. The negative of the average index of probability is associated with the macroscopic entropy

$$S = -\overline{\eta} = -k \int \eta\rho \, \mathrm{d}\mathbf{p}\mathrm{d}\mathbf{q} = -k \int \rho \ln\rho \, \mathrm{d}\mathbf{p}\mathrm{d}\mathbf{q} \tag{7.99}$$

From these equations we obtain $S = (U - F)/T$, so that F and T can be interpreted as the Helmholtz energy and temperature, respectively.

We now consider the average of $\dot{\eta}$ and find

$$\overline{\dot{\eta}} = \int \dot{\eta}\rho \, \mathrm{d}\mathbf{p}\mathrm{d}\mathbf{q} = \int \frac{\mathrm{d}}{\mathrm{d}t}(\ln\rho)\rho \, \mathrm{d}\mathbf{p}\mathrm{d}\mathbf{q} = \int \dot{\rho} \, \mathrm{d}\mathbf{p}\mathrm{d}\mathbf{q} = \frac{\mathrm{d}}{\mathrm{d}t} \int \rho \, \mathrm{d}\mathbf{p}\mathrm{d}\mathbf{q} = 0$$

We also have $\dot{\overline{\eta}} = -\dot{S}$ and therefore differentiation with respect to time and averaging over an ensemble are *not* interchangeable.

To establish the relation between $\overline{\dot{H}}$ and $\dot{\overline{H}}$ we write $\eta = (F - H)/T$ as $H = F - T\eta$. On the one hand phase averaging yields

$$\overline{H} = U = F - T\overline{\eta} = F + TS \qquad \text{and therefore} \qquad \dot{\overline{H}} = \dot{F} + \dot{T}S + T\dot{S}$$

On the other hand differentiation with respect to time yields

$$\dot{H} = \dot{F} - T\dot{\eta} - \dot{T}\eta \qquad \text{and therefore} \qquad \overline{\dot{H}} = \dot{F} - T\overline{\dot{\eta}} - \dot{T}\overline{\eta} = \dot{F} + \dot{T}S$$

since $\overline{\dot{\eta}} = 0$ and $\overline{\eta} = -S$. Combining leads to

$$\blacktriangleright \qquad \dot{\overline{H}} = \overline{\dot{H}} + T\dot{S} \tag{7.100}$$

showing again that phase averaging and differentiation with respect to time are not interchangeable.

From $\eta = (F - H)/T$ we have $\rho = \exp[(F - H)/kT]$ and in view of the normalisation condition

$$\int \rho(\mathbf{p},\mathbf{q}) \, \mathrm{d}\mathbf{p}\mathrm{d}\mathbf{q} = 1$$

we obtain

$$\exp(-F/kT) = \int \exp(-H/kT) \, \mathrm{d}\mathbf{p}\mathrm{d}\mathbf{q}$$

Differentiation yields

$$\exp(-F/kT)\left[-\frac{1}{kT}\mathrm{d}F + \frac{F}{(kT)^2}\mathrm{d}(kT)\right]$$
$$= \frac{1}{(kT)^2}\mathrm{d}(kT) \int H\exp(-H/kT)\mathrm{d}\mathbf{p}\mathrm{d}\mathbf{q} - \frac{1}{kT}\mathrm{d}\mathbf{a} \cdot \int \frac{\partial H}{\partial \mathbf{a}} \exp(-H/kT)\mathrm{d}\mathbf{p}\mathrm{d}\mathbf{q}$$

Evaluating this expression a bit further one obtains

$$\mathrm{d}F - \frac{F}{kT}\mathrm{d}(kT)$$
$$= -\frac{1}{kT}\mathrm{d}(kT) \int H\exp\left(\frac{F-H}{kT}\right)\mathrm{d}\mathbf{p}\mathrm{d}\mathbf{q} + \mathrm{d}\mathbf{a} \cdot \int \frac{\partial H}{\partial \mathbf{a}} \exp\left(\frac{F-H}{kT}\right)\mathrm{d}\mathbf{p}\mathrm{d}\mathbf{q}$$
$$= -\frac{1}{kT}\mathrm{d}(kT)\,\overline{H} + \mathrm{d}\mathbf{a} \cdot \overline{\left(\frac{\partial H}{\partial \mathbf{a}}\right)}$$

Finally we obtain, using the internal energy $U=\overline{H}$, the (reversible or quasi-conservative) macro-forces $\mathbf{A}^{(\mathrm{q})}=\overline{(\partial H/\partial \mathbf{a})}$ and $S=(U-F)/T$

$$\mathrm{d}F-\frac{F}{kT}\mathrm{d}(kT)=-\frac{U}{kT}\mathrm{d}(kT)+\mathbf{A}^{(\mathrm{q})}\cdot\mathrm{d}\mathbf{a} \qquad \text{or} \qquad \mathrm{d}F=-S\mathrm{d}T+\mathbf{A}^{(\mathrm{q})}\cdot\mathrm{d}\mathbf{a} \tag{7.101}$$

From $F=U-TS$, taking the differential and eliminating dF we obtain

$$\mathrm{d}U=T\mathrm{d}S+\mathbf{A}^{(\mathrm{q})}\cdot\mathrm{d}\mathbf{a} \tag{7.102}$$

recovering the *Gibbs equation*. From the expressions for dF (dU) the macro-force and entropy (temperature) can be obtained by partial differentiation in the usual way.

The use of the superscript (q), indicating that the forces $\mathbf{A}^{(\mathrm{q})}$ are the quasi-conservative or reversible forces, can be justified as follows. On the one hand, from $\mathrm{d}H=(\partial H/\partial \mathbf{a})\cdot\mathrm{d}\mathbf{a}$ we have $\dot{H}=(\partial H/\partial \mathbf{a})\cdot\dot{\mathbf{a}}$ and substituting this expression together with $U=\overline{H}$ in $\dot{\overline{H}}=\overline{\dot{H}}+T\dot{S}$ we find $\dot{U}=\overline{(\partial H/\partial \mathbf{a})}\cdot\dot{\mathbf{a}}+T\dot{S}$. On the other hand, from $U=F+TS$ we obtain $\dot{U}=\dot{F}+\dot{T}S+T\dot{S}$ and comparing the result is $\dot{F}+\dot{T}S=\overline{(\partial H/\partial \mathbf{a})}\cdot\dot{\mathbf{a}}$. A process is called irreversible or reversible depending on whether entropy is produced $(\dot{S}\neq 0)$ or not $(\dot{S}=0)$. We thus may associate $\dot{F}+\dot{T}S=\overline{(\partial H/\partial \mathbf{a})}\cdot\dot{\mathbf{a}}$ with the reversible part and $T\dot{S}$ with the irreversible part. Consider now the first law of thermodynamics, in rate form given by $\dot{U}=\mathbf{A}\cdot\dot{\mathbf{a}}+\dot{Q}$ where the heat input rate $\dot{Q}$ is used. Decomposing the macro-forces in $\mathbf{A}=\mathbf{A}^{(\mathrm{q})}+\mathbf{A}^{(\mathrm{d})}$ we can write $\dot{U}=\mathbf{A}^{(\mathrm{q})}\cdot\dot{\mathbf{a}}+\mathbf{A}^{(\mathrm{d})}\cdot\dot{\mathbf{a}}+\dot{Q}$ and comparing this with $\dot{U}=\mathbf{A}^{(\mathrm{q})}\cdot\dot{\mathbf{a}}+T\dot{S}$ we have

$$T\dot{S}=\dot{Q}+\mathbf{A}^{(\mathrm{d})}\cdot\dot{\mathbf{a}} \qquad \text{or} \qquad \dot{S}\equiv\dot{S}^{(\mathrm{r})}+\dot{S}^{(\mathrm{i})}=\frac{1}{T}\dot{Q}+\frac{1}{T}\mathbf{A}^{(\mathrm{d})}\cdot\dot{\mathbf{a}} \tag{7.103}$$

where the reversible and irreversible parts of the entropy rate are introduced. The term $\Phi\equiv\mathbf{A}^{(\mathrm{d})}\cdot\dot{\mathbf{a}}$, often designated as *dissipation function*, is due to the flux in phase space.

In any given state of the macro-system Φ depends only on the velocities $\dot{\mathbf{a}}$. We noted that the velocity $\dot{\mathbf{a}}$ does not play an explicit role in the micro-system. This implies that small variations in $\dot{\mathbf{a}}$ do not affect the motion of the micro-system. This is not generally true for the macro-system but for variations that do not affect the expression $\Phi(\dot{\mathbf{a}})=\mathbf{A}^{(\mathrm{d})}\cdot\dot{\mathbf{a}}$, $\dot{S}=-\overline{\dot{\eta}}$ is the same. Therefore we require that the variation of the velocity $\dot{\mathbf{a}}+\delta\dot{\mathbf{a}}$ is also compatible with the dissipation function of the macro-system in its present state. More formally $\Phi(\dot{\mathbf{a}}+\delta\dot{\mathbf{a}})=\Phi(\dot{\mathbf{a}})$. Evaluating $\Phi(\dot{\mathbf{a}}+\delta\dot{\mathbf{a}})$ we obtain

$$\Phi(\dot{\mathbf{a}}+\delta\dot{\mathbf{a}})=\mathbf{A}^{(\mathrm{d})}\cdot\dot{\mathbf{a}}+\mathbf{A}^{(\mathrm{d})}\cdot\delta\dot{\mathbf{a}} \qquad \text{and} \qquad \Phi(\dot{\mathbf{a}}+\delta\dot{\mathbf{a}})\cong\Phi(\dot{\mathbf{a}})+\frac{\partial\Phi}{\partial\dot{\mathbf{a}}}\cdot\delta\dot{\mathbf{a}}$$

Comparing these two expressions we conclude that $\mathbf{A}^{(\mathrm{d})}\sim(\partial\Phi/\partial\dot{\mathbf{a}})$ or that the irreversible force is perpendicular to the momentary dissipation surface. Using $\Phi(\dot{\mathbf{a}})=\mathbf{A}^{(\mathrm{d})}\cdot\dot{\mathbf{a}}$ again we obtain the formal expression for the *orthogonality principle*

► $$\mathbf{A}^{(\mathrm{d})}=\left(\frac{\partial\Phi}{\partial\dot{\mathbf{a}}}\cdot\dot{\mathbf{a}}\right)^{-1}\Phi\frac{\partial\Phi}{\partial\dot{\mathbf{a}}} \tag{7.104}$$

Summarising, we can say that we recovered on the basis of statistical arguments the macroscopic description of irreversible thermodynamics including the orthogonality principle as given in Chapter 6. The statistical interpretation of the reversible and irreversible macro-forces has become clear. We have refrained from discussing complications due to gyroscopic forces but it can be shown that the final result remains the same. For large deviations from reversible behaviour the theory discussed above is insufficient but fortunately in many cases in thermomechanics it is.

7.6 Bibliography

General reference

Tolman, R.C. (1938), *The principles of statistical mechanics*, Oxford University Press, Oxford (also Dover, 1979).

Classical mechanics

Jeffreys, H. and Jeffreys, B. (1956), *Methods of mathematical physics*, 3rd ed., Cambridge University Press, Cambridge.

Goldstein, H. (1981), *Classical mechanics*, 2nd ed., Addison-Wesley, Amsterdam.

Lanczos, C. (1970), *The variational principles of mechanics*, 4th ed., University of Toronto Press, Toronto (also Dover, 1986).

Quantum mechanics

Pilar, F.L. (1968), *Elementary quantum chemistry*, McGraw-Hill, London.

McWeeny, R. and Sutcliffe, B.T. (1976), *Methods of molecular quantum mechanics*, Academic Press, London.

Merzbacher, E. (1970), *Quantum mechanics*, 2nd ed., Wiley, New York.

Schiff, L.I. (1955), *Quantum mechanics*, 2nd ed., McGraw-Hill, New York.

Statistical mechanics

Callen, H. (1985), *Thermodynamics and an introduction to thermostatistics*, Wiley, New York.

Fowler, R.H. and Guggenheim, E.A. (1936), *Statistical thermodynamics*, Oxford University Press, Oxford, UK.

Khinchin, A.I. (1949), *Mathematical foundations of statistical mechanics*, Dover, New York.

Landsberg, P.T. (1978), *Thermodynamics and statistical mechanics*, Oxford University Press, Oxford, UK (also Dover, 1990).

Reif, F. (1965), *Fundamentals of statistical and thermal physics*, McGraw-Hill, New York.

Schrödinger, E. (1952), *Statistical thermodynamics*, 2nd ed., Cambridge University Press, Cambridge (also Dover, 1989).

Slater, J.C. (1939), *Introduction to chemical physics*, McGraw-Hill, New York (also Dover, 1970).

Waldram, J.R. (1985), *The theory of thermodynamics*, Cambridge University Press, Cambridge.

8

Structure and bonding

After having discussed in the previous chapter some basic concepts of C, Q and S mechanics, we deal in this chapter with the structure and bonding. By structure we mean crystallographic, molecular and defect structure as well as microstructure, i.e. structural aspects in the widest sense. We start with some lattice concepts and deal subsequently with ideal (crystalline and non-crystalline) structures and chemical bonding. We continue with an overview of defects in structures and end with a summary of the geometrical description of microstructures.

8.1 Lattice concepts

In this section some important concepts related to crystalline lattices are presented that pervade all of solid-state science (Ziman, 1972; Seitz, 1940).

The direct lattice

In many cases we need a labelling of the atoms in a crystal. We recall that a crystal can be considered as a regular stacking of *unit cells*, whose (not necessarily orthogonal) non-coplanar basis vectors are denoted by $\mathbf{a}_1$, $\mathbf{a}_2$ and $\mathbf{a}_3$. In ideal solids

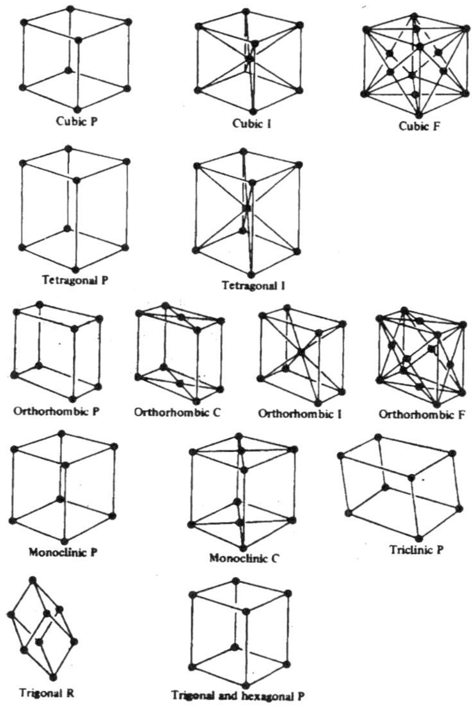

Fig. 8.1: The 14 Bravais lattices in 7 crystal systems.

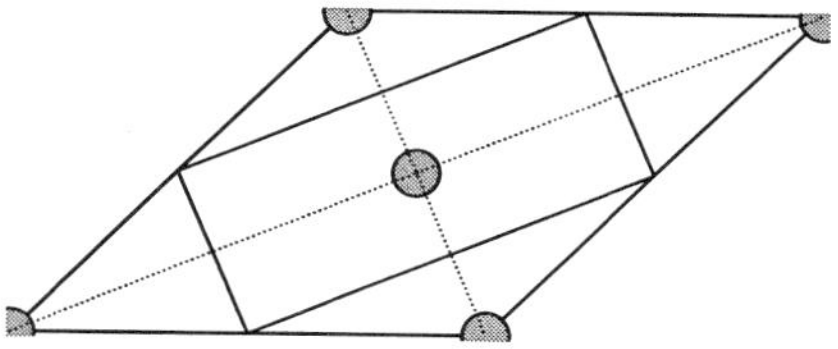

Fig. 8.2: The Wigner-Seitz cell in 2D.

each crystallographic unit cell contains the same amount of matter in the same configuration. To create a crystal structure one must arrange points in space in such a way that each of these points has an identical neighbourhood. Stacking of the unit cells in this way can be done in only 14 ways in 7 crystal systems, resulting in the *Bravais*[a] *lattices* (Fig. 8.1). If these points are actually occupied by atoms and no other atoms are present in the cell, we have a *primitive lattice*. In case more atoms occupy the unit cell, we speak of a *lattice with a basis*. If we have N unit cells, each with r atoms, the atoms are labelled according to the unit cell they occupy by the vector $\mathbf{n} = (n_1,n_2,n_3)$ and within the unit cell by α, β, … . The position of cell $\mathbf{n}$ is $\mathbf{r}_\mathbf{n} = n_1\mathbf{a}_1+n_2\mathbf{a}_2+n_3\mathbf{a}_3$ with $\mathbf{a}_1$, $\mathbf{a}_2$ and $\mathbf{a}_3$ the basis vectors while the position of atom α in cell $\mathbf{n}$ is denoted by $\mathbf{r}_\alpha$. The position of an atom with respect to the origin is thus $\mathbf{r}_{\mathbf{n}\alpha} = \mathbf{r}_\mathbf{n} + \mathbf{r}_\alpha$. The components of the position vector $\mathbf{r}_{\mathbf{n}\alpha}$ are $r_{\mathbf{n}\alpha i}$.

The reciprocal lattice

Instead of the conventional crystallographic unit cell, as discussed in the previous section, it is often convenient to take the *Wigner*[b]*-Seitz*[c] *cell*, i.e. the cell obtained by drawing the perpendicular bisectors of the translation vectors from a chosen centre to the nearest equivalent lattice site (Fig. 8.2). Using a basis of three non-coplanar unit vectors $\mathbf{a}_1$, $\mathbf{a}_2$ and $\mathbf{a}_3$, obviously different from the set used before, the volume of the Wigner-Seitz cell is given by $V_{\text{WS}} = \mathbf{a}_1\cdot(\mathbf{a}_2\times\mathbf{a}_3)$. Each lattice point can be addressed by a *lattice vector* $\mathbf{l} = l_1\mathbf{a}_1+l_2\mathbf{a}_2+l_3\mathbf{a}_3$, where l_1, l_2 and l_3 are integers. If $\mathbf{r}$ denotes a position in the zeroth cell, periodicity requires that for any function $f(\mathbf{r})$ it holds that $f(\mathbf{r}+\mathbf{l}) = f(\mathbf{r})$. We write $f(\mathbf{r})$ as

$$f(\mathbf{r}) = \sum\nolimits_{\mathbf{g}} A_{\mathbf{g}} \exp(\mathrm{i}\mathbf{g}\cdot\mathbf{r}) \tag{8.1}$$

where $A_\mathbf{g}$ is the gth Fourier component of the function $f(\mathbf{r})$ given by

$$A_{\mathbf{g}} = \frac{1}{V_{\text{WS}}}\int_{\text{cell}} f(\mathbf{r})\exp(-\mathrm{i}\mathbf{g}\cdot\mathbf{r})\,\mathrm{d}\mathbf{r}$$

The vector $\mathbf{g}$ is a *reciprocal lattice vector* for which the basis is given by

$$\mathbf{b}_1 = \frac{\mathbf{a}_2\times\mathbf{a}_3}{V_{\text{WS}}} \qquad \mathbf{b}_2 = \frac{\mathbf{a}_3\times\mathbf{a}_1}{V_{\text{WS}}} \qquad \text{and} \qquad \mathbf{b}_3 = \frac{\mathbf{a}_1\times\mathbf{a}_2}{V_{\text{WS}}} \tag{8.2}$$

---

[a] Auguste Bravais (1811-1863). French scientist who contributed to astronomy, meteorology, physics, botany and crystallography.

[b] Eugene Paul Wigner (1902-1995). Born in Hungary but American naturalised physicist who received the Noble Prize for physics in 1963 for his contributions to the theory of the atomic nucleus and the elementary particles, particularly through the discovery and application of fundamental symmetry principles.

[c] Frederick Seitz (1911-2…). American physicist who made important contributions to solid-state physics. His most influential book *The modern theory of solids* was published in 1940.

It is easy to see that $\mathbf{a}_i \cdot \mathbf{b}_j = \delta_{ij}$, i.e. the bases $\mathbf{a}_i$ and $\mathbf{b}_j$ form orthonormal sets. Any reciprocal lattice vector thus can be expressed as a linear combination of the basis vectors in reciprocal space, i.e. $\mathbf{g} = g_1\mathbf{b}_1+g_2\mathbf{b}_2+g_3\mathbf{b}_3$. Using Eq. (8.1) periodicity requires that $\exp[i\mathbf{g}\cdot(\mathbf{r}+\mathbf{l})] = \exp(i\mathbf{g}\cdot\mathbf{r})$ for any value of $\mathbf{g}$ or $\exp(i\mathbf{g}\cdot\mathbf{l}) = 1$ for all lattice vectors $\mathbf{l}$. This implies that $\mathbf{g}\cdot\mathbf{l} = 2\pi\cdot$integer or defining $\mathbf{g} = 2\pi\mathbf{h}$

$$\mathbf{g}\cdot\mathbf{l} = (g_1\mathbf{b}_1+g_2\mathbf{b}_2+g_3\mathbf{b}_3)\cdot(l_1\mathbf{a}_1+l_2\mathbf{a}_2+l_3\mathbf{a}_3)$$
$$= g_1l_1+g_2l_2+g_3l_3 = 2\pi(h_1l_1+h_2l_2+h_3l_3) = 2\pi\cdot\text{integer}$$

so that h_1, h_2 and h_3 are integers[d]. We note that

- each vector of the reciprocal lattice $\mathbf{g}$ is normal to a set of lattice planes $\mathbf{l}$,
- the volume of the Wigner-Seitz cell in reciprocal space, usually addressed as *Brillouin*[e] *zone* (BZ), is given by $V_B = (2\pi)^3/V_{WS}$ and
- the direct lattice is the reciprocal of its own reciprocal lattice.

Of course, a reciprocal lattice can also be defined using the crystallographic description of the previous section.

Bloch's theorem

Let us now turn our attention to the effect of translational invariance for the description of bonding. This invariance requires that $H(\mathbf{l}) = H(\mathbf{0})$ where $H(\mathbf{0})$ and $H(\mathbf{l})$ are the Hamilton operators before and after a lattice translation $\mathbf{l}$ is applied. Similarly $\Psi(\mathbf{r})$ and $\Psi(\mathbf{r}+\mathbf{l})$ denote the wavefunction before and after the application of a lattice translation $\mathbf{l}$. Since $H(\mathbf{0})\Psi(\mathbf{r}) = \varepsilon\Psi(\mathbf{r})$ and $H(\mathbf{l})\Psi(\mathbf{r}+\mathbf{l}) = \varepsilon\Psi(\mathbf{r}+\mathbf{l})$ we also have $H(\mathbf{0})\Psi(\mathbf{r}+\mathbf{l}) = \varepsilon\Psi(\mathbf{r}+\mathbf{l})$ where ε is the eigenvalue of $\Psi(\mathbf{r})$. Moreover $\Psi(\mathbf{r}+\mathbf{a}_1)$ is indistinguishable from $\Psi(\mathbf{r})$ and thus $\Psi(\mathbf{r}+\mathbf{a}_1) = \lambda\Psi(\mathbf{r})$. From normalisation we have $|\lambda|^2 = 1$ so that $\lambda = \exp(ik_1)$ where k_1 is a constant. Similar results arise for $\mathbf{a}_2$ and $\mathbf{a}_3$. Since for a general translation it holds that $\Psi(\mathbf{r}+\mathbf{l}) = \Psi(\mathbf{r}+l_1\mathbf{a}_1+l_2\mathbf{a}_2+l_3\mathbf{a}_3)$ we obtain

► $$\Psi(\mathbf{r}+\mathbf{l}) = \exp(i\mathbf{k}\cdot\mathbf{l})\Psi(\mathbf{r}) \quad (8.3)$$

where $\mathbf{k} = k_1\mathbf{b}_1+k_2\mathbf{b}_2+k_3\mathbf{b}_3$ is a vector in reciprocal space, the *wave vector*. This equation holds generally also for degenerate cases although here only indicated for a non-degenerate case. Eq. (8.3) represents *Bloch's theorem*[f], which is of significant importance throughout solid-state physics. It states that for any wavefunction satisfying the Schrödinger equation a wave vector $\mathbf{k}$ exists such that a translation by a lattice vector $\mathbf{l}$ is equivalent to a phase factor $\exp(i\mathbf{k}\cdot\mathbf{l})$. Hence we can label wavefunctions by their wave vector $\mathbf{k}$ and write

$$\Psi_{\mathbf{k}}(\mathbf{r}+\mathbf{l}) = \exp(i\mathbf{k}\cdot\mathbf{l})\Psi_{\mathbf{k}}(\mathbf{r}) \quad (8.4)$$

The wave vector $\mathbf{k}$ is only defined up to a reciprocal lattice vector and we take the smallest possible values as representatives. In this so-called *reduced zone scheme* we have $-\frac{1}{2}|\mathbf{b}_i| < k_i \le \frac{1}{2}|\mathbf{b}_i|$ and the wave vector is restricted to the (first) Brillouin zone.

To count the number of allowed $\mathbf{k}$-vectors and avoid surface effects *periodic boundary* (or *Born-von Kármán*) *conditions* are invoked. These conditions imply that the last cell in all three lattice directions is connected to the first, i.e. (Born and Huang, 1954)

---

[d] Sometimes the factor 2π is incorporated in the definition of the reciprocal lattice vectors.

[e] Léon Brillouin (1889-1969). French scientist who made many contributions to quantum science.

[f] Felix Bloch (1905-1983). American physicist who received the Nobel Prize in 1952 for the measure of the magnetic fields in atomic nuclei.

$\Psi(\mathbf{r}+L_1\mathbf{a}_1) = \Psi(\mathbf{r})$ $\quad\Psi(\mathbf{r}+L_2\mathbf{a}_2) = \Psi(\mathbf{r})$ $\quad$ and $\quad\Psi(\mathbf{r}+L_3\mathbf{a}_3) = \Psi(\mathbf{r})$

where L_i is the number of cells in the i-direction. Using Bloch's theorem we have $\Psi_\mathbf{k}(\mathbf{r}+L_1\mathbf{a}_1) = \exp(\mathrm{i}\mathbf{k}\cdot L_1\mathbf{a}_1)\Psi_\mathbf{k}(\mathbf{r})$ and similar relations for $\mathbf{a}_2$ and $\mathbf{a}_3$. This implies that $\exp(\mathrm{i}\mathbf{k}\cdot L_1\mathbf{a}_1) = \exp(\mathrm{i}\mathbf{k}\cdot L_2\mathbf{a}_2) = \exp(\mathrm{i}\mathbf{k}\cdot L_3\mathbf{a}_3) = 1$, which can only be obeyed if

$$\mathbf{k} = k_1\mathbf{b}_1 + k_2\mathbf{b}_2 + k_3\mathbf{b}_3 = 2\pi\left(\frac{m_1}{L_1}\mathbf{b}_1 + \frac{m_2}{L_2}\mathbf{b}_2 + \frac{m_3}{L_3}\mathbf{b}_3\right)$$

where m_1, m_2 and m_3 are integers. For the (first) Brillouin zone $-\frac{1}{2}L_i < m_i \le \frac{1}{2}L_i$ and because the number of cells in the crystal is $L_1\cdot L_2\cdot L_3 = N$, the number of $\mathbf{k}$-vectors is N. Since $V_{\mathrm{WS}} = V/N$, where V is the volume of the crystal, the volume per $\mathbf{k}$-vector is

$$\frac{V_\mathrm{B}}{N} = \frac{1}{N}\frac{(2\pi)^3}{V_{\mathrm{WS}}} = \frac{(2\pi)^3}{V}$$

In practice N is large and for later reference we note that the sum over $\mathbf{k}$-vectors can be replaced by an integral, i.e.

$$\sum\nolimits_\mathbf{k} \rightarrow \int \mathrm{d}\mathbf{k} = \frac{V}{(2\pi)^3}\iiint \mathrm{d}k_1\mathrm{d}k_2\mathrm{d}k_3$$

It can be shown that this result is independent of the shape of the crystal.

Problem 8.1

Show that $V_\mathrm{B} = (2\pi)^3/V_{\mathrm{WS}}$.

8.2 Crystalline structures

The crystal structure of many compounds is based on a limited number of relatively simple stackings. These are the simple cubic (SC), face centred cubic (FCC), body centred cubic (BCC) and hexagonal close packed (HCP) structures. Most metals actually crystallise in one of these structures. Inorganic materials usually crystallise in a more complex lattice. This is true for ionic as well as for covalent materials. Here a basic structure is the diamond structure. Molecules, small ones such as CO_2 and larger ones such as aromatic compounds, usually crystallise in a relatively simple lattice, e.g. in the orthorhombic lattice. The same is true for polymers but in this case crystallisation is usually incomplete.

The simplest structure one can consider is the *SC structure*. In this lattice atoms

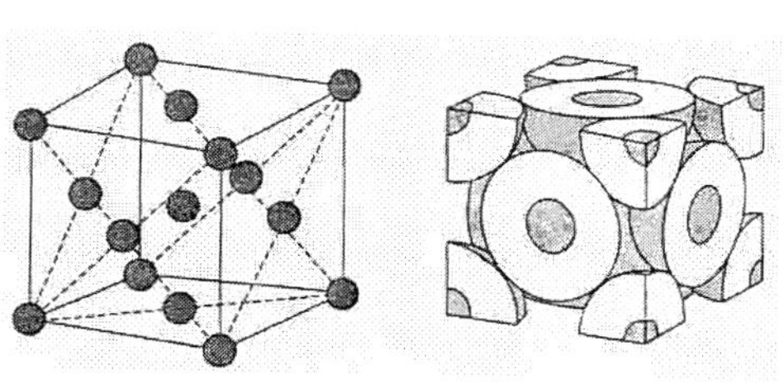

Fig. 8.3: The FCC structure.

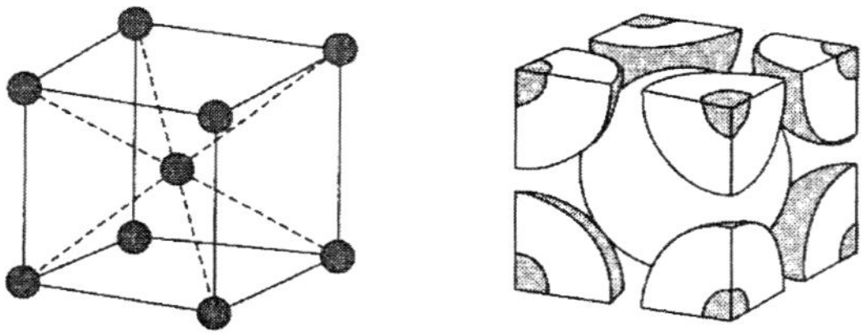

Fig. 8.4: The BCC structure.

are located at the corners of a cubic unit cell with edge length a. The lattice points are given by (n_1a,n_2a,n_3a) where n_1, n_2 and n_3 are integer. The only material known to crystallise in a SC structure is the element Po.

In the *FCC structure* (Fig. 8.3) the atoms are located at the corners of a cubic unit cell with edge length a and at the centres of its faces. Each atom has 12 equidistant nearest neighbours at a distance of $\frac{1}{2}a\sqrt{2}$. The six next nearest neighbours are located at a distance a. If one considers the atoms to be located at the corners of a polyhedron, the space inside that polyhedron is usually called (interstitial) *hole*. There are two types of holes in this lattice: an octahedral hole and a tetrahedral hole. The lattice points can be addressed by (n_1a,n_2a,n_3a) with integers n_1, n_2 and n_3 of which one or all are even, i.e. $n_1 + n_2 + n_3$ is even. The Bravais lattice points are generated by $\mathbf{l} = l_1\mathbf{a}_1+l_2\mathbf{a}_2+l_3\mathbf{a}_3 = l_1[\frac{1}{2}a(0+\mathbf{e}_y+\mathbf{e}_z)]+ l_2[\frac{1}{2}a(\mathbf{e}_x+0+\mathbf{e}_z)]+ l_3[\frac{1}{2}a(\mathbf{e}_x+\mathbf{e}_y+0)]$ where $\mathbf{e}_x$, $\mathbf{e}_y$ and $\mathbf{e}_z$ are the Cartesian unit vectors.

In the *BCC structure* (Fig. 8.4) the atoms are located at the corners of a cubic unit cell with edge length a and at the centre of that cube. Each atom has eight nearest neighbours at a distance of $\frac{1}{2}a\sqrt{3}$. The six next nearest neighbours are located at a distance a. Also in this lattice there are octahedral and tetrahedral holes. The lattice points can be denoted by either (n_1a,n_2a,n_3a) or $((n_1+\frac{1}{2})a, (n_2+\frac{1}{2})a, (n_3+\frac{1}{2})a)$. The first set represents the corners of the cubic unit cell and the second set the centres. The Bravais lattice points are generated by $\mathbf{l} = l_1\mathbf{a}_1+l_2\mathbf{a}_2+l_3\mathbf{a}_3 = l_1[\frac{1}{2}a(-\mathbf{e}_x+\mathbf{e}_y+\mathbf{e}_z)]+ l_2[\frac{1}{2}a(\mathbf{e}_x-\mathbf{e}_y+\mathbf{e}_z)]+ l_3[\frac{1}{2}a(\mathbf{e}_x+\mathbf{e}_y-\mathbf{e}_z)]$.

In the *HCP structure* (Fig. 8.5) the atoms are arranged in hexagonal layers such that each atom has six nearest neighbours in the same layer, three in the layer above and three in the layer below. The distance between the atoms in the layer is referred to as a and the height of the cell as c. The six nearest neighbour atoms in the layer above and below are also located at a distance a, if the ratio c/a has the value $(8/3)^{1/2}$ = 1.633. In that case the lattice is said to be *ideally close-packed.* Most HCP crystals have c/a ratios in the range from 1.56 to 1.63, i.e. below the ideal ratio. In the HCP lattice also octahedral and tetrahedral holes are present. The lattice points can be denoted by $(\frac{1}{2}n_1a, \frac{1}{2}\sqrt{3}n_2a, n_3c)$ and $(\frac{1}{2}(n_1+1)a, \frac{1}{2}(n_2+\frac{1}{3})\sqrt{3}a, (n_3+\frac{1}{2})c)$. The HCP lattice is *not* a Bravais lattice but a lattice with a basis in which one atom is at (0,0,0)

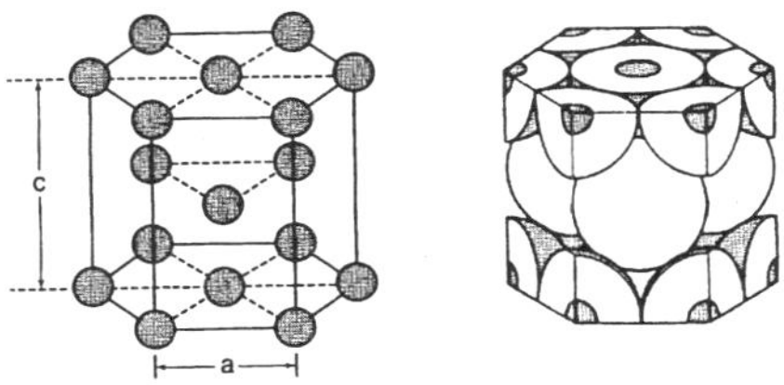

Fig. 8.5: The HCP structure.

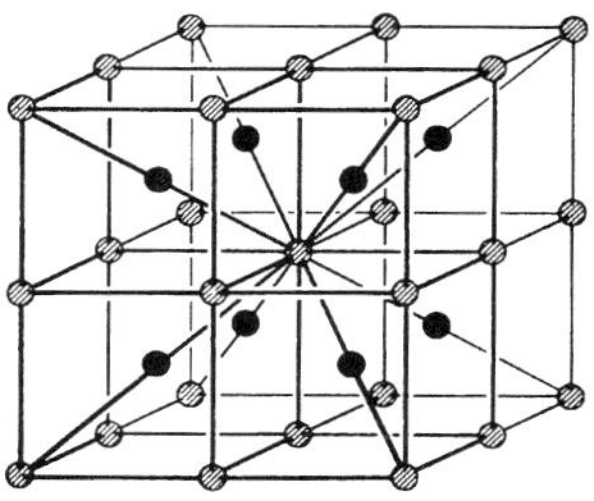

Fig. 8.6: The CsCl structure.

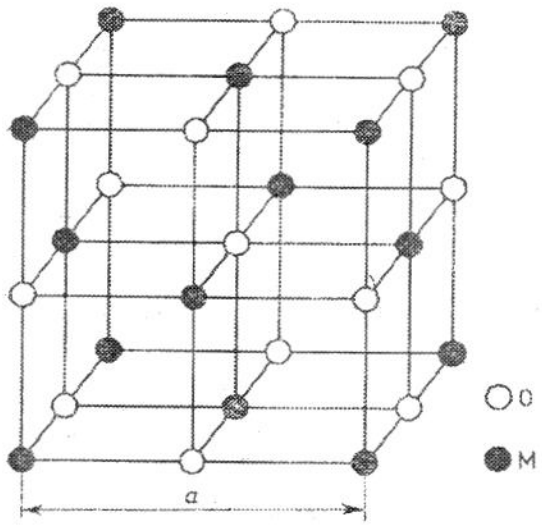

Fig. 8.7: The NaCl structure.

while the other is at (⅓,⅔,½).

Both FCC and HCP lattices may be seen as a stacking of hexagonally close-packed layers. While in the HCP lattice the lattice stacking is ABAB…, in the FCC lattice the stacking sequence is ABCABC… .The energy difference between these two lattice types is small (according to their lattice sums, see Section 8.6) and therefore stacking faults, a deviation from the normal stacking pattern, easily occur.

Similar descriptions can be given for more complex lattices, usually containing more than one atom per unit cell. We provide only two examples. The first is the CsCl structure (Fig. 8.6). In essence this is a body-centred lattice in which the alternate lattice points are occupied by positive and negative ions, respectively. The sublattice of each type of ion is the SC lattice. As the second example we mention the NaCl structure (Fig. 8.7). This is a simple cubic lattice whose lattice points are alternately occupied by the positive and negative ions, respectively. The sublattice of each type of ion is the FCC lattice. Apart from metal halides such as NaCl and KCl, oxides also crystallise in this structure, e.g. MgO. For a complete description of these and other lattices we refer to e.g. Greenwood (1968).

The structure of covalently bonded materials is generally highly complex. An important basic type is the diamond lattice in which each atom is bonded in a tetrahedral configuration with another atom (Fig. 8.8). As is well known using solely C atoms this leads to the hardest and stiffest existing material, diamond. This tetrahedral configuration can be realised by covalent bonding of sp^3 hybridised C atoms. The structure can also be described as connected network of puckered six-rings of carbon atoms with a distance of 0.154 nm in the so-called chair configuration. The structure of many compounds is a variation on this basic structure. We mention only β-SiC, where each Si atom is coordinated by four C atoms and vice versa.

The van der Waals interaction is often important in layered structures, the most

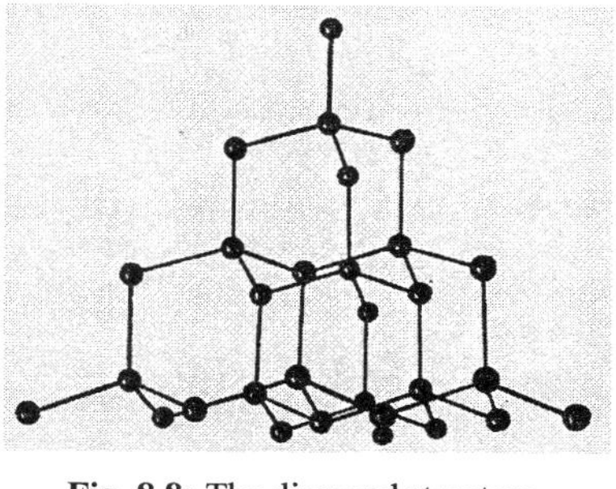

Fig. 8.8: The diamond structure.

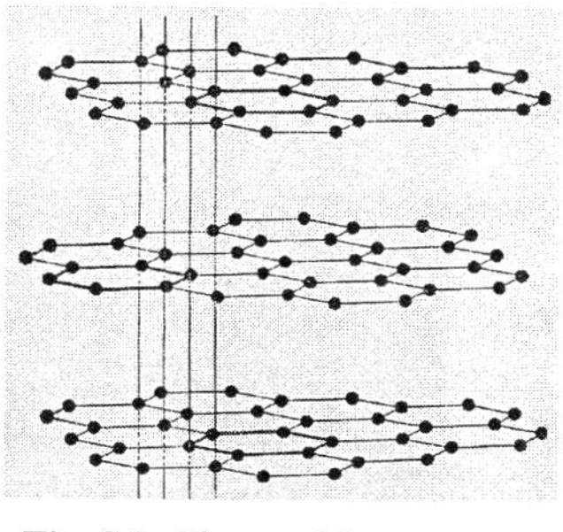

Fig. 8.9: The graphite structure.

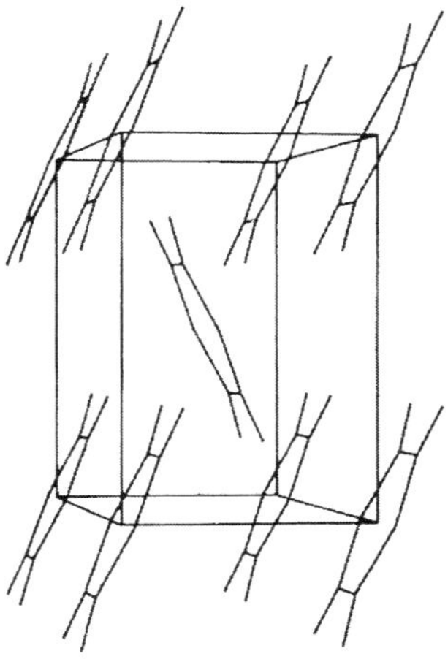

Fig. 8.10: The crystallographic structure of pyrazine.

well-known example of which is graphite (Fig. 8.9). In this compound sp^2 hybridised C atoms bond covalently to form layers while the bonding between the layers is provided by the van der Waals interaction. The layers can also be described by a connected network of flat six-rings of C atoms with a distance of 0.142 nm. The distance between the layers is about 0.335 nm. Due to the relatively weak van der Waals interaction graphite is easily deformed. Moreover, the thermal as well as electrical conductivity is high in the layers but low perpendicular to the layers.

Crystallisation can also result solely due to the van der Waals interaction. The most well-known example is provided by the noble gas crystals. They generally crystallise in the FCC lattice. The van der Waals interaction is also responsible for the bonding in so-called *molecular crystals*. As an example we take pyrazine[g], $C_4H_4N_2$, which crystallises in an orthorhombic unit cell (space group P_{mnn}), as illustrated in Fig. 8.10. The molecules occupy the corners of the unit cell and the centre. Owing to the weak van der Waals interaction a molecular crystal is usually relatively soft and shows a low melting point, in this case about 33 °C. For this particular crystal it has been predicted from crystal considerations that upon contraction of the *c*-axis, i.e. lowering the temperature, the *a*-axis expands. This is due to the rotation of the molecules to a more aligned configuration. Indeed the thermal expansion at room temperature is given by $\alpha_a = -14\times10^{-6}$ K^{-1}, i.e. small negative, while $\alpha_b = 114\times10^{-6}$ K^{-1} and $\alpha_c = 246\times10^{-6}$ K^{-1}, i.e. both large positive. Generally the unit cell for molecular crystals has a low symmetry, the most common space group being $P2_1/c$.

Also hydrogen bonding plays its role. An interesting example is oxamide[h], $C_2H_4N_2O_2$, whose crystallographic structure is shown in Fig. 8.11. The space group is $P\bar{1}$. All molecules are planar within experimental accuracy and this plane deviates only about 0.65° from the *b*-*c* crystallographic plane. The molecular 'sheets' are hydrogen bonded while the van der Waals interaction is present between the sheets. It will come as no great surprise that the optical anisotropy, as reflected in the refractive index difference Δn, is quite large, i.e. $\Delta n \cong 0.3$, and that the thermal expansion is

[g] de With, G., Harkema, S. and Feil, D. (1976), Acta Cryst. B **32**, 3178; de With, G. (1976), J. Appl. Cryst. **9**, 502.

[h] de With, G. and Harkema, S. (1977), Acta Cryst. B **33**, 2367; de With, G. (1977), J. Appl. Cryst. **10**, 353.

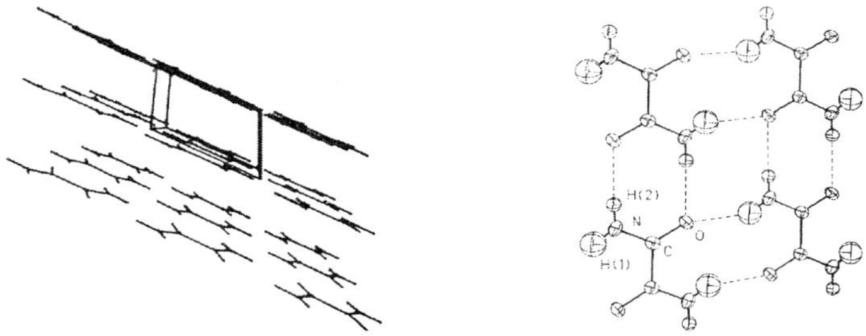

Fig. 8.11: The crystallographic structure of oxamide.

mainly perpendicular to the sheets. The thermal expansion tensor at room temperature is given by $\alpha_a = 207\times10^{-6}\ K^{-1}$ versus $\alpha_b = 27\times10^{-6}\ K^{-1}$ and $\alpha_c = -3\times10^{-6}\ K^{-1}$, respectively. Hydrogen bonding can also occur in three dimensions, the most well known example being ice (solid H_2O). The structure resembles the diamond structure in the sense that each H atom is linked with the lone pair electrons of an O atom of a neighbouring molecule.

Problem 8.2

Calculate the second nearest-neighbour distance for the FCC and BCC lattices.

8.3 Non-crystalline structures

Apart from crystalline solids considerable interest exists in non-crystalline solids, i.e. solids lacking strict translational invariance. The simplest deviation is the *modulated structure*, which can be described by underlying periodic lattices with the actual atomic positions displaced with respect to these lattices. The displacement is also periodic in space. In case the ratio of lattice and displacement periods is rational, one speaks of *commensurate structures*. In case that ratio is non-rational we have *incommensurate structures*. In the former case the overall structure is still periodic but with a larger period while in the latter case the overall structure becomes non-periodic. These concepts are illustrated in Fig. 8.12.

While the above-mentioned structures are still based on lattices, we also have *quasi-crystals*, which are non-crystalline materials with perfect long-range order but with no 3D periodicity ingredient whatsoever[i]. The initial compounds of this class,

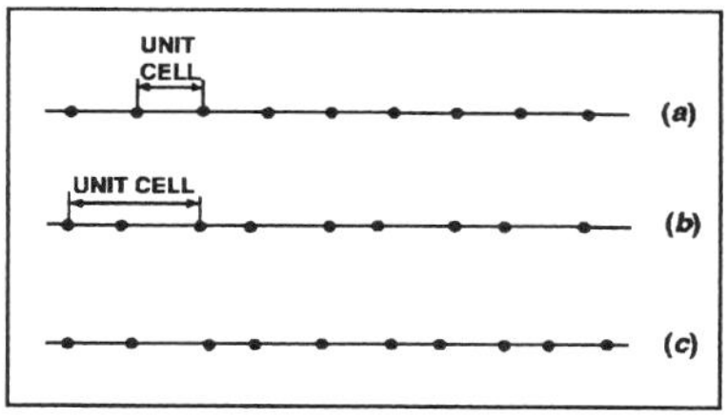

Fig. 8.12: A periodic chain of atoms (a) transfers into a longer period structure upon commensurate modulation (b) but into a non-periodic structure upon incommensurate modulation (c).

[i] For more details, see Janot, C. (1994), *Quasicrystals* , 2nd ed., Oxford Science Publications, Oxford.

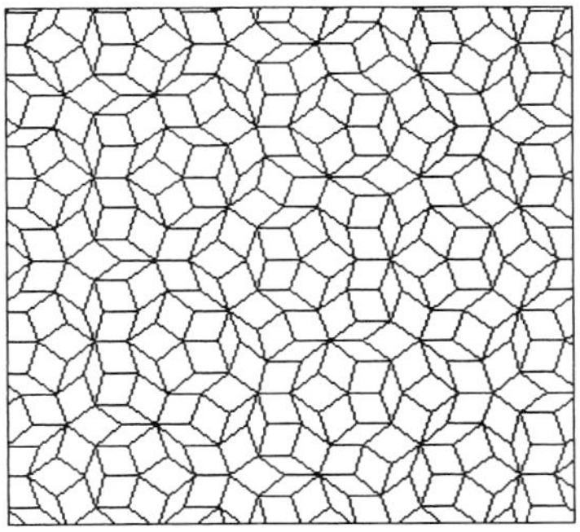

Fig. 8.13: Two-dimensional Penrose tiling showing two rhombi with acute angles of 72° and 36°.

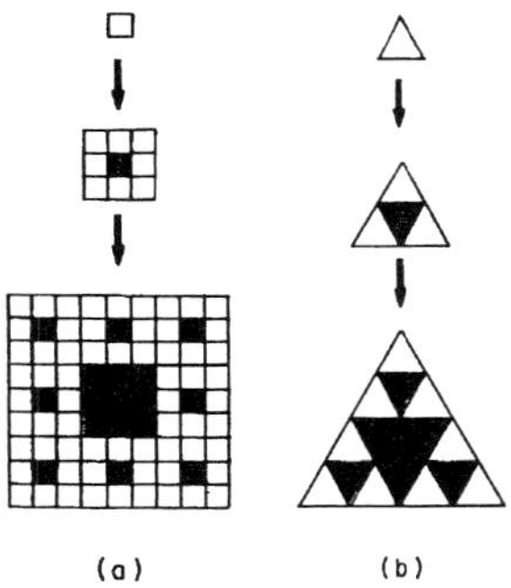

Fig. 8.14: The construction of a square and triangular Sierpinsky carpet.

such as rapidly quenched $MnAl_6$ alloy as discovered in 1984 and the more stable alloys AlCuFe and Al_6CuLi_3 discovered somewhat later, exhibited five-fold symmetry in their diffraction patterns. Since a 3D periodic lattice cannot show five-fold (or seven-fold) rotations, these materials are not periodic, yet show orientational order. A 2D mathematical example is the Penrose tiling (Fig. 8.13), which is an aperiodic structure showing five-fold rotation symmetry, and that can be realised by an appropriate arrangement of two rhombi with acute angles of 72° and 36°, respectively.

As another type of non-crystalline structures we mention *fractals*. These structures are self-similar, i.e. the structure looks identical at various length scales. A good daily life example is a tree with a trunk, which splits into branches, which on their turn split again in twigs, etc. A 2D mathematical example is the Sierpinsky carpet, which is constructed by repetitive extension of a basic unit and filling in of the resulting structure as illustrated in Fig. 8.14. In real materials self-similarity is restricted to a certain range of length scales. The basic unit from which the structure is build determines the lower limit. The upper limit is determined by either the specimen size or inhomogeneity in the specimen. The most well-known real material showing a fractal structure is silica aerogel, with a density of about 0.1 g/cm^3, and which can be

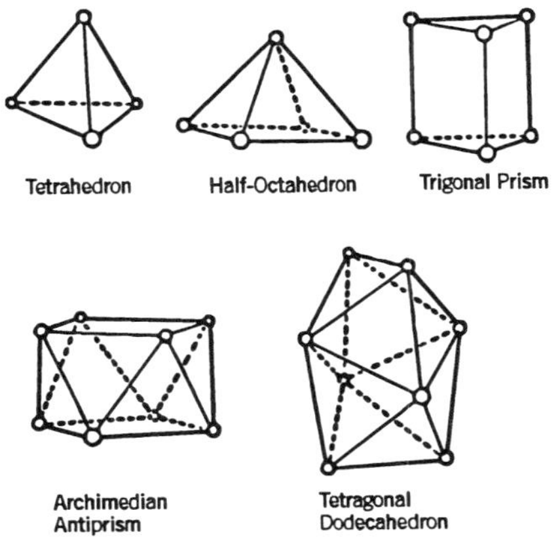

Fig. 8.15: The five canonical holes of Bernal[j].

[j] Bernal, J.D. (1964), Proc. Roy. Soc. A **280**, 299.

envisaged as an ever-branching tree of connected silica particles.

However, from our point of view the most important non-crystalline structures are the *amorphous structures* in which the atomic packing is irregular throughout the solid. We distinguish between structures based on atomic or small molecule packing and on long chain molecule packing. While the first type of structures occurs mainly in inorganics and, to a lesser extent, in metals, the second type occurs mainly for polymers. The characteristics of first type are discussed below while the next section deals with polymer characteristics.

Amorphous inorganics and metals

The simplest amorphous structures occur in metals, in which a more or less random close packing of (atomic) spheres exists. These materials are sometimes addressed as metallic glasses or *metglasses*. A useful description of this random packed structure for pure metals, as first presented by Bernal[k], is given in terms of polyhedral shapes of the interstitial holes. Five basic types of holes could be discerned (Fig. 8.15), sometimes referred to as the *canonical holes*. The actual holes are somewhat distorted but a random packing contains a certain volume fraction of each type (Table 8.1). Their frequency represents the frozen structure. The average coordination number of each atom is about 11.2, to be compared with a coordination number of 12 in the FCC lattice. This already points at another way to characterise a glassy structure, namely the distribution of the volume per atom. In an ideal crystal the volume per atom is fixed. In an amorphous material there are some atoms with a

Table 8.1: Relative frequency of canonical holes.

Hole	Frequency (%)	Hole	Frequency (5)
Tetrahedron	60	Archimedian antiprism	3
Half-octahedron	30	Tetragonal dodecahedron	2
Trigonal prism	5		

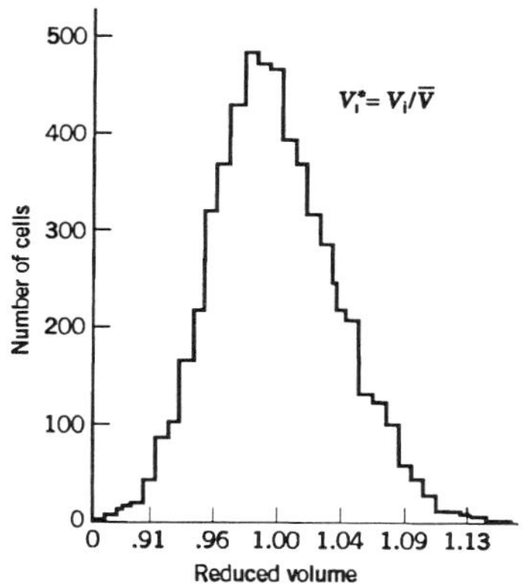

Fig. 8.16: The volume per atom distribution in an atomic glass.

[k] James Desmond Bernal (1901-1971), prominent Irish scientist, who did pioneering work in X-ray crystallography. Later he did pioneering work in social studies of science or 'science of science'. Being a marxist in philosophy and a communist in politics, he led a complicated life, sitting on hundreds of committees and playing a leading role in many scientific and political organisations. He also led a somewhat unconventional domestic life of a notoriously non-monogamous nature.

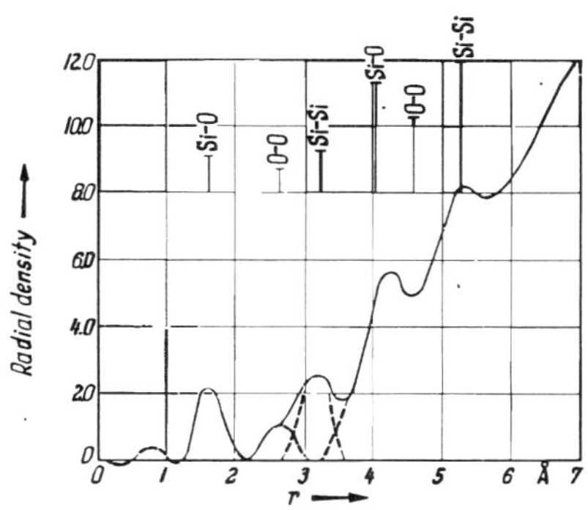

Fig. 8.17: The RDF for SiO_2.

surrounding for which the volume per atom is less than in a reference close-packed crystal while others have a larger volume and there is a considerable width for this distribution (Fig. 8.16). On average the volume per atom is larger than the volume per atom in the reference crystal, corresponding to a lower coordination number. The excess volume is referred to as the *free volume*. Of course, these types of descriptions become more complex in alloys.

The last way to characterise the random closed-packed solid we will mention is the *radial distribution function* (RDF). In a RDF the number of atoms in a spherical shell around a reference atom is given. If $\rho(r)$ denotes the number density, the RDF $N(r)$ is given by

$$N(r) = 4\pi \int \rho(r) r^2 \, dr \tag{8.5}$$

The average distance between the reference atom and its first co-ordination shell can easily be detected as well as the next nearest distances. The number of co-ordinating atoms is related to area under the first co-ordination shell peak while the correlation or coherence length is the length where essentially the peaks disappear. In alloys and inorganic glasses more than one type of atoms is present so that different types of distances can be discerned. This complicates the interpretation again considerably. A typical example showing the RDF of amorphous silica as determined by X-ray diffraction is given in Fig. 8.17. In this case the Si–O, the O–O and the Si–Si atomic distances of the first coordination shell can be discerned as well as some further distances. The structure of inorganic glasses is also often described as a connected network of co-ordination shells of anions around a cation. We refer to the literature for a discussion.

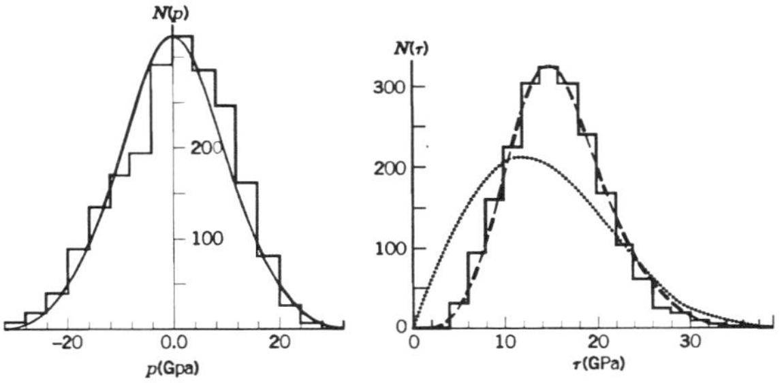

Fig. 8.18: The mean and equivalent stress distribution in an atomic glass.

Finally, we want to emphasise the following point. Although in a random packed structure each atom is in mechanical equilibrium, because of the disorder in the structure, the atoms are under a large balance of local forces as they interact with the surroundings. In fact, an atomic site stress tensor can be defined. As for the normal stress tensor the mean stress σ_{ave} and equivalent stress σ_{equ} can be calculated. The calculated distribution for σ_{ave} and σ_{equ} for an atomic glass is shown in Fig. 8.18[1]. Obviously the magnitude of both the mean stress and the equivalent stress can be huge. Similar distributions have been calculated for polymeric glasses.

8.4 Polymer characteristics

The characterisation of polymers requires a different approach. In Chapter 1 we have noted that polymers consist of long chains of covalently bonded atoms. The lengths of the molecular chains are not all equal and can be described by either the number or the weight distribution. Denoting the number distribution by $p(M)$, the number of molecular weights between M and M+dM is given by $p(M)\mathrm{d}M$. The *number average* of the molecular weight M_n is given by (Gedde, 1995)

$$M_\mathrm{n} = \frac{\int p(M)M\,\mathrm{d}M}{\int p(M)\,\mathrm{d}M} = \int p(M)M\,\mathrm{d}M$$

where the second step can be made since $p(M)$ is assumed to be normalised. Strictly speaking we should treat the distribution as discrete but since the degree of polymerisation involved is generally very high the difference with a continuous distribution is negligible. The *weight average* M_w is given by

$$M_\mathrm{w} = \frac{\int p(M)M^2\,\mathrm{d}M}{\int p(M)M\,\mathrm{d}M}$$

and is always larger than M_n. Since the second moment of the distribution function is given by

$$\langle\Delta M^2\rangle = \int p(M)[M - M_\mathrm{n}]^2\,\mathrm{d}M = \int p(M)M^2\,\mathrm{d}M - M_\mathrm{n}^2 = M_\mathrm{w}M_\mathrm{n} - M_\mathrm{n}^2$$

we have

$$U \equiv \frac{\langle\Delta M^2\rangle}{M_\mathrm{n}^2} = \frac{M_\mathrm{w}}{M_\mathrm{n}} - 1$$

and the *polydispersity coefficient* U describes the width of the distribution. In practice, instead of U, the ratio M_w/M_n is often used as an indicator for the width. The shape of $p(M)$ can vary widely, dependent on the polymerisation process. For the *step-growth mechanism* (see Chapter 1) typically wide distributions with $M_w/M_n \cong 1.5$-2 result, empirically often well described by the *gamma distribution*, but in polymer science often referred to as the *Flory-Schulz* (or *Schulz-Zimm*) *distribution*. The gamma distribution density in terms of the number of monomers in the chain N is given by

[1] Egami, T. and Vitek, V. (1983), page 127 in *Amorphous materials: Modeling of structure and properties*, V. Vitek, ed., AIME, New York.

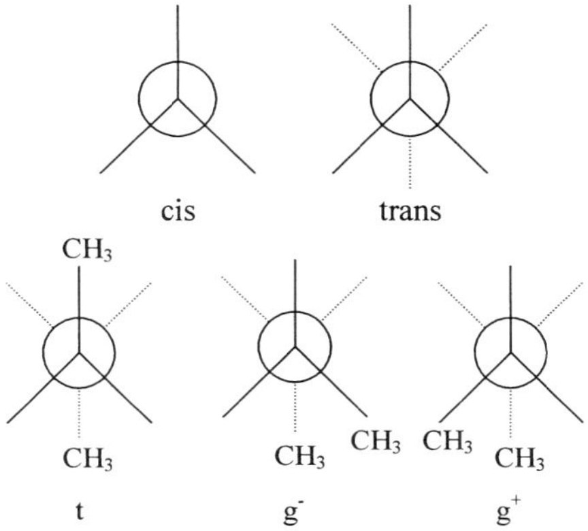

Fig. 8.19: The *cis* and *trans* conformations in ethane (upper) and the t, g^+ and g^- conformation in butane (lower), as shown by the Newman projection.

$$p(N) = \frac{1}{\Gamma(\beta)}\left(\frac{\beta}{N_n}\right)^{\beta} N^{\beta-1}\exp(-\frac{\beta N}{N_n})$$

where β is a shape parameter, N_n denotes the average number of monomers and Γ indicates the gamma function. For this distribution it holds that $U = 1/\beta$ and in practice a value of $\beta \cong$ 1-2 often results in a good fit. For the *chain-growth mechanism* (see Chapter 1) usually narrower distributions are obtained, typically described by the *Poisson distribution* given by

$$p(N) = \exp(-N_n)\frac{(N_n)^N}{\Gamma(N+1)} \cong \exp(-N_n)\left(\frac{N_n e}{N}\right)^N$$

and fully characterised by the average N_n. For this distribution it holds that $U = 1/N_n$ and thus effectively becomes monodisperse for high N_n values. The molecular weight distribution is of importance for all properties of polymers and can in fact be used to optimise the material behaviour. For further discussion, see Boyd and Phillips (1993).

After having discussed the molecular weight distribution we turn to the molecular conformation. Briefly indicated in Chapter 1 are the *gauche* and *trans* conformations. To elaborate a bit, consider first the central bond between two C atoms in ethane, C_2H_6. Fig. 8.19 shows the two extremes in conformation, namely *cis* and *trans*, in a view along the C–C bond axis. In ethane three equivalent minimum energy or *trans* conformations are present. To rotate the two CH_3 groups with respect to each other, energy has to be spent and an energy barrier exists between two *trans* states. Substituting on each C atom one H atom by a CH_3 group, so that we get butane, C_4H_{10}, the equivalence between the *trans* states is lost and we obtain one *trans* (t) conformation and two equivalent (g^+, g^-) *gauche* conformations with dihedral angles $\phi = 0°$ and $\phi = +120°$ and $\phi = -120°$, respectively, for the minimum energy conformations (Fig. 8.19). Continuing with substitution of end H atoms with CH_3 groups results in polyethylene (PE). Although the details for each C–C bond for this molecule may slightly differ, one *trans* and two *gauche* conformations are present for each C–C bond. They all have to be specified for a complete description of the molecule. For PE the lowest energy conformation is the all-*trans* conformation with a zig-zag structure of the C–C bonds.

This is no longer true for other polymers where the H atoms have been replaced by other atoms or groups. Consider for concreteness polytetrafluorethylene (PTFE) where all H atoms have been replaced by F atoms. Since the F atoms are larger than the H atoms, the non-bonded repulsive interactions between CF_2 groups of second nearest carbon atoms become much more important (Fig. 8.20) and repulsive energy can be gained by rotating a bit along the C–C axis of each bond. Of course, this increases the bond rotation energy and in this way equilibrium is reached. For the case of PTFE an optimum dihedral angle of $\phi \cong 16.5°$ is obtained. The result of all this is that the molecule forms a helix along its axis in which the positions of the side groups (the F atoms in the case of the PTFE) rotate along the molecular axis. After n screws along the axis the position of the mth monomer regains the position of the first monomer apart from a shift along the axis. Described in this way we refer to them as m/n helices. For example, PE has a 2/1 helix and PTFE has a 13/6 helix below 19 °C and a 15/7 helix above 19 °C. The description is not as exact as it appears though, since the 'periodicity' along the chain may vary slightly (Boyd and Phillips, 1993).

Due to the regular chain structure of these types of molecules, they crystallise, if cooled down either from the melt or from solution. Since the cross-section of such molecules is more or less rectangular, they tend to crystallise in an orthorhombic crystal structure. A feature closely related to the helix structure is *polymorphism*, i.e. more than one crystal structure can be observed. As an example we take polyoxymethylene (POM, $[-CH_2-O-]_n$). For this molecule the *gauche* conformation is the most stable. The energy difference between *gauche* and *trans* is about 8 kJ/mol. The all-*gauche* conformation ($\ldots g^+g^+g^+\ldots$ or $\ldots g^-g^-g^-\ldots$) with a torsion angle of 60° generates a 2/1 helix with the aforementioned rectangular cross-section leading to an orthorhombic unit cell. However, a small change in the torsion angle to about 77° leaves the chain essentially in an all-*gauche* conformation but leads to a 9/5 helix with a more or less circular cross-section. This leads to hexagonal packing (Fig. 8.21).

So we see that polymers may crystallise given sufficient regularity along the chain, which usually implies linear, isotactic or syndiotactic molecules (see Chapter 1). Generally crystallisation is incomplete though, i.e. amorphous regions exist between the crystallites, contrary to inorganics and metals. The origin of this effect can be found in the chain-like nature of polymers, which generally precludes full orientation of all the molecules. In fact, the amorphous region typically contains the non-regular parts, e.g. the chain ends, the defective parts of the chains and the crossovers to other crystals. The density is correspondingly between the theoretical density of the crystals and that of the amorphous regions and on X-ray diffraction patterns, apart from relatively sharp diffraction rings, also diffuse halos appear. Originally this semi-crystalline behaviour was described by the fringed micelle model (Fig. 8.22). In this model the molecular chains alternate between regions of order (the crystallites) and disorder (the amorphous regions). The lateral dimensions of the

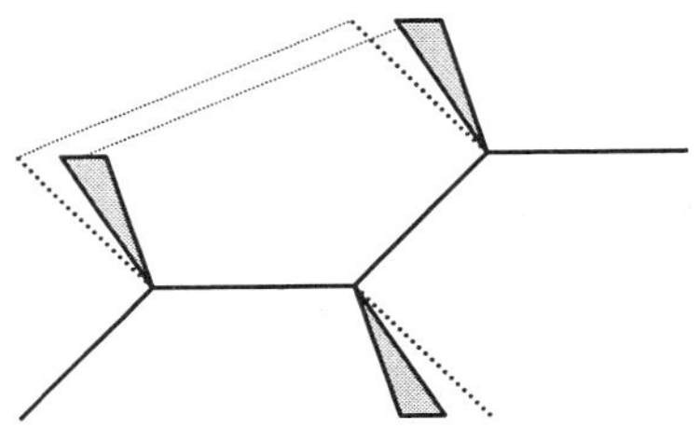

Fig. 8.20: Non-bonded interactions between CX_2 groups of the second nearest C atoms.

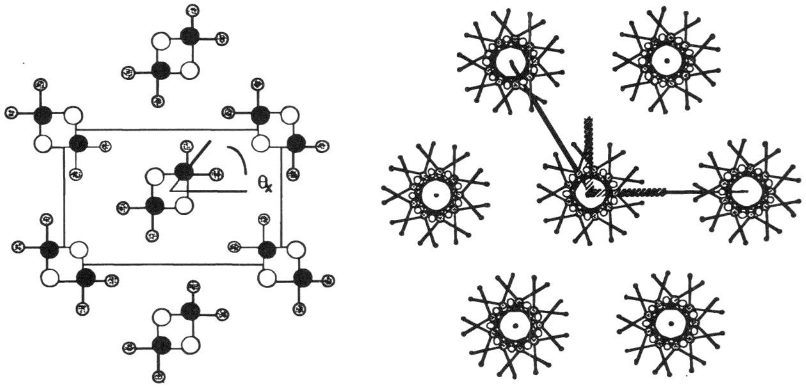

Fig. 8.21: Orthorhombic packing for the 2/1 helix (left) and hexagonal packing for the 9/5 helix (right) of polyoxymethylene as seen along the axis of the helix.

crystals so formed can be several tens of micrometres while the thickness is about 10 nm. In view of this shape these crystallites are often referred to as *lamellae*. Later developments, in particular using solution grown crystals, suggest that lamellae are formed with regular folds (Fig. 8.22). The question whether the folds at the surface of lamellae are sharp and regular or that there is some deviation from regular re-entrance, is complex. From small angle scattering the end-to-end distance in solution grown crystals has been determined to be much smaller than in the liquid, leading to a high 'regular fold' fraction. From infrared measurements an estimate of 75% regular folds in linear PE was made, essentially in agreement with scattering data. However, the end-to-end distance upon melt crystallisation is not dramatically changed and therefore it seems logical to conclude that the entangled structure of the melt is largely preserved in the semi-crystalline state. Nevertheless, also in this case a high fraction of regular folds is present as e.g. can already be assessed from density measurements. Likely, in general there is a (varying) mixture between the 'pure' micellar and 'pure' lamellar structure, dependent on crystallisation conditions and type of polymer.

Typically in a melt-crystallised polymer the lamellae are organised further and the description of this organisation, addressed as *morphology*, contains generally three levels (Gedde, 1995):

- the lamellae of folded chains,
- stacks of nearly parallel lamellae separated by amorphous material and
- superstructures the most important one of which is the spherulite.

A *spherulite* (Fig. 1.12) is a part of the material in which all the lamellar stacks have grown radially leading to a spherical shape. This feature requires a mechanism for

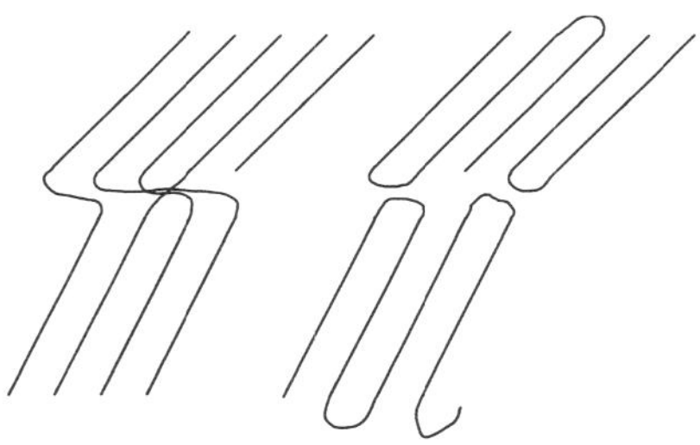

Fig. 8.22: The fringed micelle and regular fold model of lamellae in polymer crystals.

branching and splaying of the lamellae, which is different for different polymers. While for linear PE a screw dislocation is suggested as the initiating factor, for polystyrene (PS) a sheaf-like central initiating part is identified.

After this excursion to crystallinity of polymers, we turn to the amorphous state. So far we have discussed only isotactic or syndiotactic molecules but in many syntheses atactic molecules result. These molecules, like the isotactic and syndiotactic molecules, have their own preferred conformation, depending on the sequence of preferred local conformations along the chain. This irregularity of the sequence leads to a coil-like conformation, different for each molecule since the sequence is different for each molecule, and prevents regular packing in a lattice. Atactic polymers are thus generally *amorphous*. We deal with amorphous solids obtained from polymer melts only but in order to discuss them we have to say a few words on solutions. After that we deal with various aspects of long chain coils, ending with the equivalent chain.

A polymer molecule, whether in solution or in the melt, can be characterised by the *end-to-end distance*[m] r. In a good solvent the polymer-solvent attractions prevail, the coil expands and r increases. The effective monomer-monomer interaction is always repulsive. In a poor solvent the polymer-polymer interactions, irrespective of whether they are due to parts from the same or from different chains, prevail. The coil shrinks and r decreases until the effective monomer-monomer repulsion due to excluded volume forces sets in. Under certain conditions the intramolecular interactions are similar in magnitude to the intermolecular interactions. In other words the enthalpy and entropy contributions from solvent-monomer and monomer-monomer interactions to the Helmholtz energy of the assembly of molecules under consideration compensate and one part of the molecule does not seem to 'notice' other parts of the molecule. The molecules behave like phantoms and are sometimes referred to as phantom chains. The temperature for which this happens is the *Flory temperature* θ and one speaks of *theta conditions*. Under theta conditions the coil neither shrinks nor expands and has unperturbed dimensions. If $\langle r^2\rangle$ denotes the mean of r^2 (second moment), the influence of the solvent can be described by

$$\langle r^2\rangle = \alpha\langle r^2\rangle_\theta \tag{8.6}$$

where the subscript denotes the theta conditions and α a parameter dependent on solvent, temperature and molecular weight. At the Flory temperature $T = \theta$, theta conditions hold and $\alpha = 1$. Important for solids is now the *Flory theorem*: in a dense polymeric system theta conditions prevail. Describing theta conditions as the configuration where intra- and intermolecular interactions compensate and since the 'solvent' is the polymer melt itself, the theorem is highly plausible. Rephrasing, on the one hand, the monomers of a certain reference chain are subjected to a repulsive potential due to the excluded volume effect of its own monomers and this leads to an expansion of the coil. On the other hand, the other chains, interpenetrating the reference chain, generate a counteracting attractive potential acting inwards on the reference chain and under theta conditions the two effects cancel leading to (pseudo)-unperturbed chains. Small angle neutron scattering experiments support the theorem. Since we deal only with solids we omit the subscript θ in $\langle r^2\rangle_\theta$ from now on.

Focusing on the chains themselves, a first estimate of the end-to-end distance[n] is made via the *freely jointed chain* model: n bonds, each of length l, connected without

[m] Sometimes also the root mean square distance of the atoms from the centre of gravity, the *radius of gyration* s, is used. It holds that $\langle s^2\rangle = \langle r^2\rangle/6$.

[n] For derivations we refer to the literature, see e.g. Gedde (1995), which we have taken as guide.

any restriction. The probability distribution of the end-to-end vectors for long chain molecules is described by the random walk model resulting in

$$P(\mathbf{r})\mathrm{d}\mathbf{r} = \left(\frac{3}{2\pi nl^2}\right)^{3/2} \exp\left[-\frac{3r^2}{2nl^2}\right]\mathrm{d}\mathbf{r} \tag{8.7}$$

For such a model chain one obtains, in the limit of a large number of atoms,

$$\langle r^2\rangle = \int r^2 P(\mathbf{r})\,\mathrm{d}\mathbf{r} = 4\pi\int_0^\infty r^4 P(\mathbf{r})\,\mathrm{d}r = nl^2$$

where $\langle r^2\rangle = \langle x^2\rangle + \langle y^2\rangle + \langle z^2\rangle$ is the mean square end-to-end distance of the chains. The end-to-end distance $r = \langle r^2\rangle^{1/2}$ is thus proportional to $n^{1/2}$.

However, we know that the bonds are not freely connected but have a certain bond angle τ. Leaving the bonds otherwise unrestricted we obtain the *freely rotating chain* model for which it holds in the limit of a large number of bonds that

$$\langle r^2\rangle = nl^2\left[\frac{1-\cos\tau}{1+\cos\tau}\right] \tag{8.8}$$

As expected the square root dependence on n is preserved but the proportionality factor is changed. For sp^3 hybridised carbon atoms, e.g. in a polyethylene (PE) chain, with a bond angle of $\tau = 109.5°$, we have approximately $\langle r^2\rangle = 2.0nl^2$.

A further improvement is obtained by introducing the *independent hindered rotation model*, i.e. a rotating chain but with a preferential orientation for the dihedral (bond rotation) angle ϕ between two groups connected by a bond. In this model one obtains

$$\langle r^2\rangle = nl^2\left[\frac{1-\cos\tau}{1+\cos\tau}\right]\left[\frac{1+<\cos\phi>}{1-<\cos\phi>}\right] \tag{8.9}$$

Again the square root dependence on n is preserved and the proportionality factor changes. For the PE chain we have one *trans* (t) configuration with a dihedral angle $\phi = 0°$ and two equivalent *gauche* (g^+,g^-) configurations with a dihedral angle $\phi = 120°$ and $\phi = -120°$, respectively (compare Fig. 8.19). The latter have a higher energy by an amount E_{gau}. Denoting the Boltzmann factor by $\sigma = \exp(-E_{gau}/RT)$ we obtain for the average dihedral angle

$$\langle\cos\phi\rangle = \frac{1+\sigma\cos(120°)+\sigma\cos(-120°)}{1+\sigma+\sigma} = \frac{1-\sigma}{1+2\sigma} \tag{8.10}$$

For the end-to-end distance we thus have

$$\langle r^2\rangle = nl^2\left[\frac{1-\cos\tau}{1+\cos\tau}\right]\left[\frac{2+\sigma}{3\sigma}\right] \tag{8.11}$$

For PE at 140 °C, using $E_{gau} = 2.1$ kJ/mol, we find $\sigma = 0.54$ leading to $\langle r^2\rangle \cong 3.4nl^2$.

Finally we recognise that the hindered rotation around a bond is correlated and this is taken into account in the *correlated hindered rotation model*. The final expression becomes

► $$\langle r^2\rangle = Cnl^2 \tag{8.12}$$

where the *characteristic ratio C* is a function of the correlation of the rotations along the chain and therefore a measure of the stiffness of the chain. For PE Flory

Table 8.2: Values for characteristic ratio C for various polymers.

Material	C	Material	C
PEO	4.0	a-PMMA	8.4
PE	6.7	i-PMMA	10
a-PS	10.0	s-PMMA	7
i-PS	10.7	PVC	13
a-PP	5.5	a-PVAc	8.9
i-PP	5.8	PDMS	6.2
s-PP	5.9	a-PiB	6.6

PE polyethylene, PEO polyoxyethylene, PS polystyrene, PP polypropylene, PMMA poly(methyl methacrylate), PVC poly(vinyl chloride) PVAc poly(vinyl acetate), PDMS poly(dimethylsiloxane), PiB poly(isobutylene), a atactic, i isotactic, s syndiotactic.

calculated, taking into account the correlation up to two bonds away, that $C = 6.7 \pm 0.2$, in good agreement with experiment. For other polymers other values of C are obtained (see Table 8.2).

The above considerations lead to the introduction of the *equivalent chain*, in which a real chain, containing n correlated and rotation hindered bonds of length l, is described as a freely jointed chain of n' segments of length l'. Each of the *segments* thus represents a number of real bonds but since the correlation along the chain is limited to a few bonds, these segments can be considered as freely jointed. Hence for this description we match $\langle r^2\rangle$ with $n'l'^2$ and the maximum projected length of the chain r_{max} with $n'l'$. This can be done in a unique way leading to $l' = \langle r^2\rangle/r_{max}$ and $n' = r_{max}^2/\langle r^2\rangle$. Let us take again PE as an example. The maximum projected length of the PE chain r_{max} is $r_{max} = nl \sin(\tau/2) \cong 0.83\ nl$ and using $\langle r^2\rangle = 6.7nl^2$ leads to $l' \cong 8l$ and $n' \cong 0.1n$. The segment thus contains about 10 (real) bonds and its length, often addressed as *Kuhn length,* reflects the stiffness of the molecular chain. In discussing the properties of polymers frequent use is made of the equivalent chain model. For other polymers, of course, other equivalent lengths are obtained.

Finally, we have to say a few words about *cross-linking* or the *network formation* of polymers, i.e. the formation of bonds at certain points along a particular molecular chain to neighbouring chains. Cross-linking can be random or controlled implying the formation of bonds at random points or at well-controlled points along the chain. Such a bond is often referred to as a *junction* and the part of the original chain between junctions is called a *sub-chain* (or *network chain*). We denote the total number of chains and junctions in a network by ν and μ, respectively. The parameter M_{sub} indicates the molecular weight of a sub-chain, while the number of chains meeting at a junction is called the *functionality*, indicated by ϕ. A chain connected to a junction at only end is a *dangling chain* and the one that is attached to the same junction at both ends is called a *loop* (see also Section 8.12). A dangling chain does not contribute to the elasticity of the network, neither does a loop that is not penetrated by another chain that itself elastically active. A network with no dangling bonds or loops and no junctions with functionality less than 3 is a *perfect network.*

A network can be thought as being formed in two steps. In the first step all chains are joined at the junctions to macromolecule in the form of a *tree*. There are $\nu+1 \cong \nu$ junctions in such a tree. To some of the junctions chains are connected that can react with one another in the second step to form a network. In this process the number of junctions is reduced to $\nu + 1 - \xi \cong \nu - \xi$, where we introduced the number of independent paths ξ, generally addressed as *cycle rank* of the network. It is the number of bonds that has to be cut to change the network to a tree. For an ideal network of the five parameters that characterise the network (ν, μ, M_{sub}, ϕ and ξ) only two are independent. It can be shown that

$$\mu = 2\nu/\phi \qquad \xi = (1 - 2/\phi)\nu \qquad \text{and} \qquad \xi/V_0 = (1 - 2/\phi)\rho N_A/M_{sub}$$

where V_0 is the volume of the network in the formation stage, ρ is the corresponding density and N_A is Avogadro's number. For imperfect networks we have to identify the *active* (or *effective*) chains and junctions. Flory defined an active chain as the one that contributes to the elasticity of the network and related their number ν_{eff} to the cycle rank ξ by $\nu_{eff} = 2\xi$. It appears that general expressions relating ν_{eff} to other network parameters are not available at present. However, for an imperfect tetrafunctional network the number of effective chains is (see Chapter 9) $\nu_{eff} = \nu(1 - 2M_{sub}/M)$, where M is the average molecular weight of the primary molecules.

To conclude we note that for an elastomer a typical sub-chain contains between 100 and 1000 skeletal bonds. Below 100 bonds the material is likely to be a thermoset while above 1000 bonds long times are required to reach equilibrium under load. In a typical elastomer a sub-chain containing 500 bonds has a root mean square end-to-end distance $\langle r^2\rangle^{1/2}$ of about 7 to 8 nm. Such a domain contains about 40 cross-links and the associated chains. Thus a sub-chain shares its available space with many other sub-chains, resulting in entanglements permanently trapped in the network structure.

Paul Flory (1910-1985)

Born in Sterling, Illinois, Flory earned a doctorate in physical chemistry from the Ohio State University in 1934. He then went to work for DuPont, where he became involved in polymer chemistry under the direction of Carothers. Since Carothers was an organic chemist, Flory's abilities in physical chemistry and mathematics complemented well those of his mentor. Flory worked with Carothers to develop the basic principles of polymerization kinetics and the statistics of molecular mass distribution in polymer samples, among other things. A year after the death of Carothers in 1937, Flory left DuPont for an academic career, returning to industry for a few years during World War II to work on the development of synthetic rubber for the war effort. In postwar academia he continued to develop his theories on the conformation of polymer chains in solution and produced his well-known book *Principles of Polymer Chemistry*, which is still the classic reference for polymer chemistry. For the rest of his career Flory worked out rigorous mathematical theories on the thermodynamics of polymer solutions and of rubber elasticity, and statistical treatment of polymer-chain conformations. He received the Nobel Prize in chemistry in 1974 for his fundamental achievements, both theoretical and experimental, in the physical chemistry of the macromolecules, an honor which he felt would and should have gone to Wallace Carothers had he lived longer. Always a person of conscience, Flory used the prestige of the award to campaign for international human rights, especially with regard to the treatment of scientists in the Soviet block.

Problem 8.3

Show that for PE $l' \cong 8l$ and $n \cong 0.1\, n'$.

8.5 Bonding in solids

Having described in broad terms the configuration of atomic/molecular structure, we now turn our attention to bonding. We present a brief outline of Hartree-Fock theory and start with the formal theory as applicable to all types of systems, i.e. atoms, molecules and solids. In a later stage we specialise to solids and discuss two extremes, namely the nearly free electron approximation and the tight-binding approximation. Finally density functional theory is briefly addressed.

General theory

Even using the Born-Oppenheimer approximation an exact solution for molecules exists only for the H_2^+ ion-molecule, albeit a (partial) numerical one. For bonding in many-electron molecules or for solids no exact solutions are known. The description of bonding in molecules and solids relies heavily on orbital theories. Making allowance for Pauli's principle, a many-electron wavefunction can be expanded in anti-symmetrised products of one-electrons wavefunctions (or *spin orbitals*) ϕ_j, each ϕ_j consisting of a spatial part (or *orbital*) χ_j and a spin function[o] $\sigma_j = \alpha$ or $\sigma_j = \beta$. A convenient form for such an anti-symmetrised product is the *Slater*[p] *determinant*, in shorthand for N electrons written as (Sutton, 1993; Seitz, 1940)

$$|\phi_a(1)\,\phi_b(2)\ldots\phi_n(N)| \equiv \frac{1}{\sqrt{N!}}\begin{vmatrix} \phi_a(1) & \phi_a(2) & \ldots & \phi_a(N) \\ \phi_b(1) & \phi_b(2) & \ldots & \phi_b(N) \\ \ldots & \ldots & \ldots & \ldots \\ \phi_n(1) & \phi_n(2) & \ldots & \phi_n(N) \end{vmatrix} \tag{8.13}$$

and where (i) refers to the co-ordinates of an electron including the spin.

An electronic wavefunction must also be an eigenfunction of the total spin operator S^2. The determinants, constructed from the same orbitals, can be grouped in such a way that each group is an eigenfunction of S^2. Such a group is called a *configuration* Δ. Approximating the wavefunction by more than one configuration is denoted by *configuration interaction* (CI). Hence we can write

$$\Psi_i^{(e)}(\mathbf{x};\mathbf{X}) = \sum\nolimits_j C_{ij}\Delta_j \tag{8.14}$$

If all possible configurations for a given set of spin orbitals are included one speaks of complete CI, otherwise of a limited CI. The choice of the type of orbitals determines the further development. Choosing spin orbitals that are centred throughout a molecule, e.g. on the various nuclei, one obtains what is called *molecular orbital* (MO) theory. In solids this becomes the crystal orbital theory although it is usually called the *tight-binding approximation*. Within this choice, one frequently tries to

---

[o] Sometimes the spin function is indicated by the overbar notation for orbitals using an overbar for the function β and no overbar for the function α, e.g. $\chi\alpha \to \chi$ and $\chi\beta \to \bar{\chi}$.

[p] John Clarke Slater (1900-1976). American physicist who did work on the application of quantum mechanics to the chemical bond and the structure of substances.

describe the system with only one configuration. For a closed shell ground state this configuration contains only one determinant.

$$\Psi_i^{(e)}(\mathbf{x};\mathbf{X}) = \Delta_i = |\phi_a(1)\,\phi_b(2)\ \dots\ \phi_n(N)| \tag{8.15}$$

for which the total energy $E = E_{ele}+E_{nuc}$ in Ha is given by

$$E = E_{ele} + E_{nuc} = \left[2\sum\nolimits_i h_i + \sum\nolimits_{i,j}(2J_{ij} - K_{ij})\right] + \tfrac{1}{2}\sum\nolimits_{A,B}\frac{Z_A Z_B}{R_{AB}} \quad \text{where} \tag{8.16}$$

$$h_i = \langle\phi_i\,|-\tfrac{1}{2}\nabla^2(1) - \sum\nolimits_A \frac{Z_A}{r_{1A}}\,|\,\phi_i\rangle \tag{8.17}$$

$$J_{ij} = \langle\phi_i^*(1)\phi_j^*(2)\,|\,\frac{1}{r_{12}}\,|\,\phi_i(1)\phi_j(2)\rangle \quad \text{and} \tag{8.18}$$

$$K_{ij} = \langle\phi_i^*(1)\phi_j^*(2)\,|\,\frac{1}{r_{12}}\,|\,\phi_j(1)\phi_i(2)\rangle \tag{8.19}$$

In these equations r_{1A} denotes the distance between electron 1 and nucleus A with charge Z_A while r_{12} and R_{AB} denote the distances between electrons 1 and 2 and nuclei A and B, respectively. The first term in Eq. (8.16) indicates the sum over the contributions of the electrons in the field of the nuclei, the second the Coulomb repulsion between electrons, the third the exchange interaction between electrons (due to the antisymmetry of the wavefunction) and the last term the nuclear repulsion between the various nuclei A and B. The notation for a Coulomb integral J and exchange integral K is often abbreviated to $J = \langle ij|ij\rangle$ and $K = \langle ij|ji\rangle$, respectively.

Applying the variation principle to the electronic energy expression to determine the orbitals of the one-determinant approximate wavefunction under the constraint $\langle\phi_i|\phi_j\rangle = \delta_{ij}$ leads to the *Hartree-Fock*[q] *self-consistent field* (HF-SCF) equations

$$f(1)\phi_i(1) = \varepsilon_i\phi_i(1) \qquad i = 1,\dots,N \qquad \text{with} \tag{8.20}$$

$$f(1) = -\tfrac{1}{2}\nabla^2(1) - \sum\nolimits_A \frac{Z_A}{r_{1A}} + \int\frac{\rho(2)}{r_{12}}\,d\mathbf{x}_2 - \int\frac{\rho_{Xi}(1,2)}{r_{12}}\,d\mathbf{x}_2 \quad \text{where} \tag{8.21}$$

$$\rho(1) = \sum n_j\phi_i^*(1)\phi_i(1) \quad \text{and} \tag{8.22}$$

$$\rho_{Xi}(1,2) = \sum\nolimits_j n_j\frac{\phi_i^*(1)\phi_j^*(2)\phi_j(1)\phi_i(2)}{\phi_i^*(1)\phi_i(1)} \tag{8.23}$$

Here the number n_j is the occupation number, 1 for an occupied orbital and 0 for an unoccupied one. The *Fock operator* $f(j)$ is an effective one-electron operator, describing the behaviour of the electrons in the field of the others. The first two terms in Eq. (8.21) represent the kinetic energy and nuclear attraction energy, respectively. The last two terms describe the Coulomb repulsion and exchange energy, both thought to be due to a charge density, $\rho(1)$ and $\rho_{Xi}(1,2)$, respectively. The latter quantity is non-local and different for each spin orbital. The HF equations are

[q] Vladimir Aleksandrovich Fock (1898-1974). Russian scientist from St. Petersburg where he was teaching at the University for more than 40 years and who introduced the antisymmetrised mean field approximation in quantum mechanics, independent of D.R. Hartree.

(pseudo-)eigenvalue equations where the eigenvalues (or Lagrange multipliers) ε_i are to be interpreted as orbital energies. The orbital energies ε_i are given by Eq. (8.20) applying the Fock operator $f(j)$ upon each spin orbital ϕ_j. The result is

$$\varepsilon_i = h_i + \sum_j (2J_{ij} - K_{ij}) \tag{8.24}$$

Note that the electronic energy *not* equals the sum of orbital energies, i.e. $E_{\text{ele}} \neq 2\Sigma_i\, \varepsilon_i$.

Only for atoms Eq. (8.20) can be solved numerically to any desired accuracy. In molecules and solids the orbital χ_j is usually approximated by a linear combination of basis functions ξ_k. If the basis functions are largely localised at atomic sites, they are conventionally addressed as atomic orbitals (although they are not necessarily solutions of the atomic Schrödinger equation) so that one speaks of a *linear combination of atomic orbitals* (LCAO). Hence for a spin orbital ϕ_j,

$$\phi_j = \sigma_j \chi_j = \sigma_j \sum_k c_{jk} \xi_k \tag{8.25}$$

The set of basis functions ξ_k is usually called the *basis set*. A *minimal basis set* is a set, which employs only one basis function of the proper symmetry per occupied orbital for each atom. For example, for H_2 we use one basis function, say $1s_A$, at nucleus A and one, say $1s_B$, at nucleus B. Similarly for e.g. CO we have at each atom the 1s, 2s, $2p_x$, $2p_y$ and $2p_z$ functions. For large sets with functions of sufficient flexibility the results approach the results which would be obtained from a numerical solution of Eq. (8.20). The latter is called the *Hartree-Fock* (HF) *limit*. Restricting ourselves to closed shell configurations, as before, each orbital is doubly occupied and we may write $\boldsymbol{\phi} = \boldsymbol{C\xi}$ using the notation of Section 7.2. Given a set of basis functions ξ_k (equivalently $\boldsymbol{\xi}$) the optimum coefficients c_{jk} are obtained, conform Eq. (7.44), from

$$\boldsymbol{FC}-\boldsymbol{ESC} = \boldsymbol{0} \tag{8.26}$$

The *overlap integrals* S_{rs} and *Fock integrals* F_{rs}, taking the place of the Hamilton matrix elements, are given by

$$S_{rs} = \int \xi_r^* \xi_s \,\mathrm{d}\mathbf{x} \qquad \text{and} \qquad F_{rs} = \int \xi_r^* f(1) \xi_s \,\mathrm{d}\mathbf{x} \tag{8.27}$$

where the integration is over all space and $f(1)$ again indicates the Fock operator.

Mainly because of the non-local character of the exchange potential and because it is different for each spin orbital, the scheme just outlined is rather computing time consuming. An approximate exchange potential energy formula same for all spin orbitals and of local character

$$-\int \frac{\rho_{\mathrm{X}i}(1,2)}{r_{12}} \mathrm{d}\mathbf{x}_2 \quad \Rightarrow \quad V_{\mathrm{X}}(1) = -3\alpha \left[\frac{3}{8\pi}\rho(1)\right]^{1/3}, \quad \alpha = 1 \tag{8.28}$$

was proposed by Slater in 1950. It is based on the average exchange energy in the homogeneous electron gas model. An alternative derivation replacing the exchange energy in the gas before the variational procedure results in the same expression but with $\alpha = 2/3$. Clearly, the approximations do not commute. Subsequently the factor α has been considered as a parameter. For atoms and molecules $\alpha = 0.75$ seems to yield the most reliable result. The Hartree-Fock equations with the exchange part replaced by Eq. (8.28) are referred to as the *Hartree-Fock-Slater* (HFS) equations.

We note two important aspects of the Fock integrals: their evaluation assumes the LCAO coefficients to be known and involves the calculation of a large number of one- and two-electron integrals. The former aspect makes the LCAO-MO-SCF method an iterative procedure. The latter aspect results in large computing time since the number of integrals increases as the fourth power of the basis set size.

In view of the computing time various approximation schemes have been applied for the calculation of the integrals. Usually these schemes contained empirical parameters, to be gauged against experimental data, and therefore these methods are indicated as *semi-empirical* calculations (Harrison, 1980). Nowadays *ab-initio* calculations, i.e. calculations without empirical parameters, can be done for a medium number of atoms. The maximum number that can be handled is largely limited by the computing power and memory available. From the mechanical point of view the most important aspect is that the energy can be calculated as a function of the nuclear co-ordinates yielding not only the optimum geometrical configuration but also the force constants for extension, bending and torsion of bonds between (groups of) atoms.

The HF solution obviously does not represent an exact solution. It takes into account the exchange correlation, i.e. electrons with the same spin repel each other, and one obtains in this approximation a diminished probability of finding another electron near the reference electron. This diminished probability for one electron being in the neighbourhood of another is the so-called *exchange hole* or Fermi hole. Integrating over all space this hole contains exactly one electron. On the other hand the Coulomb correlation, i.e. the repulsion of electrons with different spins, is not taken properly into account although each electron is moving in the average field of the others. One can show that a diminished probability near and an enhanced probability somewhat further away from the reference electron should be present, which, when integrated over all space, yield exactly zero electron. However, within the HF framework a *Coulomb hole*, corresponding to the exchange hole, is absent. The difference between the HF limit energy E_{HF} and the exact energy E_0 is called the *correlation energy* (Pettifor, 1995; Sutton, 1993).

Example 8.1: The minimal basis set H_2 molecule

The (nearly) simplest molecule is H_2, containing two nuclei A and B separated by the internuclear distance R, and two electrons, indicated by 1 and 2. Within the MO model the simplest description of the H_2 molecule is obtained by using a minimal basis set of a 1s function at atom A ($\xi_A = \pi^{-1/2}\exp[-(\mathbf{r}-\mathbf{r}_A)]$) as well as on atom ($\xi_B$). These basis functions overlap and the overlap integral S is given by $S = \int \xi_A(\mathbf{r})\xi_B(\mathbf{r})\,d\mathbf{r}$. From these two basis functions we can construct two spatial MOs, namely the symmetrical $\chi_1 = c_1\xi_A + c_2\xi_B$ and the anti-symmetrical $\chi_2 = c_3\xi_A - c_4\xi_B$. In this particular case the expansion coefficients c_i are determined fully by the symmetry of the molecule and the normalisation condition $\int \chi_1^2\,d\mathbf{r} = \int \chi_2^2\,d\mathbf{r} = 1$. In general the coefficients c_i have to be carried through the calculation in order to determine them via Eq. (8.20). The symmetry and normalisation considerations lead to $c_1 = c_2 = [2(1+S)]^{-1/2}$. Similarly for χ_2 we obtain $c_3 = c_4 = [2(1-S)]^{-1/2}$. From the spatial orbitals χ_1 and χ_2 we can form four spin orbitals ϕ by multiplication with the spin functions α or β. Hence

$\phi_1 = \chi_1\alpha$ or χ_1 $\phi_2 = \chi_1\beta$ or $\overline{\chi}_1$ $\phi_3 = \chi_2\alpha$ or χ_2 $\phi_4 = \chi_2\beta$ or $\overline{\chi}_2$

The spin orbitals ϕ_1 and ϕ_2 are degenerate and appear to have the lower energy; hence they describe bonding. Similarly ϕ_3 and ϕ_4 are degenerate and describe the first excited state. The HF ground state in this model is given by

$$\Psi = |\,\chi_1\overline{\chi}_1\,|$$

In this particular case the determinant can be split into a spatial and spin part according to

$$\Psi = [\xi_A(1)+\xi_B(1)][\xi_A(2)+\xi_B(2)][\alpha(1)\beta(2)-\beta(1)\alpha(2)]/\sqrt{2}[2(1+S)]$$

leading, after a somewhat lengthy, straightforward calculation, to the energy

$E = -1 + [2(h_{AA}+h_{AB})/(1+S)] + J - 1/R$ where

$h_{AA} = \langle\xi_A|-1/\mathbf{r}_A+1/R|\xi_A\rangle$ $h_{AB} = \langle\xi_A|-1/\mathbf{r}_B+1/R|\xi_B\rangle$ and

$J = \langle\chi_1(1)\chi_1(2)|/r_{12}|\chi_2(1)\chi_2(2)\rangle$

known as the molecular Coulomb integral. The term –1 Ha represents the energy of two hydrogen atoms forming a H_2^+ ion-molecule while the term $(h_{AA}+h_{AB})/(1+S)$ represents the bonding contribution to a H_2^+ ion-molecule when a calculation is made for that species in the same approximation. The total energy E is thus twice the energy of the H_2^+ ion-molecule except for the term J, correcting for the electron-electron repulsion, and for the term $-1/R$, correcting for counting the nuclear repulsion twice. The full expression for E is somewhat lengthy (see Pilar, 1968) and we omit it here. This model leads to a binding energy of –1.097 Ha or –29.85 eV, corresponding to a dissociation energy of 2.65 eV, at an equilibrium distance of 1.57 a_0 or 0.084 nm. These predictions are to be compared with the experimental values of 4.75 eV and 0.0741 nm.

So far the theory is general and can be applied to molecules as well as to solids. We now specialise to solids. The most important feature of a crystalline solid is that the potential energy is also periodic. The interaction between the electrons and the lattice can be considered as weak, in which case we obtain the nearly free electron approximation, or strong, in which case we obtain the tight-binding approximation.

The nearly free electron approximation

We first recall that the Schrödinger equation for a free electron of mass m is given by (Pettifor, 1995; Sutton, 1993; Mott and Jones, 1936)

$$-\frac{\hbar^2}{2m}\nabla^2\psi(\mathbf{r}) = E\psi(\mathbf{r}) \tag{8.29}$$

where $\hbar = h/2\pi$ with h Plancks's constant and E the energy. The free electrons can be described by plane waves

$$\psi_{\mathbf{k}}(\mathbf{r}) = L^{-3/2}\exp(i\mathbf{k}.\mathbf{r}) \tag{8.30}$$

with position vector $\mathbf{r} = (x,y,z)$ and wave vector $\mathbf{k} = (2\pi/L)(n_1,n_2,n_3)$. The numbers n_1, n_2 and n_3 are the (integer) quantum numbers and periodic boundary conditions over a length L are used. The corresponding eigenvalues are

$$E_\mathbf{k} = \frac{\hbar^2}{2m}k^2 \qquad \text{where} \quad k = \|\mathbf{k}\| \tag{8.31}$$

Each state corresponding to a given wave vector can contain two electrons of opposite spin according to Pauli's principle. Therefore at 0 K for N electrons the lowest energy is obtained by filling a sphere of volume $V = L^3$ with electrons. The radius k_F of that sphere is the *Fermi sphere*. This results in $(4/3)\pi k_F^3 2V/(2\pi)^3 = N$ since a unit volume of **k**-space can contain $V/(2\pi)^3$ states. Hence

$$k_F = (3\pi^2 N/V)^{1/3}$$

The corresponding energy $E_F = \hbar^2 k_F^2/2m$ is the *Fermi energy*. At elevated temperature one has to take into account the occupation of the density of states as described by the Fermi distribution. We refer to the literature (e.g. Pettifor, 1995) for details.

For nearly free electrons[r] (NFE) one approach is to expand the wavefunction, taking also into account the Bloch condition, in a Fourier series as

$$a^{3/2}\Psi_\mathbf{k} = e^{i\mathbf{k}\cdot\mathbf{r}}u(\mathbf{r}) = e^{i\mathbf{k}\cdot\mathbf{r}}\sum_{\mathbf{k}-\mathbf{g}} c_{\mathbf{k}-\mathbf{g}}e^{-i\mathbf{g}\cdot\mathbf{r}} = \sum_{\mathbf{k}-\mathbf{g}} c_{\mathbf{k}-\mathbf{g}}e^{i(\mathbf{k}-\mathbf{g})\cdot\mathbf{r}}$$

where a is the unit cell size and $\mathbf{g}$ is a reciprocal lattice vector. On substituting these expressions in the Schrödinger equation $H\psi_\mathbf{k} = [-(\hbar^2/2m)\nabla^2+V]\psi_\mathbf{k} = E\psi_\mathbf{k}$, multiplying by $\exp[-i(\mathbf{k}-\mathbf{g}')\cdot\mathbf{r}]$ and integrating, meanwhile realising that the integral expression $a^{-3}\int \exp[i(\mathbf{k}-\mathbf{g})\cdot\mathbf{r}]\exp[-i(\mathbf{k}-\mathbf{g}')\cdot\mathbf{r}]\,d\mathbf{r} = \delta_{\mathbf{g}'\mathbf{g}}$, we obtain an infinite set of coupled, linear equations

$$\left[\frac{\hbar^2|\mathbf{k}-\mathbf{g}|^2}{2m} - E\right]c_{\mathbf{k}-\mathbf{g}} + \sum_{\mathbf{g}'} V_{\mathbf{g}'-\mathbf{g}}c_{\mathbf{k}-\mathbf{g}'} = 0 \tag{8.32}$$

where $V_\mathbf{k} = \int \exp(i\mathbf{k}\cdot\mathbf{r})V(\mathbf{r})\,d\mathbf{r}$ denotes the **k**th Fourier transform coefficient of the potential energy V. To solve this infinite set we have to approximate. As a first approximation we assume that all contributions of the potential energy are zero except those for which $\mathbf{g} = \mathbf{0}$ with $\mathbf{g}' = \mathbf{0}$ and $\mathbf{g}' = \mathbf{g}$ and those for which $\mathbf{g} = \mathbf{g}$ with $\mathbf{g}' = \mathbf{0}$ and $\mathbf{g}' = \mathbf{g}$. These two contributions are the dominant ones when Bragg reflection of an electron with wave vector **k** to an electron with wave vector **k**–**g** at the zone boundary occurs (Fig. 8.23). Due to the periodicity of the lattice these vectors are equivalent and both have to be included. Neglecting all other contributions we then obtain

$$\left[\frac{\hbar^2|\mathbf{k}|^2}{2m} - E + V_0\right]c_\mathbf{k} + V_\mathbf{g}c_{\mathbf{k}-\mathbf{g}} = 0 \quad \text{and} \quad V_{-\mathbf{g}}c_\mathbf{k} + \left[\frac{\hbar^2|\mathbf{k}-\mathbf{g}|^2}{2m} - E + V_0\right]c_{\mathbf{k}-\mathbf{g}} = 0$$

Solving for E as a function of **k** yields

$$E^\pm = V_0 + \frac{\hbar^2}{4m}\left[|\mathbf{k}|^2 + |\mathbf{k}-\mathbf{g}|^2\right] \pm \frac{1}{2}\left[\frac{\hbar^2}{2m}(|\mathbf{k}|^2 - |\mathbf{k}-\mathbf{g}|^2)^2 + 4|V_\mathbf{g}|^2\right]^{1/2}$$

When $\mathbf{k} << \mathbf{g}/2$, we obtain by expanding the square root according to $(1+x)^{1/2} \cong 1+x/2$,

[r] For a lucid introduction, see Pettifor, D.G. (1983), *Electron theory of metals*, page 73 in *Physical metallurgy*, R.W. Cahn and P. Haasen, eds., North-Holland, Amsterdam.

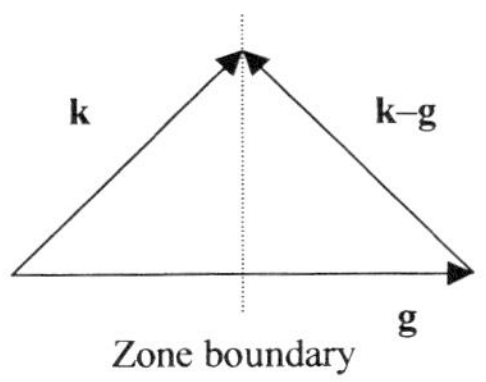

Fig. 8.23: Bragg reflection at the zone boundary.

$$E^{\pm}=V_0+\frac{\hbar^2\,|\mathbf{k}|^2}{2m}\pm\frac{2m\,|V_{\mathbf{g}}|}{\hbar^2\left(|\mathbf{k}|^2-|\mathbf{k}-\mathbf{g}|^2\right)^2}$$

and this reflects a small correction to the free electron solution. When the condition for reflection is fulfilled exactly, $\mathbf{k}=\mathbf{g}/2$, the energy expression reduces to

$$E^{\pm}=V_0+\frac{\hbar^2\,|\mathbf{g}/2|^2}{2m}\pm|V_{\mathbf{g}}|$$

At the zone boundary the energy therefore has a discontinuity and the allowed energy levels are clustered in *energy bands*. The two allowed energy band values are separated by a *band gap*, in this approximation of size $2|V_{\mathbf{g}}|$. If more than two electron waves contribute significantly, the energy expression becomes correspondingly complex but the discontinuities remain at the zone boundary. Plotting the energy as a function of the component k_1, Fig. 8.24 is obtained in which the allowed energy bands can be distinguished. Although in 1D this implies always the presence of an energy gap, in 3D the bands may overlap in the interior of the zone and generally they will do so. Therefore, in principle in the NFE approximation the electrons have access to a quasi-continuous energy spectrum and since the bands are typically only partially filled, materials in this description are electrically conducting.

The symmetry of the lattice determines which set of reciprocal lattice vectors is present and therefore also determines the shape of the Brillouin zone (BZ). For a SC lattice the shape of the first BZ is a cube. The usual lattice types FCC and BCC can be described as a superposition of several SC lattices. For example, the BCC lattice can be considered as two interpenetrating SC lattices of lattice constant a displaced by (½a, ½a, ½a) relative to each other. It follows that the Fourier coefficients V_{100}, V_{010} and V_{001} of the potential energy are all zero and the first BZ becomes a dodecahedron. Similarly the FCC lattice can be considered as four interpenetrating SC lattices and all

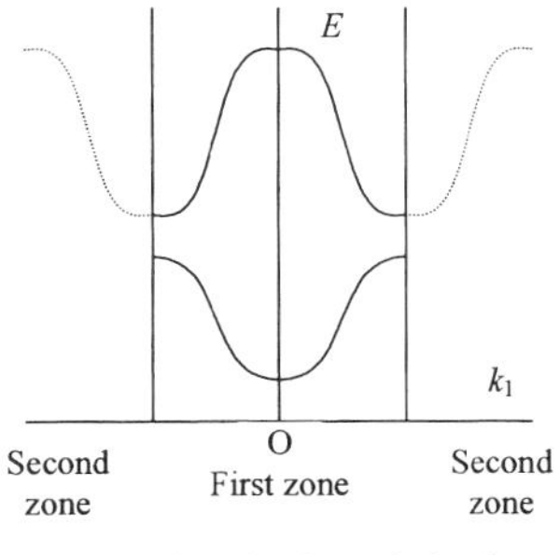

Fig. 8.24: The energy gap between NFE bands along k_1 in the reduced (—) and extended (····) zone representation.

the Fourier coefficients such as V_{100} and V_{110} are zero and V_{111} is the first non-zero coefficient. The first BZ becomes a cubo-octahedron. The shapes of the BZ for the BCC and FCC lattices are shown in Fig. 8.25.

Neglecting the details of the band structure for simple metals a convenient approach is to use the NFE model in which the Z valence electrons with mass m and charge e are distributed spatially uniformly but the ions are represented by localised point charges Ze. We denote the *atomic volume* by $\Omega = V/N$, where N represents the number of atoms in a volume V. Equivalently, the size of the atomic volume may be expressed by the *Wigner-Seitz radius* r which is the radius of a sphere of the same volume as the atomic volume; hence

$$\Omega = \frac{4}{3}\pi r^3 \tag{8.33}$$

All results in the remainder of this section will be given as a function of the *dimensionless parameter* r_s which is a measure of the electron number density and represents the radius of a sphere with one electron. The product $r_s a_0$ thus indicates the radius of a sphere with volume Ω/Z, where Z is the number of electrons donated by each ion and a_0 is the Bohr radius. Consequently,

$$\frac{4\pi}{3}(r_s a_0)^3 = \frac{\Omega}{Z} = \frac{V}{ZN} = \frac{4\pi r^3}{3Z} \quad \text{or} \quad r_s^3 = \frac{r^3}{Za_0^3} = \frac{3V}{4\pi NZa_0^3} \tag{8.34}$$

The average *kinetic energy* U_{kin} for the free electron gas[s] can be shown to be $3E_F/5$ or

$$U_{\text{kin}} = \frac{3me^4}{10(4\pi\varepsilon_0\hbar)^2}\left(\frac{9\pi}{4}\right)^{2/3} Zr_s^{-2} = 2.210Zr_s^{-2}\ [\text{Ry}] \tag{8.35}$$

The *Coulomb energy* U_{Cou} is given by

$$U_{\text{Cou}} = -\frac{Ze^2}{(4\pi\varepsilon_0)}\int \rho(\mathbf{r})\frac{1}{\mathbf{r}}\,\mathrm{d}\mathbf{r} + \tfrac{1}{2}\frac{e^2}{(4\pi\varepsilon_0)}\int \frac{\rho(\mathbf{r})\rho(\mathbf{r}')}{|\mathbf{r}-\mathbf{r}'|}\,\mathrm{d}\mathbf{r}\,\mathrm{d}\mathbf{r}' \tag{8.36}$$

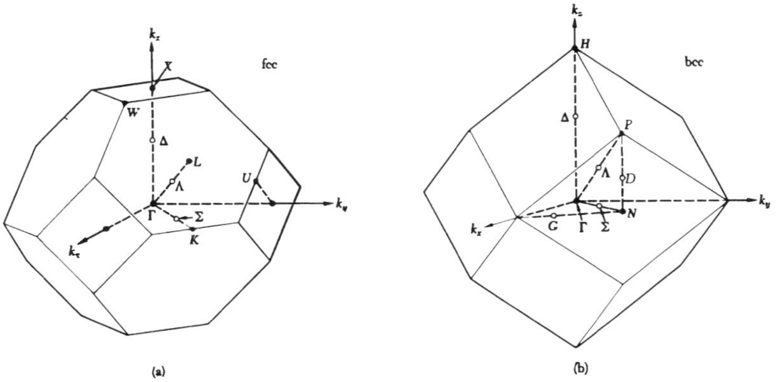

Fig. 8.25: The shape of the Brillouin zone for the BCC and FCC lattices. High symmetry points are indicated.

---

[s] For a derivation of the various contributions we refer to the literature. The second expression is in Rydberg (see Chapter 7).

where the first and second contributions result from the electron-ion and electron-electron interactions, respectively. The integration of the electron density $\rho(\mathbf{r}) = Z/\Omega$ is over the Wigner-Seitz sphere. Evaluating U_{Cou} results in

$$U_{\mathrm{Cou}} = \frac{me^4}{2(4\pi\varepsilon_0\hbar)^2}\left(-3+\frac{6}{5}\right)Z^2r^{-1} = -\frac{9}{5}Z^{5/3}r_{\mathrm{s}}^{-1} = -\alpha_{\mathrm{C}}Z^{5/3}r_{\mathrm{s}}^{-1}\,[\mathrm{Ry}] \qquad (8.37)$$

The parameter α_{C} is similar to the Madelung constant for ionic solids but here for the interaction of a uniform distribution of a negative charge with positive point ions and of the interaction of the electrons with themselves. For various lattice types calculation leads to only slightly different values for the parameter α_{C} (Table 8.3).

Table 8.3: Values of the constant α_{C} for various lattices.

Lattice	Single point ion	Simple cubic	BCC	FCC	HCP			C	White Sn
					c/a=1.5	c/a=1.6	c/a=1.8		c/a=0.56
α_{C}	1.800	1.76	1.792	1.792	1.790	1.792	1.789	1.671	1.773

In this simple model the total energy per atom is given by $U = U_{\mathrm{kin}} + U_{\mathrm{Cou}}$, which has a minimum when $\partial U/\partial r_{\mathrm{s}} = 0$, which occurs at $r_{\mathrm{bin}} = (4.42/\alpha_{\mathrm{C}})Z^{-2/3}$ resulting in a binding energy $U_{\mathrm{bin}} = 0.113\alpha_{\mathrm{C}}^2Z^{7/3}$. Since α_{C} is to a large extent insensitive to the structure, both the equilibrium distance and binding energy vary only slightly. Moreover, the value of r_{bin} is typically too small by a factor of 2. Obviously, this simple model is inadequate.

To improve we will need three more contributions. First, the *exchange energy* U_{exc},

$$U_{\mathrm{exc}} = -\frac{3}{2\pi}\left(\frac{9\pi}{4}\right)^{1/3}r_{\mathrm{s}}^{-1} = -0.916r_{\mathrm{s}}^{-1}\,[\mathrm{Ry}] \qquad (8.38)$$

which is attractive since parallel spins are kept apart by Pauli's exclusion principle leading to weaker mutual Coulomb repulsion. Second, the *correlation energy* U_{cor}

$$U_{\mathrm{cor}} = 0.0313\ln r_{\mathrm{s}} - 0.115\,[\mathrm{Ry}] \qquad (8.39)$$

which is also attractive due to the dynamic correlation between the electrons and is, even in this simplified model, only an approximate interpolation formula. Finally, we need to add a structure independent term resulting since the ions have been treated as point charges but in reality the valence electrons tend to stay away from the core electrons. This gives rise to core pseudo-potential with corresponding energy U_{pse},

$$U_{\mathrm{pse}} = 2\pi Ze^2r_{\mathrm{c}}^2\frac{Z}{\Omega} = 3Z^2\left(\frac{r_{\mathrm{c}}^2}{r_{\mathrm{s}}^3}\right)[\mathrm{Ry}] \qquad (8.40)$$

Here r_{c} is a parameter, indicating the radius of the core, which is in the order of 1 a_0. The improved total energy U obviously is

$$U = U_{\mathrm{kin}} + U_{\mathrm{Cou}} + U_{\mathrm{exc}} + U_{\mathrm{cor}} + U_{\mathrm{pse}}$$

Similarly as before minimising this equation with respect to r_{s} yields the equilibrium distance r_{bin} and binding energy U_{bin} as a function of the parameter r_{c}. The core radius

r_c can be chosen in such a way that the experimental Wigner-Seitz radius r_0 is matched[t]. This results in the relation

$$\left(\frac{r_c}{r_0}\right)^2 = \frac{1}{5} + \frac{0.102}{Z^{2/3}} + \frac{0.0035r_0}{Z} - \frac{0.491}{Z^{1/3}r_0} \tag{8.41}$$

Using the values so obtained (see Table 11.5) results in a binding energy as a function of r_s. It appears that the contribution of the exchange-correlation energy cancels almost exactly with the electron-electron Coulomb term so that we have

$$U \cong -\frac{3}{Z^{1/3}r_s}\left[1-\left(\frac{r_c}{Z^{1/3}r_s}\right)^2\right] + \frac{2.210}{r_s^2} \tag{8.42}$$

For simple metals the above-described theory results in reasonable estimates for many properties. For non-simple metals, e.g. the transition metals, similar but more complex theories are necessary.

The success of the NFE theory is remarkable if one realises that the essential assumption of a not too severely fluctuating potential cannot be fulfilled in any lattice. In fact, it is the screening of the nuclear charge by the inner electrons which provides a more or less flat potential. The screened effect of the core, i.e. the inner electrons and nucleus, can be incorporated via the pseudo-potential, as briefly indicated.

Summarising, the metallic bond may be described as an attractive force between an assembly of positive ions and the 'nearly free electron' gas. The electrons are not bound to specific atoms and the bonding is non-directional. In the NFE approach allowed energy bands and non-allowed energy gaps arise. Generally the bonding potential ϕ_{met} can be described by

$$\phi_{met} = -\frac{C}{\Omega^{1/3}} + \frac{B}{\Omega^{2/3}} + \frac{A}{\Omega} \tag{8.43}$$

where $\Omega = V/V_0$ with V_0 the equilibrium volume. The first term represents the Coulomb energy, the second the kinetic energy of the electrons and the third a contribution from the interaction of the valence electrons with the inner electrons. In the NFE approximation we have $C = 3/Z^{1/3}$, $B = 2.210$ and $A = 3r_c^2/Z$. Finally, it should be remarked that the quantities A, B and C could also be considered as parameters (see Chapter 11).

Problem 8.4

Using the approximation $U = U_{kin}+U_{Cou}$ show that at $r_{bin} = (4.42/\alpha_C)Z^{-2/3}$ the minimum in binding energy $U_{bin} = 0.113\alpha_C^2Z^{7/3}$.

Problem 8.5

a) Show that Eq. (8.41) holds.
b) Plot the various contributions to the binding energy U_{bin} as a function of r_s.
c) Derive Eq. (8.42).

---

[t] Girifalco, L.A. (1976), Acta Metall. **24**, 759.

The tight-binding approximation

In the case of strong interaction of the electrons with the lattice it is more appropriate to consider the atomic states as the zero approximation. Therefore in this case we expand the wavefunction for lattice with lattice constant a as

$$a^{3/2}\Psi_{\mathbf{k}} = \exp(i\mathbf{k}\cdot\mathbf{r})u(\mathbf{r}) = \exp(i\mathbf{k}\cdot\mathbf{r})\sum\nolimits_{\mathbf{m}} \exp[-i\mathbf{k}\cdot(\mathbf{r}+\mathbf{r}_{\mathbf{m}})]\phi(\mathbf{r}-\mathbf{r}_{\mathbf{m}})$$

where the labelling $\phi(\mathbf{r}-\mathbf{r}_{\mathbf{m}})$ or $\phi_{\mathbf{m}}$ is used to identify the atomic orbital localised at $\mathbf{r}_{\mathbf{m}}$, supposing for simplicity a Bravais lattice. We also suppose that the Schrödinger equation for the individual atoms

$$\frac{\hbar^2}{2m}\nabla^2\phi_{\mathbf{n}} + (E_0 - V_{\mathbf{n}})\phi_{\mathbf{n}} = 0 \tag{8.44}$$

is solved, assuming that the potential $V_{\mathbf{n}}$ is large near the lattice points and decreases rapidly with distance. As zero-order approximation we suppose that each electron is in the neighbourhood of one particular lattice point and we neglect the effect of other atoms. This state is highly degenerate since positioning the electrons in a similar orbit around each of the ions at the various lattice point results in the same energy for each atom and, and assuming one valence electron per ion, the degeneracy is equal to the number of atoms in the crystal. Allowing now for interaction the degeneracy is removed and each level splits into a band. We explicitly assume that the presence of the other atoms introduces an energy change small as compared to the levels associated with Eq. (8.44). This implies that the electrons are *tightly bound* (TB). Calculating the expectation value of this wavefunction results in

$$\langle E_{\mathbf{k}}\rangle = \frac{\int \sum_{\mathbf{m}'}\exp(-i\mathbf{k}\cdot\mathbf{r}_{\mathbf{m}'})\phi_{\mathbf{m}'}^{*} H \sum_{\mathbf{m}}\exp(i\mathbf{k}\cdot\mathbf{r}_{\mathbf{m}})\phi_{\mathbf{m}}\,d\mathbf{r}}{\int \sum_{\mathbf{m}'}\exp(-i\mathbf{k}\cdot\mathbf{r}_{\mathbf{m}'})\phi_{\mathbf{m}'}^{*} \sum_{\mathbf{m}}\exp(i\mathbf{k}\cdot\mathbf{r}_{\mathbf{m}})\phi_{\mathbf{m}}\,d\mathbf{r}} = \frac{\sum_{\mathbf{m}''}\exp(i\mathbf{k}\cdot\mathbf{r}_{\mathbf{m}''})H_{\mathbf{m}''}}{\sum_{\mathbf{m}''}\exp(i\mathbf{k}\cdot\mathbf{r}_{\mathbf{m}''})S_{\mathbf{m}''}}$$

where $H_{\mathbf{m}''} = \langle\phi_{\mathbf{m}'}|H|\phi_{\mathbf{m}}\rangle$ and $S_{\mathbf{m}''} = \langle\phi_{\mathbf{m}'}|\phi_{\mathbf{m}}\rangle$. Evaluating $H_{\mathbf{m}''}$ we have

$$\langle\phi_{\mathbf{m}'}|H|\sum\nolimits_{\mathbf{m}}\exp(i\mathbf{k}\cdot\mathbf{r}_{\mathbf{m}})\phi_{\mathbf{m}}\rangle = E_{\mathbf{k}}\langle\phi_{\mathbf{m}'}|\sum\nolimits_{\mathbf{m}}\exp(i\mathbf{k}\cdot\mathbf{r}_{\mathbf{m}})\phi_{\mathbf{m}}\rangle$$

Using $H = -\tfrac{1}{2}\nabla^2 + \sum_{\mathbf{n}} V_{\mathbf{n}}$ we obtain

$$\langle\phi_{\mathbf{m}'}| -\tfrac{1}{2}\nabla^2 + V_{\mathbf{m}'} + \sum\nolimits_{\mathbf{m}\neq\mathbf{m}'} V_{\mathbf{m}} | \sum\nolimits_{\mathbf{m}}\exp(i\mathbf{k}\cdot\mathbf{r}_{\mathbf{m}})\phi_{\mathbf{m}}\rangle$$
$$= E_{\mathbf{k}}\langle\phi_{\mathbf{m}'}|\phi_{\mathbf{m}'}\exp(i\mathbf{k}\cdot\mathbf{r}_{\mathbf{m}'}) + \sum\nolimits_{\mathbf{m}\neq\mathbf{m}'}\phi_{\mathbf{m}}\exp(i\mathbf{k}\cdot\mathbf{r}_{\mathbf{m}})\rangle \qquad \text{or}$$

$$E_0\exp(i\mathbf{k}\cdot\mathbf{r}_{\mathbf{m}'}) + \sum\nolimits_{\mathbf{m}\neq\mathbf{m}'}\langle\phi_{\mathbf{m}'}|\phi_{\mathbf{m}}\exp(i\mathbf{k}\cdot\mathbf{r}_{\mathbf{m}})\rangle\times$$
$$\sum\nolimits_{\mathbf{m}\neq\mathbf{m}'}\sum\nolimits_{\mathbf{m}'''\neq\mathbf{m}'}\langle\phi_{\mathbf{m}'}|V_{\mathbf{m}'''}|\phi_{\mathbf{m}'}\exp(i\mathbf{k}\cdot\mathbf{r}_{\mathbf{m}'}) + \phi_{\mathbf{m}}\exp(i\mathbf{k}\cdot\mathbf{r}_{\mathbf{m}})\rangle$$
$$= E_{\mathbf{k}}\left[\exp(i\mathbf{k}\cdot\mathbf{r}_{\mathbf{m}'}) + \sum\nolimits_{\mathbf{m}\neq\mathbf{m}'}\langle\phi_{\mathbf{m}'}|\phi_{\mathbf{m}}\exp(i\mathbf{k}\cdot\mathbf{r}_{\mathbf{m}})\rangle\right]$$

Setting $\sum_{\mathbf{m}'''\neq\mathbf{m}'}\langle\phi_{\mathbf{m}'}|V_{\mathbf{m}'''}|\phi_{\mathbf{m}'}\rangle = V_0$ we finally obtain

$$E_{\mathbf{k}} = E_0 + \frac{V_0 + \sum_{\mathbf{m}\neq\mathbf{m}'}\sum_{\mathbf{m}'''\neq\mathbf{m}'}\exp(i\mathbf{k}\cdot\mathbf{r}_{\mathbf{m}''})\langle\phi_{\mathbf{m}'}|V_{\mathbf{m}'''}|\phi_{\mathbf{m}}\rangle}{1 + \sum_{\mathbf{m}''}\exp(i\mathbf{k}\cdot\mathbf{r}_{\mathbf{m}''})S_{\mathbf{m}''}} \tag{8.45}$$

where $\mathbf{r}_{\mathbf{m}''} = \mathbf{r}_{\mathbf{m}} - \mathbf{r}_{\mathbf{m}'}$ is a non-zero lattice vector. Now several assumptions, addressed as the *Hückel approximations*, are often made. First, since in the tight-binding method one assumes little overlap between different sites, i.e. the basis functions $\phi_{\mathbf{n}}$ decrease rapidly with distance from $\mathbf{r}_{\mathbf{n}}$, it is usual to take $\langle\phi_{\mathbf{m}'}|V_{\mathbf{m}'''}|\phi_{\mathbf{m}}\rangle = 0$ so that only the simpler two-centre integrals remain. Second, in fact one even assumes that only the nearest-neighbour two-centre integrals are non-zero. Finally, in the same spirit, it is often assumed that $\langle\phi_{\mathbf{n}}|\phi_{\mathbf{m}}\rangle = \int\phi_{\mathbf{n}}^*\phi_{\mathbf{m}}\,\mathrm{d}\mathbf{r} = \delta_{\mathbf{nm}}$. In this approximation the energy expression becomes

$$E_{\mathbf{k}} = E_0 + V_0 + \sum\nolimits_{\mathbf{m}'=\mathbf{m}-1,\mathbf{m},\mathbf{m}+1} \exp(\mathrm{i}\mathbf{k}\cdot\mathbf{r}_{\mathbf{m}'})\langle\phi_{\mathbf{m}'}|V_{\mathbf{m}}|\phi_{\mathbf{m}}\rangle \tag{8.46}$$

Moreover, we write $\alpha_{\mathbf{n}} = -\langle\phi_{\mathbf{n}}|V_0 - V_{\mathbf{n}}|\phi_{\mathbf{n}}\rangle$ and $J(\mathbf{n}-\mathbf{m}) = -\langle\phi_{\mathbf{n}}|V_0 - V_{\mathbf{n}}|\phi_{\mathbf{m}}\rangle$, where $J(\mathbf{n}-\mathbf{m})$ depends on $\mathbf{n}$ and $\mathbf{m}$ only through their difference $|\mathbf{n}-\mathbf{m}|$. The values of α and J are generally positive, provided $\phi_{\mathbf{n}}$ has only nodes inside the atomic cores.

Let us discuss a few examples. For the SC lattice there are six nearest neighbours at a distance a. Using only s-orbitals with energy E_1 and taking $J(100) = J(010) = J(001) = \beta_1$, the energy expression becomes

$$E = E_1 - \alpha_1 - 2\beta_1[\cos(ak_1) + \cos(ak_2) + \cos(ak_3)]$$

The degenerated levels, all with energy E_1, thus split into a band centred at $E_1 - \alpha_1$ and of width $12\beta_1$.

In a BCC lattice there are eight nearest neighbours at a distance $\tfrac{1}{2}a\sqrt{3}$ along the cube diagonals (see Section 8.2). The corresponding values for J are indicated by γ_1 and we obtain, again only using s-orbitals,

$$E = E_1 - \alpha_1 - 2\gamma_1\begin{bmatrix}\cos\tfrac{1}{2}a(k_1 + k_2 + k_3) + \cos\tfrac{1}{2}a(k_1 + k_2 - k_3) \\ + \cos\tfrac{1}{2}a(k_1 - k_2 + k_3) + \cos\tfrac{1}{2}a(k_1 - k_2 - k_3)\end{bmatrix}$$

In a FCC lattice there are twelve nearest neighbours at a distance $\tfrac{1}{2}a\sqrt{2}$ along the face diagonals. Writing δ_1 for the corresponding J values, the energy becomes

$$E = E_1 - \alpha_1 - 2\delta_1\begin{bmatrix}\cos\tfrac{1}{2}a(k_1 + k_2) + \cos\tfrac{1}{2}a(k_1 - k_2) \\ + \cos\tfrac{1}{2}a(k_2 + k_3) + \cos\tfrac{1}{2}a(k_2 - k_3) \\ + \cos\tfrac{1}{2}a(k_1 + k_3) + \cos\tfrac{1}{2}a(k_1 - k_3)\end{bmatrix}$$

For p-states on the SC lattice the levels are triply degenerate but these states, which can be represented by $xf(\mathbf{r})$, $yf(\mathbf{r})$ and $zf(\mathbf{r})$, will not split up in a SC lattice. We consider only the p_x-states for the moment so that $\phi_{\mathbf{n}} = x_{\mathbf{n}}\,f(\mathbf{r}_{\mathbf{g}})$ is the wavefunction with energy E_2. Since $\phi_{\mathbf{n}}$ is not spherically symmetric, there are now more integrals involved. As before we set $\alpha_2 = -\langle\phi_{\mathbf{n}}|V_0 - V_{\mathbf{n}}|\phi_{\mathbf{n}}\rangle$ but also use $\beta_2 = -\langle\phi_{n_1,n_2+1,n_3}|V_0 - V_{\mathbf{n}}|\phi_{\mathbf{n}}\rangle$ (for π overlap) and $\gamma_2 = -\langle\phi_{n_1,n_2,n_3+1}|V_0 - V_{\mathbf{n}}|\phi_{\mathbf{n}}\rangle$ (for σ overlap). This leads to

$$E = E_2 - \alpha_2 - 2\gamma_2\cos(ak_1) - 2\beta_2[\cos ak_2 + \cos ak_3]$$

Similar expression results with permutation of k_1, k_2 and k_3 to account for the $yf(\mathbf{r})$ and $zf(\mathbf{r})$ contributions. In all cases thus a band arises due to the interaction of the atoms with each other. Although not very satisfactory quantitatively, the tight-binding approximation is extensively used in semi-quantitative and qualitative discussions.

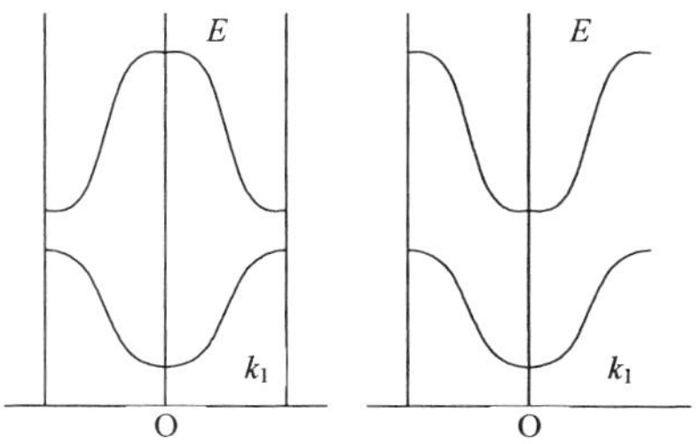

Fig. 8.26: The energy bands in tight-binding approximation for s-states and p-states.

The energy levels in the TB approximation resemble to some extent those of the NFE approximations. Expanding the cosines we have for small value of $|\mathbf{k}|$

$$E = E_1 - \alpha_1 - 6\beta_1 + \beta_1 a^2|\mathbf{k}|^2 \quad \text{for the SC lattice,}$$

$$E = E_1 - \alpha_1 - 8\gamma_1 + \gamma_1 a^2|\mathbf{k}|^2 \quad \text{for the BCC lattice and}$$

$$E = E_1 - \alpha_1 - 12\delta_1 + \delta_1 a^2|\mathbf{k}|^2 \quad \text{for the FCC lattice.}$$

In this approximation the electron moves as if it were free but with effective mass m^* given by an expression such as $m^* = \hbar^2/2a^2\beta_1$ obtained by comparing with the corresponding free electron equation.

Overall for the TB approximation to be valid, the value of β_1 (γ_1, δ_1,...) must be small and therefore it is necessary that the levels E_1, E_2, ... should be far apart. In that case $m^* > m$. The maximum of the s-states is obtained at $k_1 = k_2 = k_3 = \pi/a$ and $E = E_1 - \alpha_1 + 6\beta_1$. Similarly for the p-states the maximum is obtained at $k_1 = \pi/a$, $k_2 = k_3 = 0$ and $E = E_2 - \alpha_2 - 4\beta_2 - 2\gamma_2$. We must therefore have

$$E_2 - \alpha_2 - 4\beta_2 - 2\gamma_2 > E_1 - \alpha_1 + 6\beta_1$$

All the energies of the second band lie above those of the first band and there are forbidden energy ranges, in contrast to the NFE case. The schematic energy curves for the s-state and p-states are shown in Fig. 8.26. In conclusion, in the TB approximation the atomic levels spread to bands, which do not overlap. Therefore, in principle in the TB approximation the materials are electrically non-conducting if the gap is wide or semi-conducting if the gap is narrow as compared to kT.

Example 8.2: The minimal basis set band structure of Si

Silicon is an important covalently bonded material for which the band structure as calculated with the TB approximation holds reasonably well. In this model we take only 2s and 2p atomic orbitals as basis functions. The nearest-neighbour matrix elements are estimated semi-empirically as

$$J(\text{ss}\sigma) = -1.94 \text{ eV} \quad J(\text{sp}\sigma) = 1.75 \text{ eV} \quad J(\text{pp}\sigma) = 3.10 \text{ eV} \quad J(\text{pp}\pi) = -1.08 \text{ eV}$$

where σ and π denote the symmetry and are illustrated in Fig. 8.27. The on-site elements are estimated as

$$\alpha_s = -5.25 \text{ eV} \quad \text{and} \quad \alpha_p = 1.20 \text{ eV}$$

The diamond lattice consists of two interpenetrating FCC lattices, separated by ¼a[1,1,1] where a is the lattice constant. The WS cell can be defined by the

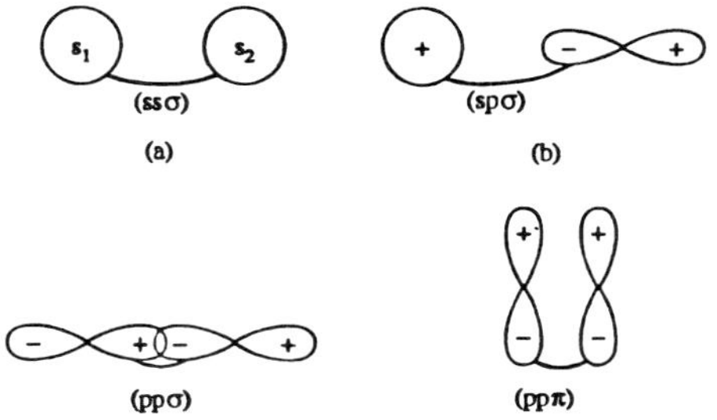

Fig. 8.27: The symmetry of the four basic types of interactions between s- and p-states.

vectors ½*a*[1,1,0], ½*a*[1,0,1] and ½*a*[0,1,1] which contains two atoms, one at (0,0,0) and the other at ¼*a*(1,1,1). Since we have a basis set of four orbitals at each atomic site, each WS cell contains eight basis functions and therefore we have eight bands.

Let $|\mathbf{m}j\alpha>$ denote an atomic state at $\mathbf{r}_{\mathbf{m}j}$ and α the label for s, p_x, p_y or p_z. Further we denote with $|n\mathbf{k}>$ an eigenstate with band index n ($1 \le n \le 8$) and wave vector **k**. Using Bloch's theorem we expand $|n\mathbf{k}>$ as

$$| n\mathbf{k} > = \sum_{\mathbf{m},j,\alpha} \exp(i\mathbf{k}\cdot\mathbf{r}_{\mathbf{m}j})\, c_{j,\alpha}^{(n)}\, | \mathbf{m}j\alpha >$$

The expansion coefficients are obtained in the usual way by inserting the expansion in the Schrödinger equation $H|n\mathbf{k}> = E^{(n)}|n\mathbf{k}>$ and multiplying by $\langle \mathbf{0}j'\alpha'|$. In this way we obtain

$$\sum_{j,\alpha} H_{j\alpha j'\alpha'}(\mathbf{k})\, c_{j\alpha}^{(n)}(\mathbf{k}) = E^{(n)}(\mathbf{k})\, c_{j'\alpha'}^{(n)}(\mathbf{k}) \qquad \text{where}$$

$$H_{j\alpha j'\alpha'}(\mathbf{k}) = \sum_{\mathbf{m}} \exp[i\mathbf{k}\cdot(\mathbf{r}_{\mathbf{m}} + \mathbf{r}_j \text{-} \mathbf{r}_{j'})] < \mathbf{0}j'\alpha' | H | \mathbf{m}j\alpha >$$

For each **k**-value an 8 by 8 matrix eigenvalue equation has to be solved. In this approximation when $j = j'$ the matrix element $\langle \mathbf{0}\, j'\alpha'|H|\mathbf{m}j\alpha\rangle$ is zero unless $\mathbf{m} = \mathbf{0}$ and $\alpha = \alpha'$. Hence the diagonal elements are either α_s or α_p depending on whether α denotes an s or p state. For the off-diagonal terms with $j \neq j'$ the expression above reduces to just four terms, one for each neighbour. Fig. 8.28 shows the resulting band structure along a few directions in the BZ as well as the density of states. In the latter the lower part contains the four valence bands and the upper part the conduction bands, separated by a small band gap

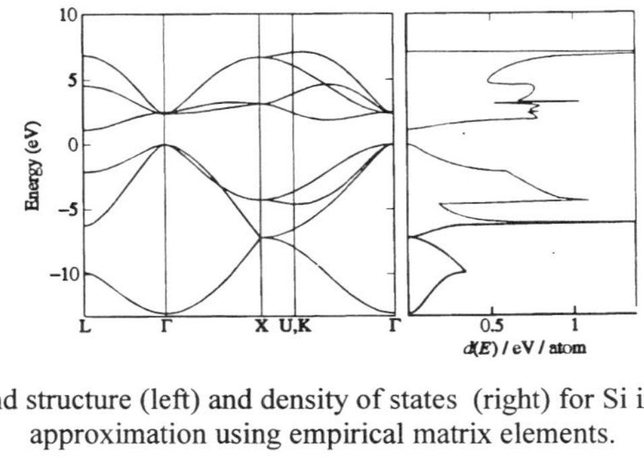

Fig. 8.28: The band structure (left) and density of states (right) for Si in the tight-binding approximation using empirical matrix elements.

of about 1.1 eV, indicating that Si a semi-conductor. Also the total energy can be estimated but since the matrix elements are constants, the energy as a function of lattice constant cannot be obtained. Using matrix elements dependent on the distance between neighbours, either calculated semi-empirically or *ab-initio*, will allow this. The semi-empirical approach, used extensively by Harrison (1980), is capable via proper parameterisation of various occurring integrals to reproduce many features of the electronic structures and the resulting properties of materials. The *ab-initio* approach in its simplest form does not provide a particularly realistic starting point for calculations but it provides a basis for more accurate models.

Summarising we can say that, although the NFE and TB approaches represent extreme views of bonding in solids, both produce a band structure with similar features. In general a full self-consistent calculation not neglecting any elements has to be employed for reliable predictions and to determine whether for a particular material an NFE-like or TB-like approach is the most appropriate. Generally for metals one uses as a first approximation the NFE approach while for non-metals typically the TB approach is preferred as a first approximation. In many cases the simple schemes as outlined here have to be refined considerably before reliable predictions can be made. This is true for metals as well as non-metals. Several more sophisticated theories have been developed. For details of these theories we refer to the literature, e.g. Harrison (1980), Pettifor (1995) and Sutton (1993).

Density functional theory

The solution of the Schrödinger equation for a particular system, even in an approximate way, is generally an enormous task. For an N-electron system the wavefunction contains $4N$ (or $3N$ omitting spin) variables. Therefore directly after the discovery of quantum mechanics attempts were made to employ the (observable) electron density ρ directly. In fact there are good arguments to do so since

- the electron density $\rho(\mathbf{r})$ integrates to the number of electrons, i.e. $\int\rho(\mathbf{r})\,\mathrm{d}\mathbf{r} = N$,
- $\rho(\mathbf{r})$ has maxima, actually cusps, at the positions of the nuclei R_A and
- the density at the position of the nucleus contains information about the nuclear charge Z via $\lim_{r_{iA}\to 0}\left[\frac{\partial}{\partial r}+2Z_A\right]\bar{\rho}(r)=0$, where $\bar{\rho}(r)$ denotes the spherically averaged electron density at the nuclear position of atom A.

Thus $\rho(\mathbf{r})$ already provides all information necessary for specifying a Hamilton operator H for a specific structure.

The earliest attempt was the Thomas-Fermi(-Dirac) model (around 1928) in which the kinetic energy T and exchange energy V_X expressions for a homogeneous electron gas were used and which was applied primarily to atoms but also molecules and solids. In that approximation both T and E_X become a functional of $\rho(\mathbf{r})$

$$T=\frac{3}{10}(3\pi^2)^{2/3}\int\rho^{5/3}(\mathbf{r})\,\mathrm{d}\mathbf{r} \quad \text{and} \quad E_X=-\alpha\frac{9}{8}\left(\frac{3}{\pi}\right)^{1/3}\int\rho^{4/3}(\mathbf{r})\,\mathrm{d}\mathbf{r} \qquad (8.47)$$

The parameter $\alpha = ⅔$ for the homogeneous electron gas, but was used as a parameter in the HFS method, which can be seen as a second example to reduce from wavefunction to electron density $\rho(\mathbf{r})$.

The next important step was made only in 1964 by the discovery of two important theorems by Hohenberg[u] and Kohn[v] and which has led to *density functional theory* (DFT). The first theorem states that 'the external potential $V_{ext}(\mathbf{r})$ is (within a constant) a unique functional of $\rho(\mathbf{r})$; since in turn $V_{ext}(\mathbf{r})$ fixes H, we see that the full many particle ground state is a unique functional of $\rho(\mathbf{r})$'. Symbolically we have

$$\rho_0 \rightarrow (n, Z_A, R_A) \rightarrow H \rightarrow \Psi_0 \rightarrow E_0$$

where ρ_0 and Ψ_0 denote the exact electron density and wavefunction, respectively. Since the ground state energy E_0 is a functional[w] of $\rho(\mathbf{r})$, so must be its individual components and we write $E_0[\rho_0] = T[\rho_0]+E_{ee}[\rho_0]+E_{ext}[\rho_0] \equiv F_{HK}[\rho_0]+\int V_{ext}[\rho_0(\mathbf{r})]\,\mathrm{d}\mathbf{r}$. The last term contains the system-specific terms, i.e. the potential energy $V_{ext} = -\sum_A Z_A/r_{1A}$ due to nuclear-electron attraction. The first terms, i.e. the kinetic energy T and electron-electron interaction E_{ee}, are system independent, i.e. independent of N, Z_A and R_A, and define $F_{HK}[\rho]$, the *Hohenberg-Kohn functional*. This implies that, if $F_{HK}[\rho]$ is calculated where ρ is an arbitrary density[x], it yields

$$F_{HK}[\rho] = T[\rho] + E_{ee}[\rho] = \langle \Psi | T+V_{ee} | \Psi \rangle$$

i.e. the proper expectation value of $T+V_{ee}$. Unfortunately, the form of F_{HK} is unclear. The only contribution that is evident is the classical Coulomb part $E_C[\rho]$ so that we have

$$E_{ee}[\rho] = \tfrac{1}{2}\int \rho(\mathbf{r}_1)V_C\,\mathrm{d}\mathbf{r}_1 + E_{ncl}[\rho] \equiv E_C[\rho] + E_{ncl}[\rho]$$

where the Coulomb potential $V_C = \int[\rho(\mathbf{r}_2)/r_{12}]\mathrm{d}\mathbf{r}_2$ is used. The term $E_{ncl}[\rho]$ is a non-classical contribution containing all effects of self-interaction, exchange and correlation. The major challenge of DFT is to find explicit expressions for $T[\rho]$ and $E_{ncl}[\rho]$. However, the important point to note is that the ground state electron density $\rho(\mathbf{r})$ determines all properties.

Now suppose that a trial density $\rho(\mathbf{r})$ is employed. This density defines its own Hamilton operator H_ρ and therefore Ψ_ρ. This wavefunction can be taken as a trial wavefunction in the variational principle using the true Hamilton operator H generated by V_{ext}. Hence we have

$$\langle \Psi_\rho | H_\rho | \Psi_\rho \rangle = T[\rho] + E_{ee}[\rho] + \int \rho(\mathbf{r})V_{ext}\,\mathrm{d}\mathbf{r} = E[\rho] > E_0[\rho] = \langle \Psi_0 | H | \Psi_0 \rangle$$

This is the second theorem of Hohenberg and Kohn which states that the functional $F_{HK}[\rho(\mathbf{r})] = T[\rho(\mathbf{r})]+E_{ee}[\rho(\mathbf{r})]$ that delivers the ground state energy of the system provides the lowest energy if (and only if) $\rho(\mathbf{r}) = \rho_0(\mathbf{r})$. This theorem thus provides a variational principle for $\rho(\mathbf{r})$. However, this variational principle applies only to the exact functional with approximate densities. For approximate functionals the variational principle is not valid.

[u] Hohenberg, P. and Kohn, W. (1964), Phys. Rev. **136**, B864.

[v] Walter Kohn (1923-2…). American physicist who received the Noble Prize in Chemistry in 1998 for his development of the density-functional theory.

[w] A function maps a number on a number. A functional maps a function on a number. A functional x dependent on the function y is indicated by $x[y(z)]$ while $y(z)$ denotes the function y dependent on the variable z.

[x] Arbitrary in this respect means n-representable. See e.g. Parr, R.G. and Yang, W. (1982).

Another major step was made by Kohn and Sham[y]. They realised that for a system of non-interacting electrons a Slater determinant of spin orbitals is an exact solution of the Schrödinger equation. In such a case the Hamilton operator H_S is given by

$$H_S = -\tfrac{1}{2}\sum_i [\nabla_i^2 + V_S(\mathbf{r}_i)] \equiv \sum_i f_{KS}(i) \tag{8.48}$$

where V_S is a one-electron potential to be identified later. In this connection the one-electron operator f_{KS} is referred to as the *Kohn-Sham operator*. The Slater determinant, indicated by Θ_S to distinguish it from the conventional Ψ, reads

$$\Theta_S = N^{-1/2}|\theta_1(\mathbf{r}_1)\ \theta_2(\mathbf{r}_2)\ \ldots\ \theta_N(\mathbf{r}_N)|$$

and the spin orbitals $\theta_i(\mathbf{r}_i)$, addressed as *Kohn-Sham orbitals* to distinguish them from their HF counterparts, are determined from

$$f_{KS}(1)\ \theta_i(\mathbf{r}_1) = \varepsilon_i\ \theta_i(\mathbf{r}_1)$$

Since we are interested in a real system with interacting electrons the potential V_S is chosen in such a way that the electron density $\rho(\mathbf{r})$ equals the exact density $\rho_0(\mathbf{r})$ or

$$\rho(\mathbf{r}) = \sum_i \theta_i^2(\mathbf{r}) = \rho_0(\mathbf{r}) \tag{8.49}$$

Next Kohn and Sham suggested to calculate the kinetic energy T_S of the non-interacting, reference system with the same density as the real, interacting system

$$T_S = -\tfrac{1}{2}\sum_i \langle\theta_i|\nabla^2|\theta_i\rangle \tag{8.50}$$

It can be shown that $T_S \leq T$ where T is the kinetic energy of the real system. The functional $F_{HK}[\rho(\mathbf{r})]$ is separated according to

$$F_{HK}[\rho] = T_S[\rho] + E_C[\rho] + E_{xc}[\rho]$$

defining the *exchange-correlation energy* as

$$E_{xc}[\rho] \equiv \{T[\rho] - T_S[\rho]\} + \{E_{ee}[\rho] - E_C[\rho]\} = T_c[\rho] + E_{ncl}[\rho]$$

The residual part T_c of the true kinetic energy not covered by T_S is thus added to the non-classical electrostatic contributions. Therefore E_{xc} is a functional that contains everything that is unknown, i.e. the self-interaction, exchange, correlation and the remainder of the kinetic energy. Since the total energy is a functional of $\rho(\mathbf{r})$, T_S must also be a functional of $\rho(\mathbf{r})$.

The total energy of the real system is thus

$$E[\rho] = T_S[\rho] + E_C[\rho] + E_{xc}[\rho] + E_{ext}[\rho] \tag{8.51}$$

where the only term for which no explicit form is present is E_{xc}. Now the variational theorem is applied in order to minimise this energy expression under the usual constraint of $\langle\theta_i|\theta_j\rangle = \delta_{ij}$. It can be shown that this procedure results in

$$\{-\tfrac{1}{2}\nabla^2 + [V_C + V_{xc} + V_{ext}]\}\theta_i(\mathbf{r}_1) = [-\tfrac{1}{2}\nabla^2 + V_S]\ \theta_i(\mathbf{r}_1) = \varepsilon_i\theta_i(\mathbf{r}_1)$$

identifying the potential V_S as introduced in Eq. (8.48). The term V_{xc} is simply the potential due to the exchange-correlation energy E_{xc} and given by the functional derivative $V_{xc} = \delta E_{xc}/\delta\rho$.

[y] Kohn, W. and Sham, L.S. (1965), Phys. Rev. **140**, A1133.

As indicated before, an important practical goal of DFT is to approximate the functional E_{xc} as best as possible. Unfortunately, in contrast to the HF approach where systematic improvement can be realised by extending the basis set, no such systematic improvement is possible for the DFT functionals and a strong trial-and-error component is involved. We discuss briefly three generally accepted approaches. The first is the local density approximation, the second is the generalised gradient approximation while the third is the hybrid functional approximation.

The basis for the *local density approximation* (LDA) is the homogeneous electron gas, referred to before, because the homogeneous electron gas is the only system for which the exchange and correlation energy are known exactly or to a high degree of accuracy. It is assumed that

$$E_{xc}^{LDA}[\rho] = \int \rho(\mathbf{r})\varepsilon_{xc}[\rho(\mathbf{r})]\,d\mathbf{r} = E_x^{LDA} + E_c^{LDA} = \int \rho\{\varepsilon_x[\rho] + \varepsilon_c[\rho]\}\,d\mathbf{r} \quad (8.52)$$

where ε_{xc} is the exchange-correlation energy per particle for the homogeneous electron gas. The exchange part ε_x is well known and given by Eq. (8.47). For the correlation part ε_c such an explicit expression is not available but accurate Padé approximant interpolations to highly accurate calculations for the homogeneous electron gas are available. The LDA overestimates the correlation energies typically by 100% and therefore overestimates bonding. With reference to a standard set of about 50 experimental values of atomisation energies for small molecules, the so-called G2 set, the average error is about 150 kJ/mol, to be compared with about 330 kJ/mol for HF calculations. This does not make this approximation a thermochemical tool since 'chemical accuracy' requires about 0.1 eV or 10 kJ/mol absolute error. On the other hand, properties such bond angles, bond lengths, vibrational frequencies, etc. tend to agree surprisingly well with experiments for normal bonds, i.e. hydrogen and van der Waals bonds excepted, so that it can be considered as a useful structural tool.

To improve upon this approximation, it is assumed that not only $\rho(\mathbf{r})$ but also $\nabla\rho(\mathbf{r})$ is important, i.e. the LDA is considered as a first term in a Taylor expansion of the density. Hence we write

$$E_{xc} = \int \rho(\mathbf{r})\varepsilon_{xc}\,d\mathbf{r} + \int C_{xc}(\rho)s(\mathbf{r})\,d\mathbf{r}$$

where the dimensionless gradient $s = |\nabla\rho(\mathbf{r})|/\rho^{4/3}(\mathbf{r})$ is used. The parameter s assumes a large value for a large value of the gradient $\nabla\rho(\mathbf{r})$ but also for a small value of the density $\rho(\mathbf{r})$. However, this approximation does not lead to the desired improved accuracy. This is due to the fact that the exchange-correlation hole in this approximation has lost several useful properties. In particular, the integrals of the Fermi and Coulomb hole do not yield the required values of –1 and 0 electrons, respectively. These requirements can be enforced by putting these values as constraints for the holes. The gradient approximation extended in this way is known as the *generalised gradient approximation* (GGA). The same separation is made as in the LDA, Eq. (8.52), and we write $E_{xc}^{GGA} = E_x^{GGA} + E_c^{GGA}$. The exchange part E_x^{GGA} is

$$E_x^{GGA} = E_x^{LDA} - \int F(s)\rho^{4/3}(\mathbf{r})\,d\mathbf{r}$$

For $F(s)$ several complex expressions are in use, either semi-empirical or derived from first principles. The corresponding Coulomb expressions are even more complex. Generally they are, as many of the exchange expressions, devoid of direct physical interpretation but mainly chosen to represent mathematically correct properties as well as to be a useful result. The accuracy of the GGA is considerably better than that for the LDA. For the G2 set an average error of 25 kJ/mol results.

A more recent approach uses the so-called adiabatic switching-on of the Coulomb correlation as indicated by

$$E_{\text{xc}} = \int_0^1 E_{\text{ncl}}(\lambda)\,\text{d}\lambda$$

At $\lambda = 0$ we have a non-interacting system and the Coulomb correlation is absent. Only the exchange correlation, which can be calculated exactly from the determinant of Kohn-Sham orbitals, is present. At $\lambda = 1$ we have a fully interacting system and the Coulomb correlation is fully present. The value of E_{xc} thus can be approximated as a sum of contributions for various values of λ, of which one is the exact exchange correlation. Therefore we can focus on the Coulomb correlation. Using $E_{\text{xc}} = \frac{1}{2}(E_{\text{xc}}^{\lambda=0} + E_{\text{xc}}^{\lambda=1})$, where for $E_{\text{xc}}^{\lambda=1}$ the LDA form is used, the G2 set results in an error of about 30 kJ/mol. Using an empirical expression for $E_{\text{xc}}^{\lambda=1}$ with three parameters fitted to the G2 data leads to an accuracy of about 13 kJ/mol. Functionals of this sort, where a certain amount of exact exchange is mixed in, are called *hybrid functionals* and provide the most promising route for improving accuracy at present.

For actual calculation of the Kohn-Sham orbitals typically an expansion in basis functions is used, quite similar to the HF case, although a full numerical solution is feasible. For molecules similar basis sets as used in the HF method are employed. For solids plane waves are also employed. Finally we have to mention that many complexities have not been addressed. We only mention degenerate systems, open shells, excited states and finite-temperature calculations. For these aspects we refer to e.g. Parr and Yang (1989), Dreizler and Gross (1990) or Koch and Holthausen (2001).

Summarising we have for DFT that:

- For an external potential V_{ext} all properties are determined by $\rho(\mathbf{r})$. For a given ρ the functional $F_{\text{HK}}[\rho]+\int\rho(\mathbf{r})V_{\text{ext}}\,\text{d}\mathbf{r}$ yields the corresponding ground-state energy and this functional attains a minimum value with respect to ρ if $\rho = \rho_0$.
- A reference system of non-interacting n electrons described by a Slater determinant Θ_{S}, containing orbitals $\theta_i(\mathbf{r}_i)$ and obeying $\rho = \rho_0$, can be defined. For this reference system an effective single-particle potential V_{S} can be defined and the kinetic energy T_{S} can be calculated according to Eq. (8.50).
- The energy of the real, interacting system is given by $E = T_{\text{S}}+E_{\text{C}}+E_{\text{xc}}+E_{\text{ext}}$ according Eq. (8.51). From the variational principle it appears that $V_{\text{S}} = V_{\text{C}}+V_{\text{xc}}+V_{\text{ext}}$.
- Provided we know the explicit form of the potentials we know V_{S} and solving the one-electron equations provides the Kohn-Sham orbitals $\theta_i(\mathbf{r}_i)$, which yield on their turn the electron density $\rho(\mathbf{r})$ according to Eq. (8.49). Since the potentials are defined in terms of $\rho(\mathbf{r})$, an iterative procedure has to be used. This procedure yields the exact energy and density, provided the exact functional for V_{xc} is used.

Since the exact functional is not known, an approximate functional has to be used. DFT is the physicists' method of choice for the calculation of electronic properties of solids. More recently chemists also used it extensively. For small molecules and high degree of accuracy the conventional methods are preferred. For large systems and more modest accuracy, DFT is preferable. While for conventional methods a systematic improvement yielding in principle arbitrary accuracy can be reached, there is no systematic way to achieve an arbitrary high level of accuracy in DFT.

Example 8.3: Spinel $MgAl_2O_4$ in DFT

AB_2O_4 compounds with the spinel structure are important prototype structures in inorganics. In this structure the oxygen atoms form an FCC lattice. The A atoms are situated at the tetrahedral and the B atoms at the octahedral interstitial sites of the FCC network. However, for certain spinel compounds the A atoms exchange partially with the B atoms to form $[A_{8-x}B_x](B_{16-x}A_x)O_{32}$ where [] indicate the tetrahedral and () the octahedral sites. For $MgAl_2O_4$ the so-called inversion parameter x is typically about 2. There is also a high-pressure (HP) form of $MgAl_2O_4$, which is supposed to be important in the deep mantle of the earth.

For MgO, Al_2O_3, $MgAl_2O_4$ and the HP-form of $MgAl_2O_4$, DFT calculations[z] within the LDA were done with full relaxation of the unit cell size and the atomic parameters. So-called 'vanderBilt' pseudo-potentials were used and the Kohn-Sham orbitals were expanded in plane waves with a kinetic energy cut-off of 36 Ry. The results in Fig. 8.29 show that $MgAl_2O_4$ is stable with respect to MgO + Al_2O_3 and has a larger unit cell volume than the sum of MgO + Al_2O_3. The HP-form is unstable at normal pressure and has a smaller unit cell volume with respect to the forming compounds. The calculations indicate that with increasing pressure $MgAl_2O_4$ first decomposes at pressure P_1 into MgO + Al_2O_3 while upon further increase at pressure P_2 these compounds combine again to the HP-form. Taking into account the zero-point vibrational energy and the inversion contribution to the entropy it is estimated that at 1800 K, $P_1 \cong 4$ GPa and $P_2 \cong 33$ GPa. The ranges for the experimental data are $8 < P_1 < 15$ GPa and $25 < P_2 < 40$ GPa. The theoretical estimates are rather sensitive to the precise values of the various contributions involved and the difference with the experimental result is probably mainly due to the relatively simple estimate for the inversion contribution. However, the unit cell volumes of both forms are in good agreement with the experimental results as are the estimates for the bulk modulus at 0 K using a fit to energy-volume data using the second-order Murnaghan equation of state (Chapter 9).

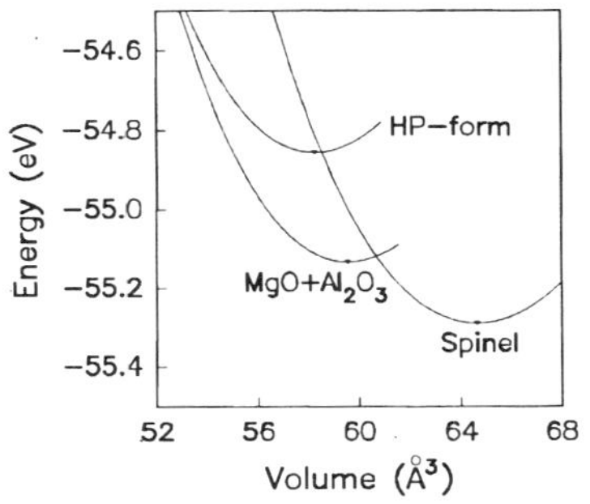

Fig. 8.29: The energy-volume relationship for MgO+Al_2O_3, $MgAl_2O_4$ and the HP-form of $MgAl_2O_4$.

[z] Fang, C.M. and de With, G. (2002), Phil. Mag. A **82**, 2885.

8.6 Bonding in solids: other approaches

Although all bonding characteristics and the resulting structures are determined by quantum mechanics, historically one distinguished between ionic, covalent, metallic and van der Waals bonding. The associated descriptions are based on various concepts such as atomic potentials, (partial) charges, etc., only indirectly based on quantum mechanics. In view of their usefulness they are widely used nevertheless.

The ionic bond

One way to gain energy during the formation of solids is to transfer an electron from one atom to another, e.g. Na → Na^+ and Cl → Cl^-. The resulting ions attract and repel each other and the *ionic bond* is based on the balance between these electrostatic forces, described by the electrostatic or *Coulomb energy* V_C, and the short-ranged repulsion forces described by V_{rep} due to orbital overlap. Obviously the total potential energy V is given by $V = V_C+V_{rep}$. The equilibrium distance r_0 is given by the solution of $\partial V/\partial r = 0$ while the force constants are related to $\partial^2 V/\partial r^2|_{r=r0}$.

The Coulomb[aa] potential energy ϕ_C between two charges $Q_A e$ and $Q_B e$ separated by a distance r is given by (Born and Huang, 1954; Slater, 1939)

$$\phi_C = \frac{Q_A Q_B e^2}{4\pi\varepsilon_0 r} = \frac{Q_A Q_B a}{r} \qquad a \equiv \frac{e^2}{4\pi\varepsilon_0} \tag{8.53}$$

where e is the unit charge and ε_0 is the dielectric permittivity of the vacuum. The electrostatic forces are centrally symmetric but their range is large so that in a lattice all ion pair interactions up to a (very) large distance have to be taken into account. Writing this sum in terms of the (momentary) nearest-neighbour distance r it appears to be attractive and we have

$$V_C = -\tfrac{1}{2}\sum\nolimits'_{\mathbf{n},\alpha} \phi_C(\mathbf{r}_{\mathbf{n},\alpha}) = -\tfrac{1}{2}\frac{a}{r}\sum\nolimits'_{\mathbf{n},\alpha} Q_0 Q_{\mathbf{n},\alpha}\left(\frac{r}{r_{\mathbf{n},\alpha}}\right) = -M\frac{a}{r} \tag{8.54}$$

where the prime over the summation sign excludes the atom at the origin. The lattice sum M is a pure number with a value characteristic for the crystal structure and usually addressed as the *Madelung constant*. This expression yields the total interaction energy per ion pair. In view of the long-range nature of the interaction special techniques are necessary to calculate accurately the value of M. In Table 8.4 values for M for several lattice types are given. For structures with more than one magnitude of charge, e.g. as in CaF_2, it is convenient to define a reduced Madelung constant M'. For a compound A_xB_y charge neutrality requires that $xQ_A = yQ_B$. The electrostatic energy per unit $x+y$ atoms is then given by[bb]

$$V_C = -M'\frac{(x+y)}{2}\frac{Q_A Q_B e^2}{(4\pi\varepsilon_0)r} = -M'\frac{(x+y)}{2}Q_A Q_B\frac{a}{r} \tag{8.55}$$

For $x = y = 1$ this expression reduces to the previous definition. Defined in this way the Madelung constant M' does not change if we define a molecular unit to be an integral number of primitive cells of a simpler structure. Moreover, M' is nearly constant (see Table 8.4).

[aa] Charles Augustin de Coulomb (1736-1806). French scientist who contributed to both electromagnetism and mechanics.

[bb] Johnson, Q.C. and Templeton, D.H. (1961), J. Chem. Phys. **34**, 2004.

Table 8.4: Madelung constants M for various crystals.

Structure	M	M'
Rock salt (NaCl)	1.748	1.75
Cesium chloride (CsCl)	1.763	1.76
Zinc blende (ZnS)	1.638	1.64
Wurtzite (ZnS)	1.641	1.64
Fluorite (CaF_2)	5.039	1.68
Rutile (TiO_2)	4.816	1.60
Anatase (TiO_2)	4.800	1.60
Corundum (Al_2O_3)	25.03	1.68

The repulsion is described by

$$\phi_{\text{rep}} = b\exp(-r/\rho) \qquad \text{or} \qquad \phi_{\text{rep}} = b/r^n \tag{8.56}$$

where (b,ρ) or (b,n) are parameters. The range of the repulsion is relatively short so that generally only nearest-neighbour interaction has to be taken into account. However, here also lattice sums can be defined. For example, for the power law interaction

$$V_{\text{pow}} = \tfrac{1}{2}\sum\nolimits_{\mathbf{n},\alpha}\phi(\mathbf{r}_{\mathbf{n},\alpha}) = \tfrac{1}{2}\frac{b}{r^n}\sum\nolimits_{\mathbf{n},\alpha}\left(\frac{r}{r_{\mathbf{n},\alpha}}\right)^n = \tfrac{1}{2}\frac{bS_n}{r^n} = \tfrac{1}{2}\frac{B_n}{r^n}$$

where r is again the nearest-neighbour distance and $\mathbf{r}_{\mathbf{n},\alpha}$ is the distance to atom α in cell $\mathbf{n}$. Values for S_n from $n = 4$ to $n = 30$ have been calculated[cc] for the SC, BCC, FCC and HCP structure. In Table 11.1 selected values are given. Since the overall structure must be neutral and the size of the various ions may be rather different, the co-ordination generally differs in different structures and the co-ordination of positive and negative ions may be different. For example, in the NaCl structure both the Na and Cl ions have a six-fold co-ordination while in the CsCl structure each ion has an eight-fold co-ordination. In the CaF_2 structure the Ca ions are surrounded by eight F ions while each F ion is co-ordinated by four Ca ions.

Example 8.4: The Born model

In simple binary salts such as NaCl, CsCl and ZnO each ion is coordinated with Z ions of the opposite charge. Using the exponential form and neglecting all but the nearest-neighbour contributions the energy per unit cell $u(r)$ is

$$u(r) = -\frac{A}{r} + B\mathrm{e}^{-r/\rho} \qquad A = aM \qquad \text{and} \qquad B = Zb$$

where the symbols have the meaning defined previously. Since $\mathrm{d}\ln V = 3\mathrm{d}\ln r$ or $(1/V)\mathrm{d}V = (3/r)\mathrm{d}r$ one easily finds for the pressure p and the bulk modulus K

$$p = -\frac{\mathrm{d}u}{\mathrm{d}V} = -\frac{1}{3V}\left[\frac{-A}{r} + B\left(\frac{r}{\rho}\right)\mathrm{e}^{-r/\rho}\right] \qquad \text{and} \tag{8.57}$$

[cc] Lennard-Jones, J.E. and Ingham, A.E. (1925), Proc. Roy. Soc. (London), A **107**, 636 and Kihara, T. and Kuba, S. (1952), J. Phys. Soc. Japan **7**, 348.

$$K=-V\frac{\mathrm{d}p}{\mathrm{d}V}=\frac{1}{9V}\left[\frac{-A}{r}-B\left(\frac{r}{\rho}\right)\mathrm{e}^{-r/\rho}+B\left(\frac{r}{\rho}\right)^2\mathrm{e}^{-r/\rho}\right]+p$$

For static equilibrium at $p = 0$ we have from Eq. (8.57)

$$\frac{A}{r_0}=B\left(\frac{r_0}{\rho}\right)\mathrm{e}^{-r_0/\rho} \tag{8.58}$$

Taking $r = r_0$, $p = 0$ and eliminating B using Eq. (8.58) we obtain for K and u

$$K=\frac{A}{9V_0r_0}\left(-2+\frac{r_0}{\rho}\right) \quad \text{and} \quad u(r_0)=\frac{-A}{r_0}\left(1-\frac{\rho}{r_0}\right) \tag{8.59}$$

In this way the parameters (b,ρ) can be determined from the bulk modulus K and equilibrium distance r_0. For example, for LiCl r_0 = 0.2572 nm and K = 29.3 GPa and we obtain $r_0/\rho = 7.75$ and $b = 0.782\times10^{-16}$ J resulting in energy u = 822 kJ/mol, to be compared with $u_{\exp}$ = 844 kJ/mol. In general, the error is about a few percent and because $r_0/\rho \cong 10$, most of the lattice energy is provided by the attraction.

Problem 8.6

Taking the power-law form of the repulsion energy $\phi_{\text{rep}} = B/r^n$, show that the corresponding equations to Example 8.4 are

$$\frac{A}{r_0}=\frac{nB}{r_0^n} \qquad K=\frac{A}{9V_0r_0}(-1+n) \quad \text{and} \quad u(r_0)=\frac{-A}{r_0}\left(1-\frac{1}{n}\right)$$

The covalent bond

Another way to gain energy during the formation of molecules or solids is to share an electron between atoms. For example, each Cl atom has seven electrons and by combining two Cl atoms a closed shell configuration with a lower energy results. This linkage is known as the *covalent bond*. The covalent bonds are generally highly directional. This type of bonding has to be described by quantum mechanics but nevertheless to simplify the matter attempts have been presented based on atomic potentials, usually neglecting the directional dependence. In this case also an attractive and a repulsive part are introduced. An often-used expression is the *Mie potential* given by

$$\phi_{\mathrm{M}}=\frac{b}{r^n}-\frac{c}{r^m} \tag{8.60}$$

where b, c, n and m are parameters and $n > m$. In crystal lattices the interactions have to be summed over all atom pairs and using the lattice sums, again indicated by S_n, we have

$$V_{\mathrm{M}}=½\sum\nolimits_{\mathbf{n},\alpha}\phi_{\mathrm{M}}(\mathbf{r}_{\mathbf{n},\alpha})=½\left(\frac{bS_n}{r^n}-\frac{cS_m}{r^m}\right)=½\left(\frac{B_n}{r^n}-\frac{C_m}{r^m}\right) \tag{8.61}$$

The parameters are generally determined by fitting on macroscopic quantities such as equilibrium distance, lattice energy and compressibility. An often-used choice, mainly for convenience, is the 12-6 or *Lennard-Jones potential* with $n = 12$ and $m = 6$. In this case an alternative form for Eq. (8.60) is

$$\blacktriangleright \quad \phi_{\rm LJ} = 4\varepsilon\left[\left(\frac{\sigma}{r}\right)^{12} - \left(\frac{\sigma}{r}\right)^{6}\right]$$

The parameter ε describes the depth of the potential energy curve. At $r = \sigma$, $\phi_{\rm LJ} = 0$.

While the power-law relationship with $m = 6$ can be rationalised for the attractive part of the potential, the repulsion is on quantum-mechanical grounds expected to be better described by an exponential, like in Eq. (8.56). Combining leads to the *exp-6 potential*

$$\phi = b\exp(-r/\rho) - cr^{-6}$$

Another often-used expression is the *Morse potential*[dd] given by

$$\phi_{\rm M} = \phi_0\{\exp[-2\alpha(r-r_0)] - 2\exp[-\alpha(r-r_0)]\} \qquad (8.62)$$

with ϕ_0, r_0 and α parameters. The dissociation energy ϕ_0 is obtained at the equilibrium distance r_0. The parameter α describes the curvature at the minimum of the curve. Apart from describing the potential energy curve reasonably, it has the additional advantage that the quantum-mechanical oscillator problem using this potential can be solved nearly exactly, resulting in

$$u_n = \hbar\omega(n+½) - x\hbar\omega(n+½)^2 \quad \text{with}$$

$$\omega = \frac{1}{2\pi}\sqrt{\frac{k}{\mu}} = \frac{\alpha}{\pi}\sqrt{\frac{\phi_0}{2\mu}} \qquad \text{and} \qquad x = \frac{\hbar\omega}{4\phi_0}$$

Here μ is the reduced mass and k is the force constant given by $k = 2\alpha^2\phi_0$. The Morse potential is sometimes also given as

$$\phi_{\rm M} = \phi_0\{1 - \exp[-\alpha(r-r_0)]\}^2$$

which is an equivalent form but with a different zero energy level.

The van der Waals interaction

The origin of the *van der Waals interaction* lies in the polarising effect caused by the influence of the electric field associated with the electrons moving around an atom upon the electrons moving around the nucleus of a neighbouring atom. Although relatively weak, this bond is responsible for the bonding in inert gas and diatomic molecular solids. Moreover, the van der Waals interaction plays an important role in polymers in which case the interaction is generally addressed as secondary interaction, the primary interaction being the covalent bonds.

To find the proper expression let us consider first an arbitrary charge distribution ρ of point charges $e^{(i)}$ at position $\mathbf{r}^{(i)}$ from the origin O located at the centre of mass (Fig. 8.30). The potential energy ϕ at a certain point P located at $\mathbf{r}$ outside the sphere containing all charges is given by

[dd] Morse, P.M. (1929), Phys. Rev. **20**, 57.

Johan Diderik van der Waals (1837-1923)

Born in Leyden, The Netherlands, he became a schoolteacher and later a director of a secondary school in The Hague. In 1873 he obtained his doctor's degree for a thesis entitled *Over de Continuïteit van den Gas - en Vloeistoftoestand* (On the continuity of the gas and liquid state). In 1876 the old Athenaeum Illustre of Amsterdam became university and Van der Waals was appointed the first professor of physics. Together with Van't Hoff and Hugo de Vries, the geneticist, he contributed to the fame of the university, and remained faithful to it until his retirement, in spite of enticing invitations from elsewhere. The immediate cause of his interest in the subject of his thesis was Clausius' treatise considering heat as a phenomenon of motion. In order to explain T. Andrews' experiments (1869) revealing the existence of 'critical temperatures' in gases, he did see the necessity of taking into account the volumes of molecules and the intermolecular forces in establishing the relationship between the pressure, volume and temperature of gases and liquids. In 1880 he enunciated the Law of Corresponding States, which served as a guide during experiments, which ultimately led to the liquefaction of hydrogen by J. Dewar in 1898 and of helium by H. Kamerlingh Onnes in 1908. In 1890 the first treatise on the *Theory of Binary Solutions* appeared. Van der Waals and Ph. Kohnstamm subsequently assembled lectures on this subject in the *Lehrbuch der Thermodynamik*. His thermodynamic theory of capillarity first appeared in 1893. In this he accepted the existence of a gradual, though very rapid, change of density at the boundary layer between liquid and vapour. He received the Nobel Prize for Physics in 1910 for his work on the equations of state of gases and fluids.

$$\phi(\mathbf{r}) = \sum\nolimits_i \frac{e^{(i)}}{s^{(i)}}$$

where $s^{(i)} = |\mathbf{s}^{(i)}| = |\mathbf{r}-\mathbf{r}^{(i)}|$ is the distance of charge e_i to P. Developing $1/s^{(i)}$ in a Taylor series with respect to $\mathbf{r}_i$, we may write $1/s^{(i)} = 1/r + \mathbf{r}^{(i)}(\nabla 1/r)_\mathrm{O} + \cdots$ and therefore

$$\begin{aligned}\phi(\mathbf{r}) &= \sum\nolimits_i \frac{e^{(i)}}{r} + \sum\nolimits_i e^{(i)}\mathbf{r}^{(i)} \cdot \left(\nabla\frac{1}{r}\right)_\mathrm{O} + \tfrac{1}{2}\sum\nolimits_i e^{(i)}\mathbf{r}^{(i)}\mathbf{r}^{(i)} : \left(\nabla\nabla\frac{1}{r}\right)_\mathrm{O} + \cdots \\ &= e\phi(\mathbf{0}) + \mathbf{m}\cdot\phi'(\mathbf{0}) + \mathbf{Q}:\phi''(\mathbf{0}) + \cdots \\ &= \frac{e}{r} - \mathbf{m}\cdot\left(\nabla\frac{1}{r}\right)_\mathrm{P} + \mathbf{Q}:\left(\nabla\nabla\frac{1}{r}\right)_\mathrm{P} - \cdots\end{aligned} \tag{8.63}$$

where $r = |\mathbf{r}|$ is the distance of the point P to the origin O of ρ. In the second line we defined $e = \sum_i e^{(i)}$ the *total charge*, $\mathbf{m} = \sum_I e^{(i)}\mathbf{r}^{(i)}$ the *dipole moment* and $\mathbf{Q}$ =

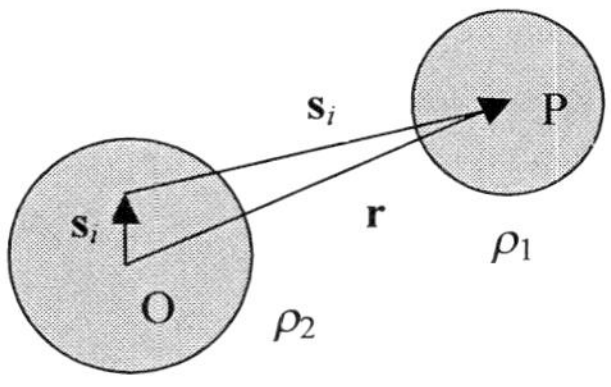

Fig. 8.30: The interaction of two charge distributions ρ_1 and ρ_2 with common origin O.

$\frac{1}{2}\sum_i e^{(i)}\mathbf{r}^{(i)}\mathbf{r}^{(i)}$ the *quadrupole moment*[ee]. Moreover we abbreviate $1/r$ by $\phi(\mathbf{0})$, $[\nabla(1/r)]_\mathrm{O}$ by $\phi'(\mathbf{0})$, etc. The minus sign in the last step for the second, fourth, etc. terms arises because the differentiation is now made at P and not at the origin of the potential O. This follows since for an arbitrary vector $\mathbf{x}$ it holds that $\nabla(1/x) = -\mathbf{x}/x^3$ and therefore $[\nabla(1/r)]_\mathrm{P} = -[\nabla(1/r)]_\mathrm{O}$. Eq. (8.63) is usually called the *multipole expansion* of the potential.

The *interaction energy* W of a point charge q at $\mathbf{r}$ in the potential ϕ of a charge distribution is $W = q\phi(\mathbf{r})$. Hence to calculate the interaction energy between two charge distributions ρ_1 and ρ_2, separated by $\mathbf{r}$ (Fig. 8.30), we express this energy as the energy of $\rho_1 = \sum_i e_1^{(i)}$ in the field of $\rho_2 = \sum_i e_2^{(i)}$. The interaction energy $W_{12} = \rho_1\phi_2$ ($= \rho_2\phi_1$) is then given by

$$W_{12} = \sum_i e_1^{(i)}\phi_2(r^{(i)}) = \sum_i e_1^{(i)}\left[\phi_2(\mathbf{0}) + \mathbf{r}_1^{(i)}\cdot\phi_2'(\mathbf{0}) + \tfrac{1}{2}\mathbf{r}_1^{(i)}\mathbf{r}_1^{(i)} : \phi_2''(\mathbf{0}) + \cdots\right]$$

$$= e_1\phi_2(\mathbf{0}) + \mathbf{m}_1\cdot\phi_2'(\mathbf{0}) + \mathbf{Q}_1 : \phi_2''(\mathbf{0}) + \cdots$$

If we take the centre of mass P of ρ_1 as common origin, the potential ϕ_2 due to ρ_2 is given by the last line of Eq. (8.63) and substituting in the previous equation results in

$$\blacktriangleright \quad W_{12} = \frac{A}{r} + B\cdot\left(\nabla\frac{1}{r}\right)_0 + \Gamma : \left(\nabla\nabla\frac{1}{r}\right)_0 + \cdots \qquad \text{where} \qquad (8.64)$$
$$A = e_1e_2 \qquad B = \left(e_1\mathbf{m}_2 - e_2\mathbf{m}_1\right) \qquad \Gamma = \left(e_1\mathbf{Q}_2 - \mathbf{m}_1\mathbf{m}_2 + e_2\mathbf{Q}_1\right)$$

This is the general expression for the electrostatic interaction energy expressed in terms of multipole moments of the charge distributions with respect to their own centre of mass[ff]. At long distance r the charge-charge interaction term, proportional to r^{-1}, will dominate. With decreasing distance the next terms will start to contribute. First, the charge-dipole interaction, proportional to r^{-2}, then the dipole-dipole and charge-quadrupole interaction, each proportional to r^{-3}, and so on. For neutral molecules the leading term is the dipole-dipole interaction.

In quantum mechanics we consider Eq. (8.64) as an operator to be used in perturbation theory. The zero-order wavefunctions are the product functions of the non-interacting molecules. If we denote the electronic ground states in molecules 1 and 2 by i and j, respectively, and similarly the excited states by i' and j', the result of second-order perturbation theory using terms up to the dipole-dipole interaction is

---

[ee] Since Laplace's equation $\nabla^2(1/r) = 0$ holds the tensor $\mathbf{Q}$ can be reduced to $\boldsymbol{\Theta} = \frac{1}{2}(3\mathbf{Q}-\mathrm{tr}(\mathbf{Q})\mathbf{I})$ and most authors refer to $\boldsymbol{\Theta}$ as the quadrupole moment. Since we will not really use either $\boldsymbol{\Theta}$ or $\mathbf{Q}$ but in this section to elucidate the origin of the van der Waals forces, we introduced only $\mathbf{Q}$ in the main text.

[ff] Böttcher, C.J.F. (1973), *Theory of dielectric polarization,* Elsevier, Amsterdam. See also Hirschfelder et al. (1964).

$$\phi = \phi_{\rm C} + \phi_{\rm ind} + \phi_{\rm vdW} \qquad \phi_{\rm C} = \langle i | W_{12} | i \rangle \langle j | W_{12} | j \rangle$$

$$\phi_{\rm ind} = \sum\nolimits'_{i'} \frac{\left| \langle i | \mathbf{m}_1 | i' \rangle \langle j | \mathbf{m}_2 | j \rangle \phi'(\mathbf{0}) \right|^2}{(E_i - E_{i'})} + \sum\nolimits'_{j'} \frac{\left| \langle i | \mathbf{m}_1 | i \rangle \langle j | \mathbf{m}_2 | j' \rangle \phi'(\mathbf{0}) \right|^2}{(E_j - E_{j'})} \tag{8.65}$$

$$\phi_{\rm vdW} = \sum\nolimits'_{i',j'} \frac{\left| \langle i | \mathbf{m}_1 | i' \rangle \langle j | \mathbf{m}_2 | j' \rangle \phi''(\mathbf{0}) \right|^2}{(E_i - E_{i'}) + (E_j - E_{j'})}$$

where the prime in the summation excludes $i = i'$ and $j = j'$. The quantities E_i, etc. are the eigenvalues (energies) and $\langle i|\mathbf{m}_1|i'\rangle$ the matrix elements of the dipole operator of molecule 1 between states i and i'. The first-order term $\phi_{\rm C}$ represents the (classical) *Coulomb* (or *electrostatic*) interaction. For neutral molecules the A- and B-terms are zero. If the molecules are polar ($\mathbf{m}_1,\mathbf{m}_2 \neq 0$ and therefore $\Gamma \neq 0$), we have the *Debye* interaction. It depends strongly on the orientation of the two molecules. We refrain from further discussion of this term since for spherically averaged interactions the net result is usually small. The second-order term contains the *induction* interaction $\phi_{\rm ind}$ and the *van der Waals* (or *dispersion*) interaction $\phi_{\rm vdW}$. The term $\phi_{\rm ind}$ is only non-zero if at least one of the molecules has a permanent dipole moment $\langle i|\mathbf{m}|i\rangle$ and usually also is small when spherically averaged. This is not true for the second-order term $\phi_{\rm vdW}$ in Eq. (8.65), which, after some calculation[gg], reduces to

$$\blacktriangleright \qquad \phi_{\rm vdW} = \frac{2}{3} \sum\nolimits'_{i',j'} \frac{\left(\sum_\alpha \left| \langle i | m_{1\alpha} | i' \rangle \right|^2 \right) \left(\sum_\beta \left| \langle j | m_{2\beta} | j' \rangle \right|^2 \right)}{(E_i - E_{i'}) + (E_j - E_{j'})} \frac{1}{r^6} \equiv -\frac{C}{r^6} \tag{8.66}$$

where $m_{1\alpha}$ and $m_{2\beta}$ indicate the components of $\mathbf{m}_1$ and $\mathbf{m}_1$, respectively. Since $(E_i - E_i') < 0$, $(E_j - E_j') < 0$ and the enumerator > 0, the van der Waals interaction is always negative and therefore attractive. For us the main feature of this complex expression to be noticed is the r^{-6} dependence.

Although nowadays *ab-initio* calculations based on this equation are feasible for realistic systems, approximations are generally still made. A frequently used approximation, due to London (1930), between molecules 1 and 2 yields

$$C = \frac{3}{2} \frac{\Delta_1 \Delta_2}{\Delta_1 + \Delta_2} \alpha_1 \alpha_2 \tag{8.67}$$

where α and Δ denote the polarisibility and the oscillation frequency, respectively. This approximation is exact for the so-called Drude model[hh] of an atom with only one characteristic frequency. For real atoms the energies Δ should be chosen in accordance with the strongest absorption frequencies of the molecules. In the absence of this information the ionisation energies can be taken. For the noble gases, however, about twice the ionisation potential has to be taken in order to match more reliable calculations.

Similar considerations on the dipole-quadrupole and quadrupole-quadrupole interactions lead to

---

[gg] For a spherical average $T_{\rm sph}$ of a second-rank tensor $\mathbf{T}$ we have $T_{\rm sph} = ⅓{\rm tr}\mathbf{T}$. If $\mathbf{T}$ is given as the dyadic product $\mathbf{T} = \mathbf{rr}$ of a vector $\mathbf{r}$, this reduces to $(r_{\rm sph})^2 = ⅓{\rm tr}(\mathbf{rr}) = ⅓\mathbf{r}.\mathbf{r} = ⅓\mathbf{r}^2 = ⅓r^2 = x^2 = y^2 = z^2$.

[hh] In the Drude model one assumes that an atom or molecule can be considered as a set of particles with charge e_i and mass m_i. Each of these particles is harmonically and isotropically bound to its equilibrium position.

$$-\frac{C'}{r^8} \quad \text{and} \quad -\frac{C''}{r^{10}} \tag{8.68}$$

where the full expressions for C' and C'' are even more complex. A similar approximation as for C leads to

$$C' = \frac{45}{8}\frac{\Delta_1\Delta_2\alpha_1\alpha_2}{e^2}\left(\frac{\alpha_1\Delta_1}{2\Delta_1+\Delta_2}+\frac{\alpha_2\Delta_2}{\Delta_1+2\Delta_2}\right) \quad \text{and} \quad C'' = \frac{315}{16}\frac{\Delta_1^2\Delta_2^2\alpha_1^2\alpha_2^2}{e^4(\Delta_1+\Delta_2)} \tag{8.69}$$

where e is the unit charge and the expressions are again exact for the Drude atoms. Although estimates for the various coefficients thus can be made, C, C' and C'' are usually considered as parameters. In that case usually only the leading r^{-6} term is used.

Problem 8.7

Prove Eq. (8.66) making use of the associated footnote. Note that $\langle i|\mathbf{m}|i\rangle$ behaves like a vector and $[\nabla\nabla(1/r)]_0$ like a second-rank tensor. For the latter use $\nabla r^n = nr^{n-1}\nabla r = nr^{n-2}\mathbf{r}$ and $\nabla\mathbf{r} = \mathbf{I}$.

8.7 Defects in solids

In this section an overview of the important defects in solids is given. For crystalline solids one can distinguish between zero-, one-, two- and three-dimensional defects. While for a zero-dimensional defect only a single atom deviates from the ideal crystallographic order, for a one-, two- and three-dimensional defect this is a (connected) line, plane or volume of atoms (Flynn, 1972; Henderson, 1972).

Upfront we note that zero-dimensional defects, better known as *point defects*, are thermodynamically stable and thus intrinsically present. This is not true for the other type of defects. A simple model will illustrate this point. Suppose we have a cube with N atoms containing p line defects with enthalpy ε per atom in the line defect. One line defect contains $N^{1/3}$ atoms so that the total enthalpy of the defects H is given by

$$H = pN^{1/3}\varepsilon.$$

Each line defect can be positioned in $N^{2/3}$ ways so that p line defects can be positioned in $W = (N^{2/3})^p/p!$ indistinguishable ways. Using for the entropy $S = k\ln W$ we obtain for the Gibbs energy

$$G = H - TS = pN^{1/3}\varepsilon - kT\ln[(N^{2/3})^p/p!] \cong pN^{1/3}\varepsilon - pkT\,[\ln(N^{2/3}/p)-1]$$

where for the last step use has been made of Stirling's approximation. As usual equilibrium is obtained when $\partial G/\partial p = 0$, which results in

$$p = N^{2/3}\exp(-\varepsilon N^{1/3}/kT).$$

From this expression we conclude that if $N \to \infty$, the number of defects $p \to 0$. Similarly for q planar defects or r volume defects we find

$$q = N^{1/3}\exp(-\varepsilon N^{2/3}/kT) \quad \text{and} \quad r = \exp(-\varepsilon N/kT)$$

from which we draw the same conclusion, namely at infinite crystal size the thermodynamically required number of planar and volume defects vanishes. Let us repeat this operation for n point defects, again with enthalpy ε. Here we find for the number of indistinguishable arrangements $W = N^n/n!$. The expression for G becomes

$$G = n\varepsilon - nkT\ln(N^n/n!) \cong n\varepsilon - nkT\,[\ln(N/n)-1]$$

Again using $\partial G/\partial n = 0$ results in

$$n = N\exp(-\varepsilon/kT)$$

In this case only at $T = 0$ K the number of point defects n becomes zero and at finite temperature point defects are thermodynamically stable. The number of point defects depends exponentially on temperature. The above expression appears to be generally true but more detailed models are required to make an estimate of the pre-factor for both simple and more complex point defects. Making an estimate for the defect energy requires different models.

In the next sections we deal in some detail with the various types of defects in crystalline lattices and conclude with a few brief remarks on polymer defects.

8.8 Zero-dimensional defects

Of the zero-dimensional defects several types are known. The basic types are (Fig. 8.31) *vacancies* (a missing atom at a lattice point), interstitials (atoms positioned *not* at lattice points), *substitutional impurities* (a foreign atom at a lattice point) and *interstitial impurities* (foreign atoms positioned *not* at lattice points). These defects occur in metals as well as inorganic materials and in principle also in polymeric crystals, although in the latter case they play a minor role. In all cases locally a deformation of the lattice arises. With this deformation an energy term is associated that contributes, in addition to electrostatic terms, to the formation energy of the defect. Obviously a larger energy value is associated with interstitial atoms than that with substitutional atoms (Henderson, 1972; Mott and Gurney, 1940).

In a metal a defect of the basic type, as indicated in the previous paragraph, can occur by itself since charge compensation can easily be achieved due to the relatively high electronic conduction. In inorganic materials, on the other hand, the electrical conductivity is typically relatively small and the basic defects are present in combination with each other to provide charge compensation. Here one can distinguish between Frenkel[ii] defects and Schottky defects (Fig. 8.32). For the *Frenkel defect* a (positive or) cation is displaced while for a *Schottky defect* a pair of ions is missing (or rather moved to the surface[jj]). For completeness we mention that also anti-defects can be present although their occurrence is less frequent than that of the normal defects. For an anti-Frenkel defect a (negative) anion is displaced while for an anti-Schottky defect an extra pair of ions is present interstitially.

[ii] Yakov Ilich Frenkel (1894-1952). Russian scientist well known for his contributions to many fields of physics, e.g. the defect named after him and the mobility of dislocations. Of his many books *Kinetic theory of liquids*, published in 1946, is still very useful.

[jj] In the case of a missing ion pair the volume is constant and the equilibrium condition is minimal Helmholtz energy F or $(\mathrm{d}F)_{T,V} = 0$ while for transport to the surface the pressure is constant and the equilibrium condition is minimal Gibbs energy G or $(\mathrm{d}G)_{T,p} = 0$. In theoretical calculations the conditions T,V constant are usually easier to handle than T,p constant, as is the case here.

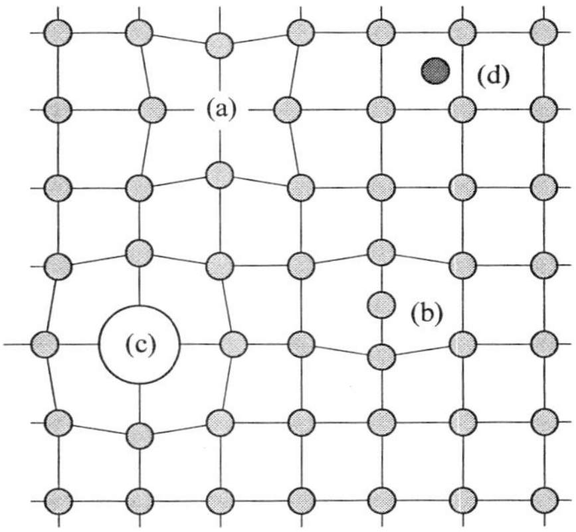

Fig. 8.31: A schematic representation of a vacancy (a), interstitial (b), substitutional impurity (c) and interstitial impurity atom (d).

Walter Schottky (1886-1976)

Born in Zurich, Switzerland, he studied physics at the Humboldt University in Berlin starting in 1904. In 1912 he was awarded a doctorate in Berlin for his thesis on the special theory of relativity. For the next 15 years his career consisted of movements between university and industrial research. He began with a couple of years with Max Wien at Jena where he began his work on the interaction of electrons and ions in vacuum and solid bodies. Then he joined the Siemens industrial research laboratories in Berlin, staying there until 1919. In 1920 he returned to the university, where he worked under Wilhelm Wien at Wurzburg and became qualified as a university lecturer. After 3 years he advanced his academic career by becoming the Professor of Theoretical Physics at Rostock. Never been a enthusiastic university lecturer, at the age of 41 he moved for the last time back to industrial research, rejoining Siemens AG. He remained at Siemens until his retirement in 1958. His research in solid-state physics and electronics yielded many effects and devices that now bear his name (Schottky defect, Schottky barrier, Schottky diode). He was one of the first to point out the existence of electron 'holes' in the valence-band structure of semiconductors. In 1935 he noticed that a vacancy in a crystal lattice results when an ion from that site is displaced to the crystal's surface, now known as a Schottky defect. In 1938 he created a theory that explained the rectifying behaviour of a metal-semiconductor contact as dependent on a barrier layer at the surface of contact between the two materials. Throughout the 1920s Schottky gathered material, which eventually appeared in 1929 in his influential book *Thermodynamik*. It presented the thermodynamic theory of solids with very low impurity content or with small deviations from stoichiometry.

The number of Schottky defects (Fig. 8.32) can be estimated in a similar way as in the introduction to this section. Consider a binary crystal AB having N ion pairs and containing n_S vacancies A and n_S vacancies B. If ε_S denotes the energy to displace an AB ion pair from the crystal, the energy difference U_S between the perfect and the defective crystal is given by $U_S = n_S\varepsilon_S$. Further we denote the number of possible configurations in which n_S pairs of vacancies can be distributed over N positions by W. If the A and B vacancies are independent, n_S A vacancies can be distributed indistinguishably in $w = N!/(N-n_S)!n_S!$ ways over N sites. Since a similar estimate can be made for the B vacancies, we have $W = w^2$. For the Helmholtz energy $F = U - TS$ we thus obtain

$$F = n_S\varepsilon_S - kT\ln W = n_S\varepsilon_S - 2kT\ln[N!/(N-n_S)!n_S!].$$

Using the equilibrium condition $\partial F/\partial n_S = 0$ in connection with Stirling's approximation we obtain

► $$n_S = (N-n_S)\exp(-\varepsilon_S/2kT) \cong N\exp(-\varepsilon_S/2kT)$$

The last step can be made since the validity of this expression extends to only a few tenth of a percent for n_S.

A similar calculation can be made for a Frenkel defect (Fig. 8.32). Consider again the binary crystal AB, now with N regular and N^* interstitial sites. The number of interstitial atoms is indicated by n_F and their energy is ε_F. Since the number of indistinguishable distributions of n_F vacancies over N sites is $w = N!/(N-n_F)!n_F!$ and the number of ways of distributing n_F interstitial atoms over N^* sites is $w^* = N^*!/(N^*-n_F)!n_F!$, the total number of ways of distributing a Frenkel defect W is given by $W = ww^*$. The Helmholtz energy F is then

$$F = U - TS = n_F\varepsilon_F - kT\ln W$$

$$= n_F\varepsilon_F - kT\ln[N!/(N-n_F)!n_F!] - kT\ln[N^*!/(N^*-n_F)!n_F!]$$

The equilibrium condition $\partial F/\partial n_F = 0$, using Stirling's approximation, leads to

► $$n_F^2 = (N-n_F)(N^*-n_F)\exp(-\varepsilon_F/kT) \cong NN^*\exp(-\varepsilon_F/kT)$$

if $n_F \ll N$ or N^*. A similar range of validity as for Schottky defects applies.

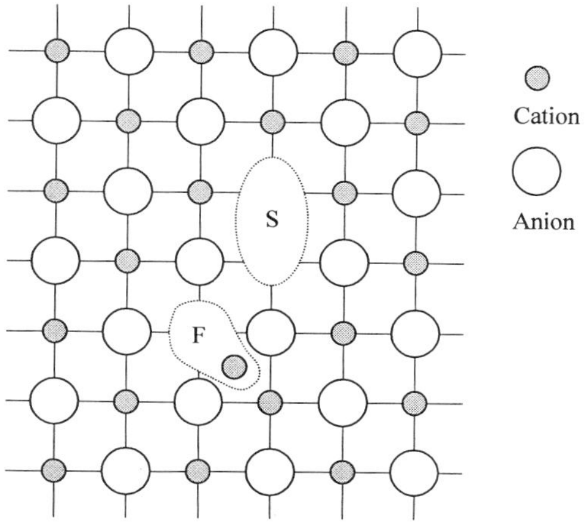

Fig. 8.32: Frenkel (F) and Schottky (S) defects.

We see that both for Frenkel and Schottky defects their concentration increases exponentially with temperature. Their existence is used in the explanation of e.g. electrical conductivity, creep and diffusion. Consequently, on the basis of these models a great deal of insight is obtained in spite of the fact that many assumptions have been made. We note that, apart from that we assumed the defect energy to be known, we also assumed that

- Either Schottky or Frenkel defects are present but not both. This can be remedied relatively easily. Since their energy of formation is typically about 1 eV (see later) the type of defect with the lower ε will generally dominate significantly and we have to consider generally this dominant type only.
- The concentration of point defects is low. No interaction is supposed to be present. However, a positive ion vacancy has an effective negative charge and vice versa so that complex formation can lower the Helmholtz energy. This complicates the analysis slightly but the main problem is that the value of the association energy is usually unknown.
- The volume is constant while actually the macroscopic volume changes and the pressure remains constant. This will influence the activation energy ε.
- Only the change in entropy due to mixing is important. However, the vibrations in the neighbourhood of the defects change thereby also contributing to the entropy.

In the following we discuss briefly some relevant improvements.

Let us deal with the last two aspects first for the case of Schottky defects. The simplest model to deal with lattice vibrations is the Einstein model (see Chapter 11) where a single vibration frequency ω_E characterises the behaviour. The vibrational Helmholtz energy at sufficiently high temperature is $3NkT\ln(\hbar\omega_E/kT)$ so that the total Helmholtz energy of an ideal crystal becomes

$$F = N[U(V) + 3kT\ln(\hbar\omega_E/kT)]$$

where N is the number of atoms and $U(V)$ is the energy of the crystal at volume V. The energy of the vacancy also becomes volume dependent and we write $\varepsilon(V)$. We assume that each atom has z neighbours and that the vibration frequency of the z neighbours of the vacancy decreases in the direction of the vacancy to $\omega_E{}'$ and that it remains constant at ω_E in the two other directions. In this case F is given by

$$F = NU + n_S\varepsilon_S + kT(3N - n_S z)\ln\frac{\hbar\omega_E}{kT} + n_S zkT\ \ln\frac{\hbar\omega_E{}'}{kT} - kT\ln\frac{N!}{(N-n_S)!n_S!}$$

From the equilibrium condition $\partial F/\partial n_S = 0$ we obtain in the usual way

$$\varepsilon_S(V) + zkT\ln\frac{\omega_E{}'}{\omega_E} + kT\ln\frac{n_S}{N-n_S} = 0$$

which can for $n_S \ll N$ and with $\gamma = (\omega_E/\omega_E{}')^z$ be written as

$$n_S = N\gamma\exp(-\varepsilon_S(V)/kT) \tag{8.70}$$

In this equation the value for $\varepsilon_S(V)$ has to be taken at the volume V for the actual temperature considered. We assume that the change in defect energy $\mathrm{d}\varepsilon$ by a change in temperature is given by

$$\mathrm{d}\varepsilon = \frac{\partial\varepsilon}{\partial T}\mathrm{d}T = \frac{\partial\varepsilon}{\partial V}\frac{\partial V}{\partial T}\mathrm{d}T = 3\alpha V\frac{\partial\varepsilon}{\partial V}\mathrm{d}T$$

where the (cubic) thermal expansion coefficient $3\alpha = V^{-1}(\partial V/\partial T)$ is used. Hence

$$\varepsilon_S(T) = \varepsilon_S(0) + \int_0^T d\varepsilon_S = \varepsilon_S(0) + 3\alpha VT\frac{\partial \varepsilon_S}{\partial V}$$

if α and $\partial\varepsilon_S/\partial V$ are temperature independent. Inserting in Eq. (8.70) yields

$$\blacktriangleright \quad n_S = N\gamma B\exp(-\varepsilon_S(0)/kT) \tag{8.71}$$

where $\varepsilon_S(0)$ is the energy to form a Schottky defect at absolute zero and

$$B = \exp[-3\alpha Vk^{-1}(\partial\varepsilon_S/\partial V)]$$

Since $\partial\varepsilon_S/\partial V < 0$, $B > 1$, as is γ. To make an order of magnitude estimate we write for $3\alpha Vk^{-1}(\partial\varepsilon_S/\partial V)$ the equivalent expression $3\alpha\varepsilon_S R^{-1}(\partial\ln\varepsilon_S/\partial\ln V)$ using molar quantities. Using NaCl as an example we have ε_S = 170 kJ/mol and 3α = 120×10^{-6} K^{-1}. Estimating $\partial\ln\varepsilon_S/\partial\ln V = 2$, we obtain $B = \exp(4.9) \cong 130$. The co-ordination number is $z = 6$ and estimating $\omega_E/\omega_E' = 2$, we obtain $\gamma = 2^6 = 64$. Hence for γB we find a value of 10^3 to 10^4. Because

$$\alpha_S = \frac{1}{V}\frac{\partial V_S}{\partial T} = \frac{a^3}{V}\frac{\partial n_S}{\partial T} = \frac{N}{V}\frac{a^3\gamma B\varepsilon_S}{kT^2}\exp(-\varepsilon_S/kT)$$

there is a considerable contribution of the defects to the overall thermal expansion coefficient. X-ray diffraction (XRD) yields the average lattice constant, which is in view of the low value of n_S very nearly equal to the ideal lattice constant. Hence the difference in α from XRD and dilatometry gives an indication for the vacancy density.

For Frenkel defects similar considerations yield

$$\blacktriangleright \quad n_F = \sqrt{NN^*}\gamma B\exp(-\varepsilon_F(V)/2kT)$$

with in this case $\gamma = \omega_E^{z+z'+1}/\omega_{E,i}(\omega_{E,i}')^{z'}(\omega_E')^z$ where $\omega_{E,i}$ is the frequency of an ion in the interstitial position, $\omega_{E,i}'$ that of its z' neighbours and ω_E', as before, that of the z neighbours of a vacancy. Since we expect that $\omega_{E,i} > \omega_E$ and $\omega_E' < \omega_E$, γ can be greater or smaller than 1. The values for γB are thus expected to be much smaller than those for the Schottky defects. Clearly the pre-exponential factor is loaded with uncertainties.

Defect energetics

For a Schottky defect in a van der Waals crystal an estimate for the defect formation energy can be made as follows. If ε_H represents the energy to remove an atom (or molecule) from the interior to the surface, leaving a vacancy in the bulk, and ε_L the lattice energy per atom, we have $\varepsilon_S = \varepsilon_H - \varepsilon_L$. In the nearest-neighbour approximation $\varepsilon_L = \frac{1}{2}Z\phi(r_k)$, where Z denotes the co-ordination number and $\phi(r_k)$ the bond energy for the nearest-neighbour distance r_k. In the same spirit $\varepsilon_H = Z\phi(r_k)$ and it follows that $\varepsilon_H = 2\varepsilon_L$ or $\varepsilon_S = \varepsilon_L$. This estimate neglects relaxation and thus generally yields a value too high value by a factor of ⅔ to ¾. However, the relaxation energy is insufficient to account for the full discrepancy. The explanation is provided by three body interactions, which make the bond energy in a simple bond picture dependent on the environment of the bond. For details we refer to e.g. Phillips (2001).

For inorganic materials, in particular alkali halides, a theoretical estimate for the defect formation energy can be made by using the Born model. The first attempt was made by Jost in 1933 who assumed that a vacancy could be represented as a hole of radius r in a dielectric medium with dielectric constant ε. From the polarisation, given by $P = (D-\varepsilon_0 E)$ with the electric field $E = e/4\pi\varepsilon\varepsilon_0 r^2$ and the dielectric displacement $D = \varepsilon E$, the potential ϕ at the centre of the hole is calculated as

$$\phi = \int_R^{\infty} \frac{P}{4\pi\varepsilon_0 r^2} 4\pi r^2 \, dr = \left(1 - \frac{1}{\varepsilon}\right) \frac{e}{4\pi\varepsilon_0 R}$$

The choice for R is not *a priori* clear but has to be in the order of interatomic distance.

The next step was made by Mott[kk] and Littleton in 1938 who indicated that the problem of the calculation of the potential ϕ due to the missing ion could be separated in two halves the first of which is the calculation of the potential with all ions kept at their original equilibrium positions and the second when the ions are allowed to relax to their new equilibrium positions. The first step can be solved by electrostatics. The second step requires a more sophisticated structure model in which the atoms can move. These more exact calculations indicate that for alkali halides with lattice constant a the radius R in the Jost model can be estimated as $R^- \cong 0.9a$ and $R^+ \cong 0.6a$ according to whether a negative (R^-) or positive ion (R^+) is missing. The second step is complex since it requires consideration of the forces acting on the displaced atoms but show that atoms adjacent to the hole can be displaced considerably, e.g. in NaCl about $0.07a$ outwards. In total these estimates yield energy values of the order of 2 eV, which is the proper order of magnitude. Flynn (1972) provides a concise review of point defects predating the era of *ab-initio* simulations. Detailed atomistic calculations of defect energies have now become available, an early overview of which is given by Stoneham (1975). See also Phillips (2001).

There exist a number of correlations known that can be useful in estimating unknown defects energies. For metal halides a strong correlation of the Schottky defect enthalpy h_S with the melting point T_m exists which can be represented by $h_S\text{ (eV)} = 2.14\times10^{-3}T_m$, as illustrated in Fig. 8.33. For metal halides and oxides also a

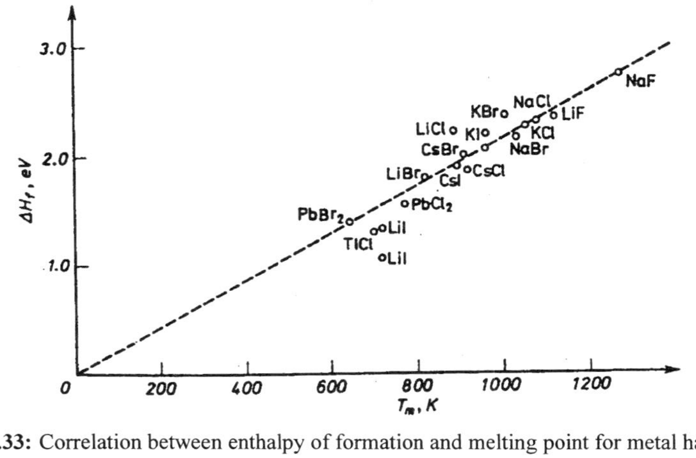

Fig. 8.33: Correlation between enthalpy of formation and melting point for metal halides.

[kk] Neville Francis Mott (1905-1996). English physicist who received the Nobel Prize for physics in 1977 together with Philip Warren Anderson (1923-2...) and John Hasbrouck van Vleck (1899-1980) for their fundamental theoretical investigations of the electronic structure of magnetic and disordered systems.

correlation with the atomisation energy[ll] E_{ato} is observed. However, while for metal halides the relation $h_S \cong 1.1E_{ato}$ is observed, for metal oxides this empirical relation reads $h_S \cong 0.63E_{ato}$. For metals estimates based on continuum elasticity are sometimes used (see Section 10.3).

8.9 One-dimensional defects

The most important one-dimensional defect is the dislocation. Dislocations occur in all the categories of materials and are largely responsible for the plastic deformation of materials, in particular in metals. There are two basic types: the *edge dislocation*, essentially an extra half-plane of atoms inserted in the lattice and the *screw dislocation*, essentially a 'staircase' in the lattice (Fig. 8.34). The line representing the end of the half-plane and the centre of the staircase is addressed as *dislocation line* and indicated by the unit tangent vector **l**. The dislocation can have a mixed character implying that it is neither a pure edge dislocation nor a pure screw dislocation. An important characteristic of dislocations is their *Burgers vector*. This vector **b** can be found by making a closed loop around a dislocation line and counting the mismatch in lattice vectors. While for a pure edge dislocation **b** is perpendicular to **l**, for a pure screw dislocation **b** is parallel to **l**. In fact the Burgers vector **b** is an invariant characteristic. Dislocations have an excess energy that can be considered as a sum of the energy of the core and the outer region. The core is the inner part of a dislocation the energy of which has to be estimated by atomic models. For the outer region elasticity theory can be used because the displacements of the atoms from their equilibrium configuration in the ideal crystal are small. The associated energy for isotropic materials is estimated as

$$U = \alpha \frac{Gb^2}{4\pi} \ln \frac{R}{r_0}$$

where G is the shear modulus, b is the length of the Burgers vector and α is a factor, 1

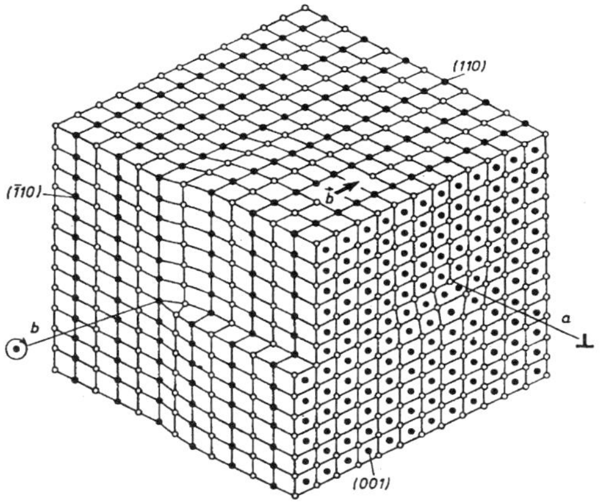

Fig. 8.34: Schematic of a mixed dislocation with Burgers vector b in a hypothetical crystal. At point 'a' the dislocation has a pure edge character while at point 'b' the character is purely screw.

[ll] For oxides Me_aO_b the atomisation energy E_{ato} is given by $E_{ato} = \Delta H + aL_{Me} + (b/2)D_{O2}$ where ΔH is the heat of formation, L_{Me} is the sublimation energy and D_{O2} is the dissociation energy of O_2 (D_{O2} = 498.2 kJ/mol).

for a screw dislocation and $(1-\nu)^{-1}$ for an edge dislocation. Further r_0 denotes the dislocation core radius and R the range of influence of a dislocation estimated as $R = \rho^{-1/2}$, where ρ is the dislocation density. The dislocation density can vary widely for different materials and conditions. For example, for a Si or Ge single crystal the dislocation density can be as low as $\rho \cong 10$ cm^{-2} while for a worked metal such as Au or Cu it can be as high as $\rho \cong 10^{11}$ cm^{-2}. Because of the logarithmic dependence of the energy on R and r_0, this factor is often neglected and the expression is simplified to $U = \frac{1}{2}\alpha Gb^2$. For the core energy U_{cor} a rough estimate yields about $0.1U$ so that the total energy $U_{dis} = U + U_{cor}$ is still $U_{dis} \cong \frac{1}{2}\alpha Gb^2$. To estimate the order of magnitude of the energy of the atoms in the dislocation line we take $\rho \cong 10^{11}$ cm^{-2}, $r_0 \cong 10^{-9}$ m, which leads to $U \cong 170$ kJ/mol, comparable to the energy of a point defect. We discuss dislocations and the associated energy relations in more detail in Chapters 15 and 16.

8.10 Two-dimensional defects

Also of the two-dimensional defects several types are known. We deal with stacking faults, grain boundaries and surfaces (Henderson, 1972).

Stacking faults

Let us consider a crystal as a stacking of lattice planes, e.g. the FCC lattice as the sequence ...ABCABC... and the HCP lattice as ...ABAB... . Now it is possible that the order of these lattice planes is not maintained throughout the crystal but that a stacking error occurs, e.g. in the FCC lattice ...ABCABC... → ...ABCBCA... or ...ABCABC... → ...ABCACBCA...where the total number of lattice planes of each type is conserved in the first example but not in the second. These two-dimensional defects are generally indicated as *intrinsic* and *extrinsic stacking faults*, respectively. With these stacking faults, energy is associated, for metals typically 0.05 to 0.2 J/m^2. These defects occur in all three material classes but are again of limited importance in polymers. In well-annealed metals the number of stacking faults is small since these faults cost relatively a large amount of energy. In some inorganic materials the required energy is small and stacking faults can be abundantly present, e.g. in SiC.

Grain boundaries

In polycrystalline materials grain boundaries delineate areas of a different crystallographic orientation. These grain boundaries can be considered as two-dimensional defects with associated grain boundary energy, ranging from about 0.01 to about 1 J/m^2. Several types of boundaries can be present. We distinguish between small-angle grain boundaries and wide-angle grain boundaries. For *small-angle grain boundaries* the crystallographic orientation difference between two grains is small and can be characterised by the difference in angle θ between the normal vectors associated with both grains. This can occur by a simple *tilt* (no rotation of grain boundary plane, only tilting the lattice) or by a *twist* (no tilt, only rotation in the grain boundary plane). Since a small-angle grain boundary can be considered as an array of dislocations an estimate for the energy of small-angle grain boundaries can be made based on the behaviour of arrays of dislocations (Fig. 8.35). This leads to (Section 15.8)

$$U_{gb} = \frac{Gb}{4\pi(1-\nu)}\theta(\ln\theta_0 - \ln\theta)$$

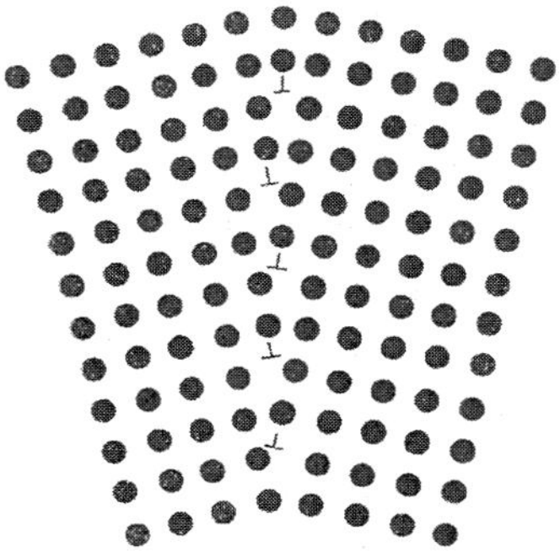

Fig. 8.35: Schematic of a simple tilt small-angle grain boundary.

where G denotes the shear modulus, v Poisson's ratio, b the length of the Burgers vector and θ_0 a constant representing the contribution of the dislocation core energy. This expression appears to be in agreement with experiment up to rather high values of θ (Fig. 15.26), in fact much better and also extending to a much larger tilt angle than can be anticipated from this dislocation model.

This description cannot be made for *wide-angle grain boundaries*, which are generally considered as the non-matching area between the lattices of two grains (Fig. 15.26). In some particular orientations though the lattices can match, each with a different repetitive unit. The superlattices that result are addressed as coincident site lattices (CSLs). The structure of real grain boundaries is frequently based on the CSL. Moreover, these CSLs play an important role in twinning phenomena and for coherent inclusions. For wide-angle boundaries in general theoretical estimates are much harder to make and a simple model is not available. Experimentally the values are determined from dihedral angle measurements at triple junctions. If γ_{12} indicates the grain boundary energy between grains 1 and 2 and φ_3 indicates the dihedral angle in grain 3, in equilibrium a force balance leads to

$$\frac{\gamma_{12}}{\sin\varphi_3} = \frac{\gamma_{23}}{\sin\varphi_1} = \frac{\gamma_{31}}{\sin\varphi_2} \qquad (8.72)$$

Since in a cross-section the true dihedral angles are not displayed, a correction to obtain these true angles is in principle required. It has been shown that the median angle of a set of measurements on a cross-section yields a reasonably accurate estimate of the true angle.

Surfaces

Surfaces can also be considered as defects, in the sense that the ideal repetition of lattice cells is interrupted. We can distinguish several types of surfaces. The simplest is the *bulk-like surface*, the structure of which is more or less alike to that of the corresponding lattice plane in the bulk, although large atomic displacement are generally present. They occur primarily in metals. For example, the {111}, {110} and {100} planes in the FCC structure, the {111} and {110} planes in the BCC structure and the {0001} and $\{10\bar{1}0\}$ planes in the HCP structure. They also occur in oxides, e.g. {100} NiO and {110} in III-V compounds (e.g. GaAs) or II-VI compounds (e.g. ZnS). A simple bond pair model can estimate the surface energy. Consider a lattice in which only nearest-neighbour interactions are present. Furthermore, the co-ordination

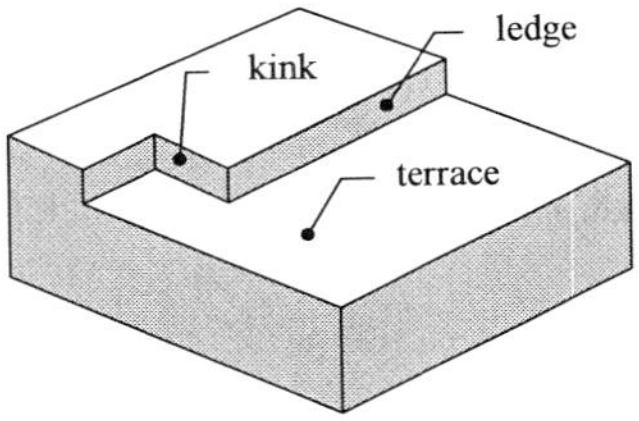

Fig. 8.36: The TLK surface model.

number is z and we assume that the bond energy B is temperature independent. In this approach the heat of sublimation ΔH_S can be estimated as $\Delta H_S = \frac{1}{2}N_A zB$, where N_A denotes Avogadro's number. Now consider the (111) plane in a FCC lattice where z = 12. There are six nearest neighbours in the plane, three above and three below. Hence after cleavage three bonds are broken yielding two surfaces. Hence the surface energy per atom $3B/2 = 3\Delta H_S/(2N_A\times12\times\frac{1}{2}) = \Delta H_S/4N_A$. For the number of atoms per unit surface N we take the estimate $N = (N_A/V_{mol})^{2/3}$, where V_{mol} is the molar volume, so that the final result for the surface energy E_{sur} becomes

$$E_{sur} = N\Delta H_S/4N_A = \Delta H_S/4V_{mol}{}^{2/3}N_A{}^{1/3}$$

For metals this simple estimate yields reasonable results (see Table 8.5).

Generally surfaces are not really planar but *stepped*. The steps are also referred to as *ledges*. While in FCC and BCC metals the step height typically is monoatomic, for HCP metals it is generally doubled. In the steps *kinks* (Fig. 8.36) are present representing a deviation from the overall step direction. The planar area between the steps is referred to as *terrace*. Together these features represent the terrace-ledge-kink or *TLK model* of surfaces. For non-metals information on steps is limited.

Atoms in the surface region exhibit in general a relatively large *relaxation* from the ideal lattice positions, leading in a number of cases even to *reconstruction*. For a relaxed surface the overall structure is still similar to a bulk-like surface while for a reconstructed surface a clear symmetry break occurs.

For unreconstructed surfaces of metals typically a bond contraction of a few percent occur between the first and second layer. The contraction is larger for the more open surfaces. For example, generally a contraction of ~0% occurs for the {111} FCC and {100} FCC (or even a widening as for Pt and Pd) and a contraction of ~10% for {110} FCC (to as large as 16% for Pb). Deeper layers generally alternate in contraction and widening. In semiconductors both the bond angle and bond length vary. Contraction as well as elongations of ~5% can occur in bond length while the bond angle can range from 90° to 120°, deviating considerably from the ideal angle of 109.5°. For oxides the situation is rather mixed. While e.g. for {100} NiO, MgO, CaO and CoO little contraction occurs, for {111} CoO a contraction of about 17% is present. Even more extreme, for {100} $SrTiO_3$ buckling of the surface occurs.

Table 8.5: Experimental and estimated surface energies for a few metals.

Material	E_{exp} (J/m^2)	E_{the} (J/m^2)
Ag	1.68	1.14
Au	2.01	1.55
Cu	2.35	1.38

Table 8.6: Surface energy of spinel.

Surface	γ (non-relaxed)	γ (relaxed)	γ (hydrated)
(100)	4.0	2.5	0.6 (5)
(110)	5.6	2.7	0.2 (8)
(111)	8.4	3.1	0.1 (7)

For each of the orientations the most stable type is given. In parentheses the number of adsorbed water molecules is indicated.

Reconstruction of a surface leads to a unit cell with a different symmetry as the corresponding lattice plane. Frequently also the size of the surface unit cell increases, e.g. for Si a well-known reconstruction is to a 7×7 surface unit cell. For *displacive* reconstruction the displacements are small but a symmetry break occurs, e.g. in 1×1 {100} Mo and W. In a *missing row* surface a row of atoms is missing from the surface leading to a 2×1 unit cell, e.g. in {110} Ir, Pt and Au. Since decreasing co-ordination often leads to a lower bond length, top layers can be *closer-packed* so that an approximate hexagonal structure results. This occurs e.g. in {100} Ir, Pt and Au.

External agents generally enhance these processes, the most prominent agent being H_2O, present nearly everywhere. For metals the influence of gases such as CO, NO, etc. are extensively studied. This relaxation and/or reconstruction leads to a lowering of the surface energy. As an example we quote the results of pair potential calculations[mm] including polarisation effects on spinel $MgAl_2O_4$. In Table 8.6 the surface energies of non-relaxed, relaxed and hydrated surfaces of (100), (110) and (111) orientations are given. We note, apart from a large change in surface energy upon relaxation, also a further decrease upon hydration leading to a reversed order of preferred planes. We note also that, contrary to simple crystallographic structures, several terminations of a certain crystallographic plane exist, which complicates the matter considerably. An edge-on view of the (100) surface is given in Fig. 8.37.

Finally, we note that segregation of certain impurity elements to grain boundaries and surfaces may occur. In crystalline materials the driving force is related to the difference in size and/or charge. In inorganic materials really small amounts of impurities may segregate dramatically to the interfaces so that at these positions a considerable concentration of that element may be present, influencing the material properties to a large extent. The presence of just a few ppm of CaO in polycrystalline Al_2O_3 provides an example[nn]. In this case the segregation factor is about 600 leading to several atomic percent of Ca at the grain boundaries. This in turn leads, besides to a minor degradation in mechanical properties, also to a large decrease in sodium corrosion resistance. In metals segregation is usually less pronounced. The segregation of P in Cu[oo] provides an example. In this case the segregation factor is relatively large leading to strongly enhanced grain boundary diffusion, relevant in solid-state reactions. For example, it has been shown that minute impurities of P in Cu lead to dramatic changes in reaction layer morphology and kinetics of reactions with Si, thus influencing the adherence.

An overview of surface structure of single crystals is provided by van Hove (1993).

[mm] Fang, C.M., de With, G. and Parker, S.C. (2001), J. Am. Ceram. Soc. **84**, 1553.

[nn] de With, G., Vrugt, P.J. and van de Ven, A.J.C. (1985), J. Mater. Sci. **20**, 1215.

[oo] Becht, J.G.M. (1987), *The influence of phosphorus on the solid state reactions between copper and silicon or germanium*, Thesis, Eindhoven University of Technology.

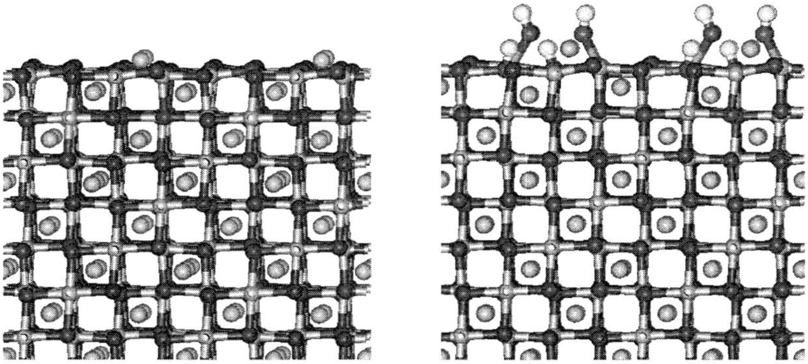

Fig. 8.37: A view along the [010] direction for the {100}b surface of spinel: clean relaxed (left) and clean partially hydrated (right).

8.11 Three-dimensional defects

For three-dimensional defects two basic types are present: *pores* (a relatively large cluster of unoccupied lattice points) and *inclusions* (a relatively large cluster of foreign atoms). In the latter case the lattice of the inclusion may match the lattice of the host material fully, only partially or not at all. These cases are addressed as *coherent*, *semi-coherent* and *incoherent*, respectively (Fig. 1.5).

Inclusions in a matrix are accompanied by a stress field since the inclusions are typically formed at a higher temperature than the temperature in use and the inclusion and matrix have a different coefficient of thermal expansion. For simplicity we assume isotropic, spherical inclusions in an isotropic matrix material. Employing the bulk modulus K and shear modulus G and denoting the values for the inclusions with an asterisk, the total elastic energy U is the sum of elastic energy for the inclusion U_{inc} and that of the matrix U_{mat} and is given by

$$U = U_{\text{mat}} + U_{\text{inc}} = \frac{2}{3}G\frac{\Delta V^2}{V} + \frac{1}{2}\frac{(4G\Delta V)^2}{9K^*V} = \frac{2}{3}G\frac{3K^*}{4G+3K^*}\frac{\Delta V^2}{V}$$

where the volume difference between free inclusion and free hole in the matrix is indicated by ΔV. Note that this expression is independent of K. This elastic energy expression can be used to estimate the elastic constants of a material with a small volume fraction of inclusion (see Chapter 12).

Volume defects occur in all three material classes. In metals they are important for hardening, in particular if they are of small size. This is true for inorganic materials as well but in this case the larger inclusions can also act as a flaw from which fracture can originate. In polymers the stress field due to mismatch is usually less important since plastic or viscous relaxation can occur relatively easy.

Accepting that continuum theory may also be applied to atomic phenomena, for which there is of course no physical justification at all apart from simplicity, this model can also be used for an order of magnitude estimate for the elastic energy of a point defect. In that case we assume that $G^* = G$ so that we obtain

$$U = \frac{2}{3}G\frac{1+\nu}{1-\nu}\frac{\Delta V^2}{V}$$

with ν Poisson's ratio. Let us take as an example Cu for which $G = 40$ GPa, $\nu = 1/3$ and an atom has a radius r of about 0.15 nm or a volume V of 12×10^{-30} m^3, leading to $U \cong 1.4(\Delta V/V)^2$ eV. For a substitutional atom $\Delta V/V \cong 0.1$ so that $U_{\text{sub}} \cong 0.014$ eV, a negligible contribution. For an interstitional atom $\Delta V/V \cong 0.5$ so that $U_{\text{int}} \cong 0.35$ eV, which is a far from negligible contribution. Since in reality atoms may relax more than allowed by elasticity theory, a more realistic estimate yields lower energy values. For FCC metals[pp] it has been estimated that an extra factor of 0.54 for impurity atoms has to be applied, altogether not too surprising in view of the crude assumptions.

8.12 Defects in polymers

Polymers can also crystallise[qq] and in crystalline polymers the same defects can arise as in inorganics and metals. Their significance is considerable less though.

More important are the following considerations. Two other types of defects can be distinguished: defects in the chain and in the network. In the chain irregularity may occur with respect to e.g. the tacticity or isomerism and this may hinder crystallisation. With respect to the network, defects may arise in the connectivity. The cross-links ought to be homogeneously distributed but may be clustered so that the molecular weight of the chain parts between the cross-links, the so-called *sub-chains*, is an important characteristic for many properties. Ideally all the sub-chains are also identically cross-linked implying identical configurations at all cross-links, e.g. at all cross-links four sub-chains are bonded together. However, defects can be present and we can distinguish between *loose ends* (sub-chains connected only at one side with a cross-link) and *loops* (sub-chains connected with both sides at the same cross-link) (Fig. 8.38). Although of a quite different nature as point and line defects, these defects play a certain role in molecular models for elastic and viscous behaviour of polymers. *Entanglements* (sub-chains wriggled into each other without bond but difficult to separate) can also be considered as a kind of defect and are quite important in the deformation of both semi-crystalline and amorphous polymers.

8.13 Microstructure

In Chapter 1 we have indicated that, apart from atomic and molecular aspects, microstructural (or morphological) aspects are also of importance. In that chapter the definition of microstructure was also given. Here we discuss a number of geometrical aspects of microstuctures, relevant for our main purpose.

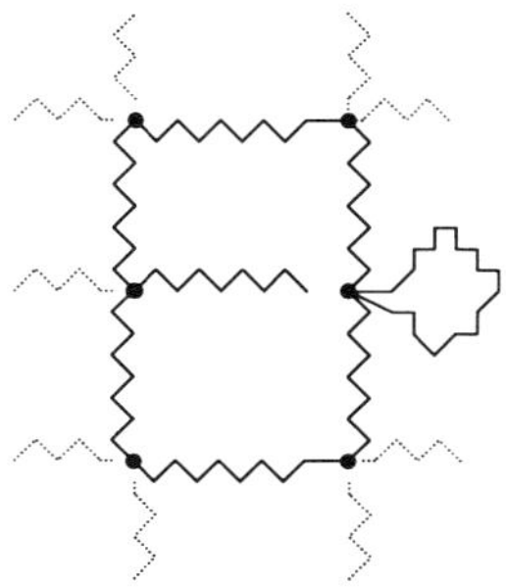

Fig. 8.38: A loop and a loose end in a polymer network with on average four cross-links.

[pp] Flinn, P.A. and Maradudin, A.A. (1962), Ann. Phys. **18**, 81.

[qq] For a review of early work, see P.H. Geil (1963), *Polymer single crystals*, Wiley, New York.

Stereology

The microstructure of any material contains a great deal of information. The attempt to characterize quantitatively the geometrical aspects of microstructures is called *stereology*. It is described somewhat more formally as a body of mathematical methods relating three-dimensional (3D) parameters defining a structure to two-dimensional (2D) measurements obtainable on sections or projections of the structure. Roughly the discipline of stereology can be divided into two parts: a part in which no assumptions, particularly on shape, are made and which results in statistically exact relations between various parameters of a microstructure and a part in which some assumptions on shape are made because otherwise no further progress is possible. In this section some basic parameters and relations for both are discussed. Several textbooks are available, e.g. DeHoff and Rhines (1968), Underwood (1969) and Weibel (1979, 1980), which vary in approach and contents and of which the one by Underwood is particularly recommended. Generally it is assumed that the measured section is an unbiased subset of all possible sections and that Euclidean geometry is applicable. The 2D image or *micrograph* used for analysis is normally obtained by optical or electron microscopy. 3D, 2D and 1D features in a microstructure result in 2D, 1D and 0D features in a micrograph, known as profiles, lines and transsections, respectively.

General relations

The characteristics of the microstructure are quantitatively described by various microstructural parameters. Each microstructure parameter consists of a ratio of a microstructural quantity M over the chosen test quantity T, indicated by the symbol M_T. If necessary the relevant phase, say α, is indicated, either between brackets or as a subscript. Here we use the bracket notation, i.e. $M_T(\alpha)$. For the determination of the microstructural parameters various test methods exist. The test methods use areas, lines or points and are called areal, lineal or point testing, respectively. They are illustrated in Fig. 8.39. For example, in point counting a set of points is distributed over the 2D image, either in a random or in a systematic way, and the point fraction for, say, the grey phase can be determined. In a similar way a set of lines can be distributed over the 2D image and the ratio of test line length in the grey phase over the total length results in the line fraction. Measuring the areas of the grey profiles and the total area of the micrograph results in the area fraction. Also the number of

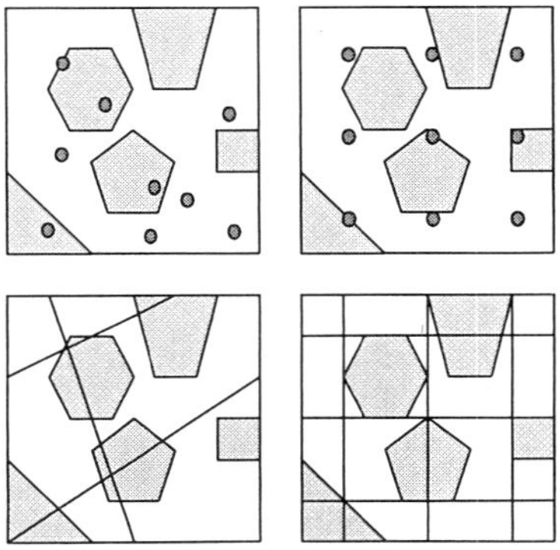

Fig. 8.39: Random and systematic point count (upper row) and random and systematic lineal analysis (lower row). Areal analysis measures the areas.

features, either linear ones such as intercepts, or areal ones such as profiles, can be determined.

The notation for quantities in stereology is more or less standardised and the more important quantities are indicated below.

- P: number of points on profiles or total number of test points
- L: length of profiles on test lines or total test line length
- A: planar area of profiles or total test area
- V: volume of a feature in the microstructure
- S: 3D surface area of a feature in the microstructure
- J: length of a linear feature in the microstructure
- N: number of features in the microstructure or micrograph
- I: number of intersections of a test line with features in the micrograph
- B: length of the boundary of a profile in the micrograph (perimeter)
- Q: number of transsections of linear features in the microstructure with the micrograph plane

Some typical microstructural parameters are as follows:

- $I_L = I/L$, the number of intersections per unit length (intersection line density)
- $S_V = S/V$, the amount of surface per unit volume (surface density)
- $V_V = V'/V''$, the amount of volume per volume or the *volume fraction* (volume density)
- $N_L = N/L$, the number of features (and thus intercepts[π]) per unit length (intercept line density)

Between the various microstructure parameters that can be determined from a micrograph, there exist relations to 3D features, not accessible to direct determination. These relations are statistically exact and some of them are (Underwood, 1969)

$$V_V = A_A = L_L = P_P \qquad S_V = (4/\pi)B_A = 2I_L \tag{8.73}$$

$$L_V = 2Q_A \qquad P_V = \tfrac{1}{2}J_V S_V = 2Q_A I_L$$

By far the most important relation is $V_V = A_A = L_L = P_P$ and it can be shown that it remains true even if the section chosen is not planar, but irregular like a fracture surface. The only requirement is that the direction of observation is normal to the 'average' plane. These relations provide an estimate of the volume fraction V_V by measurement of the point fraction P_P, lineal fraction L_L or area fraction A_A. The specific surface area can be estimated simply by counting doubly the number of intersections per unit length of test line and using $S_V = 2I_L$. The measurement of the number of intercepts per unit length N_L requires some special care. Consider a two-phase material of α-particles in a β-matrix. The α-particles can be entirely separated, entirely connected or partly connected. In the case of contiguous particles $N_L = I_L$ while in the case of separated particles $N_L = I_L/2$. In general we define

$$N_L(\alpha) = [2I_L(\alpha\alpha) + I_L(\alpha\beta)]/2$$

where $I_L(\alpha\alpha)$ and $I_L(\alpha\beta)$ denote the number of intersections between α-particles and α-particles and α-particles and the β-matrix, respectively. In this respect it is also useful to consider the so-called *contiguity* C describing the amount of connectivity between the particles. If the interface between material α and α (β and β, α and β) is denoted by $S(\alpha\alpha)$ ($S(\beta\beta)$, $S(\alpha\beta)$), the contiguity is defined by

[π] Note the difference between intercept and intersection.

$$C = 2S_V(\alpha\alpha)/(2S_V(\alpha\alpha) + S_V(\alpha\beta)) = 2I_L(\alpha\alpha)/(2I_L(\alpha\alpha) + I_L(\alpha\beta))$$

Unfortunately it is impossible by the measurement of C to determine whether a continuous 3D network exists or not. For this purpose real 3D information is necessary. The contiguity is used, however, in the description of the mechanical properties of two-phase materials.

The calculation of N_V for a particulate system from 2D images is of great interest, but unfortunately there does not exist a simple general equation in terms of quantities that can be obtained from a micrograph alone. For convex particles it holds that

$$N_A = N_V \overline{H}$$

where $\overline{H}$ is the mean projected height of the randomly oriented particles. The value of $\overline{H}$ is obviously different for different shapes. For a sphere of radius a, $\overline{H} = 2a$ and for a system of polydispersed spherical particles we have

$$N_V = N_A / 2\overline{a}$$

where $\overline{a}$ is the average sphere radius. The latter can be obtained from $\overline{a} = \pi/4\overline{m}$ with $\overline{m}$ the mean value of the reciprocals of the circle diameters on the test section.

Example 8.5

A simple example will illustrate the use of the test methods and relevant equations. Suppose we have a matrix β in which another phase α is embedded and we want to determine its volume fraction $V_V(\alpha)$. Hereto we determine the point ratio for the α-phase, $P_P(\alpha)$. Now $P_P = p'/p''$, where p' is the number of grid points on the α-phase and p'' is the total number of points. Further assume 43 hits in the α-phase in 100 applications of a nine-point grid. From the relation $V_V = P_P$ and the numbers mentioned above it follows that $V_V = P_P = 43/900 = 0.048$. The error s in V_V (95% confidence level) is given by

$$s^2(V_V) = s^2(P_P) = 4P_P(1-P_P)/\,p''$$

where p'' is the total number of test points. The above-mentioned example corresponds to a relative error of 0.014/0.048 = 30%. In comparing figures from different determinations one should be aware of the relative large errors involved. For the determination of the volume fraction one could also use the lineal or areal method but systematic point count is the most efficient method.

For an unbiased estimate of the various quantities some care is necessary in analysing a sample. In order to avoid this bias, only that part of the sampled microstructure has to be considered for which all features are completely visible (Fig. 8.40[ss]). This is usually achieved by cutting off the edges of the micrograph with about the average feature size. This part defines the area to be used for normalisation, e.g. in N_A or A_A. Furthermore, particles intersecting the so-called forbidden line are not taken

[ss] Gundersen, H.J.G. (1977), J. Microscopy **111**, 219.

into account. This procedure should be used for all analyses. Other non-biasing procedures[tt] exist but the one described is simple and efficient.

Size and size distribution

For the determination of the grain size many measures are in use. The use of the *mean intercept length* $l = V_V/N_L$ is advised because

- the measurement is simple;
- no assumptions are made on the convexity of the particles;
- this quantity is related to the other stereological quantities, contrary to other grain size definitions, e.g. $S_V = 2I_L = 2V_VC/l$ and
- no transformation is necessary for calculating the 3D value: $l(3D) = l(2D)$.

The value of the mean intercept is well known for many shapes, e.g. for

- a sphere of radius a, $l = 4a/3 = 1.333a$,
- a cube of edge a, $l = 2a/3 = 1.333(a/2)$,
- an octahedron with edge a, $l = 1.545a$,
- a cubo-octahedron with edge a, $l = 1.690a$ and
- a pentadodecahedron with edge a, $l = 1.485a$.

For these regular shapes in all cases $l \cong 1.5a$. Sometimes a grain size d is calculated from the mean intercept l by the so-called Mendelson correction[uu]. A lognormal size distribution (see Appendix D and later in this section) with a width given by $\sigma' = 0.23$ is assumed. This number results from assuming a maximum size of 2.5 times the mean size, which occurs at 4 standard deviations. Further assuming all grains to be cubo-octahedra, one can show that $d = 1.56l$. The constant 1.56 is, however, dependent on the shape of the grain and width of the distribution assumed and nothing is gained by multiplying by this constant number.

For a microstructure containing partly connected α-particles in a β-matrix, l measures the intercept through the individual α-particles. If one is interested in L, the mean intercept through clumps of connected α-particles, the effect of the contiguity should be taken into account. This results in

$$l = L(1-C)$$

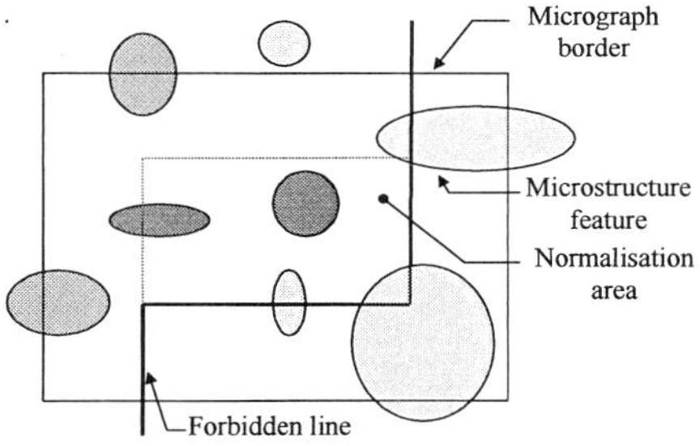

Fig. 8.40: Schematic microstructure showing the 'forbidden' line procedure. Only the dark grey areas are counted.

[tt] For space filling features in a micrograph one also frequently uses the complete area A of the micrograph and counts the number of fully visible features N_{ful} and the number of partially visible features N_{par} to make the estimate $N = N_{ful}+½N_{par}$ in the estimate for $N_A = N/A$.

[uu] Mendelson, M.I. (1969), J. Am. Ceram. Soc. **52**, 443.

Intersect counting can be done using straight test lines or circles. Using circles as test lines assures a randomisation of the data even in materials with preferred orientation.

Example 8.6

Assume that the average number of intersections for a chosen test length of 1000 μm is 120. The mean intercept length is then l = 1000/120 = 8.3 μm. In this case also an error estimate (95% confidence level) can be made, which is $s(l) = Klp^{-1/2}$ where K = 1.2 for contiguous particles, K = 4.0 for dispersed particles and p is the total number of intersections. For the example above, assuming dispersed particles, $s(l) = 4\times8.3\times(120)^{-1/2} = 3.0$ μm. Relatively large errors are thus involved.

In two-phase materials containing α-particles in a continuum matrix a useful quantity, directly related to the mean intercept, is the *mean free distance* λ between the α-particles. It is defined by

$$\lambda = [1-V_V(\alpha)]/N_L = l[1-V_V(\alpha)]/V_V(\alpha)$$

The mean free distance is used in the description of the relation between microstructure and mechanical properties of two-phase materials. The parameters l, L, λ and C are not independent. They are related by

$$l = L(1-C) = \lambda V_V/[1-V_V(\alpha)]$$

Apart from checking the consistency of various measurements, this relation is sometimes useful when one of the parameters is difficult to measure.

Many other grain size measures are in use. The most important ones are the equivalent sphere diameters and the Feret or caliper diameter. For the *equivalent sphere diameters* the size of the particle is equated to the size of a circle with equal perimeter (equivalent boundary) B or equal area (equivalent area) A

$$d_B = B/\pi \quad \text{or} \quad d_A = (4A/\pi)^{1/2}$$

Complex transformations are necessary for the calculation of the 3D distribution from the 2D distribution (contrary to the mean intercept in 3D which is equal to the one in 2D). The *Feret* or *caliper diameter* $d_F(\theta)$ is defined as the length of the projection of the particle in a fixed direction θ. The value of this parameter is dependent on the direction of projection. For convex particles the average caliper diameter $d_{F,ave}$ is equal to the equivalent boundary diameter d_B. Moreover $l(\theta) = A/d_F(\theta)$. Hence the average intercept l is related to the average caliper diameter $d_{F,ave}$ by $l = A/d_{F,ave} = \pi A/B$. Such general relations are not present for the equivalent area diameter. Only for a circular particle the equivalent area diameter d_A is equal to d_F. Otherwise $d_F > d_A$, e.g. for a hexagon $d_F/d_A = 1.05$, for a square $d_F/d_A = 1.13$, for an equilateral triangle $d_F/d_A = 1.29$ while for ellipses of axial ratio $q > 1$ the ratio is $d_F/d_A = 1[(1+q^2)/2q]^{1/2}$.

We should also mention that statistical diameters are frequently used (though usually unrecognised). Generally statistical parameters are defined by

$$D_{m,n} = \left(\sum_i N_i D_i^m \Big/ \sum_i N_i D_i^n\right)^{1/(m-n)}$$

where N_i is the number of particles with size D_i. Typical examples are provided by the mean area diameter $D_{2,0}$,

$$D_{2,0} = \left(\sum_i N_i D_i^2 \Big/ \sum_i N_i\right)^{1/2}$$

in essence the mean 'equivalent area' diameter, and by the volume surface diameter

$$D_{3,2} = \sum_i N_i D_i^3 \Big/ \sum_i N_i D_i^2$$

The latter is in fact the average particle diameter as calculated for a powder from the equation $D = 6/\rho S$ with ρ the density and S the specific surface area of the powder. The problem of which size measure to choose arises again for the statistical diameters.

Distributions of intercepts (or other measures for that matter) can be determined as well. In many cases these distributions are skew which means that various characteristic 'size' parameters can be used. Here we distinguish between the *mean* or *average* x_{ave}, the *median* x_{50} and *mode* x_{mod} (see Appendix D). Skew distributions are often described by the lognormal distribution. The lognormal distribution is obtained by replacing the running co-ordinate x in the normal (or Gauss) distribution by $y = \ln x$

$$f(y) = (2\pi)^{-1/2}\,(\sigma')^{-1}\exp\{-(y-\mu')^2/2(\sigma')^2]$$

where μ' and σ' denote the mean and standard deviation for the variable y. One can show that between μ', the mode x_{mod}, the median x_{50} and the mean x_{ave} the relations

$$\ln x_{mod} = \mu' - (\sigma')^2 \qquad \ln x_{50} = \mu' \qquad \text{and} \qquad \ln x_{ave} = \mu' + (\sigma')^2/2$$

exist. To illustrate the influence of the choice of the characteristic size measure for a lognormal distribution with $\mu' = \ln x_{50} = \ln 10$ and a (typical) value $\sigma' = 0.6$, we note that the mode $x_{mod} = 7.0$ and the mean $x_{ave} = 12.0$, respectively.

The use of various types of distribution functions should be recognised. If we denote for a size measure, say x, the normalised number distribution by $f_N(x)$, the length, area and volume distributions are given by

$$f_1(x) = k_1 x f_N(x) \qquad f_2(x) = k_2 x^2 f_N(x) \qquad \text{and} \qquad f_3(x) = k_3 x^3 f_N(x)$$

respectively. The normalization factors k_1, k_2 and k_3 are determined from

$$\int f_i(x)\,\mathrm{d}x = \int k_i x^i f_N(x)\,\mathrm{d}x = 1$$

where i is either 1, 2 or 3. For lognormal distributions a simple conversion is possible between the different types of distributions. In this case σ' is invariant for the type of distribution and the relation between calculated median value μ_c' and analysed median value μ_a' is given by

$$\mu_c' = \mu_a' + (c-a)(\sigma')^2$$

where a and c denote 0, 1, 2 or 3 depending on whether a number, length, area or volume distribution is involved. For the same example as mentioned before ($\sigma' = 0.6$), $\mu_V' - \mu_N' = 1.08$ corresponding to $x_{50,V}/x_{50,N} = 2.94$.

In general, it can be said that when one author speaks of a grain size and calls the result of his measurement d without any further comment, another author can easily end up with $3d$ or $d/3$. This is due to the use of a different definition of the grain size parameter (mean, mode, median), the use of a different measure of the grain size (intercept, equivalent area, equivalent boundary, caliper, ...) or the use of a different type of distribution (number, length, area, volume). Of course, combinations can arise. It is therefore necessary to define quite precisely the procedure followed to quantify the grain size, if one wants to avoid confusion.

8.14 Bibliography

General

Phillips, R. (2001), *Crystals, defects and microstructures*, Cambridge University Press, Cambridge.

Structure

Born, M. and Huang, K. (1954), *Dynamical theory of crystal lattices*, Oxford University Press, Oxford.

Boyd, R.H. and Phillips, P.J. (1993), *The science of polymer molecules*, Cambridge University Press, Cambridge.

Gedde, U.W. (1995), *Polymer physics*, Chapman and Hall, London.

Greenwood, N.N. (1968), *Ionic crystals, lattice defects and non-stoichiometry*, Butterworth, London.

Ziman, J.M. (1972), *Principles of the theory of solids*, 2nd ed., Cambridge University Press, London.

Bonding

Dreizler, R.M. and Gross, E.K.U. (1990), *Density functional theory*, Springer, Berlin.

Harrison, W.A. (1980), *Electronic structure and the properties of solids*, Freeman, San Francisco (also Dover, 1989).

Hirschfelder, J.O., Curtiss, C.F. and Bird, R.B. (1964), *Molecular theory of gases and liquids*, Wiley, New York.

Koch, W. and Holthausen, M.C. (2001), *A chemist's guide to density functional theory*, Wiley-VCH, Weinheim.

Mott, N.F. and Jones, H. (1936), *The theory of metals*, Clarendon press, Oxford (also Dover, 1958).

Parr, R.G. and Yang, W. (1989), *Density-functional theory of atoms and molecules*, Oxford University Press, Oxford.

Pettifor, D.G. (1995), *Bonding and structure of molecules and solids*, Clarendon, Oxford.

Seitz, F. (1940), *Modern theory of solids*, McGraw-Hill, New York.

Slater, J.C. (1939), *Introduction to chemical physics*, McGraw-Hill, New York (also Dover, 1970).

Sutton, A. (1993), *Electronic structure of materials*, Oxford Science Publishers, Clarendon, Oxford.

Defects

Flynn, C.P. (1972), *Point defects and diffusion*, Clarendon, Oxford.

Henderson, B. (1972), *Defects in crystalline solids*, Edward Arnold, London.

Van Hove, M.A. (1993), *Crystal surfaces*, page 481 in *Structure of solids*, Mater. Sci. Tech. Vol. I, R.W. Cahn, P. Haasen and E.J. Kramer, eds., VCH Verlag, Weinheim.

Mott, N.F. and Gurney, R. (1940), *Electronic processes in ionic crystals*, Clarendon, Oxford.

Stoneham, A.M. (1975), *Theory of defects in solids*, Clarendon, Oxford.

Stereology

DeHoff, R.T. and Rhines, F.N. (1968), *Quantitative microscopy*, McGraw-Hill, New York.

Underwood, E.E. (1969), *Quantitative stereology*, Addison-Wesley, Reading, MA.

Weibel, E.R. (1979), *Stereological methods*, vols. 1 and 2, Academic Press, London.

9

Continuum elasticity

In this chapter the elastic constitutive behaviour is discussed. From the general description of materials response defined by the Helmholtz energy and dissipation function as presented before, we specialise to a material without internal variables so that dissipation is absent. The behaviour for anisotropic elastic materials is derived and further simplified to isotropic materials. The various elastic constants and the elastic energy are discussed. An extension to thermo-elastic materials is made and we consider briefly large deformations in principal axes. Finally, a short introduction to potential energy formulations is given.

9.1 Elastic behaviour

Dealing with thermodynamics we have emphasised that the constitutive behaviour of materials can be described by the specific Helmholtz function f and the specific dissipation function φ. For elastic materials no internal variables are present so that the dissipation function is identically zero (Ziegler, 1983). This implies the *absence of hysteresis*. Consequently, we deal with the Helmholtz function only. Furthermore, we have emphasised that the Helmholtz function is a function of the kinematical variables a_i and the temperature T. For mechanical problems a convenient set of kinematical variables is the set of infinitesimal strains ε_{ij} and thus $f = f(\varepsilon_{ij}, T)$. The stresses and entropy are given by (see Chapter 6)

$$\blacktriangleright \qquad \sigma_{ij}^{(q)} = \rho \frac{\partial f}{\partial \varepsilon_{ij}} \qquad \text{and} \qquad s = -\frac{\partial f}{\partial T} \tag{9.1}$$

For the remainder of this chapter we omit the superscript (q) since there are no internal variables. We note again that the Helmholtz function f acts as a potential for the tensor[a] σ_{ij}/ρ and the specific entropy s.

For the moment we consider only isothermal changes and consequently $f = f(\varepsilon_{ij})$. We consider a *virgin state*, i.e. a state at the reference temperature having neither stress, nor strain. For small deviations from the equilibrium situation the Helmholtz function for a virgin state can be developed in a Taylor series resulting in

$$f(\varepsilon_{ij}) = f_0 + \left.\frac{\partial f}{\partial \varepsilon_{ij}}\right|_0 \varepsilon_{ij} + \tfrac{1}{2} \left.\frac{\partial^2 f}{\partial \varepsilon_{ij} \partial \varepsilon_{kl}}\right|_0 \varepsilon_{ij}\varepsilon_{kl} + \cdots \tag{9.2}$$

where the subscript 0 denotes the strain-free state. Using that in the strain-free state the stresses are also zero, we see from Eq. (9.1) that the coefficients for the linear

[a] One may wonder why in this case the potential yields the stress and not minus the stress, as is conventional for a mechanical potential. Remember that, if σ represents the internal stress, at equilibrium the stress due to the external loading σ^* has opposite sign, $\sigma^* = -\sigma$. While the derivative of the Helmholtz potential F provides the internal stress σ, a mechanical potential V refers to an external stress σ^* and therefore $\sigma = \partial F/\partial \varepsilon = -\sigma^* = -(-\partial V/\partial \varepsilon)$.

terms are zero (stress-free reference state). Therefore, $f-f_0$ is of second order and it follows that in the identity

$$\frac{\partial}{\partial \varepsilon_{ij}}\left[\rho\left(f-f_0\right)\right]=\rho\frac{\partial f}{\partial \varepsilon_{ij}}+\left(f-f_0\right)\frac{\partial \rho}{\partial \varepsilon_{ij}} \tag{9.3}$$

the second term on the right-hand side can be neglected and thus that ρ can be considered as a constant. Eq. (9.1) thus reduces to

$$\sigma_{ij}=\frac{\partial w}{\partial \varepsilon_{ij}} \qquad \text{where} \qquad w=\rho(f-f_0) \tag{9.4}$$

can be interpreted as the strain energy per unit volume or *strain energy density*. A similar treatment for adiabatic conditions yields the same expressions for the stress σ_{ij} and strain energy density w but with the specific Helmholtz energy f replaced by the specific internal energy u.

The general expression for the stress in an elastic solid, with a reference configuration that is stress free, is thus

► $$\sigma_{ij}=C_{ijkl}\varepsilon_{kl} \qquad \text{where} \qquad C_{ijkl}=\rho\left.\frac{\partial^2 f}{\partial \varepsilon_{ij}\partial \varepsilon_{kl}}\right|_0 \tag{9.5}$$

denotes the *elastic* (*stiffness*) *constants*. The equation is known as the (*generalised*) *Hooke's law*[b]. In symbolic notation it reads $\boldsymbol{\sigma}=\mathbb{C}:\boldsymbol{\varepsilon}$. The number of elements of the elastic stiffness tensor is $3^4=81$. However, the number of independent elements is reduced due to the symmetry of both $\boldsymbol{\sigma}$ and $\boldsymbol{\varepsilon}$. This allows for the index exchanges $ij \leftrightarrow ji$ and $kl \leftrightarrow lk$, reducing the number to $6^2=36$. Since the order of differentiation in Eq. (9.5) is immaterial, we also have the index exchange $ij \leftrightarrow kl$. This reduces the number of independent elements to $(6{\cdot}5)/2+6=21$. Crystal symmetry will reduce the number of required elastic constants further. Without discussing the details (see Nye, 1957) we merely state that for isotropic materials the number of independent elements is just two. Isotropy is important since many technical materials are to a good approximation isotropic. In that case the strain energy is a function of the three invariants $\varepsilon_{(1)}$, $\varepsilon_{(2)}$ and $\varepsilon_{(3)}$ of the strain tensor only. Also the stress becomes an isotropic function of the strain. Generally we can write by the Hamilton-Cayley theorem (see Section 3.11)

$$\sigma_{ij}=a\delta_{ij}+b\varepsilon_{ij}+c\varepsilon_{ik}\varepsilon_{kj} \tag{9.6}$$

The functions a, b and c are determined by the Helmholtz energy. Comparing with Eq. (9.5) we conclude that a = linear in $\varepsilon_{(1)}$, b = constant and $c=0$. Conventionally one writes

► $$\sigma_{ij}=\lambda\,\varepsilon_{kk}\,\delta_{ij}+2\mu\,\varepsilon_{ij} \tag{9.7}$$

which is referred to as *Hooke's law for isotropic materials* and where λ and μ are defined as *Lamé's constants*.

Decomposing the stress and strain in their isotropic parts and their deviators results in

---

[b] Remember that in this case law really means constitutive relation. The designation is conventional.

$$\sigma_{kk} = (3\lambda+2\mu)\varepsilon_{kk} \qquad \sigma_{ij}' = 2\mu\varepsilon_{ij}'$$

$$\varepsilon_{kk} = \sigma_{kk}/(3\lambda+2\mu) \qquad \text{and} \qquad \varepsilon_{ij}' = \sigma_{ij}'/2\mu,$$

respectively, from which we obtain the inverted equation

$$\varepsilon_{ij} = \frac{1}{2\mu}\left(\sigma_{ij} - \frac{1}{3}\sigma_{kk}\delta_{ij}\right) + \frac{1}{3(3\lambda+2\mu)}\sigma_{kk}\delta_{ij} \qquad \text{or more compact} \qquad (9.8)$$

► $$\varepsilon_{ij} = \frac{1}{2\mu}\left(\sigma_{ij} - \frac{\lambda}{3\lambda+2\mu}\sigma_{kk}\delta_{ij}\right) \qquad (9.9)$$

Robert Hooke (1635-1703)

Born on the Isle Wight, he was sent in 1653 to Christ Church, Oxford, where he was a chorister and studied so that in 1662 he took the degree of Master of Arts. About 1658 he worked with Boyle and this marked the beginning of experiments with springs. On recommendation of Boyle he became in 1662 curator of Royal Society. Between the years 1663-1664 Hooke became interested in microscopy and in 1665 his book *Micrographia* was published which contained information about his microscope. Based on the idea that light is a wave, he explained the interference colours of soap bubbles and the phenomenon of Newton's rings. In 1664 he became professor of geometry in Gresham College but continued to present his work to the Royal Society, amongst which there was a clear view on universal gravity. After the great fire of London in 1666 he was active in reconstruction work and designed several buildings. His paper *De Potentiâ Restitutiva* (*Of spring*) was published in 1678 and contained the results of Hooke's experiments with elastic bodies establishing the relation between the magnitude of forces and the deformations as well as the solution of some important problems. During all his life he was arguing with Newton, e.g. about the nature of light and the mirror telescope.

---

*Alternative formulations**

In the previous part the elastic behaviour was expressed in terms of stresses and strains. However, this behaviour can be expressed also entirely in terms of displacements. To that purpose we recall that $\varepsilon_{ij} = ½(u_{i,j}+u_{j,i})$ and therefore for isotropic materials the stress–strain relationship $\sigma_{ij} = \lambda\varepsilon_{kk}\delta_{ij} + 2\mu\varepsilon_{ij}$ can be written as

$$\sigma_{ij} = \lambda u_{k,k}\delta_{ij} + \mu(u_{i,j}+u_{j,i})$$

If we insert this equation in the equilibrium condition $\sigma_{ij,j} + \rho b_i = 0$ we obtain

$$\lambda u_{k,kj} + \mu(u_{i,jj} + u_{j,ij}) + \rho b_i = 0 \qquad \text{or equivalently} \qquad \mu u_{i,jj} + (\lambda+\mu)u_{j,ji} + \rho b_i = 0$$

These are the *Navier equations*, which describe (linear) elasticity entirely in terms of displacements.

In a similar, though more complex way one can obtain the *Beltrami-Michell equations*, expressing elasticity entirely in terms of stresses[c].

Problem 9.1

Show that by substituting the expression $\varepsilon_{ij} = \frac{1}{2\mu}\left(\sigma_{ij} - \frac{\lambda}{3\lambda+2\mu}\sigma_{kk}\delta_{ij}\right)$ in the compatibility equation $\varepsilon_{ij,kl} + \varepsilon_{kl,ij} - \varepsilon_{ik,jl} - \varepsilon_{jl,ik} = 0$ and using the equilibrium condition $\sigma_{ij,j} + \rho f_i = 0$, one obtains the *Beltrami-Michell equations*

$$\sigma_{ij,kk} + \frac{2(\lambda+\mu)}{3\lambda+2\mu}\sigma_{kk,ij} + \rho(f_{i,j} + f_{i,j}) + \frac{\lambda}{\lambda+2\mu}\rho f_{k,k}\delta_{ij} = 0$$

9.2 Stress states and the associated elastic constants

Let us consider a state of *uni-axial stress*, i.e. only σ_{11} is non-zero. This stress distribution occurs in, e.g. in a (slender) rod in tension or compression. In that case we obtain from Eq. (9.8)

$$\varepsilon_{11} = \frac{\lambda+\mu}{\mu(3\lambda+2\mu)}\sigma_{11} = \frac{1}{E}\sigma_{11}$$
$$\varepsilon_{22} = \varepsilon_{33} = -\frac{\lambda}{2\mu(3\lambda+2\mu)}\sigma_{11} = -\frac{\upsilon}{E}\sigma_{11} \qquad \text{and} \qquad \varepsilon_{23} = \cdots = 0 \tag{9.10}$$

where the constants E and v are known as *Young's modulus* and *Poisson's ratio*. The elastic constants E and v are usually called as the *engineering elastic constants*.

Similarly a *hydrostatic stress* $\sigma_{ij} = -p\delta_{ij} = \sigma_{kk}\delta_{ij}/3$, which occurs in bodies during immersion in a liquid under compression, yields

$$\varepsilon_{kk} = -\frac{1}{K}p \qquad \text{where} \qquad K = \lambda + \frac{2}{3}\mu \tag{9.11}$$

is the *bulk modulus*[d].

A *simple shear stress*, for which e.g. only $\sigma_{23} = \sigma_{32}$ is non-zero, results in

$$\varepsilon_{11} = \cdots = 0 \qquad \varepsilon_{31} = \varepsilon_{12} = 0$$
$$\gamma_{23} = \varepsilon_{23} + \varepsilon_{32} = \frac{1}{G}\sigma_{23} \qquad \text{where} \qquad G = \mu \tag{9.12}$$

is the *shear modulus* and $\gamma_{23} = \varepsilon_{23}+\varepsilon_{32}$ is a so-called *pseudo-vector component* (see Section 7.4). In Table 9.1 the most useful relations between the various elastic constants are given. Two other useful relations are

---

[c] Another scientist dealing with solutions of elasticity problems was the German Alfred Clebsch (1833-1872). He later turned to pure mathematics. His book *Theorie der Elasticität fester Körper* (1862) in the French, annotated translation of Saint-Venant is one of the most complete books in elasticity.

[d] The inverse of K, the *compressibility* $\beta = 1/K$, is also frequently used.

$$\frac{\mu}{\lambda+\mu}=1-2\nu \quad \text{and} \quad \frac{\lambda}{\lambda+2\mu}=\frac{\nu}{1-\nu} \tag{9.13}$$

Finally, we mention *incompressibility*. We note from the expression for Poisson's ratio $\nu = (3K–2G)/(6K+2G) = [3–(2G/K)]/[6+(2G/K)]$ that for $K \to \infty$, $\nu = ½$. Although in practice $K = \infty$ will not occur, high values of ν for certain materials can be reached, in particular for rubbers (see Section 9.8) and many biological tissues. For a typical rubber $K = 1$ GPa and $E = 1$ MPa, leading to $\nu = 0.4998$. While loosely speaking rubbers are characterised as incompressible, this in fact implies a low ratio of shear modulus versus bulk modulus.

Table 9.1: Relations between the various elastic constants for isotropic materials.

	E, G	E, ν	G, ν	λ, μ	K, G	K, E
E	E	E	$2(1+\nu)G$	$\frac{\mu(3\lambda+2\mu)}{(\lambda+\mu)}$	$\frac{9KG}{3K+G}$	E
ν	$\frac{E-2G}{2G}$	ν	ν	$\frac{\lambda}{2(\lambda+\mu)}$	$\frac{3K-2G}{6K+2G}$	$\frac{3K-E}{6K}$
K	$\frac{EG}{9G-3E}$	$\frac{E}{3(1-2\nu)}$	$\frac{2G(1+\nu)}{3(1-2\nu)}$	$\lambda+2\mu/3$	K	K
$G=\mu$	G	$\frac{E}{2(1+\nu)}$	G	G	G	$\frac{3KE}{9K-E}$
λ	$\frac{G(E-2G)}{3G-E}$	$\frac{E\nu}{(1+\nu)(1-2\nu)}$	$\frac{2G\nu}{(1-2\nu)}$	λ	$K-2G/3$	$\frac{3K(3K-E)}{9K-E}$

Gabriel Lamé (1795-1870)

Graduated from the École Polytechnique in 1818 and from the school of mines in 1820, both times together with B.P.E. Clapeyron (1799-1864). Thereafter they were recommended to assist in the work of the new Institute of Engineers of Ways in St. Petersburg in teaching applied mathematics and physics. They helped in the construction of several suspension bridges, which were the first to be built in Europe. To investigate the mechanical properties of the Russian iron, Lamé designed and built a testing machine and while conducting tests he observed that the iron began to stretch rapidly at about two-thirds of its ultimate strength. In connection with the reconstruction of the cathedral of St. Isaac in St. Petersburg, together with Clapeyron, examined the problem of stability of arches. During their time in St. Petersburg they wrote the memoir *Sur l'équilibre interieur des corps solides homogènes*, which contained not only the equilibrium equations but also applications of these general equations to problems of practical interest. In 1831 they returned to France and helped to outline and construct the railroad line between Paris and St.-Germain. However, soon Lamé became professor of physics at the École Polytechnique, which he remained until 1844. In 1843 he was elected member of the French Academy of Sciences and in 1850 became professor at the Sorbonne. In

1852 he published his book *Leçons sur la théorie mathématique de l'élasticité des corps solides*, the first book on the theory of elasticity. In that book he also describes a contribution of his former companion, which he calls (and we still know as) Clapeyron's theorem. Clapeyron was elected member of the Franch Academy in 1858 and worked until his death at the Academy and the École des Ponts et Chaussées.

9.3 Elastic energy

For an isotropic material we can write $\sigma_{ij} = K\varepsilon_{kk}\delta_{ij} + G\varepsilon_{ij}'$. From this equation it is clear that the stress associated with a change in shape is determined by the shear modulus G because it is the multiplier for the components of the deviatoric strain tensor ε_{ij}'. Similarly the stress associated with a change in volume is determined by the bulk modulus K since it is the multiplier for the isotropic part of the strain tensor ε_{kk}. This decoupling of stresses for shape and volume changes is characteristic for isotropic materials and does not occur for anisotropic materials.

Let us now consider the amount of energy associated with the deformation of a material element. If we accept the various strains as independent co-ordinates this energy is usually called the *strain energy density*. In fact we have seen that it is the Helmholtz energy density $\rho f = \rho f(\varepsilon_{ij},T)$ for isothermal conditions. From the strain energy density definition, valid for any stress-strain relationship,

$$w = \int \sigma_{ij}(\varepsilon_{kl})\,\mathrm{d}\varepsilon_{ij} \tag{9.14}$$

we obtain using the general expression for linear elasticity $\sigma_{ij} = C_{ijkl}\varepsilon_{kl}$

$$w = \int C_{ijkl}\varepsilon_{kl}\,\mathrm{d}\varepsilon_{ij} = ½C_{ijkl}\varepsilon_{ij}\varepsilon_{kl}\left(= ½\sigma_{ij}\varepsilon_{ij}\right) \tag{9.15}$$

as illustrated in Fig. 9.1. The expression $w = ½\sigma_{ij}\varepsilon_{ij}$ is sometimes referred to as *Clapeyron's equation*.

For isotropic materials inserting Eq. (9.7) in Eq. (9.14) also directly results in an expression for strain energy density

► $$w = ½\lambda\,\varepsilon_{ii}\,\varepsilon_{jj} + \mu\,\varepsilon_{ij}\,\varepsilon_{ij} = ½\lambda\,\varepsilon_{(1)}^2 + \mu\,\varepsilon_{(2)} \tag{9.16}$$

or in terms of K and G to

► $$w = ½K(\varepsilon_{kk})^2 + G\varepsilon_{ij}'\varepsilon_{ij}' = ½K\varepsilon_{(1)}^2 + G\varepsilon_{(2)}' = w_{\text{vol}} + w_{\text{sha}} \tag{9.17}$$

where as before $\varepsilon_{(1)} = \varepsilon_{kk}$ and $\varepsilon_{(2)} = \varepsilon_{ij}\varepsilon_{ij}$ denote the basic invariants of the strain tensor ε_{ij} and the prime indicates, as usual, the deviatoric nature. It is thus possible to express the strain energy density of an isotropic body as the sum of a term w_{sha} only dependent on the shape change and a term w_{vol} only dependent on the volume change. This energetic decoupling also is characteristic for isotropic materials and does not occur for anisotropic materials. We will re-encounter the shape energy density again in Chapter 11 dealing with plasticity where it will be shown that reaching a certain critical value of w_{sha} provides a criterion for the onset of plastic deformation.

The inverse of Hooke's law, Eq. (9.8), can be written slightly more condensed as

$$\varepsilon_{ij} = \frac{\sigma_{ij}'}{2G} - \frac{p}{3K}\delta_{ij} \tag{9.18}$$

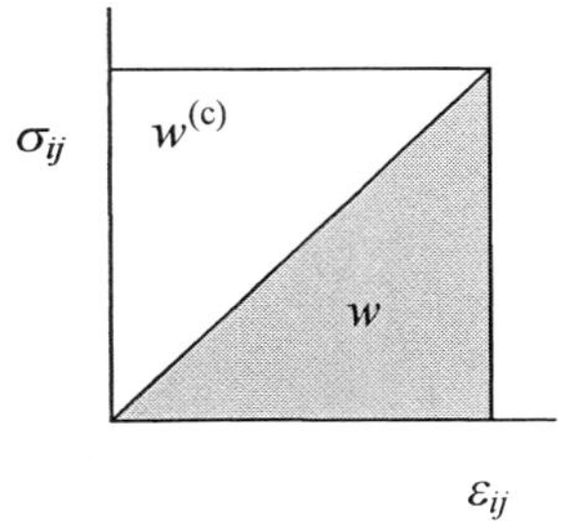

Fig. 9.1: The strain energy density w and stress energy density $w^{(c)}$.

where the bulk modulus $K = \lambda+2\mu/3$, the shear modulus $G = \mu$ and the pressure $p = -\text{tr}(\sigma_{ij})/3 = -\sigma_{jj}/3$ have been used. From Eq. (9.18) it is clear, like before, that the change in shape is determined by the shear modulus G and the deviatoric components of the stress tensor σ_{ij}'. The change in volume is determined by the bulk modulus K and the isotropic part of the stress tensor $p = -\sigma_{kk}/3$.

Let us now consider the various stresses as independent co-ordinates and discuss the *complementary strain energy density* or *stress energy density* given explicitly by

$$w^{(c)} = \int \varepsilon_{ij}(\sigma_{kl})\,\mathrm{d}\sigma_{ij}$$

From Fig. 9.1 we note that $w^{(c)} = \sigma_{ij}\varepsilon_{ij} - w$. Recognising that the Gibbs energy density ρg is given by $\rho g = \rho f - \sigma_{ij}\varepsilon_{ij}$ (similar to $G = F+pV$) we observe[e] that $w^{(c)} = -\rho g(\sigma_{ij};T)$ for isothermal conditions. For a linear elastic material $w^{(c)}$ is numerically equal to w because $w^{(c)} = \sigma_{ij}\varepsilon_{ij} - w = \sigma_{ij}\varepsilon_{ij} - ½\sigma_{ij}\varepsilon_{ij} = ½\sigma_{ij}\varepsilon_{ij}$. However, since the stresses σ_{ij} are the independent variables for $w^{(c)}$ we should write $w^{(c)} = ½S_{ijkl}\sigma_{ij}\sigma_{kl}$, where the tensor S_{ijkl} is the inverse of C_{ijkl} (or equivalently $\mathbb{S} = \mathbb{C}^{-1}$).

Limiting ourselves again to isotropic materials, inserting the inverted Hooke's law, Eq. (9.18) in $w^{(c)} = ½\sigma_{ij}\varepsilon_{ij}$, and separating the stress tensor in an isotropic and a deviatoric part yields

$$\blacktriangleright \quad \begin{aligned} w^{(c)} &= ½\left(\sigma_{ij}' + \frac{\sigma_{kk}}{3}\delta_{ij}\right)\left(\frac{\sigma_{ij}'}{2G} + \frac{\sigma_{ll}/3}{3K}\delta_{ij}\right) = ½\left(\frac{\sigma_{ij}'\sigma_{ij}'}{2G} + \frac{\sigma_{kk}\sigma_{ll}}{9K}\right) \\ &= ½\left(\frac{\sigma_{(2)}'}{2G} + \frac{\sigma_{(1)}^2}{9K}\right) = w_{\text{sha}}^{(c)} + w_{\text{vol}}^{(c)} \end{aligned} \qquad (9.19)$$

where $\sigma_{(1)} = \sigma_{kk}$ and $\sigma_{(2)} = \sigma_{ij}\sigma_{ij}$ denote the basic invariants of the stress tensor σ_{ij} and the prime indicates, as usual, the deviatoric nature. As expected, the expression for the stress energy density of an isotropic body is thus also the sum of a term w_{sha} only dependent on the shape change and a term w_{vol} only dependent on the volume change.

Finally we note that for stability reasons the strain energy w is positive definite ($w > 0$). Therefore it holds for isotropic materials that $K \geq 0$ and $G \geq 0$. From $E = 9KG/(3K+G)$ we also have $E \geq 0$. Since $G = E/2(1+\nu)$ and $K = E/3(1-2\nu)$, it follows that $-1 < \nu < ½$. For most materials though $\nu > 0$ with a value of about 0.2 to 0.4.

---

[e] The minus sign is due to the fact that the Legendre transform $g(X_i)$ of a function $f(x_i)$ in thermodynamics is given by $g(X_i) = f(X_i) - X_i x_i$ where $X_i = \partial f/\partial x_i$ (see Section 3.5) while in physics and mathematics generally the definition $g(X_i) = X_i x_i - f(X_i)$ is used.

Problem 9.2

Calculate for $\sigma_{ij} = \begin{pmatrix} \sigma & 0 & 0 \\ 0 & 0 & \tau \\ 0 & \tau & 0 \end{pmatrix}$ the relative contribution of the shape energy density and volume energy density to the elastic energy for polycrystalline Al (E = 73 GPa and ν = 0.33) and rubber (E = 1 MPa and ν = 0.49).

Problem 9.3

From a tensile test for a non-linear material the stress-strain relationship $\sigma = A\varepsilon^{1/2}$ is obtained. Calculate for this material the strain energy density w and the stress energy density $w^{(c)}$.

9.4 Some conventions

Although the notation σ_{ij}, ε_{ij} and C_{ijkl} as used so far for the stresses, strains and elastic constants is quite clear, it is convenient to reduce the number of subscripts so that the stresses and strains on the one hand and the elastic constants on the other hand can be represented as column matrices and a square matrix, respectively. Moreover, this notation is frequently used in the literature.

For the strains to that purpose the following notation is frequently used:

$$\varepsilon_1 = \varepsilon_{11} \quad \varepsilon_2 = \varepsilon_{22} \quad \varepsilon_3 = \varepsilon_{33} \quad \varepsilon_4 = \varepsilon_{23}+\varepsilon_{32} \quad \varepsilon_5 = \varepsilon_{13}+\varepsilon_{31} \quad \varepsilon_6 = \varepsilon_{12}+\varepsilon_{21}$$

This allows us to write the strain tensor as a 6 × 1 column matrix given by

$$\boldsymbol{\varepsilon} = (\varepsilon_1,\ \varepsilon_2,\ \varepsilon_3,\ \varepsilon_4,\ \varepsilon_5,\ \varepsilon_6)^{\mathrm{T}}$$

The components ε_i are denoted as the *pseudo-vector components*, while the components ε_{ij} are addressed as the *tensorial components*. The components ε_4, ε_5 and ε_6 were previously already indicated by γ_{23} (= γ_{32}), γ_{13} (= γ_{31}) and γ_{12} (= γ_{21}). While the tensorial components relate to half the change in angle between two unit vectors **n** and **m**, the pseudo-vector components, being twice as large, relate directly to the change. A similar notation can be used for the gradient ∇ and displacement **u**. Writing ∇ and **u** as

$$G_{ij} = \begin{pmatrix} \partial/\partial x & 0 & 0 \\ 0 & \partial/\partial y & 0 \\ 0 & 0 & \partial/\partial z \\ 0 & \partial/\partial z & \partial/\partial y \\ \partial/\partial z & 0 & \partial/\partial x \\ \partial/\partial y & \partial/\partial x & 0 \end{pmatrix} \quad \text{and} \quad u_j = \begin{pmatrix} u_x \\ u_y \\ u_z \end{pmatrix} \tag{9.20}$$

the relation $\boldsymbol{\varepsilon} = \nabla\mathbf{u}$ can also be written as $\varepsilon_i = G_{ij}u_j$ or, equivalently, $\boldsymbol{\varepsilon} = \boldsymbol{Gu}$.

Similar to the strain case the pseudo-vector notation for stress

$$\sigma_1 = \sigma_{11} \quad \sigma_2 = \sigma_{22} \quad \sigma_3 = \sigma_{33} \quad \sigma_4 = \sigma_{23} = \sigma_{32} \quad \sigma_5 = \sigma_{31} = \sigma_{13} \quad \sigma_6 = \sigma_{12} = \sigma_{21}$$

allows us to write the stress tensor as a 6 × 1 column matrix

$$\boldsymbol{\sigma} = (\sigma_1,\ \sigma_2,\ \sigma_3,\ \sigma_4,\ \sigma_5,\ \sigma_6)^{\mathrm{T}}$$

Note the difference in definition, as compared with the strain, for elements 4, 5 and 6.

Finally, to make full use of the advantages of matrix calculation and for easy representation, the fourth order elasticity tensor C_{ijkl} is often written as a 6×6 square matrix C_{mn} whereby the following index changes are applied:

$$11 \to 1 \quad 22 \to 2 \quad 33 \to 3 \quad 23 = 32 \to 4 \quad 31 = 13 \to 5 \quad 12 = 21 \to 6$$

To get consistent results the compliances should transform as $S_{ijkl} \to S_{mn}$ if both m and n belong to the set (1,2,3), $2S_{ijkl} \to S_{mn}$ if only one of m or n belong to the set (1,2,3) and $4S_{ijkl} \to S_{mn}$ if both m and n belong to the set (4,5,6). Although the benefits of this notation are clear from a computational point of view, it relies on a strict adherence to the ordering of the elements, not always done in the same way by different authors.

Summarising, for a homogeneous elastic material the stress-strain relation in index and direct notation reads

$$\sigma_{ij} = C_{ijkl}\varepsilon_{kl} \quad \text{and} \quad \boldsymbol{\sigma} = \mathbb{C}\mathbf{:}\boldsymbol{\varepsilon}$$

respectively, while in pseudo-vector notation it reads

► $$\sigma_i = C_{ij}\varepsilon_j \quad \text{or} \quad \boldsymbol{\sigma} = \boldsymbol{C\varepsilon} \tag{9.21}$$

The inverse relation in index and direct notation is

$$\varepsilon_{ij} = (C^{-1})_{ijkl}\sigma_{kl} = S_{ijkl}\sigma_{kl} \quad \text{and} \quad \boldsymbol{\varepsilon} = \mathbb{S}\mathbf{:}\boldsymbol{\sigma}$$

respectively, while in pseudo-vector notation we obtain

► $$\varepsilon_i = (C^{-1})_{ij}\sigma_j = S_{ij}\sigma_j \quad \text{or} \quad \boldsymbol{\varepsilon} = \boldsymbol{C}^{-1}\boldsymbol{\sigma} = \boldsymbol{S\sigma} \tag{9.22}$$

where $\boldsymbol{C}$ denotes the matrix of elastic stifnesses and $\boldsymbol{S}$ its inverse, the matrix of elastic compliances. For isotropic materials these matrices read in full

$$C_{ij} = \frac{E}{(1+\nu)(1-2\nu)}\begin{pmatrix} 1-\nu & & & & & \\ \nu & 1-\nu & & & \text{sym} & \\ \nu & \nu & 1-\nu & & & \\ 0 & 0 & 0 & \tfrac{1}{2}(1-2\nu) & & \\ 0 & 0 & 0 & 0 & \tfrac{1}{2}(1-2\nu) & \\ 0 & 0 & 0 & 0 & 0 & \tfrac{1}{2}(1-2\nu) \end{pmatrix} \tag{9.23}$$

$$C_{ij} = \begin{pmatrix} \lambda+2\mu & & & & & \\ \lambda & \lambda+2\mu & & & \text{sym} & \\ \lambda & \lambda & \lambda+2\mu & & & \\ 0 & 0 & 0 & \mu & & \\ 0 & 0 & 0 & 0 & \mu & \\ 0 & 0 & 0 & 0 & 0 & \mu \end{pmatrix} \tag{9.24}$$

$$S_{ij} = \frac{1}{E}\begin{pmatrix} 1 & & & & & \\ -\nu & 1 & & & \text{sym} & \\ -\nu & -\nu & 1 & & & \\ 0 & 0 & 0 & 2(1+\nu) & & \\ 0 & 0 & 0 & 0 & 2(1+\nu) & \\ 0 & 0 & 0 & 0 & 0 & 2(1+\nu) \end{pmatrix} \tag{9.25}$$

where $\boldsymbol{C}$ has been given in terms of Lamé's constants λ and μ as well as the engineering constants E and ν.

9.5 Plane stress and plane strain

In many situations one can simplify the problem to a two-dimensional one, i.e. to one of plane stress or plane strain. Let us first consider an example of each state. To illustrate the plane stress situation consider a structure with one dimension much smaller than the other two, e.g. a thin plate loaded in tension at the edges (Fig. 9.2). The stresses that develop due to this loading are all in-plane: stress components that are perpendicular to the plate are zero, e.g. $\sigma_{i3} = 0$. This plate is said to be in *plane stress*. For an illustration of plane strain consider a structure with one of the dimensions much larger than the other two, e.g. a retaining wall loaded perpendicular to the sides (Fig. 9.2). In this case the strain components along the wall are zero, e.g. $\varepsilon_{i3} = 0$. The wall is said to be in *plane strain*. In both cases the choice of zero components is arbitrary since the co-ordinate system can be chosen at will.

In *plane stress* with suitably chosen co-ordinate system, $\sigma_{i3} = 0$ which leads to the constitutive equation

$$\begin{pmatrix} \varepsilon_1 \\ \varepsilon_2 \\ \varepsilon_3 \\ \varepsilon_4 \\ \varepsilon_5 \\ \varepsilon_6 \end{pmatrix} = \frac{1}{E} \begin{pmatrix} 1 & & & & & \\ -\nu & 1 & & & \text{sym} & \\ -\nu & -\nu & 1 & & & \\ 0 & 0 & 0 & 2(1+\nu) & & \\ 0 & 0 & 0 & 0 & 2(1+\nu) & \\ 0 & 0 & 0 & 0 & 0 & 2(1+\nu) \end{pmatrix} \begin{pmatrix} \sigma_1 \\ \sigma_2 \\ 0 \\ 0 \\ 0 \\ \sigma_6 \end{pmatrix} \tag{9.26}$$

The only surviving components are ε_1, ε_2, ε_3 and ε_6 so that this equation reduces to

$$\begin{pmatrix} \varepsilon_1 \\ \varepsilon_2 \\ \varepsilon_3 \\ \varepsilon_6 \end{pmatrix} = \frac{1}{E} \begin{pmatrix} 1 & -\nu & -\nu & 0 \\ -\nu & 1 & -\nu & 0 \\ -\nu & -\nu & 1 & 0 \\ 0 & 0 & 0 & 2(1+\nu) \end{pmatrix} \begin{pmatrix} \sigma_1 \\ \sigma_2 \\ 0 \\ \sigma_6 \end{pmatrix} \tag{9.27}$$

Since

► $$\varepsilon_1 + \varepsilon_2 = \frac{1-\nu}{E}(\sigma_1 + \sigma_2) \tag{9.28}$$

we obtain for ε_3 the following expression in terms of ε_1 and ε_2:

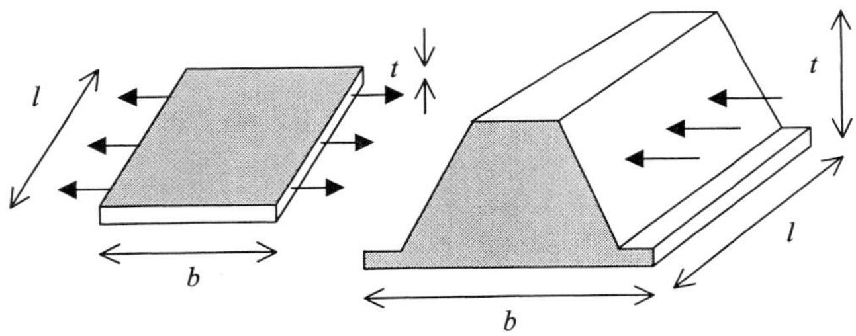

Fig. 9.2: Example of plane stress, where $t \ll b$ and $t \ll l$ (left), and plane strain, where $l \gg b$ and $l \gg t$ (right).

$$\varepsilon_3 = \frac{-\nu}{E}(\sigma_1+\sigma_2) = \frac{-\nu}{E}\frac{E}{1-\nu}(\varepsilon_1+\varepsilon_2) = \frac{-\nu}{1-\nu}(\varepsilon_1+\varepsilon_2) \quad (9.29)$$

and thus Eq. (9.27) reduces further to

► $$\begin{pmatrix}\varepsilon_1\\ \varepsilon_2\\ \varepsilon_6\end{pmatrix} = \frac{1}{E}\begin{pmatrix}1 & -\nu & 0\\ -\nu & 1 & 0\\ 0 & 0 & 2(1+\nu)\end{pmatrix}\begin{pmatrix}\sigma_1\\ \sigma_2\\ \sigma_6\end{pmatrix} \quad (9.30)$$

with Eq. (9.29) understood. Similarly for *plane strain* $\varepsilon_{i3} = 0$, which leads to

► $$\sigma_3 = \nu(\sigma_1+\sigma_2) \quad (9.31)$$

and following a similar procedure as before to

$$\sigma_3 = \frac{E}{(1+\nu)(1-2\nu)}(\varepsilon_1+\varepsilon_2) \quad \text{and} \quad (9.32)$$

► $$\begin{pmatrix}\sigma_1\\ \sigma_2\\ \sigma_6\end{pmatrix} = \frac{E}{(1+\nu)(1-2\nu)}\begin{pmatrix}1 & \nu & 0\\ \nu & 1 & 0\\ 0 & 0 & \tfrac{1}{2}(1-2\nu)\end{pmatrix}\begin{pmatrix}\varepsilon_1\\ \varepsilon_2\\ \varepsilon_6\end{pmatrix} \quad (9.33)$$

Problem 9.4

Show that for plane strain σ_3 is given by Eq. (9.31).

Problem 9.5

At a point P in an isotropic elastic body in plane stress with Young's modulus E and Poisson's ratio ν the strains in the [100], [010] and [110] directions are given by a, b and c, respectively. Calculate the strain matrix ε_{ij} and stress matrix σ_{ij}.

9.6 Anisotropic materials

So far we discussed primarily isotropic materials for which only two elastic constants remain. In many cases though, materials are anisotropic. Single crystal materials are never isotropic and with decreasing symmetry of the crystal an increasing number of elastic constants is required. A full discussion of the influence of symmetry upon the number of elastic constants is given by Nye (1957) from a physical point of view or from a group theoretical point of view by Bhagavantam (1966). The full symmetry is given in Table 9.2. The equations are augmented by the thermal expansion contributions (see Section 9.7) and then read

$$\begin{pmatrix}\boldsymbol{\varepsilon}\\ \Delta X\end{pmatrix} = \begin{pmatrix}\boldsymbol{S} & \boldsymbol{\alpha}\\ \boldsymbol{\alpha}^{\mathrm{T}} & C/T\end{pmatrix}\begin{pmatrix}\boldsymbol{\sigma}\\ \Delta T\end{pmatrix}$$

where ΔX denotes the entropy (to avoid confusion with the compliance S), $\boldsymbol{\alpha}$ the thermal expansion matrix written as a column, $\boldsymbol{\varepsilon}$ the strains, $\boldsymbol{\sigma}$ the stresses and C the heat capacity. We here just discuss two important symmetries.

Justification 9.1*

The reduction of the number of elastic constants due to crystal symmetry is based on Neumans's principle (Chapter 6), which states that for any physical property that can be represented by a tensor the symmetry elements must include the symmetry elements of the point group of the crystal. Since the elastic constants C_{ijkl} are fourth-order tensors, they transform according to $C_{ijkl}{}^* = R_{ip}R_{jq}R_{kr}R_{ls}C_{pqrs}$ and since each symmetry operation can be represented by an orthogonal matrix R_{ij}, we have according to Neumann's principle $C_{ijkl}{}^* \equiv C_{ijkl}$. As an example of symmetry operations we quote reflection, inversion and an n-fold axis of rotation (n = 2,3,4 and 6). For reflection with respect to the x_1-x_2 mirror normal to x_3 and an inversion we have

$$R_{ij} = \begin{pmatrix} 1 & 0 & 0 \\ 0 & 1 & 0 \\ 0 & 0 & -1 \end{pmatrix} \quad \text{and} \quad R_{ij} = \begin{pmatrix} -1 & 0 & 0 \\ 0 & -1 & 0 \\ 0 & 0 & -1 \end{pmatrix}$$

respectively, while for a n-fold axis around x_3 we have

$$R_{ij} = \begin{pmatrix} \cos 2\pi/n & \sin 2\pi/n & 0 \\ -\sin 2\pi/n & \cos 2\pi/n & 0 \\ 0 & 0 & 1 \end{pmatrix}$$

For cubic crystals the axes x_1-x_2-x_3 have 4-fold symmetry but in fact cubic symmetry is defined by four 3-fold axes along the $\langle 1\,1\,1 \rangle$ directions. After some straightforward but tedious algebra (as such an operation is usually called in the literature) one should obtain

$$C_{ijkl} = C_{12}\delta_{ij}\delta_{kl} + C_{44}(\delta_{ik}\delta_{jl} + \delta_{il}\delta_{jk}) + (C_{11} - C_{12} - 2C_{44})I_{ijkl}$$

where I_{ijkl} = 1 if all indices are equal and zero otherwise. Therefore cubic crystals have three independent constants, namely C_{11}, C_{12} and C_{44}. It can be shown that for isotropic materials a similar procedure leads to

$$C_{ijkl} = C_{12}\delta_{ij}\delta_{kl} + C_{44}(\delta_{ik}\delta_{jl} + \delta_{il}\delta_{jk}) \quad \text{with} \quad 2C_{44} = C_{11} - C_{12}$$

The number of independent constants is therefore reduced to two.

The positive-definiteness of the strain energy $w = \frac{1}{2}C_{ijkl}\varepsilon_{ij}\varepsilon_{kl} = \frac{1}{2}C_{ij}\varepsilon_i\varepsilon_j$ requires that $w > 0$ for all strains, implying that the eigenvectors of C_{ij} are all positive. For cubic symmetry this leads to

$$C_{44} > 0 \qquad C_{11} > |C_{12}| \qquad \text{and} \qquad C_{11} + 2C_{12} > 0$$

Similar relations can also be derived for other crystal symmetries.

The first example deals with cubic crystals as commonly encountered for metals. For cubic crystals the lattice constants are all equal, i.e. $a = b = c$, while for the lattice angles it holds that $\alpha = \beta = \gamma = 90°$. Without discussing the details (see Justification 9.1 or Nye, 1957) we quote that for cubic crystals the three necessary, independent elastic constants are C_{11}, C_{12} and C_{44} and that the complete matrix C_{ij} is given by

Table 9.2: Symmetry of the compliance matrices for the various crystal systems, $S = 2(S_{11}-S_{12})$.

Triclinic

$$\begin{pmatrix} S_{11} & & & & & & \\ S_{21} & S_{22} & & & \text{sym} & & \\ S_{31} & S_{32} & S_{33} & & & & \\ S_{41} & S_{42} & S_{43} & S_{44} & & & \\ S_{51} & S_{52} & S_{53} & S_{54} & S_{55} & & \\ S_{61} & S_{62} & S_{63} & S_{64} & S_{65} & S_{66} & \\ \alpha_1 & \alpha_2 & \alpha_3 & \alpha_4 & \alpha_5 & \alpha_6 & C/T \end{pmatrix}$$

Monoclinic

$$\begin{pmatrix} S_{11} & & & & & & \\ S_{21} & S_{22} & & & \text{sym} & & \\ S_{31} & S_{32} & S_{33} & & & & \\ 0 & 0 & 0 & S_{44} & & & \\ S_{51} & S_{52} & S_{53} & 0 & S_{55} & & \\ 0 & 0 & 0 & S_{64} & 0 & S_{66} & \\ \alpha_1 & \alpha_2 & \alpha_3 & 0 & \alpha_5 & 0 & C/T \end{pmatrix}$$

Tetragonal class 4/m

$$\begin{pmatrix} S_{11} & & & & & & \\ S_{21} & S_{11} & & & \text{sym} & & \\ S_{31} & S_{31} & S_{33} & & & & \\ 0 & 0 & 0 & S_{44} & & & \\ 0 & 0 & 0 & 0 & S_{44} & & \\ S_{61} & -S_{61} & 0 & 0 & 0 & S_{66} & \\ \alpha_1 & \alpha_1 & \alpha_3 & 0 & 0 & 0 & C/T \end{pmatrix}$$

Tetragonal class 422, 4mm, $\bar{4}$2m and 4/mmm

$$\begin{pmatrix} S_{11} & & & & & & \\ S_{21} & S_{11} & & & \text{sym} & & \\ S_{31} & S_{31} & S_{33} & & & & \\ 0 & 0 & 0 & S_{44} & & & \\ 0 & 0 & 0 & 0 & S_{44} & & \\ 0 & 0 & 0 & 0 & 0 & S_{66} & \\ \alpha_1 & \alpha_1 & \alpha_3 & 0 & 0 & 0 & C/T \end{pmatrix}$$

Trigonal class 3 and $\bar{3}$

$$\begin{pmatrix} S_{11} & & & & & & \\ S_{21} & S_{11} & & & \text{sym} & & \\ S_{31} & S_{31} & S_{33} & & & & \\ S_{41} & -S_{41} & 0 & S_{44} & & & \\ -S_{52} & S_{52} & 0 & 0 & S_{44} & & \\ 0 & 0 & 0 & 2S_{52} & 2S_{41} & 2S & \\ \alpha_1 & \alpha_1 & \alpha_3 & 0 & 0 & 0 & C/T \end{pmatrix}$$

Trigonal class 32, 3m and $\bar{3}$m

$$\begin{pmatrix} S_{11} & & & & & & \\ S_{21} & S_{11} & & & \text{sym} & & \\ S_{31} & S_{31} & S_{33} & & & & \\ S_{41} & -S_{41} & 0 & S_{44} & & & \\ 0 & 0 & 0 & 0 & S_{44} & & \\ 0 & 0 & 0 & 0 & 2S_{41} & 2S & \\ \alpha_1 & \alpha_1 & \alpha_3 & 0 & 0 & 0 & C/T \end{pmatrix}$$

Hexagonal

$$\begin{pmatrix} S_{11} & & & & & & \\ S_{21} & S_{11} & & & \text{sym} & & \\ S_{31} & S_{31} & S_{33} & & & & \\ 0 & 0 & 0 & S_{44} & & & \\ 0 & 0 & 0 & 0 & S_{44} & & \\ 0 & 0 & 0 & 0 & 0 & S & \\ \alpha_1 & \alpha_1 & \alpha_3 & 0 & 0 & 0 & C/T \end{pmatrix}$$

Orthorhombic

$$\begin{pmatrix} S_{11} & & & & & & \\ S_{21} & S_{22} & & & \text{sym} & & \\ S_{31} & S_{31} & S_{33} & & & & \\ 0 & 0 & 0 & S_{44} & & & \\ 0 & 0 & 0 & 0 & S_{55} & & \\ 0 & 0 & 0 & 0 & 0 & S_{66} & \\ \alpha_1 & \alpha_2 & \alpha_3 & 0 & 0 & 0 & C/T \end{pmatrix}$$

Cubic

$$\begin{pmatrix} S_{11} & & & & & & \\ S_{21} & S_{11} & & & \text{sym} & & \\ S_{21} & S_{21} & S_{11} & & & & \\ 0 & 0 & 0 & S_{44} & & & \\ 0 & 0 & 0 & 0 & S_{44} & & \\ 0 & 0 & 0 & 0 & 0 & S_{44} & \\ \alpha & \alpha & \alpha & 0 & 0 & 0 & C/T \end{pmatrix}$$

Isotropic

$$\begin{pmatrix} S_{11} & & & & & & \\ S_{21} & S_{11} & & & \text{sym} & & \\ S_{21} & S_{21} & S_{11} & & & & \\ 0 & 0 & 0 & S & & & \\ 0 & 0 & 0 & 0 & S & & \\ 0 & 0 & 0 & 0 & 0 & S & \\ \alpha & \alpha & \alpha & 0 & 0 & 0 & C/T \end{pmatrix}$$

Table 9.3: Elastic constants for various cubic materials.

Material	C_{11} (GPa)	C_{12} (GPa)	C_{44} (GPa)	A	E[111] (GPa)	E[100] (GPa)
Al	108.2	61.3	28.5	1.22	76.1	63.7
Mo	460	176	110	0.775	–	–
C	1076	125	576	1.21	–	–
W	501	198	151	0.997	385	385
Na	6.0	4.6	5.9	8.43	–	–
PbS	127	29.8	24.8	0.510	–	–
Ni	246.5	147.3	124.7	2.51	303	137
$Y_3Al_5O_{12}$	334	112	115	–	–	–
$(Mn,Zn)Fe_2O_4$	234	142	88.5	–	–	–

$$C_{ij} = \begin{pmatrix} C_{11} & & & & & \\ C_{12} & C_{11} & & & \text{sym} & \\ C_{12} & C_{12} & C_{11} & & & \\ 0 & 0 & 0 & C_{44} & & \\ 0 & 0 & 0 & 0 & C_{44} & \\ 0 & 0 & 0 & 0 & 0 & C_{44} \end{pmatrix} \tag{9.34}$$

Values for various cubic crystals are given in Table 9.3. The degree of anisotropy can be indicated by the anisotropy parameter $A = 2C_{44}/(C_{11}-C_{12})$. The value for A for a few materials is also given in Table 9.3. As can be seen from this table the anisotropy varies to large extent.

We can also calculate Young's modulus in any direction in the crystal. For a cubic crystal Young's modulus E in the direction $\mathbf{n} = (n_1,n_2,n_3)$ is given by

► $$\frac{1}{E} = S_{11} - 2S\left(n_1^2 n_2^2 + n_2^2 n_3^2 + n_3^2 n_1^2\right) \tag{9.35}$$

where n_1, n_2 and n_3 denote the direction cosines, $S = (S_{11}-S_{12}-\frac{1}{2}S_{44})$ and the S_{ij}'s the elastic compliance constants. For a cubic crystal the latter are explicitly given by

$$S_{11} = \frac{C_{11}+C_{12}}{C}, \qquad S_{12} = -\frac{C_{12}}{C} \qquad \text{and} \qquad S_{44} = \frac{1}{C_{44}} \tag{9.36}$$

where $C = (C_{11}-C_{12})(C_{11}+2C_{12})$. If we calculate Young's modulus in the [111] and [100] directions, as has been done for Al, W and Ni, we see that the anisotropy can result in a Young's modulus for the two directions differing more than a factor 2.

A similar relation can be derived for Poisson's ratio ν for transverse strain in the direction $\mathbf{m}$ and longitudinal strain in the direction $\mathbf{n}$ and reads

$$\nu(\mathbf{m},\mathbf{n}) = -\frac{S_{12} + S\left(n_1^2 m_1^2 + n_2^2 m_2^2 + n_3^2 m_3^2\right)}{S_{11} - 2S\left(n_1^2 n_2^2 + n_2^2 n_3^2 + n_3^2 n_1^2\right)} \tag{9.37}$$

As before, n_1, n_2, n_3, m_1, m_2 and m_3 denote direction cosines. From the expression for E and ν the shear modulus G can be calculated.

Another relatively important case is orthorhombic symmetry. In case of a crystal this implies that the lattice constants $a \neq b \neq c$, while for the lattice angles it holds that $\alpha = \beta = \gamma = 90°$. The required nine independent elastic constants are C_{11}, C_{22}, C_{33}, C_{12}, C_{13}, C_{23}, C_{44}, C_{55} and C_{66}. This symmetry is not only important for inorganic crystals but also for composites[f] and polymer crystals. An example of the latter is polyethylene for which the theoretically calculated elastic compliance constants with the chains fully ordered along the z-axis (see Section 9.5) have been estimated[g] as

$$S_{ij} = \begin{pmatrix} 14.5 & -4.78 & -0.019 & 0 & 0 & 0 \\ -4.78 & 11.7 & -0.062 & 0 & 0 & 0 \\ -0.019 & -0.062 & 0.317 & 0 & 0 & 0 \\ 0 & 0 & 0 & 31.4 & 0 & 0 \\ 0 & 0 & 0 & 0 & 61.7 & 0 \\ 0 & 0 & 0 & 0 & 0 & 27.6 \end{pmatrix} \text{GPa}^{-1}/100 \qquad (9.38)$$

In the direction of the chains we find $E_3 = S_{33}^{-1} = 312.5$ GPa while the shear modulus G_6 is equal to $G_6 = S_{66}^{-1} = 3.6$ GPa. This polymer thus shows a high stiffness in the chain direction but perpendicular to the chain it is very compliant. For comparison, Young's and shear moduli of steel are about 200 GPa and 80 GPa, respectively.

Finally, we mention that not only single crystals are anisotropic but that polycrystalline and natural materials also can show anisotropy in their elastic properties. During casting of a metal a more or less random distribution of orientation of the grains arises. In many cases the metal is deformed during which a preferential orientation, usually addressed as *texture*, may develop. Also natural materials, e.g. wood, are anisotropic. Although the trunk of a tree is more or less circular symmetric, beams cut from a trunk exhibit approximately orthorhombic symmetry. In Chapter 10 these aspects are discussed somewhat further.

Problem 9.6

Calculate the anisotropy factor and Young's modulus in the [100] and [111] directions for the garnet $Y_3Al_5O_{12}$ and the ferrite $(Mn,Zn)Fe_2O_4$.

Problem 9.7

Plot Eq. (9.35) for $(S_{11}-S_{12}-\frac{1}{2}S_{44}) > 0$ and < 0. Discuss qualitatively the difference.

Problem 9.8

Prove Eq. (9.36).

9.7 Thermo-elasticity

In many cases the conditions are neither isothermal nor adiabatic (Ziegler, 1982; Malvern, 1969). In that case the thermal response cannot be neglected and we have to assume that the Helmholtz energy f is dependent on the strains ε_{ij} and the temperature T, i.e. $f = f(\varepsilon_{ij},T)$. If we measure the strain from a stress-free reference state at a reference temperature T_0 and restrict ourselves to small temperature deviations from

[f] In this case the symmetry is often addressed as orthotropic.
[g] Tashiro, K., Kobayashi, M. and Tadakoro, H. (1978), Macromolecules **11**, 914.

T_0, we can expand $f(\varepsilon_{ij},T)$ in a power series in ε_{ij} and T and truncate the series after the quadratic terms. Since in the reference state $\sigma_{ij} = \rho\,\partial f/\partial\varepsilon_{ij} = 0$ and $T = T_0$, the linear terms in ε_{ij} are absent. Expanding ρf leads to

$$\blacktriangleright \qquad \rho f = \rho f_0 - \rho s_0(T-T_0) + \tfrac{1}{2}C_{ijkl}\varepsilon_{ij}\varepsilon_{kl} + c_{ij}\varepsilon_{ij}(T-T_0) - \frac{\rho c}{2T_0}(T-T_0)^2 \tag{9.39}$$

where the coefficients f_0, s_0, C_{ijkl}, c_{ij} and c have to be determined and the factors ρ and ρ/T_0 have been added for convenience. Taking $\varepsilon_{ij} = 0$ and $T = T_0$ shows that f_0 is the reference Helmholtz energy. Eq. (9.39) is valid for anisotropic materials. Restricting ourselves to isotropic materials, similar as before, the third term on the left-hand side can be written as

$$\tfrac{1}{2}\lambda\,\varepsilon_{ii}\,\varepsilon_{jj} + \mu\,\varepsilon_{ij}\,\varepsilon_{ij} = \tfrac{1}{2}\lambda\,\varepsilon_{(1)}^2 + \mu\,\varepsilon_{(2)} \tag{9.40}$$

In this case also c_{ij} must be isotropic and the appropriate form is $a\delta_{ij}$. It will become clear later that it is convenient to use $a = -(3\lambda+2\mu)\alpha$ and the expression for f reduces to

$$\begin{aligned}\rho f = \rho f_0 - \rho s_0(T-T_0) &+ \tfrac{1}{2}\lambda\varepsilon_{ii}\varepsilon_{jj} + \mu\varepsilon_{ij}\varepsilon_{ij} \\ &- (3\lambda+2\mu)\alpha\varepsilon_{kk}(T-T_0) - \frac{\rho c}{2T_0}(T-T_0)^2\end{aligned} \tag{9.41}$$

The stress is calculated as before from $\sigma_{ij} = \rho\,\partial f/\partial\varepsilon_{ij}$ resulting in

$$\blacktriangleright \qquad \begin{aligned}\sigma_{ij} &= \rho\frac{\partial f}{\partial\varepsilon_{ij}} = \lambda\varepsilon_{kk}\delta_{ij} + 2\mu\varepsilon_{ij} - (3\lambda+2\mu)\alpha\delta_{ij}(T-T_0) \\ &= \left[\lambda\varepsilon_{kk} - (3\lambda+2\mu)\alpha(T-T_0)\right]\delta_{ij} + 2\mu\varepsilon_{ij}\end{aligned} \tag{9.42}$$

which is the thermoelastic equivalent of Hooke's law, Eq. (9.7). Its inversion can be obtained in the same way as before and reads

$$\blacktriangleright \qquad \varepsilon_{ij} = \frac{1}{2\mu}\left\{\left(\sigma_{ij} + \left[2\mu\alpha(T-T_0) - \frac{\lambda}{3\lambda+2\mu}\sigma_{kk}\right]\delta_{ij}\right)\right\} \tag{9.43}$$

The strains are similar as before but for an additional term $\alpha(T-T_0)\delta_{ij}$, which results from the change in temperature and leads to a uniform volume change $3\alpha(T-T_0)$. The parameter α may thus be interpreted as the *(linear) thermal expansion coefficient*.

Similarly using $s = -\partial f/\partial T$ we obtain the entropy

$$\rho s = -\rho\frac{\partial f}{\partial T} = \rho s_0 + (3\lambda+2\mu)\alpha\varepsilon_{kk} + \frac{\rho c}{T_0}(T-T_0) \tag{9.44}$$

so that s_0 is the entropy in the reference state. Calculating the *heat capacity*, given by the derivative $T\partial s/\partial T$, the result is cT/T_0, so that c can be interpreted as the heat capacity at the reference temperature. This completes the interpretation of the constants in the Helmholtz expression.

Further considerations of the thermoelastic equations show that heat conduction and deformation become coupled. Generally, however, the interaction coefficient that occurs is small so that this coupling is usually neglected. We refer to the literature for details, e.g. Fung (1965) or Boley and Weiner (1960).

Problem 9.9

Prove Eq. (9.43).

Problem 9.10

Show that the internal energy $u = f + Ts$, considered as a function of ε_{ij} and T, is given by

$$\rho u = \rho u_0 + \tfrac{1}{2}\lambda\varepsilon_{ii}\varepsilon_{jj} + \mu\varepsilon_{ij}\varepsilon_{ij} + (3\lambda + 2\mu)\alpha T_0\varepsilon_{kk} + \frac{\rho c}{2T_0}(T^2 - T_0^2)$$

where $u_0 = f_0 + T_0 s_0$ is the internal energy in the reference state. We obtain a relation $u = u(\varepsilon_{ij},T)$, while the natural variables for u are ε_{ij} and s. Discuss the difference between $u = u(\varepsilon_{ij},T)$, an equation of state, and $u = u(\varepsilon_{ij},s)$, a fundamental equation.

9.8 Large deformations

The foregoing is entirely expressed in the small deformation gradient strains $\boldsymbol{\varepsilon}$. In a number of cases, however, large(r) deformations are relevant. This is particularly true for rubbers and for the accurate extraction of elastic moduli from pressure-volume relations, obtained either experimentally or theoretically. A general review of rubbers is given by Treloar (1949, 1975).

Rubbers

For rubbers in particular large deformations are possible and, although the large deformation strains could be used, it is easier to use the stretches λ. We use the approach as originated by Rivlin[h] but discuss only homogeneous deformations so that we can refer simply to principal axes. For homogeneous deformations a *stretch* is defined as the relative length after deformation. Remember that the strain is the relative length increase after deformation. In Chapter 4 we have seen that

$$\|\mathrm{d}\mathbf{x}'\|^2 = \mathrm{d}\mathbf{x}\cdot(\mathbf{I}+\mathbf{U})^{\mathrm{T}}\cdot(\mathbf{I}+\mathbf{U})\cdot\mathrm{d}\mathbf{x} \tag{9.45}$$

or equivalently

$$\lambda^2 \equiv \frac{\|\mathrm{d}\mathbf{x}'\|^2}{\|\mathrm{d}\mathbf{x}\|^2} = \frac{\mathrm{d}\mathbf{x}}{\|\mathrm{d}\mathbf{x}\|}(\mathbf{I}+\mathbf{U})^{\mathrm{T}}\cdot(\mathbf{I}+\mathbf{U})\frac{\mathrm{d}\mathbf{x}}{\|\mathrm{d}\mathbf{x}\|} \cong 1+2\boldsymbol{\varepsilon} \tag{9.46}$$

where $\mathrm{d}\mathbf{x}$ and $\mathrm{d}\mathbf{x}'$ denote the length before and after deformation, respectively, and $\mathbf{U}$ the displacement gradient matrix. The principal stretches λ_{I} are thus approximately related to the principal strains ε_{I} by

$$\lambda_{\mathrm{I}}^2 \cong 1 + 2\varepsilon_{\mathrm{I}} \tag{9.47}$$

If we assume isotropic initial behaviour, for not too large values of λ we may assume isotropic behaviour throughout and in that case the Helmholtz energy of the material F must be a function of the invariants $I_{(1)}$, $I_{(2)}$ and $I_{(3)}$ of the deformation measure, in this case the stretches. Therefore

[h] Rivlin, J. (1948), Phil. Trans. **A241**, 379.

$$F = F(I_{(1)},I_{(2)},I_{(3)})$$

where $I_{(1)} = \lambda_I^2 + \lambda_{II}^2 + \lambda_{III}^2$, $I_{(2)} = \lambda_I^2\lambda_{II}^2 + \lambda_{II}^2\lambda_{III}^2 + \lambda_{III}^2\lambda_I^2$ and $I_{(3)} = \lambda_I^2\lambda_{II}^2\lambda_{III}^2$ are the expressions for the invariants of the stretch (see Section 3.11). Rubbers can be treated as if they are incompressible. Incompressibility is represented by $I_{(3)} = 1$. In this case the second invariant reduces to

$$I_{(2)} = 1/\lambda_I^2 + 1/\lambda_{II}^2 + 1/\lambda_{III}^2$$

and the Helmholtz energy reduces to

$$F = F(I_{(1)},I_{(2)})$$

with the constraint $\lambda_I^2\lambda_{II}^2\lambda_{III}^2 = 1$. If we now expand the Helmholtz energy in a polynomial in terms of $I_{(1)}$ and $I_{(2)}$, we obtain

► $$F = \sum\nolimits_{i=0,j=0}^{\infty} F_{ij} = \sum\nolimits_{i=0,j=0}^{\infty} C_{ij}\left(I_{(1)}-3\right)^i\left(I_{(2)}-3\right)^j \tag{9.48}$$

where $F_{00} = C_{00}$ denotes the Helmholtz energy of the stretchless configuration. The quantity –3 is introduced to make C_{00} a constant, since otherwise the value of C_{00} would depend on the order of the expansion. The first-order contribution in terms of $I_{(1)}$ is

$$F_{10} = C_{10}\left(I_{(1)}-3\right) = C_{10}\left(\lambda_I^2 + \lambda_{II}^2 + \lambda_{III}^2 - 3\right) \tag{9.49}$$

The principal stretches λ_I, λ_{II} and λ_{III} give rise to principal stresses s_I, s_{II} and s_{III} (referred to the non-deformed state). The latter can be calculated in the conventional way by differentiating the total Helmholtz function of the system with respect to λ_I, λ_{II} and λ_{III}. The total Helmholtz function F' is given by the Helmholtz energy F of the material and the potential energy of the external loads and thus reads

$$F' = F_{10} - s_I\lambda_I - s_{II}\lambda_{II} - s_{III}\lambda_{III} \tag{9.50}$$

We also have to take care of the incompressibility constraint and we do this via the Lagrange multiplier technique (see Section 3.4). To that purpose we add to the function to be minimised, the Helmholtz energy in this case, the constraint function, $I_{(3)}-1 = 0$ in this case, multiplied by an undetermined multiplier, here denoted by ½p. According to Eqs. (9.50) and (9.49) we obtain

$$\frac{\partial}{\partial\lambda_I}\left[F_{10} - s_I\lambda_I - s_{II}\lambda_{II} - s_{III}\lambda_{III} - \tfrac{1}{2}p(I_{(3)}-1)\right] = 0 \quad \text{or}$$
$$s_I = 2C_{10}\lambda_I - p\lambda_I\lambda_{II}^2\lambda_{III}^2 \tag{9.51}$$

Similar expressions are obtained for s_{II} and s_{III}. Since a unit area before deformation is reduced to $1/\lambda_{II}\lambda_{III}$ after deformation, the stress in the deformed state σ_I corresponding to the stress in the non-deformed force s_I is given by

$$\sigma_I = \frac{s_I}{\lambda_{II}\lambda_{III}} = \lambda_I s_I = 2C_{10}\lambda_I^2 - p\lambda_I^2\lambda_{II}^2\lambda_{III}^2 = 2C_{10}\lambda_I^2 - p \tag{9.52}$$

where the last step can be made since $I_{(3)} = \lambda_I^2\lambda_{II}^2\lambda_{III}^2 = 1$.

Now we have to realise that λ_I, λ_{II} and λ_{III} are not independent and thus in this formalism only two of them can be chosen at will. We choose λ_I and λ_{II}. For uniaxial tension we obtain $\lambda_I = \lambda$ and $\lambda_{II} = \lambda_{III} = \lambda^{-1/2}$. Moreover, for this situation $\sigma_{II} = \sigma_{III} = 0$ and thus

$$\sigma_{II} + \sigma_{III} = 2C_{10}(\lambda_{II}^2 + \lambda_{III}^2) - 2p = 2C_{10}(\lambda^{-1} + \lambda^{-1}) - 2p = 0$$

Hence $p = 2C_{10}\lambda^{-1}$ (which, by the way, can be interpreted as the pressure) and thus

$$\sigma_I = 2C_{10}\lambda^2 - 2C_{10}\lambda^{-1} = 2C_{10}(\lambda^2 - \lambda^{-1})$$

The constant C_{10} can only be determined either from measurements or from other models. As we will see later this expression can reasonably describe the deformation behaviour of rubbers up to $\lambda \cong 1.5$. At higher values for the stretch λ more elaborate expressions for the Helmholtz function have to be used.

Another interesting form, first derived by Mooney[i] by assuming a linear stress-strain relationship in simple shear, is given by

$$F_{10} + F_{01} = C_{10}\left(I_{(1)} - 3\right) + C_{01}\left(I_{(2)} - 3\right) \qquad (9.53)$$

which is the general first-order relationship in terms of $I_{(1)}$ and $I_{(2)}$. Obviously higher order approximations can be made systematically. The value of this approach turns largely on the question whether a simple expression like Eq. (9.49) or Eq. (9.53) does represent the properties of real rubbers sufficiently.

Problem 9.11

Show that incompressibility is characterised by $I_{(3)} = \lambda_I^2\lambda_{II}^2\lambda_{III}^2 = 1$.

Problem 9.12

Show that for a biaxial stress situation λ_I, λ_{II}, $\lambda_{III} = (\lambda_I\lambda_{II})^{-1}$, the stresses are given by $\sigma_I = 2C_{10}(\lambda_I^2 - \lambda_I^{-2}\lambda_{II}^{-2})$, $\sigma_{II} = 2C_{10}(\lambda_{II}^2 - \lambda_I^{-2}\lambda_{II}^{-2})$ and $\sigma_{III} = 0$.

Problem 9.13

Show that for uniaxial tension in the limit of small λ, $C_{10} = E/6$, where E denotes Young's modulus.

Problem 9.14

a) Show that the uniaxial stress as referred to the original cross-section (the engineering stress) can be written as $s = 2C_{10}(\lambda - 1/\lambda^2)$.
b) Show that the ideal rubber as described by $s = 2C_{10}(\lambda - 1/\lambda^2)$ will not neck according to Considère's criterion.

Problem 9.15*

Show using the full first-order expression $F = F_{10} + F_{01}$ that
a) for uniaxial tension the stress becomes $\sigma_I = 2C_{10}(\lambda^2 - \lambda^{-1})(C_{10} + C_{01}/\lambda)$ and that
b) for simple shear the stress becomes $\sigma_I = -\sigma_{II} = 2(C_{10} + C_{01})\lambda$.

---

[i] Mooney, M. (1940), J. Appl. Phys. **11**, 582.

*Elastic equations of state**

So far we have considered the elastic moduli of materials as (temperature dependent) material constants, given by the second derivative of the Helmholtz energy with respect to strain. We used small deformation gradient strains $\boldsymbol{\varepsilon}$ and an expansion of the Helmholtz energy up to second order. For the accurate determination of elastic constants from experimental or theoretical pressure-volume relations, we have to elaborate a bit though and use large deformations. We limit ourselves to the bulk modulus for isotropic materials.

Let us recall that the Euler strain **E** (Eq. 4.63) is given by

$$\mathbf{E} = \tfrac{1}{2}\left[\left(\frac{\partial \mathbf{u}}{\partial \mathbf{x}}\right) + \left(\frac{\partial \mathbf{u}}{\partial \mathbf{x}}\right)^{\mathrm{T}}\right] - \tfrac{1}{2}\left(\frac{\partial \mathbf{u}}{\partial \mathbf{x}}\right)^{\mathrm{T}} \cdot \left(\frac{\partial \mathbf{u}}{\partial \mathbf{x}}\right) \equiv \varepsilon_{ij} - \tfrac{1}{2}\frac{\partial u_k}{\partial x_i}\frac{\partial u_k}{\partial x_j}$$

In the following we will need the average or isotropic part of **E** and $\boldsymbol{\varepsilon}$, i.e. $E = \tfrac{1}{3}\,\mathrm{tr}\,\mathbf{E} = \tfrac{1}{3}E_{ii}$ and $\varepsilon = \tfrac{1}{3}\,\mathrm{tr}\,\boldsymbol{\varepsilon} = \tfrac{1}{3}\varepsilon_{ii}$. By contraction we obtain $E = \varepsilon - \tfrac{1}{2}\varepsilon^2$. In the case of hydrostatic deformation an elementary volume $V = \rho^{-1} = (\mathrm{d}x_1)^3$ in the strained state has a volume $V_0 = \rho_0^{-1} = [\mathrm{d}x_1(1-\partial u_1/\partial x_1)]^3$ in the unstrained state where $\partial u_1/\partial x_1 = \varepsilon$. Therefore,

$$\frac{V_0}{V} = \frac{\rho}{\rho_0} = \left(1 - \frac{\partial u_1}{\partial x_1}\right)^3 = (1-\varepsilon)^3$$

and using $[1-\varepsilon]^3 = \left[(1-\varepsilon)^2\right]^{3/2}$ we obtain

$$\frac{\rho}{\rho_0} = \left[1 - 2\varepsilon + \varepsilon^2\right]^{3/2} = \left\{1 - 2\left[\varepsilon - \tfrac{1}{2}\varepsilon^2\right]\right\}^{3/2} = (1-2E)^{3/2}$$

For the case of small deformation this reduces to $\rho/\rho_0 \cong 1-3\varepsilon$ or $V/V_0 \cong 1+3\varepsilon$. Next we expand the Helmholtz energy F in terms of E and write, omitting as before the linear term since we expand around the strain-free reference state,

$$F - F_0 = (1/2)a(T)E^2 + (1/6)b(T)E^3 + \cdots$$

As before we limit the expansion to second order. Since the bulk modulus in the reference state $K_0 = V_0(\partial^2F/\partial V^2) = (1/9V_0)(\partial^2F/\partial\varepsilon^2)$, we directly obtain $a = 9K_0V_0$. Moreover,

$$E = \tfrac{1}{2}\left[\left(\frac{\rho}{\rho_0}\right)^{2/3} - 1\right] \quad \text{and therefore} \quad \frac{\partial E}{\partial V} = -\frac{1}{3V_0}(1-2E)^{5/2}$$

so that the pressure p is given by

$$p = -\frac{\partial F}{\partial V} = -\frac{\partial F}{\partial E}\frac{\partial E}{\partial V} = 3K_0E(1-2E)^{5/2} = \frac{3K_0}{2}\left[\left(\frac{\rho}{\rho_0}\right)^{7/3} - \left(\frac{\rho}{\rho_0}\right)^{5/3}\right] \tag{9.54}$$

an expression usually referred to as the second order *Birch-Murnaghan equation of state*. The bulk modulus follows directly from $K = -V\,\partial p/\partial V = (\rho/\rho_0)\,\partial p/\partial(\rho/\rho_0)$ as

$$K = \frac{K_0}{2}\left[7\left(\frac{\rho}{\rho_0}\right)^{5/3} - 5\left(\frac{\rho}{\rho_0}\right)^{2/3}\right] = K_0(1-7E)(1-2E)^{5/2} \tag{9.55}$$

Using this equation an accurate estimate of K_0 can be made from pressure-volume relations. The pressure derivative of the bulk modulus K_0' is given by

$$K_0' = \frac{\partial K}{\partial p} = \frac{\partial K}{\partial E}\left(\frac{\partial p}{\partial E}\right)^{-1} = (12-49E)(3-21E)^{-1} \tag{9.56}$$

Taking the limit $E \to 0$, we find $K_0' = 4$. This value is actually close to the value experimentally observed value for many close-packed minerals. Incorporation of third-order terms in the Euler strain E in the expansion of the Helmholtz energy F leads, after considerable calculation, to the third order Birch-Murnaghan equation of state

$$p = \frac{3K_0}{2}\left[\left(\frac{\rho}{\rho_0}\right)^{7/3} - \left(\frac{\rho}{\rho_0}\right)^{5/3}\right]\left\{1+\frac{3}{4}(K_0'-4)\left[\left(\frac{\rho}{\rho_0}\right)^{2/3} - 1\right]\right\}$$

Note that if $K_0' = 4$, we recover the second order equation of state. Although, with consistency in mind, the use of large deformations with the third-order expansion of F leading to the third order equation of state seems more logical, the second order equation of state is often preferred.

9.9 Potential energy formulations*

In Chapter 5 we have discussed the general equilibrium conditions and energy considerations for solids. For elastic systems, whether linear elastic or non-linear elastic, one can go one or two steps further (Teodosiu, 1982). Consider to that purpose again an isolated system consisting of two subsystems. As before we will use the Helmholtz formulation under isothermal conditions. Equilibrium is reached when the Helmholtz energy of the system $F = F_1+F_2$ is minimal or when the variation $\delta F = 0$ and thereto we have to consider variations in strain $\delta\varepsilon$. Considering system 1 indicated by a superscript as the system of interest, we have

$$\delta F^{(1)} = \int \sigma_{ij}\delta\varepsilon_{ij}\,\mathrm{d}V \tag{9.57}$$

where ε_{ij} is the strain associated with the kinematically admissible displacement field u_i. Let us also assume that the loading system, i.e. system 2, provides a loading of body forces and contact forces (tractions) independent of the deformation of system 1, neither in direction, nor in magnitude. We write for the (prescribed) body force b_i and for the (prescribed) traction[j] over the surface A_t by $\bar{t}_i$. The variation for the loading system thus becomes[k]

$$\delta F^{(2)} = -\int \rho b_i \delta u_i\,\mathrm{d}V - \int_{A_t} \bar{t}_i \delta u_i\,\mathrm{d}A = -\int \rho b_i \delta u_i\,\mathrm{d}V - \int \bar{t}_i \delta u_i\,\mathrm{d}A \tag{9.58}$$

where the last step can be made since $\delta u_i = 0$ on the surface A_u and the total surface $A = A_u+A_t$. This particular loading as provided by subsystem 2 is frequently used. It is conventionally indicated as *conservative loading*. One way to accomplish this type of

[j] As before we write $\bar{t}_i$ for prescribed tractions on A_t and use t_i for the reactions at the prescribed displacement $\bar{u}_i$ on A_u.

[k] Remember that the work is delivered *by* system 2, hence the minus sign.

loading is by using a contact force potential $T = -\bar{t}_i u_i$ and a body force potential $B = -b_i u_i$ so that $\bar{t}_i = -\partial T/\partial u_i$ and $b_i = -\partial B/\partial u_i$, respectively. This results in

$$\delta F^{(2)} = \int \rho \frac{\partial B}{\partial u_i} \delta u_i \, \mathrm{d}V + \int \frac{\partial T}{\partial u_i} \delta u_i \, \mathrm{d}A = \int \rho \, \delta B \, \mathrm{d}V + \int \delta T \, \mathrm{d}A \tag{9.59}$$

Now defining

$$\Omega = \int \rho B \, \mathrm{d}V + \int T \, \mathrm{d}A \tag{9.60}$$

we may write

$$\delta F = \delta F^{(1)} + \delta F^{(2)} = \int \sigma_{ij} \delta \varepsilon_{ij} \, \mathrm{d}V + \delta \Omega = 0 \tag{9.61}$$

We see that we may interpret $F^{(2)} = \Omega$ as the potential energy of loading or *external potential energy*[1] for subsystem 1. We have also seen that under isothermal conditions the stress σ_{ij} was given by $\sigma_{ij} = \rho\partial f/\partial\varepsilon_{ij}$ and we may consider the Helmholtz energy of the system of interest, in mechanics often indicated by W and thus $W = F^{(1)} = \int \rho f^{(1)} \, \mathrm{d}V$, as *internal potential energy* acting as a potential for σ_{ij}. For linear elastic, isotropic systems $W = \int\int \sigma_{ij} \mathrm{d}\varepsilon_{ij} \, \mathrm{d}V = \int w \, \mathrm{d}V$ where w is given by Eq. (9.16). From $w = \rho f$ we have $\sigma_{ij} = \rho\partial f/\partial\varepsilon_{ij} = \partial w/\partial\varepsilon_{ij}$. In elastic systems the quasi-conservative stress $\sigma_{ij} = \sigma_{ij}^{(q)}$ is also the only stress. Focussing on the system of interest (subsystem 1) and omitting all system superscripts we thus have the *total potential energy* $\Pi = W + \Omega$. The equilibrium condition for elastic systems

$$\delta \Pi = \delta W + \delta \Omega = 0 \tag{9.62}$$

is usually called as the *principle of minimum potential energy*. In this way reference to the environment is minimised. This theorem essentially states that from all kinematically admissible displacement fields the one, which is also statically admissible (satisfies the equilibrium conditions), renders the potential energy Π an absolute minimum. A proof can be delivered as follows. If we denote a kinematically admissible elastic state by $s = [\mathbf{u},\boldsymbol{\varepsilon},\boldsymbol{\sigma}]$, by setting $s' = s'' - s$ we have

$$\varepsilon_{ij}' = \tfrac{1}{2}(u_{i,j}' + u_{j,i}') \qquad \sigma_{ij}' = C_{ijkl}\varepsilon_{ij}' \ \text{ in } V \qquad \text{and} \qquad u_k' = 0 \ \text{ on } A_u$$

while the potential energy $\Pi = W + \Omega$ yields

$$\Pi[s''] - \Pi[s] = \tfrac{1}{2} \int \left(\sigma_{ij}''\varepsilon_{ij}'' - \sigma_{ij}\varepsilon_{ij}\right) \mathrm{d}V - \int_{A_t} \bar{t}_i u_i' \, \mathrm{d}A - \int \rho b_i u_i' \, \mathrm{d}V$$

Since $\delta u_i = 0$ on the surface A_u we may replace A_t by A for the range of surface integration. Using the boundary condition $\sigma_{ij} n_j = t_i$, the Gauss theorem $\int \sigma_{ij} n_j u_i' \, \mathrm{d}A = \int (\sigma_{ij} u_i')_{,j} \, \mathrm{d}V$ and the equilibrium condition $\sigma_{ij,j} + \rho b_i = 0$, we obtain

$$\int_A \bar{t}_i u_i' \, \mathrm{d}A + \int \rho b_i u_i' \, \mathrm{d}V = \int [(\sigma_{ij} u_i')_{,j} + \rho b_i u_i'] \, \mathrm{d}V = \int \sigma_{ij} u_{i,j}' \, \mathrm{d}V = \int \sigma_{ij} \varepsilon_{ij}' \, \mathrm{d}V$$

so that, using $\sigma_{ij}'\varepsilon_{ij} = C_{ijkl}\varepsilon_{kl}'\varepsilon_{ij} = \varepsilon_{kl}' C_{ijkl}\varepsilon_{ij} = \varepsilon_{kl}'\sigma_{kl}$ where C_{ijkl} represents the elastic constants tensor, we obtain

---

[1] Unfortunately the external potential energy is in mechanics usually indicated by V. To avoid confusion with volume V we use Ω.

$$\Pi[s''] - \Pi[s] = \tfrac{1}{2}\int \sigma_{ij}'\varepsilon_{ij}'\,\mathrm{d}V = W(\varepsilon_{ij}') \tag{9.63}$$

The strain energy W is positive definite and therefore we conclude that $\Pi[s] \le \Pi[s'']$ for any s'' and $\Pi[s] = \Pi[s'']$ only if $\varepsilon_{ij} = \varepsilon_{ij}''$ or, equivalently, if s and s'' differ by a rigid displacement. It also follows that the solution is unique and this fact is often referred to as *Kirchhoff's uniqueness theorem*. From Eq. (9.63) we also have that at equilibrium for linear elastic systems

$$W = \tfrac{1}{2}\int \sigma_{ij}\varepsilon_{ij}\,\mathrm{d}V = \tfrac{1}{2}\int \rho b_i u_i\,\mathrm{d}V + \tfrac{1}{2}\int \bar{t}_i u_i\,\mathrm{d}A \tag{9.64}$$

often referred to as the *theorem of work and energy* or *Clapeyron's theorem*. Of the work done by the external forces, half disappears and half increases the elastic energy of the system. Moreover, we used for two kinematically admissible elastic states $s = [\mathbf{u},\boldsymbol{\varepsilon},\boldsymbol{\sigma}]$ and $s^* = [\mathbf{u}^*,\boldsymbol{\varepsilon}^*,\boldsymbol{\sigma}^*]$, corresponding to the external forces $[\rho\mathbf{b},\mathbf{t}]$ and $[\rho\mathbf{b}^*,\mathbf{t}^*]$,

$$\boldsymbol{\sigma}:\boldsymbol{\varepsilon}^* = \mathbb{C}:\boldsymbol{\varepsilon}:\boldsymbol{\varepsilon}^* = \boldsymbol{\varepsilon}:\mathbb{C}:\boldsymbol{\varepsilon}^* = \boldsymbol{\varepsilon}:\boldsymbol{\sigma}^*$$

Here $\mathbb{C}$ represents the tensor of elastic constants. Therefore, we have

$$\int \boldsymbol{\sigma}:\boldsymbol{\varepsilon}^*\,\mathrm{d}V = \int \boldsymbol{\sigma}^*:\boldsymbol{\varepsilon}\,\mathrm{d}V \tag{9.65}$$

a result known as *Betti's reciprocal theorem*.

Summarising, using the strains as independent variables we discussed in Chapter 6 the principle of virtual power, Eq. (6.76), which is generally valid. Restricting ourselves to conservative loading we obtain Eq. (9.61). Further restricting ourselves to (not necessarily linear) elastic systems we finally obtain Eq. (9.62), the principle of minimum potential energy.

For completeness we mention that along similar lines the complementary strain energy can be used. In this case use is made of the principle of complementary virtual energy using the stresses as independent variables together with conservative loading. The *complementary potential energy* $\Pi^{(c)}$ is given by $\Pi^{(c)} = W^{(c)} + \Omega$ and leads to the equilibrium condition

$$\delta\Pi^{(c)} = \delta W^{(c)} + \delta\Omega = 0 \tag{9.66}$$

usually known as the *principle of minimum complementary potential energy*. This theorem states that amongst all statically admissible stress fields, the actual stress field renders $\Pi^{(c)}$ an absolute minimum. It can be proved along similar lines as the minimum potential energy theorem. The principle of minimum complementary potential energy is related to the principle of minimum potential energy similarly as the Gibbs formulation is related to the Helmholtz formulation. In Eq. (9.66) $W^{(c)}$ is given by $W^{(c)} = \int\int \varepsilon_{ij}\mathrm{d}\sigma_{ij}\,\mathrm{d}V$, which for linear elastic, isotropic systems reduces to $W^{(c)} = \int w^{(c)}\,\mathrm{d}V$ where $w^{(c)}$ is given by Eq. (9.19). We have also $w^{(c)} = \varepsilon_{ij}\sigma_{ij} - w$ and therefore $\varepsilon_{ij} = \partial w^{(c)}/\partial\sigma_{ij}$. Although for both Π and $\Pi^{(c)}$ for the external potential Ω is used, we have for Π that $\delta\Omega = (\partial\Omega/\partial\mathbf{u})\delta\mathbf{u}$ since $\mathbf{u}$ is considered as the independent co-ordinate, while for $\Pi^{(c)}$ we use $\delta\Omega = (\partial\Omega/\partial\mathbf{t})\delta\mathbf{t} + (\partial\Omega/\partial\mathbf{b})\delta\mathbf{b}$ since in this case $\mathbf{t}$ and $\mathbf{b}$ are the independent co-ordinates[m].

---

[m] In practice $\delta\Omega = (\partial\Omega/\partial\mathbf{t})\delta\mathbf{t}$ is used since the body force $\mathbf{b}$ is almost always provided by gravity and therefore difficult to vary independently.

Example 9.1

Consider a material with a power relationship between stress and strain, which is homogeneous of degree n in stress, i.e. $\varepsilon \sim \rho^n$. From $\varepsilon_{ij} = \partial w^{(c)}/\partial\sigma_{ij}$ we note that $w^{(c)}$ is homogeneous of degree $(n+1)$. Thus, using Euler's theorem on homogeneous functions, we have $w^{(c)} = (n+1)^{-1}(\partial w^{(c)}/\partial\sigma_{ij})\sigma_{ij} = (n+1)^{-1}\varepsilon_{ij}\sigma_{ij}$. Since $w^{(c)} = \varepsilon_{ij}\sigma_{ij} - w$, we also have $w = n(n+1)^{-1}\varepsilon_{ij}\sigma_{ij}$. For $n = 1$, this result reduces to Clapeyron's theorem.

The formulation $\delta F = 0$ opens a useful possibility. Suppose again that we have an isolated system consisting of two subsystems, where subsystem 1 is the system of interest and subsystem 2 the loading device. Further let us assume conservative loading. Finally, let us assume that we do not know the exact Helmholtz energy $F^{(1)}$ for subsystem 1 but that we have constructed a model which incorporates, apart from the strain ε_{ij} a parameter p, i.e. $F^{(1)} = F^{(1)}(\varepsilon_{ij},p)$. The total Helmholtz energy $F = F^{(1)}+F^{(2)}$ should be minimal at equilibrium, i.e. $\delta F = 0$ or

$$\delta F = \int \rho \frac{\partial f^{(1)}}{\partial \varepsilon_{ij}} \delta\varepsilon_{ij}\,\mathrm{d}V + \int \rho \frac{\partial f^{(1)}}{\partial p}\delta p\,\mathrm{d}V + \delta\Omega = 0 \tag{9.67}$$

Using a similar reasoning as in Section 6.6 we regain the mechanical equilibrium and boundary conditions but also the condition

$$\int \rho \frac{\partial f^{(1)}}{\partial p}\delta p\,\mathrm{d}V = 0 \tag{9.68}$$

Since δp is independent and arbitrary, we have $\partial f^{(1)}/\partial p = 0$. In fact this means that $f^{(1)}$ should be a minimum with respect to p and we denote the value of p for which this is true by p^*. This implies that we should take $\sigma_{ij} = \partial f^{(1)}(\varepsilon_{ij},p^*)/\partial\varepsilon_{ij}$ and in this way we obtain the best estimate for σ_{ij}. In case an exact solution of a thermomechanical problem cannot be found the above procedure yields a way to approximate the solution. This approximate solution can be improved systematically by taking a better trial function, e.g. by incorporating more parameters $f^{(1)}(\varepsilon_{ij},p_k)$.

Potential energy of interaction

In the previous section we have seen that the total potential energy Π is the sum of the internal potential energy $W = F^{(1)}$ and the external potential energy $\Omega = F^{(2)}$, where for a linear elastic material with volume V and surface A

$$W = \int\int \sigma_{ij}\,\mathrm{d}\varepsilon_{ij}\,\mathrm{d}V = ½\int \sigma_{ij}\varepsilon_{ij}\,\mathrm{d}V \quad \text{and} \quad \Omega = -\int \rho b_i u_i\,\mathrm{d}V - \int \bar{t}_i u_i\,\mathrm{d}A$$

and, as before, b_i and $\bar{t}_i$ refer to the prescribed external forces.

We now consider the potential energy of interaction between two elastic states $s = [\mathbf{u},\boldsymbol{\varepsilon},\boldsymbol{\sigma}]$ and $s^* = [\mathbf{u}^*,\boldsymbol{\varepsilon}^*,\boldsymbol{\sigma}^*]$ that are realised by the load systems $[\rho\mathbf{b},\mathbf{t}]$ and $[\rho\mathbf{b}^*,\mathbf{t}^*]$. Let us assume first for simplicity that both s and s^* act via the outer surface A of a volume V. The potential energy of interaction is defined by

$$\Pi_{\mathrm{int}}(s,s^*) = \Pi(s+s^*) - \Pi(s) - \Pi(s^*)$$

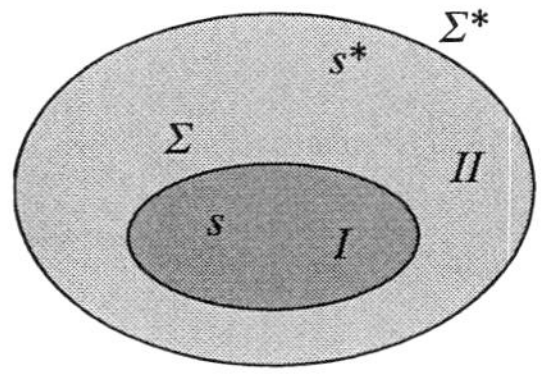

Fig. 9.3: The stress systems s acting on volume I (surface Σ) and s^* acting on volume II (surface Σ^*) of a body with total volume $V = I+II$.

from which we obtain, using the above results,

$$\begin{aligned}\Pi_{\text{int}} &= \tfrac{1}{2}\int(\boldsymbol{\sigma}:\boldsymbol{\varepsilon}^* + \boldsymbol{\sigma}^*:\boldsymbol{\varepsilon})\,\mathrm{d}V - \int(\rho\mathbf{b}\cdot\mathbf{u}^* + \rho\mathbf{b}^*\cdot\mathbf{u})\,\mathrm{d}V - \int(\bar{\mathbf{t}}\cdot\mathbf{u}^* + \bar{\mathbf{t}}^*\cdot\mathbf{u})\,\mathrm{d}A \\ &= -\int\boldsymbol{\sigma}:\boldsymbol{\varepsilon}^*\,\mathrm{d}V = -\int\boldsymbol{\sigma}^*:\boldsymbol{\varepsilon}\,\mathrm{d}V\end{aligned} \tag{9.69}$$

The interaction energy is thus given by the negative of the product of the stress due to one loading system and the strain due to the other loading system or vice versa.

For the discussion on inclusions and dislocations it is, however, useful to assume the following situation[n]. The origin of the stress system s lies wholly within the surface Σ of volume I, e.g. due to an inclusion. The source of the system s^* lies wholly outside the surface Σ but inside the surface Σ^* of the remainder volume II (Fig. 9.3). We can write $\varepsilon_{ij} = \tfrac{1}{2}(u_{i,j}+u_{j,i})$ in volume II and $\varepsilon_{ij}^* = \tfrac{1}{2}(u_{i,j}^*+u_{j,i}^*)$ in volume I but not vice versa. In this case the interaction potential energy is

$$\Pi_{\text{int}} = \int_I \sigma_{ij}\varepsilon_{ij}^*\,\mathrm{d}V + \int_{II}\sigma_{ij}^*\varepsilon_{ij}\,\mathrm{d}V$$

Using the Gauss theorem and the equilibrium condition $\sigma_{ij,j} + \rho b_i = 0$ on volume I and similarly on volume II, we obtain

$$\begin{aligned}\Pi_{\text{int}} &= \int_I \sigma_{ij}\varepsilon_{ij}^*\,\mathrm{d}V + \int_{II}\sigma_{ij}^*\varepsilon_{ij}\,\mathrm{d}V = \int_I \sigma_{ij}u_{i,j}^*\,\mathrm{d}V + \int_{II}\sigma_{ij}^*u_{i,j}\,\mathrm{d}V \\ &= \int_\Sigma \sigma_{ij}u_i^*n_j\,\mathrm{d}A + \int_{\Sigma^*}\sigma_{ij}^*u_in_j\,\mathrm{d}A - \int_\Sigma \sigma_{ij}^*u_in_j\,\mathrm{d}A\end{aligned}$$

$$\Pi_{\text{int}} = \int_\Sigma \sigma_{ij}u_i^*n_j\,\mathrm{d}A - \int_\Sigma \sigma_{ij}^*u_in_j\,\mathrm{d}A \tag{9.70}$$

where the last step is made since on the outer surface Σ^* the traction $\sigma_{ij}^*n_j$ vanishes. This expression yields the interaction energy in the form of an integral over the surface separating the two load systems.

If s^* represents a load system provided by external forces, u_i^* exists throughout the body, we may take $\Sigma = \Sigma^*$ and the interaction energy becomes

$$\Pi_{\text{int}} = \int_{I+II}\sigma_{ij}u_{i,j}^*\,\mathrm{d}V - \int_{I+II}\sigma_{ij}^*u_{i,j}\,\mathrm{d}V = \int_{\Sigma^*}\sigma_{ij}u_i^*n_j\,\mathrm{d}A - \int_{\Sigma^*}\sigma_{ij}^*u_in_j\,\mathrm{d}A$$

The first surface integral on the right-hand side, representing the elastic interaction energy between an internal stress system and an external stress system, vanishes since

[n] Eshelby, J.D. (1956), *The continuum theory of lattice defects*, Solid State Physics **3**, 79.

on the outer surface Σ^* the traction $\sigma_{ij}n_j$ vanishes. Hence this interaction energy is zero or, in other words, the response of the body is the same whether it is in a state of self-stress or not. The absence of interaction energy between an internal and an `external stress system is often named *Collonetti's theorem*. However, the second surface integral on the right-hand side, representing the interaction energy due to the load system providing the external stress (external potential energy) and the internal stress, remains so that the final result becomes

$$\Pi_{\text{int}} = -\int_{\Sigma^*} \sigma_{ij}^* u_i \, n_j \, \mathrm{d}A = -\int_{I+II} \sigma_{ij}^* \varepsilon_{ij} \, \mathrm{d}V$$

equivalent to Eq. (9.69).

As usual the force **A** associated with the potential energy Π is given by the gradient. Now if we have a inhomogeneity, e.g. a defect, a dislocation, an inclusion, etc., in an elastic matrix at position **r** with associated internal stress state s, the force exerted by an external stress state s^* on the inhomogeneity is given

$$\mathbf{A}_{\text{int}} = -\nabla \Pi_{\text{int}}(s,s^*)|_{\mathbf{r}}$$

This result will be used in the theory of inclusions as well as of dislocations.

Problem 9.16

Prove Eq. (9.64).

9.10 Bibliography

Bhagavantam, S. (1966), *Crystal symmetry and physical properties*, Academic Press, London.

Boley, B.A. and Weiner, J.H (1960), *Theory of thermal stresses*, Wiley, New York.

Fung, Y.C. (1965), *Foundations of solid mechanics*, Prentice-Hall, Englewood Cliffs, NJ.

Malvern, L.E. (1969), *Introduction to the mechanics of a continuous medium*, Prentice-Hall, Englewood Cliffs, NJ.

Nye, J.F. (1957), *Physical properties of crystals*, Oxford University Press, London.

Teodosiu, C. (1982), *Elastic models of crystal defects*, Springer, Berlin.

Treloar, L.R.G. (1949, 1975), *The physics of rubber elasticity*, 1st and 3rd ed., Clarendon, Oxford.

Ziegler, H. (1983), *An introduction to thermomechanics*, 2nd ed., North-Holland, Amsterdam.

10

Elasticity of structures

The elastic behaviour of simple structures using the small displacement gradient strains is discussed. For statically undetermined structures solving the equilibrium equations yields the stress distribution. A simplified theory, frequently defined as mechanics of materials, is used to deal with bending, torsion and buckling of bars. Thereafter, the Airy stress function approach for two-dimensional stress distributions is presented. Obtaining an approximate solution via variation theory is shown. An outline of the finite element method is given, followed by an actual example.

10.1 Preview

For almost all thermomechanical problems an initial evaluation is made on the basis of elasticity theory. In this section we first review briefly the required equations, consider some generally accepted simplifications and thereafter introduce the various ways of solving elastic problems.

Elasticity theory applications

Elasticity theory is widely used in engineering. The theory of simple beams and plates is frequently applied in the design and construction of houses. In the early days, when computers were still not widely available, and a full FEM calculation was impossible, the use of truss networks was widely used for larger structures, e.g. power relay line towers, bridges and oil rigs. Nowadays, of course, full FEM calculations are frequently done, not only for large structures but also in the design cycle of many industrial products. It is probably true that for almost any mechanical problem the first step is an elastic calculation in some form or another. Also in science extensive use is made of elasticity theory. For example, in the calculation of properties of composites such as laminates, fibre or particle reinforced materials elastic models are widely used. Moreover, elasticity theory is often applied to microscopic (atomic/molecular) and mesoscopic (microstructural) phenomena in a first approach to assess the influence of their effects. Well-known examples for the former are point defects and dislocations while for the latter grain boundaries can be mentioned.

From the discussions in part I it should be clear that in general we need four types of equations for the solution of the stress and strain distribution in a mechanically loaded structure. First, we need the continuum equations, which are general and not specifically related to the problem and material at hand. For static structures the equilibrium equations, given in terms of the stress σ_{ij} and body force b_i, read

$$\nabla\cdot\boldsymbol{\sigma} + (\rho\mathbf{b}) = 0 \qquad \text{or} \qquad \sigma_{ij,j} + (\rho b_i) = 0$$

If the body forces can be neglected, the terms in brackets are absent. We dealt with these equations in Chapter 5. Second, we need the kinematic conditions, also of a general continuum nature, and which connect the strain ε_{ij} to the displacement u_i by

$$\boldsymbol{\varepsilon} = \tfrac{1}{2}[\nabla\mathbf{u} + (\nabla\mathbf{u})^{\mathrm{T}}] \qquad \text{or} \qquad \varepsilon_{ij} = u_{(i,j)}$$

We dealt with these equations in Chapter 6. Third, we need the constitutive equations, which are specific to the material used. In the case of linear elastic materials this is *Hooke's law*

$$\sigma = \mathrm{C}{:}\varepsilon \qquad \text{or} \qquad \sigma_{ij} = C_{ijkl}\varepsilon_{kl}$$

We dealt with this equation in Chapter 7. Fourth and finally, we need the boundary conditions, which are problem specific. The boundary conditions can be divided in prescribed loading and prescribed displacement conditions. For the whole of the structure it holds that either the loading or the displacement at the surface A is prescribed but not both (at the same point in the same direction). Explicitly

$$t_i = \bar{t}_i \text{ on } A_t \qquad \text{and} \qquad u_i = \bar{u}_i \text{ on } A_u$$

where the barred parameters indicate the prescribed values and $A = A_t + A_u$.

At points where loads are applied generally locally complex stress distributions arise which hardly affect the overall deformation if the structure is large as compared with the characteristic size of the contact stress distribution. These local deformations complicate the overall solution, however, enormously. A way out is to invoke *St.-Venants's principle* which states that, if forces are acting on a small part of the surface of the structure, they can be replaced by *statically equivalent forces*, e.g. forces with the same resultant force and couple, whereby the stress state at large distance from that part is negligibly changed.

Example 10.1

A load on a small part of a large structure provides a typical example. Consider Fig. 10.1. Two tensile forces F are acting symmetrically with respect to a compressive force $2F$. Their resultant force and couple are zero. Therefore, these forces produce negligible stress in regions well removed from the point of application. According to St. Venant's principle well removed means a few times the distance between the tensile forces F.

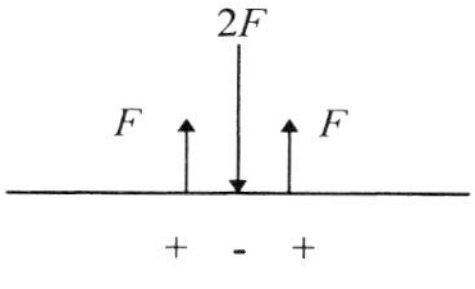

Fig. 10.1: A simple example of point loads with a near-zero remote stress field.

Example 10.2

Another more complex situation, analysed by Hertz, is the contact of two bodies of revolution. This result leads, even for simple loading like a cylinder touching a plane, to highly complex stress distributions (Fig. 10.2). However,

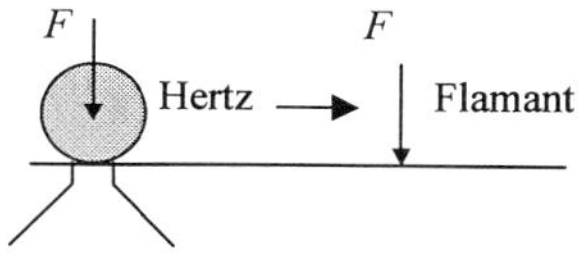

Fig. 10.2: The Hertz contact configuration for a cylinder and the Flamant approximation.

far away from the contact area, the stress is given by the Flamant solution $\sigma_{rr} = -2F\cos(\theta)/\pi r$, where F denotes the load, r the distance from the point of load application and θ the angle between the normal direction and the vector **r**. In this far field the details of the contact situation are irrelevant.

Although for a number of structures an exact solution can be obtained, generally this is impossible. Therefore approximations are made. In the following both exact and approximate solutions are given:

- In some cases the stress distribution is entirely determined by the loads only. These structures are referred to as statically determined structures. A few examples have been discussed in Chapter 5. In other cases the stresses cannot be determined with the aid of statics alone and the deformation plays a role. These structures are the statically undetermined structures.
- For statically undetermined structures a way to solve problems is to identify beforehand the important stress components and construct a simplified theory based on these assumptions.
- In some cases it is possible to solve the equations of elasticity exactly, usually using simplified one- or two-dimensional structures. The former occurs only occasionally. The latter is simply allowed for circular symmetric structures but generally requires making an abstraction of the structure in two dimensions. The solution of the two-dimensional equations, although still quite complicated, is eased by the existence of some general schemes of which the Airy stress function is one of the most important ones.
- Still another way is to approximate the stress or strain distribution in the whole structure by a function of sufficient flexibility and determine the optimum solution by variation theory, i.e. by the minimum in Helmholtz energy.
- Finally, the structure may be divided in elements and the stress or strain distribution in these elements can again be described by such a flexible function. To ensure continuity matching of these functions at the boundaries of the elements is required during the minimisation process. This is the frequently used finite element method. Apart from the simplest cases, the numerical solution of these problems requires the use of computers.

In the following sections the various approaches are discussed.

Barré de Saint-Venant (1797-1886)

Born in the Castle de Fortoiseau, Seine-et-Marne, he entered at the age of sixteen the École Polytechnique. In March 1814, with the troops of the allies approaching Paris, as a sergeant of the detachment of students of the École Polytechnique, he refused too fight, was proclaimed a deserter and forbidden to resume his study. Later the mathematician Chasles, at that time one of Saint-Venants' schoolmates, judged him indulgently. During the eight years that followed he was an assistant in the powder industry. In 1823 the government permitted him to enter the École des Ponts et Chaussées without examination. Here, for two years Saint-Venant bore the protest of the other students, who neither talked to him nor sat on the same bench with him. Disregarding all the unpleasant actions, he followed all the lectures and graduated as the first of his class. After graduating he worked for some time on the channel of Nivernais and on the channel of Ardennes. In his spare time he did theoretical work, presented two papers to the Academy of Sciences and this work made him known to French scientists. In 1837, during the illness of professor Coriolis, he was asked to give lectures on strength of materials at the École des Ponts et Chaussées. Saint-Venant became early interested in hydraulics and its application to agriculture. He published several papers in that subject for which he was awarded the gold medal of the French Agricultural Society. He was not distracted, however, from his favourite subject and published in 1855 and 1856 his two famous memoirs on torsion and bending problems. He never published a book on elasticity theory but in editing the books by Navier and Clebsch on the subject, the first became ten times and the second three times as large as the original due to his annotations and additions. In 1868 he was elected a member of the French Academy of Sciences and he was the authority of mechanics to the end of his life.

10.2 Simplified modelling

A useful approximate way to deal with elastic problems is to assume a certain stress distribution. The bending of a beam provides the most well-known and frequently used example (Derby et al., 1992). Other frequently occurring situations are torsion and buckling. These three topics are discussed briefly in this section.

Bending

An elastic beam (Fig. 10.3) of thickness h is bent to a radius ρ where $\rho >> h$. It is assumed that the deformation consists only of an extension (or contraction) in the longitudinal (in this case the x-) direction of the beam, proportional to the distance from a central extensionless surface, the so-called *neutral surface*. The relevant stress and strain components are thus σ_{11} and ε_{11}, which for convenience are denoted by σ and ε.

Consider a small element of the beam (Fig. 10.4). After bending a thin layer at a distance $\rho+y$ from the centre of curvature has a length $l+\Delta l$ while originally the length was l. At the neutral axis at a distance ρ the length is still l after bending. Accordingly

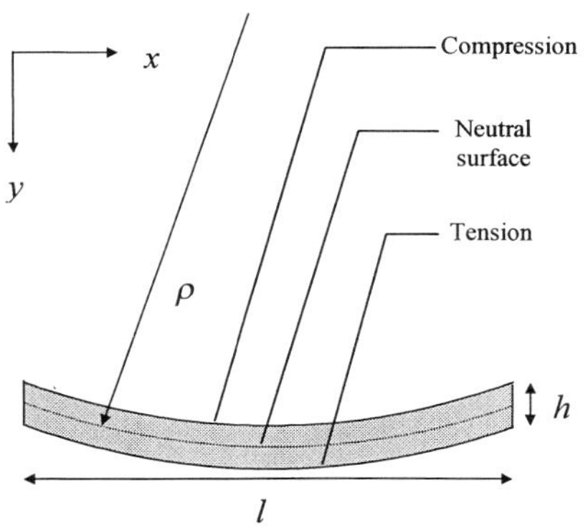

Fig. 10.3: A schematic view of a bending beam.

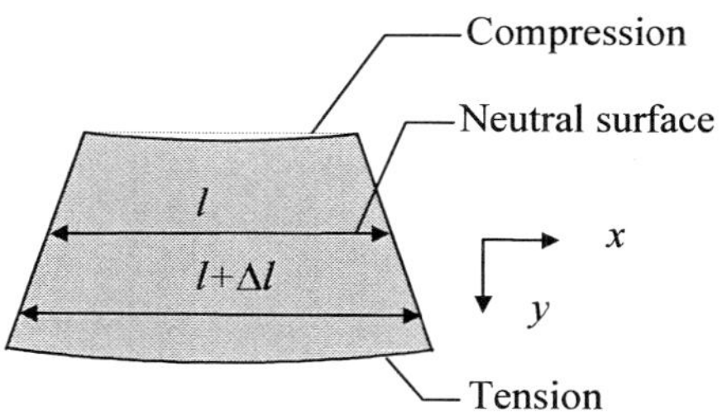

Fig. 10.4: Element of the beam.

$$(l+\Delta l)/l = 1+\Delta l/l = 1+ y/\rho$$

and the strain ε and stress σ are therefore

$$\varepsilon = y/\rho \qquad \text{and} \qquad \sigma = Ey/\rho$$

If the width of the beam at a distance y from the neutral surface is $b(y)$, the longitudinal force df in the layer between y and y+dy is $Ey\, b(y)\, dy/\rho$. In *pure bending* the total longitudinal force exerted over any cross-section of the beam is zero

$$\frac{E}{\rho}\int y b(y)\mathrm{d}y = 0 \qquad (10.1)$$

assuming a constant Young's modulus and mass density. For a symmetric cross-section the neutral axis is thus always through the centre of area of the beam. This implies also through the centre of mass. There is, however, also a moment $y\, df$ exerted by the force df around an axis in the neutral surface. The moments from the layers above and below the neutral surface co-operate and produce a total moment M

$$M = \frac{E}{\rho}\int y^2 b(y)\mathrm{d}y = \frac{EI}{\rho} \qquad (10.2)$$

where the integral I is the *second moment of inertia* of the cross-section. The product EI is known as the *flexural rigidity*. Combining the above equations yields

► $$\sigma = Ey/\rho = My/I \qquad (10.3)$$

The maximum tensile stress σ_{max} is at the outer fibre at y_{max} and yields

$$\sigma_{max}(x) = M(x)y_{max}/I = M(x)/(I/y_{max}) = M/Z_{ela}$$

where Z_{ela} is known as the (*elastic*) *section modulus*. By the way, we see that bending of bars represents a statically determined process.

Example 10.3

For rectangular beam, where the width $b(y) = b$ = constant and h denotes the height of the beam,

$$I = b\int_{-h/2}^{+h/2} y^2\mathrm{d}y = \frac{bh^3}{12} \qquad \text{and} \qquad Z_{ela} = I/y_{max} = bh^2/6$$

For a beam with circular cross-section $b(y) = 2(r^2-y^2)^{1/2}$ and radius r,

$$I = 2\int_{-r}^{+r} (r^2 - y^2)^{1/2} y^2 \mathrm{d}y = \frac{\pi r^4}{4} \qquad \text{and} \qquad Z_{\text{ela}} = I/y_{\max} = \pi r^3/4$$

Obviously, the geometry of the beam plays an important role in the resistance to bending. More details on moments of inertia are given in Appendix C.

The internal bending moment must be balanced by the external forces which produce the bend state. Let us first consider a beam subjected to symmetrical *four-point bending* (Fig. 10.5). The forces $F/2$ are applied at a distance d and result in a moment $M = (F/2)d$ for all positions between the inner load points. This moment is balanced by the moment of the internal forces. An often-used four-point configuration is the so-called *quarter point loading*. This implies that $d = l/4$. Consequently for such a configuration and using a rectangular beam, the maximum stress $\sigma_{4\text{pb}}$ at the outer fibre is given by the expression

$$\sigma_{4\text{pb}} = \frac{My}{I} = \frac{(F/2)(l/4)(h/2)}{bh^3/12} = \frac{3Fl}{4bh^2} \tag{10.4}$$

Another frequently used configuration is the *three-point bend* test (Fig. 10.5). Here the forces result in a moment $M = (F/2)(l/2-x)$, which varies over the length of the beam. The maximum outer fibre stress $\sigma_{3\text{pb}}$ present under the central load point is given by

$$\sigma_{3\text{pb}} = \frac{My}{I} = \frac{(F/2)(l/2-0)(h/2)}{bh^3/12} = \frac{3Fl}{2bh^2} \tag{10.5}$$

We can also calculate the deflection of the beam. To that purpose we recall that the radius of curvature ρ of a function $y = y(x)$ is given by

$$\frac{1}{\rho} = \frac{\mathrm{d}^2y/\mathrm{d}x^2}{\left[1+(\mathrm{d}y/\mathrm{d}x)^2\right]^{3/2}} \cong \frac{\mathrm{d}^2y}{\mathrm{d}x^2} = y'' \tag{10.6}$$

where the approximation is valid for small values of $\mathrm{d}y/\mathrm{d}x$, i.e. for large radius of curvature ρ (equivalent to small curvature $1/\rho$). Note that the differentiation, as is often done, is indicated by a prime, i.e. $y' = \mathrm{d}y/\mathrm{d}x$. It is conventional to denote the displacement in the x, y and z direction by u, v and w, respectively. Combining with $M = EI/\rho$ results in

► $$\frac{\mathrm{d}^2v}{\mathrm{d}x^2} = v'' = \frac{M}{EI} \tag{10.7}$$

which can be integrated twice to obtain $v = v(x)$ where v denotes the displacement in the y-direction as a function of x. For a specific problem we also need the boundary conditions. Typically the values of the displacement and angle of rotation at some positions are prescribed. For example, putting the origin of the axis system at the centre of the (initial position of the) beam, we have for both the four-point bend beam and the three-point beam $y = 0$ at $x = \pm l/2$.

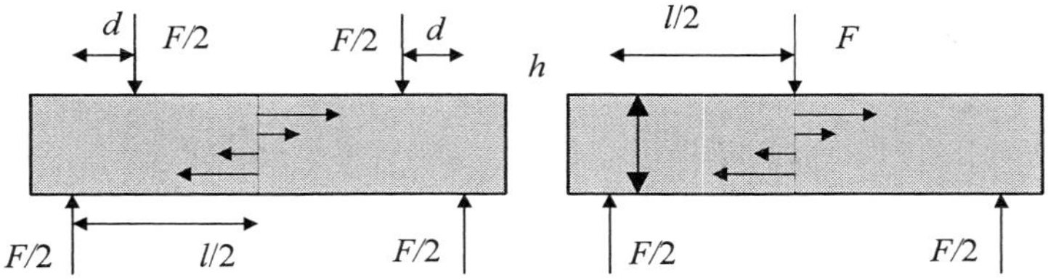

Fig. 10.5: Symmetrical four-point bending and three-point bending of a beam with Young's modulus E and moment of inertia I. The height and thickness are denoted by h and b, respectively.

For other configurations symmetry can often be used. Consider a *cantilever beam*, i.e. a beam clamped at one side and loaded at the other. This is essentially half of the three-point bend beam. Adapting the dimensions of the beam properly directly results in expressions for the cantilever beam. Similarly a four-point bend beam clamped at the outer load points can be considered as four connected cantilever beams.

Example 10.4

Consider a *cantilever beam* (Fig. 10.6). The moment M is $F(l-x)$. Integrating $\mathrm{d}^2v/\mathrm{d}x^2 = M/EI$ twice yields

$$v = (F/6EI)(l-x)^3 + a(l-x) + b$$

where a and b are integration constants. Substituting $v = v' = \mathrm{d}v/\mathrm{d}x = 0$ for $x = 0$ and solving for a and b yields

$$v = (3lx^2 - x^3)F/6EI$$

The maximum displacement $v_{\max}$ is at $x = l$ and is given by $v_{\max} = Fl^3/3EI$.

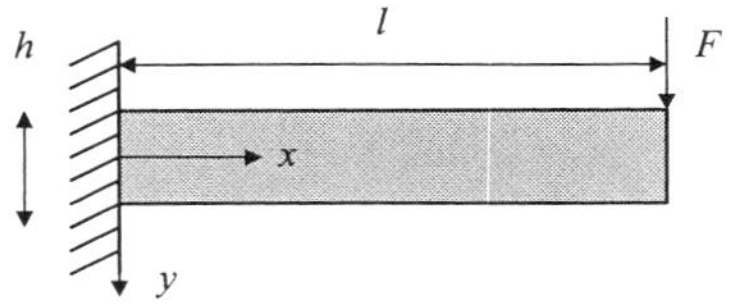

Fig. 10.6: A cantilever beam.

Many other practical important solutions can be obtained in this way. These include distributed loading, statically undetermined beams, optimisation of cross-sections, corrections for shear, etc. The theory involved is usually addressed as 'mechanics of materials' theory, although obviously the material aspects are limited. A classic reference is 'Mechanics of materials' by Timoshenko and Gere (1973). Many solutions can be found in 'Roark's formulas for stress and strain' (Young, 1989).

Also of interest is the strain energy of the beams considered. We recall that the strain energy W is given by

$$W = \int w \,\mathrm{d}V \qquad \text{with} \quad w = \int \sigma : \mathrm{d}\varepsilon = \frac{E\varepsilon^2}{2} = \frac{\sigma\varepsilon}{2} = \frac{\sigma^2}{2E} \tag{10.8}$$

Combining the 'strain expression' $w = E\varepsilon^2/2$ with $\varepsilon = y/\rho = yv''$ results in $w = \frac{1}{2}Ey^2(v'')^2$ and therefore by integration in

$$\blacktriangleright \quad W = \frac{1}{2}\int Ey^2(v'')^2\mathrm{d}V = \frac{Eb}{2}\int_{-h/2}^{+h/2} y^2\mathrm{d}y \int_0^l (v'')^2\mathrm{d}x = \frac{EI}{2}\int_0^l (v'')^2\mathrm{d}x \qquad (10.9)$$

for a beam with constant Young's modulus E and moment of inertia $I = bh^3/12$, where b and h denote the width and height, respectively. Similarly one obtains from the 'stress expression' $w = \sigma^2/2E$ with $\sigma = My/I$ by integration

$$\blacktriangleright \quad W = \frac{1}{2EI}\int_0^l M^2\mathrm{d}x \qquad (10.10)$$

It may be useful to introduce here *generalised co-ordinates*, defined as any 'stress' and 'strain' measure of which the product yields the work done. The above results, together with $M = EIv''$, a relation very similar to $\boldsymbol{\sigma} = \mathbf{C}{:}\boldsymbol{\varepsilon}$, thus can be interpreted as

$$M \Leftrightarrow \boldsymbol{\sigma} \qquad v'' \Leftrightarrow \boldsymbol{\varepsilon} \qquad EI \Leftrightarrow \mathbf{C}$$

where the symbol $\Leftrightarrow$ denotes 'corresponds with'. Finally, we mention that the bending of simple beams can be used to determine Young's modulus E of a material by measuring the deflection v using a known load F or preferably a set of deflections for a set of loads.

Problem 10.1

Compare the section modulus of a beam with circular cross-section with the one for beam with a square cross-section having the same area. Which beam is more resistant to bending stresses? Show that the section modulus Z_{ela} for an ideal I-shaped beam (for which we assume that the web contributes nothing to the moment of inertia and that $a/h << 1$, see figure) is given by $Z_{ela} = \frac{1}{2}Ah'$. Determine also the real section modulus for that beam.

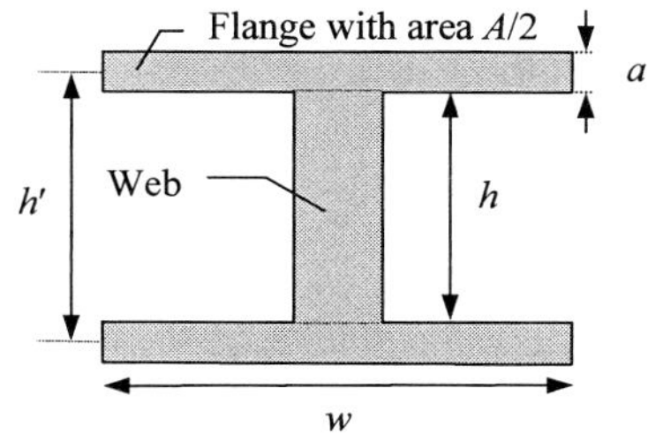

Problem 10.2

Show that for the cantilever beam (see Fig. 10.6), using the moment integration with the moment $M = F(l-x)$, one obtains $W_{can} = F^2l^3/6EI$. Show that the same result is obtained by using the deflection integration. Make use of symmetry to carry this result over in that for a three-point bending bar. Recognise that in that case we have two connected cantilever beams and that

$$F \rightarrow F/2 \quad \text{and} \quad l \rightarrow l/2$$

to obtain for the three-point bend bar

$$W_{3pb} = (F/2)^2(l/2)^3/6EI = F^2l^3/96EI$$

Problem 10.3

Show that the central deflection of a three-point and four-point bending bars are given by $v_{3pb} = \dfrac{Fl^3}{48EI}$ and $v_{4pb} = \dfrac{Fd}{48EI}(3l^2 - 4d^2)$, respectively, where the symbols have the meaning as defined in Fig. 10.5.

Torsion

Another relatively simple configuration is torsion of circular shafts. Here we discuss a shaft with length l and radius r, which is loaded by two equal but opposite torques at the ends. Using symmetry arguments we assume that during deformation a plane section remains plane and a radius remains a radius. In that case the twist angle per unit length is constant over the shaft. The deformation, denoted as *pure torsion*, thus consists of a rotation of sections with respect to each other. The only important stress is σ_{rz}, here denoted by τ. The corresponding engineering strain is denoted by γ.

Consider a cylindrical shell with thickness dr of which one surface is rotated over an angle θ (Fig. 10.7). Opening the cylinder we see that the outer surface OABC has been deformed to ODEC. For small strains we have AD = $\theta r = \gamma l$ so that the *angle of shear* γ is given by

$$\gamma = \theta r/l$$

To maintain the deformed configuration a tangential force dF is applied along DE and CO with similar forces along DO and CE to keep equilibrium. This corresponds to a circumferential force along the cylinder in the twisted cylinder shell. The area of the end face is $\mathrm{d}A = 2\pi r\,\mathrm{d}r$ so that the *shear stress* in shell is given by

$$\tau = \mathrm{d}F/\mathrm{d}A = \mathrm{d}F/(2\pi r\,\mathrm{d}r)$$

We also have $\tau = G\gamma = G\theta r/l$ according to Hooke's law, where G denotes the shear modulus. The force is thus $\mathrm{d}F = \tau\,\mathrm{d}A = (G\theta r/l)(2\pi r\,\mathrm{d}r) = 2\pi r^2(G\theta/l)\,\mathrm{d}r$ and acts tangentially at a distance r from the axis of the cylinder producing a torque $\mathrm{d}T = r\,\mathrm{d}F$. For a solid cylinder the total torque is given by

$$T = \int_A r\tau\,\mathrm{d}A = \frac{G\theta}{l}\int_0^r 2\pi r^3\mathrm{d}r = \frac{G\theta}{l}\frac{\pi r^4}{2} = \frac{G\theta}{l}I_{\mathrm{p}} \qquad (10.11)$$

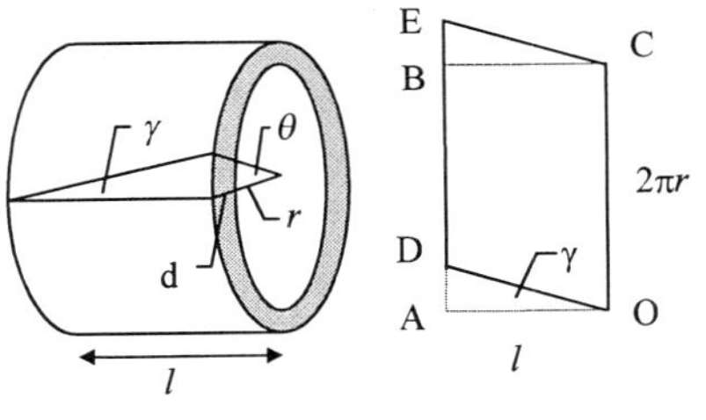

Fig. 10.7: Torsion of a circular shaft.

where I_p is the (second) *polar moment of inertia*. The stress thus can be expressed in terms of the total torque T and the polar moment I_p as

$$\tau = G\gamma = \frac{Tl}{\theta I_p}\frac{\theta r}{l} = \frac{Tr}{I_p} \tag{10.12}$$

The result is independent of the shear modulus due to the fact that the structure is statically determined.

The strain energy is calculated similar to the one for a beam. Here $w = \tau^2/2G = G\gamma^2/2$ (= $\tau\gamma/2$). Integrating the 'stress expression' $w = \tau^2/2G$ for a circular shaft results in

$$W = \int w\,\mathrm{d}V = \int \frac{\tau(r)^2}{2G}\mathrm{d}V = \int_0^l \mathrm{d}z \int_0^r \frac{(Tr/I_p)^2}{2G} 2\pi r\,\mathrm{d}r = \frac{T^2 l}{2GI_p} \tag{10.13}$$

while a similar integration of the 'strain expression' $w = G\gamma^2/2$ yields

$$W = \frac{GI_p\theta^2}{2l} \tag{10.14}$$

In this case the generalised co-ordinates are thus

$T \Leftrightarrow \sigma \qquad \theta \Leftrightarrow \varepsilon \qquad GI_p/l \Leftrightarrow \mathbb{C}$

Finally we mention that, as in bending, a material property, in this case the shear modulus G, can be determined by measuring the angle of rotation θ produced by a known torque T. Again a set of angles of rotation with corresponding torques is preferred above a single measurement. This experiment can be done relatively easy dynamically in order to determine the shear modulus as a function of frequency.

Stepan Prokof'evitch Timoshenko (1878-1972)

Born in Shpotovka, Ukrania, he graduated in 1901 from St. Petersburg Railway Engineering Institute. After one year of military service he worked since 1903 in St. Petersburg Polytechnical Institute. He was invited to join the Kiev Polytechnical Institute in 1907, where he defended a thesis, and since 1908 became a professor in materials strength and since 1909 a dean of the civil-engineering faculty. In 1907-1908 he developed and read the course of materials strength, which later was published as *Mechanics of Materials* and became a classical book on this discipline, as well as his other book on the *Theory of elasticity*. In 1911 after students' disturbances he was dismissed from Kiev Polytechnical Institute and returned to St.Petersburg where in 1915 he was elected a professor in the Polytechnical Institute. In 1918 he became one of the first academicians but had to leave in 1920 the Soviet Ukraine for Yugoslavia, where he took up a chair of materials strength in Zagreb Polytechnical Institute. In 1922 Timoshenko moved to the USA, first as an engineer in the 'Westinghouse' company, but later as a professor in the University of Michigan. His lectures on applied mechanics attracted a great number of students and young scientists and teachers. Prandtl and Westergaard e.g. went to the USA to meet with him. At that time he published a number of books on materials strength, theory of elasticity and theory of stability. Timoshenko worked since 1936 in Stanford University, where his books on technical mechanics, theory of plates and shells, dynamics were published. His book *History of strength of materials* gained a great popularity. Since 1964 Timoshenko lived in Germany. He is considered to be the founder of the technical mechanics scientific school in the USA.

Buckling

Buckling is the sudden bending of slender beams under compression. This bending is a fully elastic phenomenon and is not directly related to the failure of materials: upon unloading the material restores to its initial shape. This kind of loading can occur in many structures, e.g. in chairs, buildings, bridges, etc., and thus presents an important problem.

To calculate the force upon which buckling occurs, we consider a beam of length l loaded in compression with a force F. The clamping at the ends is such that they are considered to be able to rotate freely (Fig. 10.8). The moment upon bending is

$$M(x) = Fy(x) \tag{10.15}$$

and this moment is equal to

$$M(x) = -\frac{EI_x}{\rho} \tag{10.16}$$

where ρ is the radius of curvature, in this case equal to $\partial x^2/\partial y^2$ for small curvatures. The minus sign is used because the beam is concave to the x-axis. Hence we have

$$\frac{\partial y^2}{\partial x^2} = \frac{-Fy}{EI_x} \tag{10.17}$$

with boundary conditions $y = 0$ for $x = 0$ and $y = 0$ for $x = l$. The solution for this equation is

$$y = A\sin\left[x\sqrt{\frac{F}{EI_x}}\right] + B\cos\left[x\sqrt{\frac{F}{EI_x}}\right] \tag{10.18}$$

Making use of the first boundary condition $y(0) = 0$ results in $B = 0$, so that

$$y = A\sin\left[x\sqrt{\frac{F}{EI_x}}\right] \tag{10.19}$$

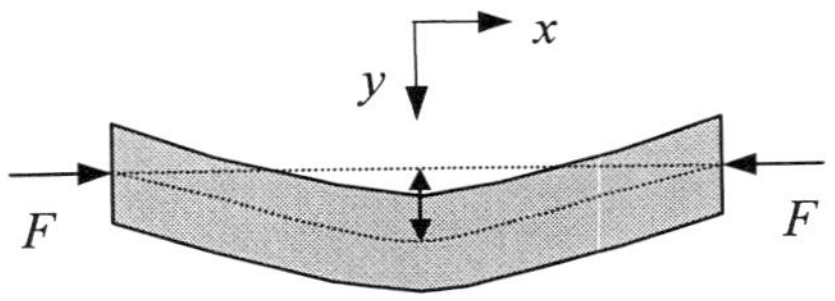

Fig. 10.8: Buckling of a beam.

and the shape of the beam is sinusoidal with A the amplitude and $2\pi/(F/EI_x)^{1/2}$ the wavelength. The second boundary condition $y(l) = A \sin[l(F/EI_x)^{1/2}] = 0$ can be satisfied by either

$$A = 0 \qquad \text{or} \qquad l(F_{\text{cri}}/EI_x)^{1/2} = n\pi$$

where n is a positive integer. While the former solution represents a beam, which is not bent, the second does represent bending. The equation is thus satisfied for every force

► $$F_{\text{cri}} = n^2 \frac{\pi^2 EI_x}{l^2} \qquad \text{where } n = 1, 2, 3, \ldots \qquad (10.20)$$

Eq. (10.20) is usually denoted as the *Euler buckling formula*. Each force corresponds to a set of shapes

$$y = A \sin\left(n\frac{\pi x}{l}\right) \qquad A = 0 \;\text{ if }\; F < F_{\text{cri}} \qquad A \neq 0 \;\text{ if }\; F = F_{\text{cri}} \qquad (10.21)$$

where n denotes the set of real numbers. This implies that the deflection in the y-direction is indeterminate for $F = F_{\text{cri}}$. Of course, this solution is only valid as long as the deflection is small enough so that the small deformation gradient approximation remains valid. The lowest load at which buckling occurs is for the lowest value of n compatible with the boundary conditions at the end of the beam. For a freely rotating beam, as assumed so far, the lowest compatible solution is $n = 1$, corresponding to wavelength of $2l$. For a beam clamped at both ends the lowest compatible solution is $n = 2$, corresponding to wavelength of l.

Summarising, the beam remains straight for $F < F_{\text{cri}}$. At $F = F_{\text{cri}}$, the beam is still in equilibrium if the shape becomes sinusoidal. For a small amount of bending, the force is independent of the deflection y, so that the bending occurs suddenly and without noticeable force increase. This is the phenomenon called *buckling*. It follows that a large value for L or a small value for I_x promotes buckling. Since hollow or I-beams have a high ratio of moment of inertia and density as compared with solid beams, the advantage of using these beams will be clear (see Appendix C).

10.3 Exact solutions*

In many cases the structure is statically undetermined and a solution of the equilibrium equations is necessary. In general a full three-dimensional solution is quite complicated. For example, even the solution for tensile loaded bar, involving St.-Venant's principle, contains some unexpected features. Therefore, abstraction of the system at hand is considered leading to one-dimensional and two-dimensional models. Nevertheless full solutions are rare. As an illustration we provide the important (effective) 1D example of a spherical inclusion in an isotropic matrix and thereafter consider some 2D solutions.

One-dimensional solutions exemplified by inclusions

The problem of a rigid or elastic spherical inclusion in a matrix provides many aspects encountered in an exact solution. Moreover, it is a highly relevant problem in materials science. We consider only isotropic materials but will discuss the effect of different moduli for the inclusion and matrix as well as the effect of a finite matrix.

By using spherical co-ordinates and assuming that the displacement $\mathbf{u}$ is spherically symmetrical radial, we have

$$\mathbf{u} = u_r(r)\mathbf{e}_r \quad \text{and} \quad u_\theta = u_\varphi = 0$$

Using Eq. (4.24) we easily find

$$\varepsilon_{rr} = \frac{\mathrm{d}u_r}{\mathrm{d}r} \qquad \varepsilon_{\theta\theta} = \varepsilon_{\varphi\varphi} = \frac{u_r}{r} \qquad \varepsilon_{r\theta} = \varepsilon_{\theta\varphi} = \varepsilon_{r\varphi} = 0 \tag{10.22}$$

and by substituting in Eq. (5.16) the non-zero components of the stress tensor are

$$\sigma_{rr} = (\lambda + 2\mu)\frac{\mathrm{d}u_r}{\mathrm{d}r} + 2\lambda\frac{u_r}{r} \qquad \sigma_{\theta\theta} = \sigma_{\varphi\varphi} = \lambda\frac{\mathrm{d}u_r}{\mathrm{d}r} + 2(\lambda+\mu)\frac{u_r}{r} \tag{10.23}$$

Now introducing this equation in the (reduced) equilibrium equation $\sigma_{ij,j} = 0$, we see that the latter equation is satisfied identically while the former results in

$$r^2\frac{\mathrm{d}^2u_r}{\mathrm{d}r^2} + 2r\frac{\mathrm{d}u_r}{\mathrm{d}r} - 2u_r = 0$$

This equation has, using α and β as constants, the general solution

$$u_r(r) = \alpha r^{-2} + \beta r \tag{10.24}$$

We now consider an infinite isotropic matrix containing a rigid inclusion with radius r_0 which is forced in a spherical cavity whose volume is δv smaller than that of the inclusion. It then holds that $\beta = 0$ and therefore $u_r^\infty = \alpha r^{-2}$. The constant α is addressed as the *strength of the singularity*. The superscript ∞ is used to indicate that the solution is for an infinite matrix. This solution may be written as

$$\mathbf{u}^\infty = \alpha r^{-2}\mathbf{e}_r = \alpha r^{-3}\mathbf{r} = -\alpha\nabla(r^{-1})$$

From the condition $4\pi r_0^2 u_r^\infty = \delta v$ we find that $\delta v = 4\pi\alpha$. For the non-zero strain and stress components we obtain

$$\varepsilon_{rr} = -2\alpha r^{-3} \qquad \varepsilon_{\theta\theta} = \varepsilon_{\varphi\varphi} = \alpha r^{-3} \qquad \sigma_{rr} = -4\mu\alpha r^{-3} \qquad \sigma_{\theta\theta} = \sigma_{\varphi\varphi} = 2\mu\alpha r^{-3}$$

Since $\operatorname{tr}\boldsymbol{\varepsilon} = \operatorname{tr}\boldsymbol{\sigma} = 0$ the volume dilatation and mean pressure vanish at any point outside the inclusion.

Taking derivatives of $\mathbf{u}^\infty$, remembering that $\nabla^2(r^{-1}) = -4\pi\delta(\mathbf{r})$, we derive

$$\nabla.\mathbf{u}^\infty = -\alpha\nabla^2(r^{-1}) = 4\pi\alpha\delta(\mathbf{r}) \quad \text{and}$$

$$\nabla^2\mathbf{u}^\infty = -\alpha\nabla^2[\nabla(r^{-1})] = -\alpha\nabla[\nabla^2(r^{-1})] = 4\pi\alpha\nabla\delta(\mathbf{r})$$

where, as usual, $\delta(\mathbf{r})$ denotes the Dirac function. By introducing these relations in the Navier equations $(\lambda+\mu)u_{m,mk} + \mu u_{k,mm} + \rho b_k = 0$ and solving for $\mathbf{b}$, we obtain

$$\rho\mathbf{b}(\mathbf{r}) = -4\pi\alpha(\lambda + 2\mu)\nabla\delta(\mathbf{r})$$

so that the effect of the inclusion can be considered as equivalent to that of a body force $\mathbf{b}(\mathbf{r})$. The volume change of a part of the matrix containing the inclusion is then

$$\delta V^\infty = \int \mathbf{u}^\infty.\mathbf{n}\,\mathrm{d}S = 4\pi\alpha = \delta v \tag{10.25}$$

where the integral is over the boundary S of the volume V and **n** is the outward normal to V. It can be proven that this change is independent of the choice of V.

If we now consider a finite matrix we have to introduce some extra displacements since the tractions at the surface of the matrix should be zero, i.e. $\beta \neq 0$. For convenience we take for the matrix a sphere of radius R with a concentric spherical inclusion with radius r_0 at the centre. As before the volume of the hole is δv smaller than that of the rigid inclusion. Since this represents again a spherically symmetrical problem, the same type of solutions can be used but now with boundary conditions

$$\sigma_{rr}(R) = 0 \qquad \text{and} \qquad 4\pi r_0^2 u_r(r_0) = \delta v$$

Therefore we find

$$\alpha = \delta v \bigg/ 4\pi\left(1 + \frac{4\mu r_0^3}{3KR^3}\right) \qquad \text{and} \qquad \beta = \frac{4\mu\alpha}{3KR^3}$$

with the bulk modulus $K = \lambda + \tfrac{2}{3}\mu$. Generally $r_0 \ll R$ so that $\delta v = 4\pi\alpha$ again holds. Substituting the general solution, Eq. (10.24), in the expression for the strain, Eq. (10.22), and stress, Eq. (10.23), we find for the non-zero components of strain and stress

$$\varepsilon_{rr} = -\frac{2\alpha}{r^3}\left(1 - \frac{2\mu}{3K}\frac{r^3}{R^3}\right) \qquad \varepsilon_{\theta\theta} = \varepsilon_{\varphi\varphi} = \frac{\alpha}{r^3}\left(1 + \frac{4\mu}{3K}\frac{r^3}{R^3}\right) \qquad \mathrm{tr}\,\boldsymbol{\varepsilon} = \frac{4\mu\alpha}{KR^3}$$

$$\sigma_{rr} = -\frac{4\mu\alpha}{r^3}\left(1 - \frac{r^3}{R^3}\right) \qquad \sigma_{\theta\theta} = \sigma_{\varphi\varphi} = \frac{2\mu\alpha}{r^3}\left(1 + \frac{2r^3}{R^3}\right) \qquad \mathrm{tr}\,\boldsymbol{\sigma} = \frac{12\mu\alpha}{R^3}$$

A uniform dilatation arises and this gives rise to a uniform pressure in the matrix. The volume change of the matrix due to the rigid inclusion is

$$\delta V = 4\pi R^2 u_r(R) = 4\pi\alpha\left(1 + \frac{4\mu}{3K}\right) = \left(1 + \frac{4\mu}{3K}\right)\delta v = \frac{3(1-\nu)}{1+\nu}\delta v$$

It can be proved that this result is independent of the shape of the matrix. For $\nu = \tfrac{1}{3}$ this expression yields $\delta V = 3\delta v/2$ and this larger expansion as compared with the infinite matrix is due to the more limited restraint of the finite matrix. Using the same procedure as before one can show that the effect of a rigid inclusion in a finite body can be represented by

$$\rho\mathbf{b}(\mathbf{r}) = -K\delta V\,\nabla\delta(\mathbf{r})$$

The interaction energy Φ_{int} between such an inclusion and an externally applied stress field characterised by $[\mathbf{u}^*, \boldsymbol{\varepsilon}^*, \boldsymbol{\sigma}^*]$ is given by

$$\Phi_{\text{int}} = -\int \rho\mathbf{b}(\mathbf{r})\cdot\mathbf{u}^*\,\mathrm{d}V$$

and substituting $\rho\mathbf{b}(\mathbf{r})$ results, via integration by parts, in

$$\begin{aligned}\Phi_{\text{int}} &= -K\delta V\int u_k^*\frac{\partial\delta(\mathbf{r})}{\partial x_k}\,\mathrm{d}V = -K\delta V\int \delta(\mathbf{r})\frac{\partial u_k^*}{\partial x_k}\,\mathrm{d}V \\ &= -K\delta V\varepsilon_{kk}^* = -\tfrac{1}{3}\delta V\sigma_{kk}^* = p^*\delta V\end{aligned}$$

where p^* is the pressure corresponding to $\boldsymbol{\sigma}^*$ evaluated at the centre of the inclusion. If $\boldsymbol{\sigma}^*$ is due to a rigid inclusion in an infinite elastic medium, we have $p^* = 0$ and the interaction energy is zero.

A better model may be obtained by replacing the rigid inclusion with an elastic inclusion of radius r_0, which, again, is forced in a hole at the centre of a spherical matrix with radius R and having a volume $\delta v'$ smaller than that of the inclusion. Let us denote by $[\mathbf{u}', \boldsymbol{\sigma}', \lambda', \mu']$ and $[\mathbf{u}, \boldsymbol{\sigma}, \lambda, \mu]$ the displacement, stress and Lamé constants of the inclusion and matrix, respectively. It follows from the general solution that, since at the origin the displacement must be zero, the only non-zero components of $\mathbf{u}'$ (within the inclusion) and $\mathbf{u}$ (outside the inclusion) are

$$u_r'(r) = \beta' r \quad \text{and} \quad u_r(r) = \alpha r^{-2} + \beta r \tag{10.26}$$

The constants β', α and β can be determined from the condition that at the interface between inclusion and matrix the radial stress must be continuous and that the radial stress vanishes at $r = R$, i.e.

$$\sigma_{rr}'(r_0) = \sigma_{rr}(r_0) \quad \text{and} \quad \sigma_{rr}(R) = 0 \tag{10.27}$$

as well as the geometric (compatibility) condition

$$4\pi r_0^2[u_r(r_0) - u_r'(r_0)] = \delta v' \tag{10.28}$$

Substitution of the displacement, Eq. (10.26), in the stress, Eq. (10.23), yields

$$\sigma_{rr}' = 3K'\beta' \quad \text{and} \quad \sigma_{rr} = -4\mu\alpha r^{-3} + 3K\beta \tag{10.29}$$

with the bulk moduli $K' = \lambda' + ⅔\mu'$ and $K = \lambda + ⅔\mu$, respectively. Substitution of Eq. (10.29) in Eq. (10.27) and Eq. (10.26) in Eq. (10.28) and solving for β', α and β results in

$$\beta = \frac{4\mu\alpha}{3KR^3} \qquad \beta' = \frac{4\mu\alpha}{3K'}\left(\frac{1}{R^3} - \frac{1}{r_0^3}\right) \qquad \alpha = \frac{\delta v'}{4\pi}\left[1 + \frac{4\mu r_0^3}{3KR^3} + \frac{4\mu}{3K'}\left(1 - \frac{r_0^3}{R^3}\right)\right]^{-1}$$

For a rigid inclusion $K' = \infty$, $\beta' = 0$ and α and β reduce to the previous result. Since in many cases $r_0^3/R^3 << 1$, we neglect r_0^3/R^3 with respect to unity and obtain for α

$$\alpha = \frac{\delta v'}{4\pi}\left[1 + \frac{4\mu}{3K'}\right]^{-1}$$

For the non-zero components of stress this leads to

$$\sigma_{rr}' = \sigma_{\varphi\varphi}' = \sigma_{\theta\theta}' = -\frac{4\mu\alpha}{r_0^3} \qquad \operatorname{tr}\boldsymbol{\sigma}' = \frac{-12\mu\alpha}{r_0^3}$$

$$\sigma_{rr} = -\frac{4\mu\alpha}{r^3}\left(1 - \frac{r^3}{R^3}\right) \qquad \sigma_{\theta\theta} = \sigma_{\varphi\varphi} = \frac{2\mu\alpha}{r^3}\left(1 + \frac{2r^3}{R^3}\right) \qquad \operatorname{tr}\boldsymbol{\sigma} = \frac{12\mu\alpha}{R^3}$$

The elastic inclusion is subject to a hydrostatic stress while the stress outside the inclusion is similar to that of the rigid inclusion. Since for a rigid inclusion the strength $\alpha = \delta v/4\pi$, it can be seen that an elastic inclusion with bulk modulus K' and strength α has the same effect as a rigid inclusion of strength $\alpha/[1+(4\mu/3K')]$. Finally, the volume change of the inclusion and matrix is given by, respectively,

$$\delta v = 4\pi r_0^2 u_r'(r_0) = -\frac{4\mu}{3K'}\frac{\delta v'}{1+(4\mu/3K')} \quad \text{and} \quad \delta V = 4\pi R^2 u_r(R) = \frac{1+(4\mu/3K)}{1+(4\mu/3K')}\delta v'$$

Two-dimensional solutions

We now turn to two-dimensional solutions. Using Greek indices α, β and γ for the summation convention over indices 1, 2 we can write the equilibrium condition in two dimensions (without body force) as

$$\sigma_{\alpha\beta,\beta} = 0 \tag{10.30}$$

In full we write

$$\sigma_{11,1} + \sigma_{12,2} = 0 \quad \text{and} \quad \sigma_{21,1} + \sigma_{22,2} = 0,$$

so that there exist functions $\phi_\alpha(x_1,x_2)$ ($\alpha = 1,2$) such that

$$\sigma_{11} = \phi_{1,2} \qquad \sigma_{12} = -\phi_{1,1} \qquad \sigma_{22} = \phi_{2,1} \qquad \text{and} \qquad \sigma_{12} = -\phi_{2,2}$$

Consequently, $\phi_{1,1} = \phi_{2,2}$ and there exists a function $\Phi(x_1,x_2)$ such that $\phi_1 = \Phi_{,2}$ and $\phi_2 = \Phi_{,1}$. In total we have

$$\sigma_{11} = \Phi_{,22} \qquad \sigma_{12} = -\Phi_{,12} \qquad \sigma_{22} = \Phi_{,11} \tag{10.31}$$

or equivalently

$$\sigma_{\alpha\beta} = \delta_{\alpha\beta}\Phi_{,\gamma\gamma} - \Phi_{,\alpha\beta} \tag{10.32}$$

The function Φ is known as the *Airy stress function*[a]. The number of compatibility equations also reduces significantly for two-dimensional systems. The only surviving equation reads

$$2\frac{\partial^2 \varepsilon_{\alpha\beta}}{\partial x_\alpha \partial x_\beta} = \frac{\partial^2 \varepsilon_{\alpha\alpha}}{\partial x_\beta^2} + \frac{\partial^2 \varepsilon_{\beta\beta}}{\partial x_\alpha^2} \tag{10.33}$$

Moreover, Hooke's law reduces and reads

$$\begin{aligned} \varepsilon_{\alpha\alpha} &= \frac{1}{E}\left(\gamma\sigma_{\alpha\alpha} - \delta\nu\sigma_{\beta\beta}\right) \quad \text{and} \\ 2\varepsilon_{\alpha\beta} &= \frac{\sigma_{\alpha\beta}}{G} = \frac{2(1+\nu)}{E}\sigma_{\alpha\beta} \qquad (\alpha \neq \beta) \end{aligned} \tag{10.34}$$

where $\gamma = \delta = 1$ for plane stress and $\gamma = 1-\nu^2$ and $\delta = 1+\nu$ for plane strain, respectively. Combining these equations leads for both plane stress and plane strain to

$$\blacktriangleright \quad \left(\frac{\partial^2}{\partial x_1^2} + \frac{\partial^2}{\partial x_2^2} + \frac{\partial^2}{\partial x_3^2}\right)^2 \Phi = (\nabla^2)^2\Phi = \Delta^2\Phi = 0 \tag{10.35}$$

---

[a] George Biddell Airy (1801-1892). English mathematician. He was professor of astronomy and director of the Cambridge Observatory before he became Astronomer Royal in 1835. When informed in 1843 by John Couch Adams (1819-1892) that he calculated the position of a new planet, later to be named Neptune, from the deviation of the other planets, Airy kept asking for more calculations. Jean-Joseph Urbain Le Verrier (1811-1877) informed the director of the Berlin observatory (!) Galle in 1846 with the same message and since Galle experimentally verified the predictions rightaway, Leverrier is also known as one of the discoverers of the planet. In 1858 Adams became professor of astronomy and in 1861 he became director of the Cambridge Observatory.

This equation is the so-called *biharmonic equation* and its solutions are the *biharmonic functions*. The solution of two-dimensional problems thus requires to guess a proper function Φ for the problem at hand, i.e. satisfying the boundary conditions. Once this function is found, it automatically satisfies the equilibrium and compatibility conditions. The stresses are obtained from Eq. (10.31) while the strains are obtained using Eq. (10.34).

The simplest solution is $\Phi = ax^2+bxy+cy^2$ and represents constant stresses in the body. The simplest solution which gives non-constant stresses is thus $\Phi = ay^3$. It yields $\sigma_{11} = 6ay$, $\sigma_{12} = \sigma_{22} = 0$ and represents pure bending of a beam about the z-axis with x the longitudinal fibre axis and with $6a = E/\rho$, where ρ is the radius of curvature. A particular useful feature is that the solution can be continuously improved by adding more terms that satisfy the biharmonic equation, e.g. $\Phi = \Phi_1 + \Phi_2 + \cdots$ An Airy function for normal bending is provided by $\Phi = axy^3+bxy$, not only yielding the bending stress but also the proper correction for shear.

Since many two-dimensional problems are cylindrically symmetrical it is convenient to use cylinder co-ordinates. In this case the Airy function is most conveniently expressed in cylinder co-ordinates. To that purpose Eq. (10.35) has to be expressed in cylinder co-ordinates. Using the equivalence

$$\nabla^2 a = \frac{\partial^2 a}{\partial x^2}+\frac{\partial^2 a}{\partial y^2}+\frac{\partial^2 a}{\partial z^2}=\frac{\partial^2 a}{\partial r^2}+\frac{1}{r}\frac{\partial a}{\partial r}+\frac{1}{r^2}\frac{\partial^2 a}{\partial \theta^2}+\frac{\partial^2 a}{\partial z^2} \tag{10.36}$$

this is easily done. For axially symmetric problems Φ is a function of r only and the general solution is

$$\Phi(r) = A\ln r + Br^2\ln r + Cr^2 + D \tag{10.37}$$

which corresponds to

$$\sigma_{rr} = \frac{A}{r^2} + B(1+2\ln r) + 2C \qquad \sigma_{\theta\theta} = -\frac{A}{r^2} + B(3+2\ln r) + 2C \qquad \sigma_{r\theta} = 0 \tag{10.38}$$

All solutions for symmetrical stress distributions in the absence of body forces can be obtained from this general solution. For finite structures the coefficients B and D are identically zero on physical grounds.

For the more general case $\Phi(r,\theta)$ also a general solution exists. For its explicit form we refer to the literature. We only present a simple but important example. Finally, we also mention that for three-dimensions a comparable but much more complex approach exists (see e.g. Fung, 1965).

Example 10.5

For a concentrated point load P on a semi-infinite plate the Airy stress function is given by $\Phi(r,\theta) = -\dfrac{P}{\pi} r\theta\sin\theta$ which results in $\sigma_{rr} = -\dfrac{2P}{\pi}\dfrac{\cos\theta}{r}$, $\sigma_{\theta\theta} = \sigma_{r\theta} = 0$. An element at a distance r from the point of application is thus

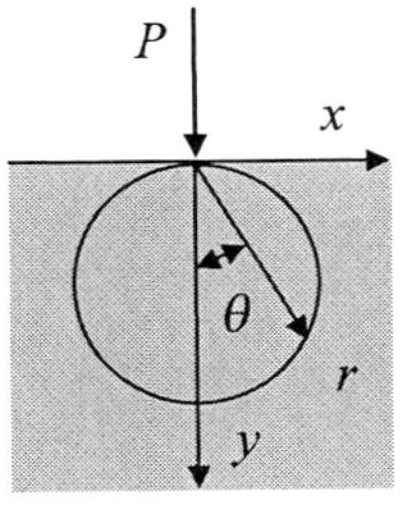

subjected to simple compression in the radial direction.

Problem 10.4

Using the boundary conditions ($x = 0$: $v = v' = 0$) for a cantilever beam with height h and width b (Fig. 10.6), show that the complete Airy function is

$$\Phi = -\frac{Fh^2}{8I}(l-x)y + \frac{F}{6I}(l-x)y^3$$

resulting in $\sigma_{11} = \frac{F}{I}(l-x)y,\ \ \sigma_{12} = \frac{F}{2I}\left(\frac{h^2}{4} - y^2\right)$ and $\sigma_{22} = 0$ while the corresponding strains for plane stress are $\varepsilon_{11} = \sigma_{11}/E$, $\varepsilon_{12} = (1+\nu)\sigma_{12}/E$, $\varepsilon_{22} = -\nu\sigma_{11}/E$. Here $I = bh^3/12$ denotes, as usual, the moment of inertia for a beam with width b and height h.

Problem 10.5

For a thick walled cylinder with uniform external pressure q and internal pressure p the boundary conditions are $\sigma_{rr} = -p$ at $r = a$ and $\sigma_{rr} = -q$ at $r = b$ (see accompanying figure). Show by applying the boundary conditions to the general solution that the stresses are given by

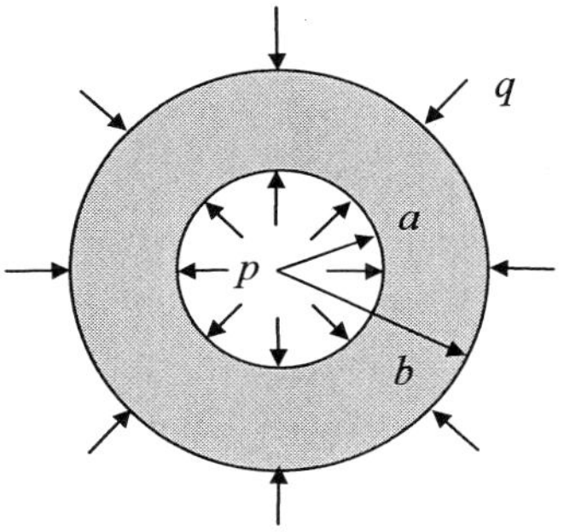

$$\sigma_{rr} = -p\frac{(b^2/r^2)-1}{(b^2/a^2)-1} - q\frac{1-(a^2/r^2)}{1-(a^2/b^2)}$$

$$\sigma_{\theta\theta} = p\frac{(b^2/r^2)+1}{(b^2/a^2)-1} - q\frac{1+(a^2/r^2)}{1-(a^2/b^2)}$$

10.4 Variational approach

A full solution of the elasticity equations in either two or three dimensions is often quite complicated. For approximate solutions we can take advantage of the virtual work formulation and the nature of elastic solids.

As discussed in Chapter 5 the principle of virtual work states that for a structure the sum of the work of the external forces W_{ext} (comprising the volume forces W_{vol} and surface forces W_{sur}) and the internal forces W_{int} equals the work of the inertia forces W_{acc} for any virtual loading, i.e $W_{int}+W_{ext} = W_{acc}$. If we neglect inertia forces and recall that the change in mechanical work W_{mec} (the negative of the internal work $W_{int} = -\int\int\sigma_{ij}\,d\varepsilon_{ij}\,dV$) equals the change in internal energy U of the system (under adiabatic conditions or Helmholtz energy F under isothermal conditions), we have

$$\delta W_{mec}\left(= \delta U = \int \sigma_{ij}\delta\varepsilon_{ij}\,dV\right) = \delta W_{ext}\left(= \int \rho b_i \delta u_i\,dV + \int t_i \delta u_i\,dA\right) \tag{10.39}$$

where ε_{ij} is the strain associated with the displacement u_i, b_i the body force and t_i the traction. We now introduce that the loading is conservative. This implies that the body force and traction can be written as the derivatives of potential functions B and T, respectively, e.g. $b_i = -\partial B/\partial u_i$ and $t_i = -\partial T/\partial u_i$. For external forces independent of the deformation of the body, neither in direction, nor in magnitude $B = -b_i u_i$ and $T = -t_i u_i$. Inserting yields

$$\delta W_{\text{ext}} = -\int \rho \frac{\partial B}{\partial u_i} \delta u_i \mathrm{d}V - \int \frac{\partial T}{\partial u_i} \delta u_i \mathrm{d}A$$
$$= -\delta \int \rho B \mathrm{d}V - \delta \int T \mathrm{d}A = \delta \int \rho b_i u_i \mathrm{d}V + \delta \int t_i u_i \mathrm{d}A = -\delta \Omega \qquad (10.40)$$

where Ω is the potential energy of the external forces. This leads us to the principle of minimum (total) potential energy (see also Chapter 7)

► $$\delta \Pi = \delta(U + \Omega) = 0 \qquad (10.41)$$

where Π denotes the (total) *potential energy* (or total Helmholtz energy). The principle states that from all statically admissible displacement fields (satisfying the boundary conditions), the real displacement field yields an extremum. It can be shown that this extremum is a minimum. In case the body forces can be neglected and the traction consists of a set of individual loads $F^{(i)}$, the potential energy expression reduces the well-known one from elementary considerations, i.e. $\Omega = -F^{(i)}u^{(i)}$.

The great utility of the principle of minimum potential energy is that it is relatively straightforward to guess an approximate displacement field. Including one or more free parameters in this approximate field offers the possibility to optimise this field by minimising the total potential energy with respect to these parameters. The following example will illustrate this.

Example 10.6

Consider a three-point bend bar and let us assume that the displacement is unknown. An approximate displacement function is $v = A \sin(\pi x/l)$ where the amplitude A is the free parameter. According to the principle of minimum potential energy we have to calculate the internal energy U and potential energy of the loading Ω. The latter is given by the central loading force F multiplied by the central displacement, which is equal to A in this case. Hence

$$\Omega = -FA$$

The internal energy, calculated from Eq. (10.9), $U = \dfrac{EI}{2}\int_0^l (v'')^2 \mathrm{d}x$, becomes

$$U = \pi^4 EIA^2/4l^3.$$

Consequently $\delta\Pi = \delta(U+\Omega) = [\partial(U+\Omega)/\partial A]\delta A = 0$. Since the variation of A is arbitrary, $\partial(U+\Omega)/\partial A = 0$. Solving results in

$$A = 2Fl^3/\pi^4 EI = 0.02053\ Fl^3/EI.$$

The exact central displacement is $Fl^3/48EI = 0.02083\ Fl^3/EI$. The difference is less than 2%. However, for the moment the result is not so good. Inserting the approximate displacement function in $M = EIv''$ results in $M = (2Fl/\pi^2)\sin(\pi x/l)$. Evaluating for $x = l/4$ and $x = l/2$ yields

	approximate	exact	difference
$x = l/4$	0.143 Fl	0.125 Fl	14%
$x = l/2$	0.203 Fl	0.250 Fl	19%

This nature of this result is generally true: the moments are calculated from the displacements by differentiation, thereby decreasing the accuracy. Since $\sigma = My/I$, the same remark applies for the stress σ.

The displacement function $v = A \sin(\pi x/l)$ contains only one parameter. The above result can be improved considerably by extending the trial function, e.g. by adding another term $C \sin(3\pi x/l)$. Problem 10.1 shows the considerable improvement upon the results of Example 10.6 for that extended trial function. Systematic improvement can be obtained in this way quite generally. However, the trial functions have to obey all the relevant boundary conditions to ensure proper convergence.

Problem 10.6

Show that adding the trial function $v^{(3)} = C \sin(3\pi x/l)$ to the trial function $v^{(1)} = A \sin(\pi x/l)$ of Example 10.6 leads to a central displacement

$$v = (2/\pi^4 - 2/81\pi^4)\, Fl^3/EI = 0.02028\, Fl^3/EI$$

Moreover show that the moment at $x = l/4$ and $x = l/2$ become 0.127 Fl and 0.225 Fl, respectively.

10.5 Discrete numerical approach

Before we discuss in the next section a continuum method to solve the elastic equations, it seems useful to discuss in this section a discrete representation that possibly shows the structure of the solution strategy more clearly (Akin, 1994). Consider to that purpose a linear elastic structure loaded with a set of forces, collectively given by the column matrix $\boldsymbol{f}$. At each of the loading points a displacement occurs, which are collectively described by the column matrix $\boldsymbol{q}$. Since the structure is linear elastic, each of the forces is linearly related to all the displacements. The force constants are collectively given by the symmetric matrix $\boldsymbol{K}$, called as *stiffness matrix*, and we have $\boldsymbol{f} = \boldsymbol{Kq}$. This equation represents Hooke's law for the structure and we may think of the structure as a system of coupled springs. We now consider the total potential energy of the structure for which two contributions can be recognised. First, the strain energy U stored in the structure given by

$$U = \int \boldsymbol{f}^{\mathrm{T}}\,\mathrm{d}\boldsymbol{q} = \tfrac{1}{2}\boldsymbol{q}^{\mathrm{T}}\boldsymbol{Kq}$$

and, second, the potential energy of loading Ω is given by

$$\Omega = -\boldsymbol{f}^{\mathrm{T}}\boldsymbol{q}$$

The solution is obtained via minimisation of the potential energy $\Pi = U + \Omega$ leading to

$$\frac{\partial \Pi}{\partial \boldsymbol{q}} = \boldsymbol{0} \qquad \text{or} \qquad \boldsymbol{Kq} = \boldsymbol{f} \qquad \text{or} \qquad \boldsymbol{q} = \boldsymbol{K}^{-1}\boldsymbol{f}$$

The $\boldsymbol{q}$'s are thus obtained by inverting $\boldsymbol{K}$ and multiplying $\boldsymbol{K}^{-1}$ with $\boldsymbol{f}$.

Now, in general, at some points the displacements are free while at some other points the displacements are constrained (i.e. prescribed and typically 0) and the corresponding forces of the latter, the *reaction forces*, have to be determined. For example, for a cantilever beam at the load point the displacement is free while at the fixation point the displacement is obviously zero. In this case a constrained optimisation is required and the easiest way to deal with this situation is via the method of Lagrange multipliers (see Chapter 3). Suppose that we have displacements $\boldsymbol{q}_1$ that are free and displacements $\boldsymbol{q}_2$ that are prescribed. The forces corresponding to the free displacements are $\boldsymbol{f}_1$. The total displacement column $\boldsymbol{q}$ is then $\boldsymbol{q}^{\mathrm{T}} = (\boldsymbol{q}_1, \boldsymbol{q}_2)^{\mathrm{T}}$ while the total load column $\boldsymbol{f}$ is $\boldsymbol{f}^{\mathrm{T}} = (\boldsymbol{f}_1, \boldsymbol{0})^{\mathrm{T}}$. The constraint can be expressed as $\boldsymbol{S}[\boldsymbol{q} - (\boldsymbol{0}, \bar{\boldsymbol{q}})] = \boldsymbol{0}$, where $\bar{\boldsymbol{q}}$ denotes the prescribed values and where $\boldsymbol{S}$ is a square matrix that selects from the total displacement column the prescribed displacements. As an example, in case the dimension of $\boldsymbol{q}$ is 5 and the last two displacements are prescribed,

$$\boldsymbol{S} = \begin{pmatrix} \boldsymbol{O} & \boldsymbol{O} \\ \boldsymbol{O} & \boldsymbol{I} \end{pmatrix} = \begin{pmatrix} 0 & & & & \\ 0 & 0 & & \text{sym} & \\ 0 & 0 & 0 & & \\ 0 & 0 & 0 & 1 & \\ 0 & 0 & 0 & 0 & 1 \end{pmatrix}$$

The function to be minimised is now the total potential energy $\Pi = U + \Omega$ plus an undetermined multiplier column $\boldsymbol{\lambda}$ times the constraint, i.e. $\boldsymbol{\lambda}^{\mathrm{T}}\boldsymbol{S}[\boldsymbol{q} - (\boldsymbol{0}, \bar{\boldsymbol{q}})]$, leading to

$$\frac{\partial}{\partial \boldsymbol{q}}\left(\tfrac{1}{2}\boldsymbol{q}^{\mathrm{T}}\boldsymbol{K}\boldsymbol{q} - \boldsymbol{f}^{\mathrm{T}}\boldsymbol{q} + \boldsymbol{\lambda}^{\mathrm{T}}\boldsymbol{S}[\boldsymbol{q} - (\boldsymbol{0}, \bar{\boldsymbol{q}})]\right) = \boldsymbol{0} \qquad \text{and}$$

$$\frac{\partial}{\partial \boldsymbol{\lambda}}\left(\tfrac{1}{2}\boldsymbol{q}^{\mathrm{T}}\boldsymbol{K}\boldsymbol{q} - \boldsymbol{f}^{\mathrm{T}}\boldsymbol{q} + \boldsymbol{\lambda}^{\mathrm{T}}\boldsymbol{S}[\boldsymbol{q} - (\boldsymbol{0}, \bar{\boldsymbol{q}})]\right) = \boldsymbol{0}$$

From the second equation we regain the constraint $\boldsymbol{S}[\boldsymbol{q} - (\boldsymbol{0}, \bar{\boldsymbol{q}})] = \boldsymbol{S}[(\boldsymbol{q}_1, \boldsymbol{q}_2) - (\boldsymbol{0}, \bar{\boldsymbol{q}})] = \boldsymbol{0}$ or $\boldsymbol{q}_2 = \bar{\boldsymbol{q}}$. The first equation yields

$$\boldsymbol{K}\boldsymbol{q} - \boldsymbol{f} - \boldsymbol{S}\boldsymbol{\lambda} = \boldsymbol{0} \qquad \text{or} \qquad \begin{pmatrix} \boldsymbol{K}_{11} & \boldsymbol{K}_{12} \\ \boldsymbol{K}_{21} & \boldsymbol{K}_{22} \end{pmatrix}\begin{pmatrix} \boldsymbol{q}_1 \\ \boldsymbol{q}_2 \end{pmatrix} - \begin{pmatrix} \boldsymbol{f}_1 \\ \boldsymbol{0} \end{pmatrix} - \begin{pmatrix} \boldsymbol{O} & \boldsymbol{O} \\ \boldsymbol{O} & \boldsymbol{I} \end{pmatrix}\begin{pmatrix} \boldsymbol{\lambda}_1 \\ \boldsymbol{\lambda}_2 \end{pmatrix} = \boldsymbol{0}$$

Solving for $\boldsymbol{q}_1$ and subsequently for the multiplier $\boldsymbol{\lambda}$ yields

► $$\boldsymbol{q}_1 = \boldsymbol{K}_{11}^{-1}(\boldsymbol{f}_1 - \boldsymbol{K}_{12}\boldsymbol{q}_2) \qquad \text{or if} \quad \boldsymbol{q}_2 = \bar{\boldsymbol{q}} = \boldsymbol{0} \quad \boldsymbol{q}_1 = \boldsymbol{K}_{11}^{-1}\boldsymbol{f}_1 \qquad (10.42)$$

► $$\boldsymbol{\lambda}_2 = \boldsymbol{K}_{21}\boldsymbol{q}_1 + \boldsymbol{K}_{22}\boldsymbol{q}_2 \qquad \text{or if} \quad \boldsymbol{q}_2 = \bar{\boldsymbol{q}} = \boldsymbol{0} \quad \boldsymbol{\lambda}_2 = \boldsymbol{K}_{21}\boldsymbol{q}_1 = \boldsymbol{K}_{21}\boldsymbol{K}_{11}^{-1}\boldsymbol{f}_1 \qquad (10.43)$$

The part $\boldsymbol{\lambda}_1$ is indetermined but this presents no problem. In fact we may write $\boldsymbol{\lambda}^{\mathrm{T}} = (\boldsymbol{0}, \boldsymbol{\lambda}_2)^{\mathrm{T}}$. The part $\boldsymbol{\lambda}_2$ can be interpreted as the reaction force $\boldsymbol{f}_2$ since it describes the force to maintain the constraint. It appears that in this case we can also write

$$\begin{pmatrix} \boldsymbol{K}_{11} & \boldsymbol{K}_{12} \\ \boldsymbol{K}_{21} & \boldsymbol{K}_{22} \end{pmatrix}\begin{pmatrix} \boldsymbol{q}_1 \\ \boldsymbol{q}_2 \end{pmatrix} - \begin{pmatrix} \boldsymbol{f}_1 \\ \boldsymbol{f}_2 \end{pmatrix} = \boldsymbol{0}$$

where solving for $\boldsymbol{q}_1$ and $\boldsymbol{f}_2$ yields the same answers as before. In our case the selection matrix $\boldsymbol{S}$ was used to select the prescribed displacements $\boldsymbol{q}_2$. Obviously it can be also be used to introduce linear relations between the displacements, if an appropriate form for $\boldsymbol{S}$ is constructed.

In summary, once the response of the system, the stiffness matrix $\boldsymbol{K}$, to a set of forces $\boldsymbol{f}_1$ and constraints $\boldsymbol{q}_2$ is given, the (free) displacements $\boldsymbol{q}_1$ are given by Eq. (10.42) and reaction forces $\boldsymbol{f}_2$ by Eq. (10.43).

10.6 Continuum numerical approach*

In the approach as described in Section 10.4 approximate continuum solutions cover the whole structure. In Section 10.5 we dealt with a discrete treatment of the complete structure. In the approach as described in this section approximate continuum solutions for the displacement of parts of the structure are used which are matched at their interfaces (Akin, 1994). These parts, generally known as *elements*, can be line, surface or volume elements depending on the representation of the structure. In each element the unknown displacement fields are represented by a linear combination of so-called *shape* or *interpolation functions*. The shape functions are functions of the displacements of certain points in the element, the so-called *nodes* (see Fig. 10.9). The nodes can be anywhere in the element but are usually located at the boundaries. This is the so-called *finite element method* (FEM) of which we describe here only the formalism generally known as the displacement formulation (other, more complex methods exists, for details see Zienkiewicz, and Taylor, 1989).

In the following we describe first the procedure of discretisation, followed by the displacements of a single element. After that we combine the results of many elements to a single equation and take care of the constraints. To this purpose we need again the discretised version of the principle of virtual work and minimum potential energy. Essentially the method specifies how the stiffness matrix $\boldsymbol{K}$, as used in the previous section, can be calculated for a continuum structure. The remainder is in principle the same.

The general procedure is as follows:

- First, we have to model the structure. A choice of elements has to be made and the material data collected. For general cases the modelling of the structure with various types of elements can be quite time consuming, in fact most of the time required for the analyst using a FEM software package. This modelling, together with the collecting of relevant material data, is often referred to as *pre-processing*.
- The actual calculation entails the solution of the equation $\boldsymbol{Kq} = \boldsymbol{f}$ where the column matrix $\boldsymbol{q}$ denotes the displacements of the nodes, the column matrix $\boldsymbol{f}$ denotes the load and the square matrix $\boldsymbol{K}$ represents the response of the structure to the load. This step is referred to as the *analysis* or *processing*.

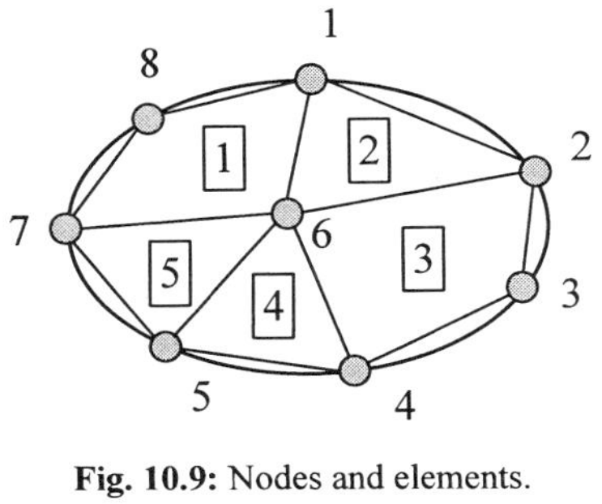

Fig. 10.9: Nodes and elements.

- After the solution for $\boldsymbol{q}$ is obtained, the strains and thus the stresses can be calculated from the displacements. It remains to be said that after calculation of the stresses, often outputted in a graphical form, some further calculations, e.g calculation of principal stresses, are often done. This is the so-called *post-processing*.

As stated, we want to approximate the displacement field in an element, indicated by a column matrix $\boldsymbol{u}^{(e)}(\boldsymbol{x})$, by the displacements of the nodes, collectively denoted by the column matrix $\boldsymbol{q}^{(e)}$. We write

$$\boldsymbol{u}^{(e)}(\boldsymbol{x}) = \boldsymbol{N}^{(e)}(\boldsymbol{x})\boldsymbol{q}^{(e)}$$

where the matrix $\boldsymbol{N}^{(e)}(\boldsymbol{x})$ contains the *shape* or *interpolation functions*. Their precise nature will become clear shortly in the example. Each node can have one, two or three degrees-of-freedom, dependent on the dimensionality of the problem, corresponding to a one-, two- or three-dimensional displacement field. We use the pseudo-vector notation for the strains $\boldsymbol{\varepsilon}^{(e)}(\boldsymbol{x})$, which we can calculate from the displacements by differentiation

$$\boldsymbol{\varepsilon}^{(e)}(\boldsymbol{x}) = \boldsymbol{N}^{(e)\prime}(\boldsymbol{x})\boldsymbol{q}^{(e)} = \boldsymbol{B}^{(e)}(\boldsymbol{x})\boldsymbol{q}^{(e)} \qquad \text{with} \quad \boldsymbol{B}^{(e)} = \boldsymbol{N}^{(e)\prime} = \partial\boldsymbol{N}^{(e)}/\partial\mathbf{x}$$

Similarly we indicate the stress by the pseudo-vector $\boldsymbol{\sigma}^{(e)}(\boldsymbol{x})$. In terms of $\boldsymbol{\varepsilon}^{(e)}(\boldsymbol{x})$ and $\boldsymbol{\sigma}^{(e)}(\boldsymbol{x})$ the variation of the strain energy is given by

$$\delta U^{(e)} = \int_e (\boldsymbol{\sigma}^{(e)})^{\mathrm{T}}\delta\boldsymbol{\varepsilon}^{(e)}\,\mathrm{d}V^{(e)} = \int_e (\boldsymbol{\sigma}^{(e)})^{\mathrm{T}}\boldsymbol{B}^{(e)}\delta\boldsymbol{q}^{(e)}\,\mathrm{d}V^{(e)} = (\boldsymbol{p}^{(e)})^{\mathrm{T}}\delta\boldsymbol{q}^{(e)} \tag{10.44}$$

where the matrix $\boldsymbol{p}^{(e)}$ is known as the (*element*) *internal load matrix*. Similarly the variation of the potential energy of the loading is given by

$$\delta\Omega^{(e)} = \int_e (\boldsymbol{N}^{(e)})^{\mathrm{T}}\rho\boldsymbol{b}^{(e)}\delta\boldsymbol{q}^{(e)}\,\mathrm{d}V^{(e)} + \int_e (\boldsymbol{N}^{(e)})^{\mathrm{T}}\boldsymbol{t}^{(e)}\delta\boldsymbol{q}^{(e)}\,\mathrm{d}A^{(e)} = (\boldsymbol{f}^{(e)})^{\mathrm{T}}\delta\boldsymbol{q}^{(e)} \tag{10.45}$$

where the matrix $\boldsymbol{f}^{(e)}$ is known as the (*element*) *external load matrix*. Equilibrium is thus reached if $\delta\Pi^{(e)} = \delta(U^{(e)} + \Omega^{(e)}) = (\boldsymbol{p}^{(e)} - \boldsymbol{f}^{(e)})\delta\boldsymbol{q}^{(e)} = 0$ or

$$\blacktriangleright \quad \boldsymbol{p}^{(e)} = \boldsymbol{f}^{(e)} \tag{10.46}$$

This is the discretised version of the principle of virtual work. The discretised version of the principle of total potential energy is obtained after inserting the constitutive equation for elasticity, $\boldsymbol{\sigma}^{(e)}(\boldsymbol{x}) = \boldsymbol{C}^{(e)}\boldsymbol{\varepsilon}^{(e)}(\boldsymbol{x}) = \boldsymbol{C}^{(e)}\boldsymbol{B}^{(e)}(\boldsymbol{x})\boldsymbol{q}^{(e)}$. One obtains for $\boldsymbol{p}^{(e)}$

$$\boldsymbol{p}^{(e)} = (\boldsymbol{q}^{(e)})^{\mathrm{T}}\int_e (\boldsymbol{B}^{(e)})^{\mathrm{T}}\boldsymbol{C}\boldsymbol{B}^{(e)}\mathrm{d}V^{(e)} = (\boldsymbol{K}^{(e)})^{\mathrm{T}}\boldsymbol{q}^{(e)} = \boldsymbol{K}^{(e)}\boldsymbol{q}^{(e)} \tag{10.47}$$

since the matrix $\boldsymbol{K}^{(e)}$, generally known as the (*element*) *stiffness matrix*, is symmetric. Solving for the unknown displacements of the nodes $\boldsymbol{q}^{(e)}$ of the element yields

$$\boldsymbol{q}^{(e)} = (\boldsymbol{K}^{(e)})^{-1}\boldsymbol{f}^{(e)}$$

However, we have many elements and we have to combine them first before solving for the $\boldsymbol{q}^{(e)}$'s since they are not independent. The displacements at the element have to matched and therefore the displacements of the nodes of adjacent elements have to be the same. To that purpose we collect the displacements of all the independent nodes in a single column $\boldsymbol{q}$ and find those of the element via

$$\boldsymbol{q}^{(e)} = \boldsymbol{A}^{(e)}\boldsymbol{q}$$

where the matrix $\boldsymbol{A}^{(\mathrm{e})}$ is a matrix that assigns the nodes to elements. The nature of this matrix will become clear in the example discussed later on. The strain energy and potential energy of the structure are the sum of the corresponding quantities for the elements so that

$$\begin{aligned}\delta\Pi &= \sum\nolimits_{\mathrm{e}}\delta\Pi^{(\mathrm{e})} = \sum\nolimits_{\mathrm{e}}(\boldsymbol{p}^{(\mathrm{e})})^{\mathrm{T}}\delta\boldsymbol{q}^{(\mathrm{e})} - (\boldsymbol{f}^{(\mathrm{e})})^{\mathrm{T}}\delta\boldsymbol{q}^{(\mathrm{e})} \\ &= \sum\nolimits_{\mathrm{e}}(\boldsymbol{q}^{(\mathrm{e})})^{\mathrm{T}}\boldsymbol{K}^{(\mathrm{e})}\delta\boldsymbol{q}^{(\mathrm{e})} - (\boldsymbol{f}^{(\mathrm{e})})^{\mathrm{T}}\delta\boldsymbol{q}^{(\mathrm{e})} \\ &= \sum\nolimits_{\mathrm{e}}(\boldsymbol{A}^{(\mathrm{e})}\boldsymbol{q})^{\mathrm{T}}\boldsymbol{K}^{(\mathrm{e})}\boldsymbol{A}^{(\mathrm{e})}\delta\boldsymbol{q} - (\boldsymbol{f}^{(\mathrm{e})})^{\mathrm{T}}\boldsymbol{A}^{(\mathrm{e})}\delta\boldsymbol{q} \\ &= \left[\sum\nolimits_{\mathrm{e}}\left(\boldsymbol{A}^{(\mathrm{e})}\right)^{\mathrm{T}}\boldsymbol{K}^{(\mathrm{e})}\boldsymbol{A}^{(\mathrm{e})}\boldsymbol{q} - \left(\boldsymbol{A}^{(\mathrm{e})}\right)^{\mathrm{T}}\boldsymbol{f}^{(\mathrm{e})}\right]^{\mathrm{T}}\delta\boldsymbol{q} = [\boldsymbol{K}\boldsymbol{q} - \boldsymbol{f}]^{\mathrm{T}}\delta\boldsymbol{q} = 0\end{aligned} \tag{10.48}$$

Now the $\delta\boldsymbol{q}$ are independent and the equation can only be satisfied by $\boldsymbol{Kq} - \boldsymbol{f} = \boldsymbol{0}$. Inverting results in $\boldsymbol{q} = \boldsymbol{K}^{-1}\boldsymbol{f}$ from which the displacement, strain and stress field can be calculated.

In most cases, though, some displacements of the independent nodes are prescribed and this has to be taken into account. Suppose we order the nodes $\boldsymbol{q}$ in such a way that those that are prescribed and those that are free are collectively denoted by $\boldsymbol{q}_2$ and $\boldsymbol{q}_1$, respectively. In that case we can write (see Section 10.5)

$$\begin{pmatrix}\boldsymbol{K}_{11} & \boldsymbol{K}_{12} \\ \boldsymbol{K}_{21} & \boldsymbol{K}_{22}\end{pmatrix}\begin{pmatrix}\boldsymbol{q}_1 \\ \boldsymbol{q}_2\end{pmatrix} = \begin{pmatrix}\boldsymbol{f}_1 \\ \boldsymbol{f}_2\end{pmatrix} \quad \text{with}$$

$$\begin{aligned}\boldsymbol{K}_{11} &= \sum\nolimits_{\mathrm{e}}(\boldsymbol{A}^{(\mathrm{e})})^{\mathrm{T}}\boldsymbol{K}_{11}^{(\mathrm{e})}\boldsymbol{A}^{(\mathrm{e})} = \sum\nolimits_{\mathrm{e}}(\boldsymbol{A}^{(\mathrm{e})})^{\mathrm{T}}\int_{\mathrm{e}}(\boldsymbol{B}^{(\mathrm{e})})^{\mathrm{T}}\boldsymbol{C}\boldsymbol{B}^{(\mathrm{e})}\mathrm{d}V^{(\mathrm{e})}\boldsymbol{A}^{(\mathrm{e})} \\ \boldsymbol{K}_{12} = \boldsymbol{K}_{21} &= \sum\nolimits_{\mathrm{e}}(\boldsymbol{A}^{(\mathrm{e})})^{\mathrm{T}}\boldsymbol{K}_{12}^{(\mathrm{e})}\boldsymbol{A}^{(\mathrm{e})} = \sum\nolimits_{\mathrm{e}}(\boldsymbol{A}^{(\mathrm{e})})^{\mathrm{T}}\int_{\mathrm{e}}(\boldsymbol{B}^{(\mathrm{e})})^{\mathrm{T}}\boldsymbol{C}\overline{\boldsymbol{B}}^{(\mathrm{e})}\mathrm{d}V^{(\mathrm{e})}\boldsymbol{A}^{(\mathrm{e})} \\ \boldsymbol{K}_{22} &= \sum\nolimits_{\mathrm{e}}(\boldsymbol{A}^{(\mathrm{e})})^{\mathrm{T}}\boldsymbol{K}_{22}^{(\mathrm{e})}\boldsymbol{A}^{(\mathrm{e})} = \sum\nolimits_{\mathrm{e}}(\boldsymbol{A}^{(\mathrm{e})})^{\mathrm{T}}\int_{\mathrm{e}}(\overline{\boldsymbol{B}}^{(\mathrm{e})})^{\mathrm{T}}\boldsymbol{C}\overline{\boldsymbol{B}}^{(\mathrm{e})}\mathrm{d}V^{(\mathrm{e})}\boldsymbol{A}^{(\mathrm{e})}\end{aligned} \tag{10.49}$$

where $\overline{\boldsymbol{B}}^{(\mathrm{e})}$ and $\boldsymbol{B}^{(\mathrm{e})}$ denote the $\boldsymbol{B}$-matrix associated with the prescribed and free displacements, respectively. Solving for $\boldsymbol{q}_1$ leads to

► $$\boldsymbol{q}_1 = \boldsymbol{K}_{11}^{-1}(\boldsymbol{f}_1 - \boldsymbol{K}_{22}\boldsymbol{q}_2) \tag{10.50}$$

The reaction forces $\bar{\boldsymbol{f}}_2$ can be calculated from

► $$\boldsymbol{f}_2 = \boldsymbol{K}_{21}\boldsymbol{q}_1 + \boldsymbol{K}_{22}\boldsymbol{q}_2 \tag{10.51}$$

The displacements are thus provided by the solution of a set of coupled, linear equations. Before illustrating the above equations with a simple example, a few general remarks are in order:

- The great advantage of the FEM method is that shapes of structures to be analysed can be chosen completely free. As can be noticed from the above equations the size of the linear equations that has to be solved is proportional to the number of degrees-of-freedom. Therefore, the FEM requires a computer for all but the smallest structures. A large selection of software packages is available, though user-friendlyness is differing widely.
- Although of great flexibility, the finite element method requires a significant amount of pre-processing, in particular in three-dimensional situations. Therefore, one nearly always tries to abstract the real structure in such a way that a two-

dimensional structure remains. Obviously this implies either an axi-symmetric or a plane (stress or strain) configuration.

- During the calculation the stiffness matrix **K** has to be inverted. This is only possible if **K** is non-singular. There is, however, as discussed in Chapter 4 apart from the deformation, also a rigid body motion of undetermined magnitude and for which the corresponding forces are zero. This leads to zero rows in the matrix **K**, hence **K** becomes singular and to avoid that the rigid body motion has to be eliminated. The usual way to accomplish this is to suppress the rigid body motion via the boundary conditions.
- In many cases the structure to be analysed contains some symmetry, e.g. in a two-dimensional calculation a three-point bend bar is symmetrical with respect to a plane perpendicular to the length of the specimen positioned at the middle of that specimen. This symmetry reflects itself in the forces and displacements implying that only the irreducible part has to be modelled. In the case of the three-point bend bar mentioned this means only half of the bar and corresponding loads.
- The flexibility of the shape functions is crucial. Generally polynomial functions, e.g. linear or quadratic functions, are used and the corresponding elements are denoted as linear and quadratic elements. In linear elements the strain is constant. Higher order polynomials provide more flexibility but at the cost of more computing. Most of the time shape functions of class C^0, which only have matching displacements at the nodes, are sufficient. In a number of cases, though, also class C^1 functions, which have also continuity of the derivative of the displacement, are required. One-dimensional modelling of a bending beam is an example. In this case cubic Hermite polynomials provide a useful alternative.
- In the above equations many integrals occur. Therefore efficient integration is required. Only in the simplest cases this is done analytically. An often-used method is *Gauss quadrature* in which the integrals are expressed as *weighted summations* over the function values (the *abscissa*) at certain points, the so-called *Gauss points*. Both the weights and the Gauss points are extensively tabulated.
- Since many integrals are similar and integration using global co-ordinates results in different expressions for each element, local (unit) co-ordinates are generally used, e.g. the co-ordinates of the nodes are scaled between 0 and 1. If for the calculation of the global co-ordinates from the local ones, the same functions are used as for the shape functions, the elements are called *isoparametric*. This choice is common nowadays.

Example 10.7

To illustrate the use of the equations presented above, we analyse a tensile bar in the absence of body forces but loaded by a force F at one of its ends. To that purpose consider Fig. 10.10. The bar is modelled by two linear, one-dimensional elements of length l, each with two nodes with one degree-of-freedom, the displacement u in the x-direction. In this case the pre-processing is trivial. Already for a two-dimensional model, the modelling has to be done with care, in particular with respect to the numbering of the nodes.

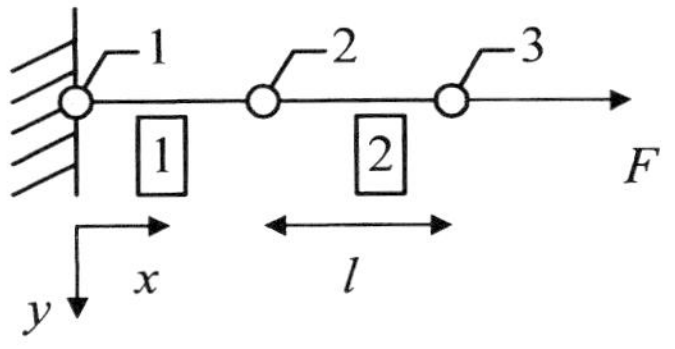

Fig. 10.10: One-dimensional model of a tensile bar.

Let us first consider the shape functions. Since we have taken the simplest one, e.g. a linear function, this implies that the displacement u is given by $u(x) = c_1+c_2x = \boldsymbol{Lc}$ with $\boldsymbol{L} = (1, x)$ and $\boldsymbol{c}^{\mathrm{T}} = (c_1, c_2)$. The displacements of the nodes are thus given by

$$\boldsymbol{q} = \begin{pmatrix} q_1 \\ q_2 \end{pmatrix} = \begin{pmatrix} u(x_1) \\ u(x_2) \end{pmatrix} = \begin{pmatrix} c_1 + c_2 x_1 \\ c_1 + c_2 x_2 \end{pmatrix} = \begin{pmatrix} 1 & x_1 \\ 1 & x_2 \end{pmatrix} \begin{pmatrix} c_1 \\ c_2 \end{pmatrix} = \boldsymbol{Gc} \quad \text{or}$$

$$\boldsymbol{c} = \boldsymbol{G}^{-1}\boldsymbol{q} = \frac{1}{x_2 - x_1} \begin{pmatrix} x_2 & -x_1 \\ -1 & 1 \end{pmatrix} \begin{pmatrix} q_1 \\ q_2 \end{pmatrix} \quad \text{and}$$

$$u(x) = \boldsymbol{LG}^{-1}\boldsymbol{q} = \boldsymbol{N}(x)\boldsymbol{q} = \frac{1}{x_2 - x_1} \big((x_2 - x) \quad (-x_1 + x) \big) \begin{pmatrix} q_1 \\ q_2 \end{pmatrix}$$

where $\boldsymbol{N}(x)$ is the shape function. The strain ε consequently is

$$\varepsilon(x) = \boldsymbol{N}'\boldsymbol{q} = \boldsymbol{Bq} = \frac{1}{x_2 - x_1} (-1 \quad 1) \begin{pmatrix} q_1 \\ q_2 \end{pmatrix} = \frac{q_2 - q_1}{x_2 - x_1} = \frac{q_2 - q_1}{l}$$

which is constant throughout the element. This is due to the choice of linear elements. Quadratic elements lead to a strain, which is linear over the elements and so forth.

The next step is the calculation of the stiffness matrix $\boldsymbol{K}^{(\mathrm{e})}$ and the load matrix $\boldsymbol{f}^{(\mathrm{e})}$ for the element. The elasticity matrix reduces in this case to the scalar Young's modulus E so that, if we take the nodes at the end-points of the element,

$$\boldsymbol{K}^{(\mathrm{e})} = \int (\boldsymbol{B}^{(\mathrm{e})})^{\mathrm{T}} E \boldsymbol{B}^{(\mathrm{e})} \mathrm{d}V^{(\mathrm{e})} = \frac{E}{l^2} \begin{pmatrix} 1 & -1 \\ -1 & 1 \end{pmatrix} \int \mathrm{d}y\mathrm{d}z \int_0^l \mathrm{d}x = \frac{EA}{l} \begin{pmatrix} 1 & -1 \\ -1 & 1 \end{pmatrix}$$

where the cross-section area of the element $A = \int \mathrm{d}x\mathrm{d}y$ has been taken constant. Since body forces are absent and the traction $\boldsymbol{t}^{(\mathrm{e})}$ is applied at node i of an element only, the load matrix $\boldsymbol{f}^{(\mathrm{e})}$ reduces to

$$\begin{aligned} \boldsymbol{f}^{(\mathrm{e})} &= \int_{\mathrm{e}} (\boldsymbol{N}^{(\mathrm{e})})^{\mathrm{T}} \rho \boldsymbol{b}^{(\mathrm{e})} \mathrm{d}V^{(\mathrm{e})} + \int_{\mathrm{e}} (\boldsymbol{N}^{(\mathrm{e})})^{\mathrm{T}} \boldsymbol{t}^{(\mathrm{e})} \mathrm{d}A^{(\mathrm{e})} \\ &= \int_{\mathrm{e}} (\boldsymbol{N}^{(\mathrm{e})})^{\mathrm{T}} F^{(i)} \delta(x - x^{(i)}) \mathrm{d}A^{(\mathrm{e})} = (\boldsymbol{N}^{(\mathrm{e})}(x^{(i)}))^{\mathrm{T}} F^{(i)} \end{aligned}$$

where $F^{(i)}$ and $x^{(i)}$ are the force applied at node i and its position, respectively, and $\delta(x)$ denotes the Dirac delta function (see Chapter 3). Before we start assembling the structure matrices from the element matrices, two remarks.

First, it is not necessary to restrict the load to be applied to the nodes only but this simplifies the example without deleting essentials. Second, in this example integration is simple but as soon as minor modifications are required, e.g. a non-constant cross-section or a quadratic shape function, unit co-ordinates and integration procedures become much more important.

The final step before solving is the assembly and reduction of the structure matrices. The independent nodes obviously are 1, 2 and 3. In element 1 nodes 1 and 2 participate while in element 2 nodes 2 and 3 play a role. The element assembly matrices $\boldsymbol{A}^{(1)}$ and $\boldsymbol{A}^{(2)}$ for element 1 and 2 are therefore defined by

$$\boldsymbol{q}^{(1)} = \boldsymbol{A}^{(1)}\boldsymbol{q} = \begin{pmatrix} 1 & 0 & 0 \\ 0 & 1 & 0 \end{pmatrix}\begin{pmatrix} q_1 \\ q_2 \\ q_3 \end{pmatrix} \quad \text{and} \quad \boldsymbol{q}^{(2)} = \boldsymbol{A}^{(2)}\boldsymbol{q} = \begin{pmatrix} 0 & 1 & 0 \\ 0 & 0 & 1 \end{pmatrix}\begin{pmatrix} q_1 \\ q_2 \\ q_3 \end{pmatrix}$$

respectively. The structure stiffness matrix $\boldsymbol{K}$ thus becomes, after some calculation,

$$\boldsymbol{K} = \sum\nolimits_{\mathrm{e}} (\boldsymbol{A}^{(\mathrm{e})})^{\mathrm{T}} \boldsymbol{K}^{(\mathrm{e})} \boldsymbol{A}^{(\mathrm{e})} = \frac{EA}{l}\begin{pmatrix} 1 & -1 & 0 \\ -1 & 2 & -1 \\ 0 & -1 & 1 \end{pmatrix}$$

Similarly, since only element 2 carries a force, the structure load matrix $\boldsymbol{f}$ becomes

$$\boldsymbol{f} = \sum\nolimits_{\mathrm{e}} (\boldsymbol{A}^{(\mathrm{e})})^{\mathrm{T}} \boldsymbol{f}^{(\mathrm{e})} = \begin{pmatrix} 0 & 0 \\ 1 & 0 \\ 0 & 1 \end{pmatrix} \frac{F}{l}\begin{pmatrix} x_3 - x \\ -x_2 + x \end{pmatrix}_{x=x_3} = \frac{F}{l}\begin{pmatrix} 0 \\ 0 \\ l \end{pmatrix}$$

The assembled equation thus is

$$\frac{EA}{l}\begin{pmatrix} 1 & -1 & 0 \\ -1 & 2 & -1 \\ 0 & -1 & 1 \end{pmatrix}\begin{pmatrix} q_1 \\ q_2 \\ q_3 \end{pmatrix} = \frac{F}{l}\begin{pmatrix} 0 \\ 0 \\ l \end{pmatrix}$$

Now it is time to remember that the displacement of node 1 is prescribed, in this case to $q_1 = 0$. Reduction of the above equation with respect to q_1 leads to

$$\frac{EA}{l}\begin{pmatrix} 2 & -1 \\ -1 & 1 \end{pmatrix}\begin{pmatrix} q_2 \\ q_3 \end{pmatrix} = \frac{F}{l}\begin{pmatrix} 0 \\ l \end{pmatrix} - \begin{pmatrix} -1 \\ 0 \end{pmatrix} q_1 = \frac{F}{l}\begin{pmatrix} 0 \\ l \end{pmatrix}$$

Finally solving for $\boldsymbol{q}$ leads to

$$\begin{pmatrix} q_2 \\ q_3 \end{pmatrix} = \frac{F}{l}\frac{l}{EA}\begin{pmatrix} 2 & -1 \\ -1 & 1 \end{pmatrix}^{-1}\begin{pmatrix} 0 \\ l \end{pmatrix} = \frac{F}{EA}\begin{pmatrix} 1 & 1 \\ 1 & 2 \end{pmatrix}\begin{pmatrix} 0 \\ l \end{pmatrix} = \frac{F}{EA}\begin{pmatrix} l \\ 2l \end{pmatrix}$$

So we find that the displacements at x_2 and x_3 are $q_2 = Fl/EA$ and $q_3 = 2Fl/EA$, respectively, conform our expectation. As a final check we calculate the reaction force, which is, as expected, given by

$$\begin{pmatrix} -1 & 0 \end{pmatrix}\frac{EA}{l}\begin{pmatrix} Fl/EA \\ 2Fl/EA \end{pmatrix} + (1)(0) = -F$$

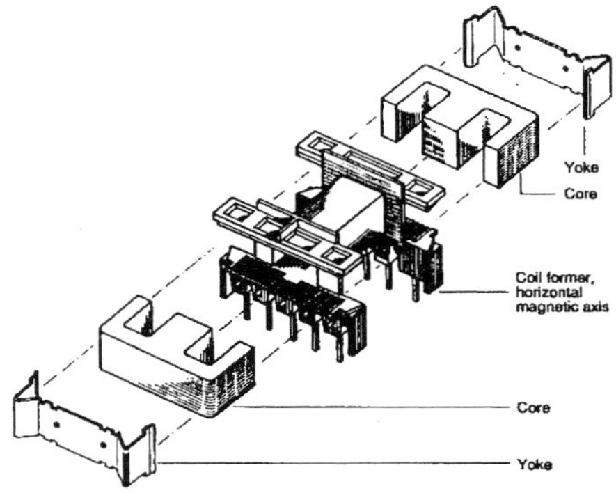

Fig. 10.11: Schematic view of a transformer containing an E-core (courtesy Philips Electronics).

Obviously for this example the FEM is like a gun for shooting a mosquito. The example presented is simple and straightforward and a solution can be obtained easy in another way. However, it should be clear that for even slightly more complex situations, which cannot be handled analytically, FEM is a powerful method for which a wide selection of software packages is available.

10.7 An example of a FEM analysis*

The previous section has outlined the most common method of FEM analysis. It is clear though that the power is only demonstrated in a real example. In this section we do so but we take an example that shows a relatively simple stress distribution so that the calculated stress distribution is easily understood intuitively.

For this purpose consider a transformer. While in the past the magnetic core frequently was made of Fe-Si alloy, nowadays ceramic parts are intensively used. These ceramic cores are typically made of MnZn-ferrite, a complex oxide with the spinel structure. The advantages of these materials are low eddy current losses (the material is electrically nearly non-conductive) and a high permeability (so that good amplification is possible). Moreover these materials are relatively cheap. The cores are produced in various shapes. As an example we take here the so-called E-core of which the assembly into a transformer is given in Fig. 10.11. Two E-cores with the legs opposite to each other form the magnetic part of the transformer. Apart from the E-cores, the structure contains copper windings on a polymer holder and some clips to

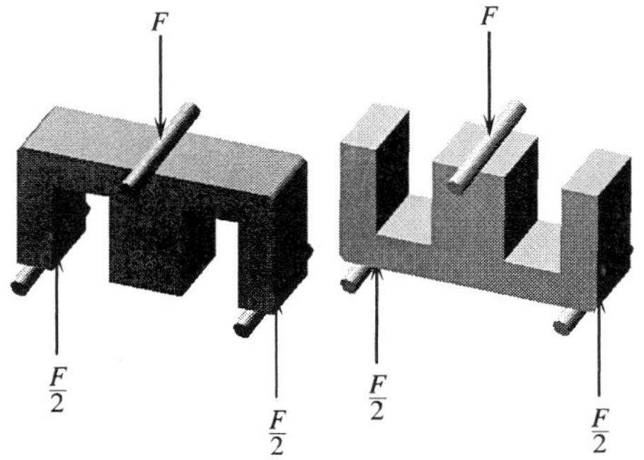

Fig. 10.12: The M- and W-test for an E-core.

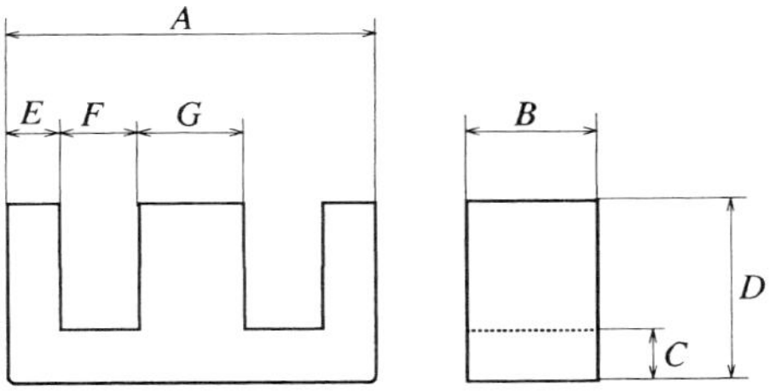

Fig. 10.13: Geometry of an E-core.

clamp the various parts together.

Although the primary function of an E-core is acting as a magnetic transformer, they have to endure thermo-mechanical stresses during production and during lifetime. First, during production the transformer is heated from the outside of the core, e.g. during re-flow soldering or during temperature cycling quality tests. This causes the exterior to expand with respect to the central part as the polymer holder containing the copper windings isolates this part. In this case maximum tensile stresses of typically 50 to 70 MPa are observed in the inner corners at the central leg of the E-core. This situation can be simulated during an M-test (Fig. 10.13) since this test also probes primarily the inner corners. Second, during lifetime power losses occur in the core and in the windings causing heating from the inside. The highest principal stresses are found in two zones at the back of the E-core and amount to about 5 MPa. This situation can be simulated during a W-test (Fig. 10.12) since in this test also the back of the E-core is stressed the most. A reasonable simulation of the loads as experienced in practice thus can be realised.

Obviously the calculation of the fracture stress from the fracture load is cumbersome analytically, if not impossible. Therefore, a FEM model of these two tests was made using plane stress conditions. The required data in this case are the Young's modulus E = 125 GPa and the Poisson's ratio ν = 0.3. Dimensions are also required and a typical set is given in Fig. 10.13 and Table 10.1. Other dimensions, smaller as well as larger are also in use. Some dimensions are not well controlled or prescribed. For example, the inner corner radius of the central leg is highly variable since it is not prescribed except for a maximum radius. This leads to varying stresses dependent on the exact value of that radius.

Iso-parametric eight node quadrilateral elements were used using a mesh, which was somewhat refined at the expected positions of high stress. The total number of elements was about 2000. The calculation[b] was done for a plane stress situation, although three-dimensional calculations, which are much more elaborate, confirm this two-dimensional analysis. It should be stated that this is not always the case, i.e. neither a check with 3D calculations, nor the agreement with 2D calculations is always present.

Table 10.1: Dimensions (mm) of E 42/21/15 cores.

A	B	C	D	E	F	G
42.7	14.8	6.1	21.1	6.1	9.2	12.0

[b] Huisman, D., de Graaf, M. and Dortmans, L. (1995), Proc. PCIM Conf., Nurnberg, 593.

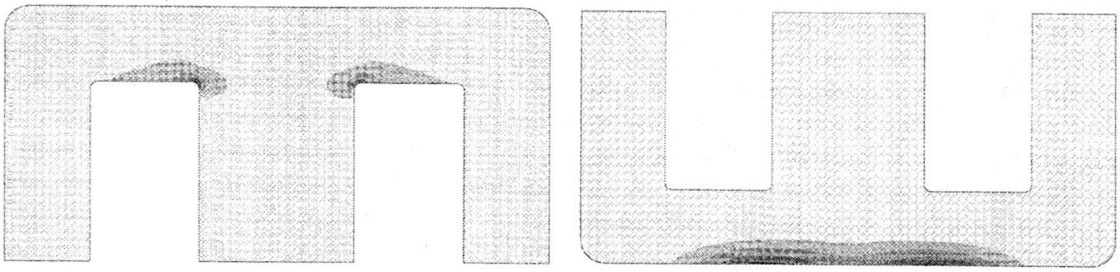

Fig. 10.14: Stress distribution as calculated by FEM for the M- and W-test. Darker grey implies a higher stress.

The stress distribution obtained for both the M- and W-test is shown in Fig. 10.14. For the M-test, using dimensions as given in Table 10.1, the maximum stress is a function of the radius of the corner of the inner leg. If this radius r is given in mm, then the maximum stress σ (MPa) appears to be related to the applied force F (N) by

$$\sigma = a(1 + b/r^{1/2})F \quad \text{with} \quad a = 0.0295 \quad \text{and} \quad b = 3.36$$

As expected the stress increases rapidly with decreasing radius of the corner. A stress singularity occurs for $r = 0$ mm and in fact the form of the above expression is chosen with this in mind. For the W-test the maximum stress is only dependent on the precise global dimensions since the lower face is essentially flat. For the dimensions as given in Table 10.1 the relation between maximum stress σ (MPa) and applied force F (N) is given by

$$\sigma = cF \quad \text{with} \quad c = 0.05895$$

The factor corresponds with the term $3l/2bh^2$ in the expression for the 3-point bend test. With these data experimental tests are calibrated for the conversion of force to stress enabling E-cores to be optimised without actual processing to complete transformers or endurance tests.

10.8 Bibliography

Akin, J.E. (1994), *Finite elements for analysis and design*, Academic Press, London.

Derby, B., Hils, D. and Ruiz, C. (1992), *Materials for engineering*, Longman Scientific & Technical, Harlow, UK.

Fung, Y.C. (1965), *Foundations of solid mechanics*, Prentice-Hall, Englewood Cliffs, NJ.

Timoshenko, S.P. and Gere, J.M. (1973), *Mechanics of materials*, Van Nostrand Reinhold, New York.

Young, W.C. (1989), *Roark's formulas for stress and strain*, 7th ed., McGraw-Hill, New York.

Zienkiewicz, O.C. and Taylor, R.L. (1989), *The finite element method*, 4th ed., McGraw-Hill, London.

11

Molecular basis of elasticity

In this chapter the molecular basis of elastic behaviour is discussed. Some general considerations are presented first. It is again convenient to divide materials into four categories, as done in Chapter 1: inorganics, metals, polymers and composites. For the first three types of material the chemical bonding and resulting forces are dealt with, resulting in order of magnitude indications (or better) for the values of the elastic parameters of these materials. Also an estimate for van der Waals crystals will be made. Thermal effects are introduced via lattice dynamics.

11.1 General considerations

Using thermodynamics it was shown that applying a strain ε_{ij} a stress σ_{ij} in a material results as a consequence of a change in internal energy u for adiabatic conditions or Helmholtz energy f for isothermal conditions

$$\blacktriangleright \qquad \sigma_{ij} = \rho\frac{\partial f}{\partial \varepsilon_{ij}} = \rho\left(\frac{\partial u}{\partial \varepsilon_{ij}} - T\frac{\partial s}{\partial \varepsilon_{ij}}\right) \tag{11.1}$$

Both terms contribute significantly but their relative contribution depends heavily on the type of material. Roughly speaking we thus can divide materials in two classes: *energy elastic materials*, for which the energy derivative is dominating, and *entropy elastic materials*, for which the entropy derivative is dominating.

To assess the relative importance of the energy and entropy contributions, we use the Maxwell relation

$$\frac{\partial s}{\partial \varepsilon_{ij}} = -\frac{\partial^2 f}{\partial \varepsilon_{ij}\partial T} = -\frac{1}{\rho}\frac{\partial \sigma_{ij}}{\partial T} \tag{11.2}$$

from which it follows that

$$\sigma_{ij} = \rho\left(\frac{\partial u}{\partial \varepsilon_{ij}} + \frac{T}{\rho}\frac{\partial \sigma_{ij}}{\partial T}\right) \tag{11.3}$$

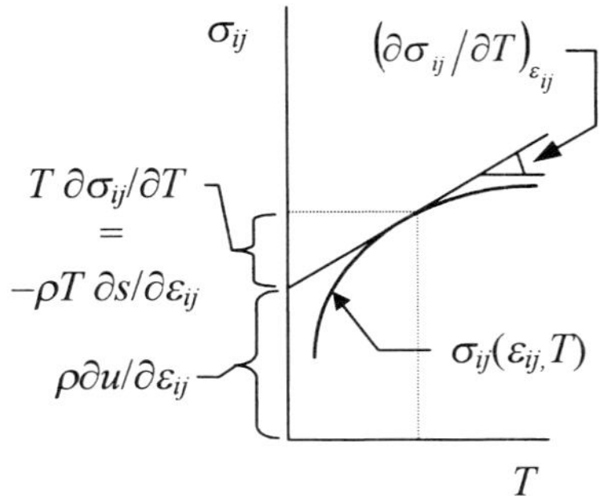

Fig. 11.1: Measuring stress as a function of temperature at constant strain.

Measuring the stress at constant strain as a function of temperature experimentally indicates the relative importance of the energy and entropy contributions, as indicated in Fig. 11.1. It appears that the energy contribution dominates for inorganic materials, metals and many polymers and that entropy is dominating for rubbers.

For energy elastic materials we have to consider the internal energy containing the potential energy Φ and kinetic energy K of the particles in the material involved. For most systems it is assumed that the dynamics of the electrons is so fast that they follow at all instances the movement of the nuclei. This is easily justified by the large difference in mass between electrons and nuclei. In this so-called *adiabatic* or *Born-Oppenheimer approximation* the internal energy is determined by the energy of the atoms as a whole only and all the atoms remain in their electronic ground state. Chemical bonding has to be treated quantum mechanically and in a quantum approach the relevant quantities in the expression for the bonding energy cannot be expressed only in simple atomic, bond or crystallographic parameters like valency, bond energy and lattice constant. We refrain from further discussion here and focus on the pair potential approach.

Leaving the kinetic energy of the atoms aside for the moment, then at low temperature the Helmholtz energy F equals the internal energy U and is only determined by the potential energy Φ of the atoms. The latter is in principle a function of the position of all atoms. Hence to make an estimate of the potential energy, we need a labelling of the atoms. If we have N unit cells, each with r atoms, the atoms are labelled according to the unit cell they occupy by the vector $\mathbf{n} = (n_1,n_2,n_3)$ and within the unit cell by α, β, ... The position of cell $\mathbf{n}$ is $\mathbf{r}_\mathbf{n} = n_1\mathbf{a}_1+n_2\mathbf{a}_2+n_3\mathbf{a}_3$, with $\mathbf{a}_1$, $\mathbf{a}_2$ and $\mathbf{a}_3$ the lattice vectors while the position of atom α in cell $\mathbf{n}$ is denoted by $\mathbf{r}_\alpha$. The position with respect to the origin is thus $\mathbf{r}_{\mathbf{n}\alpha} = \mathbf{r}_\mathbf{n} + \mathbf{r}_\alpha$. The components of the position vector $\mathbf{r}_{\mathbf{n}\alpha}$ are $r_{\mathbf{n}\alpha i}$. The potential energy is then

$$\Phi = \Phi\,(\mathbf{r}_{\mathbf{n}\alpha})$$

The theory of chemical bonding provides us with models for the potential energy Φ. Although, as stated before, in principle all chemical bonding should be treated quantum-mechanically, useful insight can be obtained by assuming that the many body potential $\Phi\,(\mathbf{r}_{\mathbf{n}\alpha})$ can be written as a sum of two body potentials ϕ so that

$$\Phi\,(\mathbf{r}_{\mathbf{n}\alpha}) = \tfrac{1}{2}\Sigma\phi_{\,\mathbf{n}\alpha,\mathbf{m}\beta}\,(\mathbf{r}_{\mathbf{n}\alpha}, \mathbf{r}_{\mathbf{m}\beta})$$

where the sum runs over $\mathbf{n}\alpha$, $\mathbf{m}\beta = 1..rN$ and rN is the total number of atoms. A further simplification is obtained by assuming a central symmetric potential, implying that the force is always directed from atom $\mathbf{n}\alpha$ to atom $\mathbf{m}\beta$ and vice versa. Along these lines a considerable amount of modelling has been done. This approach is rather successful for ionically bonded solids and using more fundamental theory can motivate the approximations involved. We will use this approach for inorganic materials in Section 11.2. The approach has also been applied to molecular crystals with reasonable success and even to metals, although obviously in that case success is not expected on the basis of the premises (Section 11.3). In the latter case the electrons provide a significant contribution to the bonding. Metals are sometimes modelled using a two-body potential for the ions plus a volume dependent potential for the delocalised electrons. Altogether this is not a two-body potential. In this book a simple quantum model for metals is presented in Section 11.4, while Section 11.5 provides some extensions. Some energetic considerations for polymers are presented in Section 11.6. By the way, a further simplification arises if for the binding energy U_{bin} only the energy contribution between bonded atoms is considered and these 'bond energies' are considered as characteristics of these bonds. In that case the energy of a molecule can be estimated by adding the bond energies (Table 1.1). This approach is sometimes used for estimating surface energies of solids and total energies of molecules. Even simpler is the estimate $U_{\text{bin}} = 2H/N_AZ$, where H denotes the atomisation energy (≅ enthalpy) replacing the sum of bond energies and Z the co-ordination number.

For entropy elastic materials we have to consider the various contributions to the entropy. Generally, the most important ones are the configurational entropy due to the positions of the atoms or molecules in space and, to a lesser extent, the vibrational entropy due to thermal motion. This approach is particularly useful for rubbers and the basics are dealt in Section 11.7 while Section 11.8 elaborates a bit. Thereafter thermal effects are discussed in Section 11.9 in the most simple terms, i.e. using independent atomic motion, while Section 11.10 treats the basics of coupled motion.

Essentially the basic procedure is thus: find an expression for the Helmholtz energy, determine its minimum by varying the independent parameters (e.g. the atomic positions) and evaluate the second derivative of the Helmholtz energy with respect to the kinematical variables (e.g. the strains) to yield the corresponding elastic modulus. We restrict ourselves mainly to bulk moduli. The reason is simple. For isotropic compression the relative positions of atoms in a lattice remain the same but for shear deformations they may change. This makes the calculation of the shear modulus more complex. In the case of incompressible materials one modulus obviously suffices.

Problem 11.1

Show by counting bond energies that polyethylene gains about 0.9 eV per monomer with respect to ethylene (see Table 1.1).

Problem 11.2

Why is the 'bond energy' model not useful for making an estimate for the Young's modulus E but sometimes used to estimate the surface energy γ?

Max Born (1882-1970)
Born in Breslau, Germany, and educated at the universities of Breslau, Heidelberg, Zurich and Göttingen, he obtained his doctorate in 1907. After working in Cambridge and Breslau he was invited to Göttingen where he began a research project with von Kármán on lattice dynamics.

In 1914 Born was offered a chair at Berlin, becoming a colleague of Planck, but he had to join the German Armed Forces meanwhile publishing his first book *Dynamik der Kristallgitter*. In 1919 he moved to a chair in Frankfurt-am-Main and in 1921 to Göttingen, at the same time as James Franck. Here he reformulated the first law of thermodynamics. During these years his most important works were created; a modernized version of his book on crystals (1923) and many investigations by him and his pupils on crystal lattices, followed by studies on the quantum theory. Among his collaborators at this time were many physicists, later to become well-known, such as Pauli, Heisenberg, Jordan, Fermi, Dirac, Hund, Hylleraas, Weisskopf, Oppenheimer and Maria Goeppert-Mayer. During the years 1925 and 1926 he published, with Heisenberg and Jordan, investigations on the principles of quantum mechanics (matrix mechanics) and soon after this, his own studies on the statistical interpretation of quantum mechanics. He was forced to emigrate in 1933 and was invited to Cambridge, where he taught for three years as Stokes Lecturer. In 1936 he was appointed Tait Professor of Natural Philosophy in Edinburgh, where he worked until his retirement in 1953 after which he returned to Germany to Bad Pyrmont, near Göttingen. He was awarded the 1954 Nobel Prize for his work in quantum mechanics and statistical interpretation of the wave equation. He received many other honours. He wrote several books, which are now considered as classics, e.g. together with K. Huang *Dynamical theory of crystal lattices* (1954), which can be considered as a quantum update of *Dynamik der Kristallgitter* (1915, 1923). His small, still very readable booklet *Natural philosophy of cause and chance* (1949) provides a concise representation of his views on natural philosophy, i.e. physics.

11.2 Inorganics

As indicated briefly in Chapter 1, bonding in inorganic materials is either by ionic bonding or by covalent bonding. We restrict ourselves in this section to ionic bonding.

Ionic bonding is generally discussed in terms of central symmetric pair potentials between atoms or molecules of which the general shape is shown in Fig. 11.2. We do so likewise but for somewhat greater generality we insert the material constants appropriate to ionic bonding later. From the bonding energy so obtained we will calculate the bulk modulus and so establish a relation with the crystallographic structure. We note that in this section we will use the term molecules as a generic

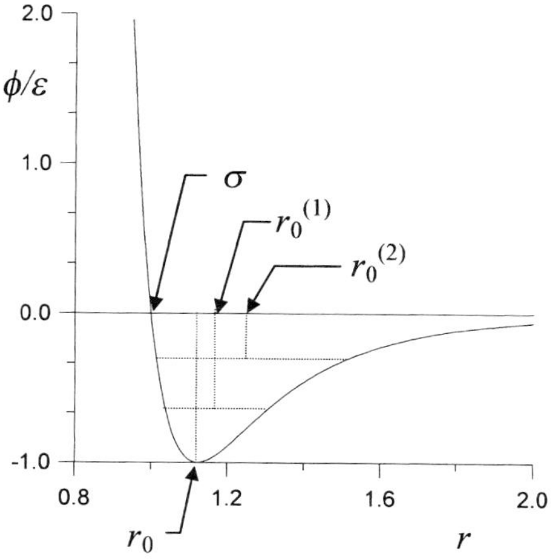

Fig. 11.2: The shape of the Mie pair potential in units of ε and σ for $m = 6$ and $n = 12$. At $T = 0$ K the equilibrium distance is r_0. This distance increases with T due to the asymmetry of the potential and the increase in population of excited vibrational states, as indicated by $r_0^{(1)}$ and $r_0^{(2)}$.

name denoting both atoms and molecules as appropriate.

A widely used potential is the *Mie potential* providing the potential energy ϕ_M for two molecules separated by a distance r. At zero temperature, neglecting zero point motion, the potential energy[a] is equal to $-\varepsilon$ at the equilibrium distance r_0. At $r = \sigma$, $\phi_M = 0$. The general expression for the Mie-potential is

$$\phi_M(r) = b/r^n - c/r^m$$

where b, c, m and n denote parameters and $n > m$. To obtain an expression in terms of r_0 and ε we use the equilibrium condition

$$\left(\frac{\partial \phi_M}{\partial r}\right)_{r=r_0} = 0 \quad \text{and} \quad \phi_M(r = r_0) = -\varepsilon$$

to obtain

$$\phi_M(r) = \frac{nm\varepsilon}{n-m}\left[\frac{1}{n}\left(\frac{r_0}{r}\right)^n - \frac{1}{m}\left(\frac{r_0}{r}\right)^m\right] \tag{11.4}$$

Alternatively, using of $\phi_M(\sigma) = 0$ we obtain $r_0 = (n/m)^{1/(n-m)}\sigma$ (Fig. 11.2) so that we can also write

► $$\phi_M(r) = \frac{n\varepsilon}{n-m}\left(\frac{n}{m}\right)^{m/(n-m)}\left[\left(\frac{\sigma}{r}\right)^n - \left(\frac{\sigma}{r}\right)^m\right] \tag{11.5}$$

By the way, if we set $n = 12$ and $m = 6$ we obtain

$$\phi_{LJ}(r) = 4\varepsilon[(\sigma/r)^{12} - (\sigma/r)^6],$$

usually referred to as the *Lennard-Jones potential*[b] or 6-12 *potential*. In this case $\sigma = (b/c)^{1/6}$ and $\varepsilon = c^2/4b$. While the choice for $m = 6$ can be rationalised from van der Waals interactions, the choice for $n = 12$ is mainly made for mathematical convenience.

Returning to Eq. (11.4) the total potential energy $\Phi(r)$ of 1 mole containing N_A molecules can be calculated by adding the various pair potential contributions. These contributions are due to nearest neighbours, next nearest neighbours and so on. Restricting ourselves for the moment to nearest neighbour interactions the potential energy is given by

$$\Phi(r) = \frac{N_A Z}{2}\phi_M(r) = \frac{N_A Z}{2}\frac{nm}{n-m}\varepsilon\left[\frac{1}{n}\left(\frac{r_0}{r}\right)^n - \frac{1}{m}\left(\frac{r_0}{r}\right)^m\right] \tag{11.6}$$

with Z the co-ordination number and r the nearest neighbour distance. In this approximation the equilibrium nearest neighbour distance in the lattice appears to be

[a] Unfortunately a number of symbols are conventionally overloaded. This applies to V for potential energy and volume. Since both properties are used in this chapter, we use Φ and ϕ for potential energy and V and v for volume. Similarly ε, normally denoting strain, is used for (potential) energy ε in connection with pair potentials. Also σ, normally denoting stress, is used in the same connection as a measure for size.

[b] John Edward Lennard-Jones (1894-1954). English theoretical chemist and physicist who contributed to the determination of interatomic and intermolecular forces, the quantum theory of molecular structure, and the statistical mechanics of liquids, gases, and surfaces. He added the suffix Jones to his name when he married with Mrs. Jones.

the equilibrium distance r_0 in a (gaseous) dimer, i.e. a two-atomic molecule or a bi-molecular complex. We now introduce the equilibrium molecular volume $v_0^* = \pi r_0^3/6$ at 0 K and momentary molecular volume $v^* = \pi r^3/6$ at 0 K. The corresponding molar volumes are $v_m = N_A v_0^*$ and $v = N_A v^*$, respectively. Using these volumes we can write

$$\Phi(v^*) = \frac{N_A Z}{2} \frac{nm}{n-m} \varepsilon \left[\frac{1}{n} \left(\frac{v_0^*}{v^*} \right)^{n/3} - \frac{1}{m} \left(\frac{v_0^*}{v^*} \right)^{m/3} \right] \tag{11.7}$$

For the calculation of the bulk modulus $K = -V(\partial p/\partial V)$ we consider hydrostatic compression under a pressure p. In that case it holds that

$$df = -p\, dv$$

so that the bulk modulus is proportional to the second derivative of the Helmholtz function with respect to the volume. Approximating the Helmholtz energy f by the internal energy u and u by the potential energy Φ we obtain

$$K = v_m \left(\frac{\partial^2 f}{\partial v^2} \right)_{v=v_m} \cong v_m \left(\frac{\partial^2 u}{\partial v^2} \right)_{v=v_m} \cong \frac{v_0^*}{N_A} \left(\frac{\partial^2 u}{\partial v^{*2}} \right)_{v^*=v_0^*} \cong \frac{v_0^*}{N_A} \left(\frac{\partial^2 \Phi}{\partial v^{*2}} \right)_{v^*=v_0^*} \tag{11.8}$$

Using for the potential energy the equation just derived results, after some calculation, in a formula due to Grüneisen[c]

$$K = \frac{mn}{9} \frac{Z}{2} \frac{\varepsilon}{v_0^*} = \frac{mn}{9} \frac{Z}{2} \frac{N_A \varepsilon}{v_m} = \frac{mn}{9} \frac{U_{sub}}{v_m} \tag{11.9}$$

where $U_{sub} = N_A Z\varepsilon/2$ is approximately the sublimation energy necessary to evaporate one mole of solid.

We now consider briefly contributions from molecules beyond the first co-ordination shell. If we take one atom at the origin, the distance to the others is given by $\mathbf{r}_{\mathbf{n}\alpha}$. Using a power law potential b/r^n the total energy per atom for various lattices can be written as

$$\phi = ½\sum\nolimits_{\mathbf{n},\alpha} \frac{b}{(r_{\mathbf{n},\alpha})^n} = ½\frac{b}{r^n} \sum\nolimits_{\mathbf{n},\alpha} \left(\frac{r}{r_{\mathbf{n},\alpha}} \right)^n$$

where r is the nearest neighbour distance in the lattice. The so-called lattice sums

$$S_n = \sum\nolimits_{\mathbf{n},\alpha} \left(\frac{r}{r_{\mathbf{n},\alpha}} \right)^n$$

Table 11.1: Lattice sums S_n for various lattices.

n	SC	BCC	FCC	HCP
4	16.532	22.639	25.338	–
6	8.402	12.253	14.454	14.555
8	6.946	10.355	12.802	12.802
10	6.426	9.564	12.311	12.312
12	6.202	9.114	12.132	12.132
14	6.098	8.817	12.059	12.059

[c] Grüneisen, E. (1926), Geiger-Scheel's Handbuch der Physik **10**, 1, Springer, Berlin.

are pure numbers that are only dependent on the type of lattice. They have been calculated[d] for various lattice types and for values from $n = 4$ to $n = 20$. A selection is given in Table 11.1. For the Mie potential the contributions b/r^n and c/r^m have to be combined and we obtain an expression similar as before but with the constants b and c replaced by B_n and C_m

$$\blacktriangleright \qquad \Phi = \frac{N_\mathrm{A}}{2}\left(\frac{bS_n}{r^n} - \frac{cS_m}{r^m}\right) \equiv \frac{N_\mathrm{A}}{2}\left(\frac{B_n}{r^n} - \frac{C_m}{r^m}\right) \tag{11.10}$$

Equilibrium is obtained if $\partial\Phi/\partial r = 0$. Evaluation leads to the equilibrium nearest neighbour distance in the lattice

$$r_0' = \sigma\left(\frac{nS_n}{mS_m}\right)^{1/(n-m)} = \left(\frac{S_n}{S_m}\right)^{1/(n-m)} r_0$$

where $r_0 = \sigma(n/m)^{1/(n-m)}$ denotes again the equilibrium distance in a dimer. For the potential energy Φ_0 we find

$$\Phi_0 = \frac{N_\mathrm{A}}{2}\frac{n}{n-m}\left(\frac{n}{m}\right)^{m/(n-m)} \varepsilon\left[S_n\left(\frac{mS_m}{nS_n}\right)^{n/(n-m)} - S_m\left(\frac{mS_m}{nS_n}\right)^{m/(n-m)}\right]$$

Specialising to the Lennard-Jones potential we obtain

$$r_0' = \sigma\left(\frac{2S_{12}}{S_6}\right)^{1/6} = \left(\frac{S_{12}}{S_6}\right)^{1/6} r_0 \qquad \text{and} \qquad \Phi_0 = -N_\mathrm{A}\varepsilon\frac{S_6^2}{2S_{12}}$$

We see that r_0' replaces r_0 and S_6^2/S_{12} replaces Z in the original expression for the potential energy. A comparison between Z and S_6^2/S_{12} is made in Table 11.2. The discrepancy is largely due to S_6. Take e.g. the FCC lattice. Since $S_{12} = 12.132$ and $Z = 12$, the next nearest contributions are only about 1%. For $S_6 = 14.454$, however, this contribution is about 20%. For the bulk modulus we obtain similar results. Since

$$\Phi = (N_\mathrm{A}Z/2)[b/r^n - c/r^m] \qquad \text{with} \qquad r_0' = \sigma(n/m)^{1/(n-m)}$$

becomes

$$\Phi = (N_\mathrm{A}/2)[B_n/r^n - C_m/r^m] \qquad \text{with} \qquad r_0' = \sigma(nS_n/mS_m)^{1/(n-m)}$$

the final result[e] is

$$\blacktriangleright \qquad K = v_\mathrm{m}\frac{\partial^2\Phi}{\partial v^2} = \frac{mn\varepsilon S_m}{18v_0^*}\left(\frac{S_m}{S_n}\right)^{\frac{m}{n-m}} \tag{11.11}$$

Table 11.2: Comparison of Z and S_6^2/S_{12}.

	SC	BCC	FCC	HCP
Z	6	8	12	12
S_6^2/S_{12}	11.38	16.47	17.220	17.222
Δ	–47%	–51%	–30%	–30%

[d] Lennard-Jones, J.E. and Ingham, A.E. (1925), Proc. Roy. Soc. (London) **A107**, 636 and Kihara, T. and Kuba, S. (1952), J. Phys. Soc. Japan 7, 348.

[e] Equivalently, making the dependence on r_0' more clear, $K = m(n-m)C_m/18(r_0')^{m+3}$.

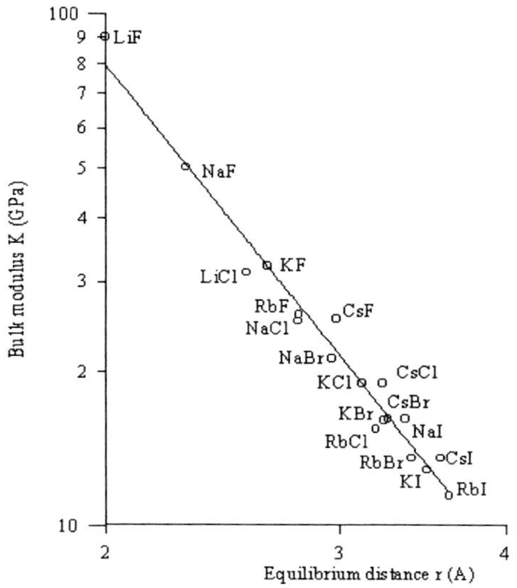

Fig. 11.3: The bulk modulus K at 0 K versus equilibrium distance r for various solids as given by Born and Huang (1954). The solid line is a least-squares fit with slope –3.2.

Specialising to the Lennard-Jones potential we obtain $K = 4\varepsilon\, S_6^{\,2}/v_0^{\,*}S_{12}$. Since for the nearest-neighbour model $K = 4\varepsilon\, Z/v_0^{\,*}$, the contribution of next-nearest neighbours to K amounts[f] up to 50% (see Table 11.2)

The parameters m and n depend on the nature of the bonding. For ionic bonding with two ions with charge Q_1 and Q_2 and separated by a distance r the attraction is given by the electrostatic or *Coulomb potential*

$$\phi_{\text{Cou}} = -M\, Q_1\, Q_2\, e^2/4\pi\varepsilon_0 r$$

where M denotes the Madelung constant, e the elementary charge and ε_0 the permittivity of the vacuum. Thus for ionic solids $C_m = MQ_1Q_2e^2/4\pi\varepsilon_0$ and $m = 1$. Therefore, the bulk modulus becomes

$$K = \frac{n-1}{18}\frac{N_{\text{A}}C_m}{(r_0')^4}$$

where it should be realised that this expression actually applies at 0 K. The above relationship is approximately observed for ionic lattices but with a slope of –3.2 instead of –4 (Fig. 11.3). Also for metallic and covalently bonded lattices a slope of –4 can be rationalised and again this relationship is only approximately obeyed.

A slightly different line of approach for ionic crystals is to use the experimentally observed bulk moduli to calculate the parameter n. It appears that this parameter varies with the solid but typically $n \cong 9$. In fact a somewhat better agreement can be obtained by taking n dependent on the electron configuration of the ions (5 for He-like, 7 for Ne-like, 9 for Ar-like, 10 for Kr-like and 12 for Xe-like and using the average value for different configurations of the cation and anion). This leads to lattice energies in fair agreement with experiment. Using an exponential expression, i.e. $\lambda\exp(-r/\rho)$, instead of the power law b/r^n also results in good agreement between

[f] Of course, in estimating K via $K = mnU_{\text{sub}}/9v_{\text{m}}$ this effect is incorporated implicitly via U_{sub}.

theoretical and experimental lattice energies. This agreement can be further improved to within experimental error by taking also into account the van der Waals attraction and zero-point energy, see e.g. Born and Huang (1954).

As stated we will not make an estimate for the shear modulus G, or equivalently Poisson's ratio ν, since it is much more difficult to make. Here it suffices to say that Poisson's ratio varies only to a small extent and is for most inorganic solids in the order of 0.25. Exceptions are, e.g. fused silica SiO_2 and polycrystalline boron carbide B_4C, both of which have a value of $\nu = 0.16$.

Problem 11.3

Derive Eq. (11.9).

Problem 11.4

Calculate the relative difference in equilibrium distance in a molecule and in the FCC lattice for the Lennard-Jones potential.

Problem 11.5

Show that the next-nearest neighbour contribution to the lattice sum S_6 and S_{12} for the FCC lattice are 0.75 and 0.094, respectively.

11.3 Van der Waals crystals

The approach as discussed in Section 11.2 can also be used for van der Waals crystals. As an example we estimate the bulk modulus for solid methane. Although methane obviously is not a centrally symmetric molecule, the deviation from spherical symmetry is small and we can expect the approach to work. Methane crystallises due to the van der Waals forces in a face centred cubic (FCC) lattice which has a co-ordination number $Z = 12$. The parameters n and m depend on the nature of the bonding. For van der Waals bonding typically $n = 12$ and $m = 6$ are taken and, as mentioned in Section 11.2, this choice is referred to as the *Lennard-Jones potential*. We also need the lattice equilibrium distance r_0 (omitting the prime from now) and bonding energy ε. From Table 11.3 we obtain

$$\varepsilon/k = 148 \text{ K} \qquad \text{and} \qquad r_0 = 0.382 \text{ nm}$$

From these data we calculate for the 6-12 potential from Eq. (9.9) $K = 4.2$ GPa.

As can be verified from the data given in Table 11.3 the values for the bulk

Table 11.3: Lennard-Jones parameters[g] for various compounds.

Material	Ne	Ar	Kr	Xe	CH_4	CCl_4	N_2	O_2	NO
$k^{-1}\varepsilon$/K	35.6	120	171	221	148	327	95.1	118	131
r_0/nm	0.275	0.341	0.360	0.410	0.382	5.88	0.371	0.346	0.317

The potential energy ε at r_0 is divided by k, Boltzmann's constant, so that the $k^{-1}\varepsilon$ is given in K.

[g] Hirschfelder, J.O., Curtiss, C.F. and Bird, R.B. (1964), *Molecular theory of gases and liquids*, Wiley, New York, Table I-A.

moduli do not vary widely. Since Poisson's ratio also is roughly constant at about 0.25, Young's modulus for methane can be estimated as

$$E = 3K(1-2\nu) = 1.5\,K = 6.3 \text{ GPa}$$

Generally we obtain for the moduli of van der Waals crystals values in the range of 1 to 5 GPa, which is indeed the experimental range.

11.4 Metals

Metallic bonding is, metaphorically speaking, provided by the 'sea' of electrons in which the ions are submerged. The valence electrons are not bound to specific atoms and the bonding is non-directional. For metals therefore, although the correlation as provided by the pair potential approach is quite good, the bonding energy has to be calculated quantum mechanically. To that purpose we consider only simple metals, which can be discussed in terms of the free electron model. For d-band and transition metals a more realistic approach is required.
Generally the bonding in simple metals can be described by

► $$\phi_{\text{met}} = -Cy + By^2 + Ay^3 \tag{11.12}$$

where $y = (V_0/V)^{1/3}$ with V the current volume[h], V_0 the equilibrium volume and A, B and C either parameters or theoretically determined functions. The first term represents the Coulomb energy, the second the electronic kinetic energy and the third a contribution from the interaction of the valence electrons with the inner electrons.

Considering A, B and C as empirical parameters one observes that in this approach the structure of the metals is entirely neglected and that the potential is *not* a two-body potential. Using the experimental values for the equilibrium volume V_0, energy U_0 and compressibility β_0 at absolute zero, a reasonable description can be obtained for simple metals. Since at absolute zero $U(V) = \phi_{\text{met}}$, from $\text{d}U = 0$ at $V = V_0$ one obtains

$$C = 3A + 2B \tag{11.13}$$

and using the energy and compressibility

$$U_0 = -2A - B \quad \text{and} \quad 3V_0/\beta_0 = 2A + 2B/3 \tag{11.14}$$

In this way Bardeen[i] obtained values for A, B and C for various alkali metals. They are given in Table 11.4. To test these results, a comparison with high-pressure data,

Table 11.4: Values for the empirical constants A, B and C and equilibrium lattice constant r_0 for the alkali (BCC) metals.

	Li	Na	K	Rb	Cs
A	1.4	4.3	5.1	4.2	3.7
B	8.4	1.2	–2.1	–0.6	0.0
C	20.8	15.4	11.2	11.4	11.2
r_0	0.346	0.424	0.525	0.562	0.605

A, B and C in 10^{-19} J/atom. r_0 in nm.

[h] Expressing V as volume per atom, ϕ_{met} is given in J/atom.

[i] John Bardeen (1908-1991). American physicist who received the Nobel Prize for physics in 1956 for the invention and development of the transistor, together with Walter H. Brattain (1902-1987) and William B. Shockley (1910-1989) and again for Physics in 1972 together with Leon. N. Cooper (1930-2...) and John Schrieffer (1931-2...) for their jointly developed theory of superconductivity, usually called BCS-theory.

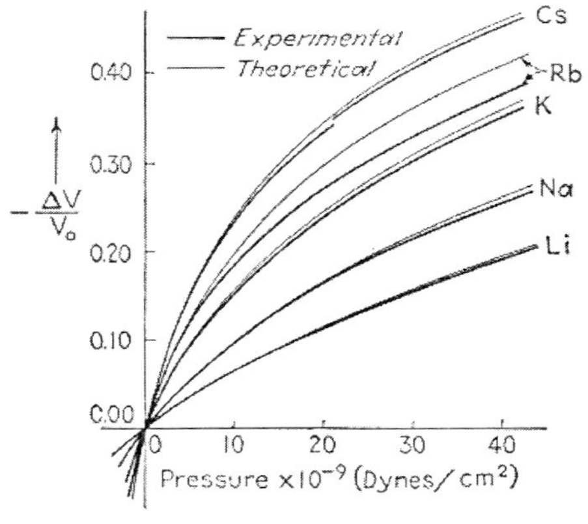

Fig. 11.4: The volume change with pressure for the alkali metals. 1 dyne = 10^{-5} N.

extrapolated to the absolute zero, was made. Determining the volume at pressure p from $p = -\partial F/\partial V = -\partial U/\partial V = -\partial \phi_{met}/\partial V$ leads to

$$pV_0 = y^4(y-1)[\,A(y+1) + 2B/3\,] \tag{11.15}$$

where, as before, $y = (V_0/V)^{1/3}$. In view of the simplicity of the approximation, the agreement with experiment is surprisingly good[j] (Fig. 11.4). The largest discrepancy occurs for Rb and amounts to about 10% at 4 GPa. The small discontinuity for Cs at about 2 GPa is due to a phase transition.

In fact Rose et al.[k] has extended this approach to metals in general. As a length scale they used the radius of the Wigner-Seitz cell r_{WS}, equal to the radius of a sphere whose volume equals the volume per atom in the material, i.e. $N{\cdot}4\pi r_{WS}^3/3 = V$, where N is the number of atoms in a volume V. In the sequel we will omit the subscript WS and write just r. They considered that the binding energy curve should be scaled. The first step is to scale the binding energy U_{bin} at any distance by the binding energy at the equilibrium distance U_0 so that $g(r) = U_{bin}/|U_0|$. The second step is to scale the interatomic distance. This could be done by the equilibrium distance r_0 but it appears to be more fruitful to use another measure. To that purpose the curvature at $r = r_0$ is used from the expansion

$$g(r) = -1 + \tfrac{1}{2}g''(r_0)(r-r_0)^2 + \cdots$$

As usual in these cases the first derivative vanishes at equilibrium. Omitting higher order terms the parabola describing $g(r)$ intersects the length axis when

$$r - r_0 = [2/g''(r_0)]^{1/2}$$

and this intersection defines a length scale. Therefore, $r - r_0$ is scaled by

$$l = \left(\frac{|U_0|}{U''(r_0)}\right)^{1/2}$$

to obtain $a = (r-r_0)/l$ as a length measure, implying that the $U_{bin}(r)$ curve is scaled as

$$U_{bin}(r) = U_{bin}(a)/|U_0|$$

[j] Bardeen, J. (1938), J. Chem. Phys. **6**, 367.
[k] Rose, J.H., Smith, J.R., Guinea, F. and Ferrante, J. (1984), Phys. Rev. **B29**, 2963.

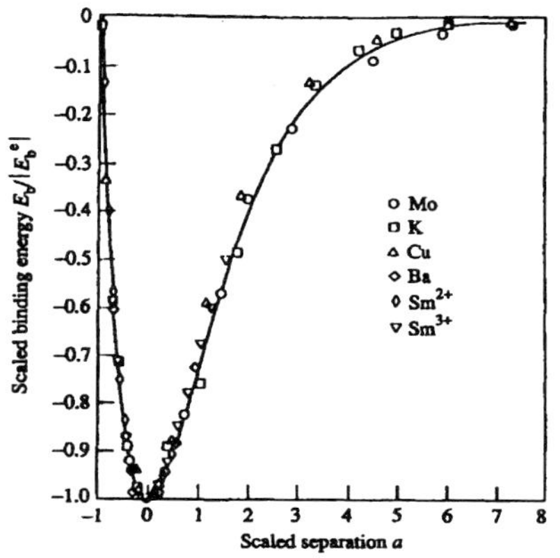

Fig. 11.5: The scaled binding energy as a function of scaled separation.

It is then found that the binding energy curves of metals lie on top of each other (Fig. 11.5). Moreover, it has been found that this scaled binding energy can be described empirically by

► $U_{\text{bin}}(r) = |U_0|g(a)$ where $g(a) = -(1 + a + 0.05a^3)\exp(-a)$ (11.16)

The consequences of the above results are large. First, as in the Bardeen treatment, the complete energy-density relationship is known whenever the equilibrium Wigner-Seitz radius r_0, the equilibrium binding energy $U_{\text{bin}}(r_0)$ and the bulk modulus at equilibrium K_0 are known. The bulk modulus is given by

$$K = V\frac{\partial^2 U_{\text{bin}}}{\partial V^2} = \frac{1}{12\pi r}\frac{\partial^2 U_{\text{bin}}}{\partial r^2}$$

so that the scaling length l becomes

$$l = \left(\frac{|U_0|}{12\pi r_0 K_0}\right)^{1/2}$$

Second, the scaling length l is proportional to the width of the binding energy curve and therefore related to the Hooke region in which the forces are linearly related to the displacements. For a so-called anharmonic crystal this region is small and as a measure for the degree of anharmonicity we use $\eta = r_0/l$. The parameters r_0, l and η are tabulated for most metals.

Given the expression above, the equation of state can be calculated as

$$p = -K_0\frac{3[(V/V_0)^{1/3} - 1]}{(V/V_0)^{2/3}}\left(1 - 0.15a + 0.05a^2\right)\exp(-a) \qquad (11.17)$$

while the variation of the bulk modulus $K = -V(\partial p/\partial V)_T$ with pressure is given by

$$(\partial K/\partial p)_T = 1 + 2.3\eta/3 \qquad (11.18)$$

In summary, although an estimate of the bulk modulus at equilibrium cannot be made by these approaches, the pressure dependence of the bulk modulus is quite well described. This is true in the Bardeen method for the alkali metals and for all metals in the Rose method. The calculation of the bulk modulus at equilibrium requires a more sophisticated approach, of which an outline is given in the refinement section.

Problem 11.6

Derive Eqs. (11.13) and (11.14).

Problem 11.7

Calculate the relative volume and compressibility for Na at 2 GPa.

Problem 11.8

Derive Eqs. (11.17) and (11.18). For the latter use $\partial K/\partial p = (\partial K/\partial r)/(\partial p/\partial r)$ and the limit $r \rightarrow r_0$.

11.5 First principles calculations for metals*

In Section 11.4 a simple approach to the elasticity of metals was given. Unfortunately, this approach was unable to provide an ab-initio estimate of the bulk modulus at equilibrium. In this section a more quantitative extension of this approach is dealt with that can provide this estimate using the nearly-free electron model as discussed in Chapter 8.

We recall that the *atomic volume* is given by $\Omega = V/N$, where N represents the number of atoms in a volume V, so that the *Wigner-Seitz radius* r reads

$$\Omega = \frac{4}{3}\pi r^3 \tag{11.19}$$

Further, the *dimensionless parameter* r_s is used where $r_s a_0$ indicates the radius of a sphere with volume Ω/Z, where Z is the number of electrons donated by each ion and a_0 the Bohr radius. Consequently,

$$\frac{4\pi}{3}(r_s a_0)^3 = \frac{\Omega}{Z} = \frac{V}{ZN} = \frac{4\pi r^3}{3Z} \quad \text{or} \quad r_s^3 = \frac{r^3}{Za_0^3} = \frac{3V}{4\pi NZa_0^3} \tag{11.20}$$

The average *kinetic energy* U_{kin} and the *Coulomb energy* U_{Cou} are given by

$$U_{kin} = 2.210 Z r_s^{-2} \quad \text{and} \quad U_{Cou} = -\alpha_C Z^{5/3} r_s^{-1} \tag{11.21}$$

Values for the Madelung-like constant α_C are given in Chapter 8. In first approximation the total energy per atom is given by $U = U_{kin} + U_{Cou}$ which has a minimum when $\partial U/\partial r_s = 0$, which occurs at $r_{bin} = (4.42/\alpha_C)Z^{-2/3}$ resulting in a binding energy $U_{bin} = 0.113\alpha_C^2 Z^{7/3}$. For the bulk modulus, given by $K = V\partial^2 U/\partial V^2$ and evaluated at $r_s = r_{bin}$, we then obtain

$$K/K_{non} = 0.200\ Z$$

where K is normalised by the bulk modulus of non-interacting electrons $K_{non} = 0.586\ r_s^{-5}$. As mentioned before, α_C is to a large extent insensitive to the structure, so that both the equilibrium distance and binding energy vary only slightly. Moreover, the value of r_{bin} is typically too small by a factor of 2. We concluded that this simple model is inadequate. This conclusion reinforced by the observation that the value of the bulk modulus is independent of structure.

To improve we will need the *exchange energy* U_{exc}, the *correlation energy* U_{cor} and the *pseudo-potential energy* U_{pse} given by

$$U_{\text{exc}} = -0.916 r_s^{-1} \qquad U_{\text{cor}} = 0.0313 \ln r_s - 0.115 \qquad \text{and} \qquad U_{\text{pse}} = 3Zr_c^2 r_s^{-3} \tag{11.22}$$

respectively. The core radius r_c was chosen in such a way that the experimental Wigner-Seitz radius r is matched (see Table 11.5). The improved total energy U is

$$U = U_{\text{kin}} + U_{\text{Cou}} + U_{\text{exc}} + U_{\text{cor}} + U_{\text{pse}}$$

Similar as before minimising this equation with respect to r_s yields the equilibrium distance r_{bin} and binding energy U_{bin} as a function of the parameter r_c. The contribution of the exchange-correlation energy cancels almost exactly with the electron-electron Coulomb term resulting in the approximate relation

$$U \cong -\frac{3}{Z^{1/3} r_s}\left[1 - \left(\frac{r_c}{Z^{1/3} r_s}\right)^2\right] + \frac{2.210}{r_s^2} \tag{11.23}$$

The bulk modulus, calculated similarly as before, is given by

$$\frac{K}{K_{\text{non}}} = 0.200Z + 0.815\frac{r_c^2}{r_s} \tag{11.24}$$

where K is again normalised by the bulk modulus of non-interacting electrons $K_{\text{non}} = 0.586\, r_s^{-5}$. The contribution of the correlation energy is neglected since it contributes just a few percent.

Table 11.5: Girifalco parameters and the bulk moduli of simple metals.

Metal	r/a_0	r_s/-	r_c/a_0	K/K_{non} theory	K/K_{non} experiment
Li	3.27	3.27	1.32	0.63	0.50
Na	3.99	3.99	1.75	0.83	0.80
K	4.86	4.86	2.22	1.03	1.10
Rb	5.31	5.31	2.47	1.14	1.55
Cs	5.70	5.70	2.76	1.29	1.43
Be	2.36	1.87	0.76	0.45	0.27
Mg	3.35	2.66	1.31	0.73	0.54
Ca	4.12	3.27	1.73	0.95	0.66
Sr	4.49	3.57	1.93	1.05	0.78
Ba	4.67	3.71	2.03	1.11	0.84
Zn	2.91	2.31	1.07	0.60	0.45
Cd	3.26	2.59	1.27	0.71	0.63
Hg	3.35	2.66	1.31	0.73	0.59
Al	2.99	2.07	1.11	0.69	0.32
Ga	3.16	2.19	1.20	0.74	0.33
In	3.48	2.41	1.37	0.83	0.39
Tl	3.58	2.49	1.43	0.87	0.39
Cu	2.67	2.67	0.91	0.45	2.16
Ag	3.02	3.02	1.37	0.71	2.94
Au	3.01	3.01	1.35	0.69	4.96

A comparison with experimental data is given in Table 11.5. The results for the Z=1 and Z=2 metals are quite good but for the Z=3 and noble metals the agreement is not so good. This is as far as the model can be pushed. For the Z=3 metals the second order contribution[1] to the bulk modulus becomes important while for the noble metals and for the transition metals the more complex band theory is essential. However, the procedure to estimate the bulk moduli is illustrated nicely and shows that a basic understanding is obtained. In principle a calculation for the shear modulus G can be made in a similar way.

Problem 11.9

Show that:

a) for non-interacting electrons, i.e. for electrons with only kinetic energy, the bulk modulus is given by $K_{\text{non}} = 0.586\, r_s^{-5}$,

b) the bulk modulus $K = V\partial^2 U/\partial V^2$ can also be written as $K = 1/(12\pi r_s)\partial^2 U/\partial r_s^{\,2}$,

c) using the approximation $U = U_{\text{kin}}+U_{\text{Cou}}$ the bulk modulus is given by $K/K_{\text{non}} = 0.200\, Z$ and that

d) Eq. (11.24) holds.

11.6 Polymers

As briefly discussed in Chapter 1, polymers consist of long molecular chains of covalently bonded atoms, which are inter- and intra-molecularly bonded to either crystalline or amorphous solids (Strobl, 1997). We discuss the deformation of such chains in terms of spring constants representing the various forces. Deformation of the chains occurs via deformation of the bonds (spring constant k_r), deformation of the bond angles (spring constant k_θ), internal rotations along the bonds (spring constant k_ϕ) and van der Waals interaction between the chains (spring constant k_i).

The spring constants k_r and k_θ are determined by the covalent bonding and are high while the constants k_ϕ and k_i are related to the intra- and inter-molecular interaction, respectively, and are thus much lower (Fig. 11.6). Approximately the ratio between these constants is $k_r : k_\theta : k_\phi : k_i = 100 : 10 : 1 : 1$. Polymers do not crystallise

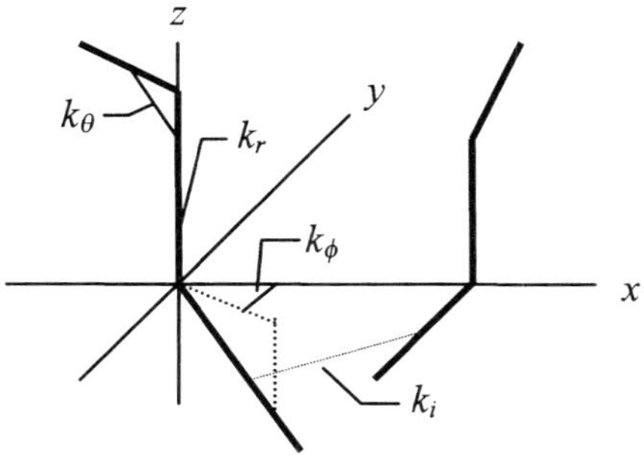

Fig. 11.6: Various spring constants representing the interactions in polymer chains.

---

[1] Ashcroft, N.W. and Langreth, D.C. (1967), Phys. Rev. **155**, 682.

more or less completely like metals and inorganics. As indicated in Chapter 1, the chains fold to small crystallised packages, the *folded chain crystal(lite)s* or *lamellae*. Stacks of these lamellae associate to sphere shaped aggregates, the *spherulites*. Space between the crystallites in the spherulite is filled with amorphous material. The crystallites are anisotropic: along the chain the stiffness is determined by the covalent bond, resulting in relatively high stiffness values, while in the other directions the dispersive, van der Waals interactions are decisive with relatively low stiffness values as a consequence.

In random or non-oriented materials the moduli at low temperature are thus also largely determined by the secondary or van der Waals interactions, in spite of the relatively high density of covalent bonds. This implies moduli of about 1 to 5 GPa. The relatively small differences between amorphous, semi-crystallised and cross-linked polymers at low temperature are caused by differences in density and dispersive interactions.

In the next section we discuss first amorphous and oriented polymers, where energy considerations prevail, and second in Section 11.7 rubber elasticity, where entropy considerations are the most important.

Amorphous polymers

Although amorphous polymers strictly speaking do not qualify for application of the van der Waals interaction model, it appears that reasonable predictions nevertheless can be made. For amorphous polymers it can be expected that the low temperature bulk modulus is higher as compared to that of actual van der Waals crystals, such as Ar or CH_4, because the compressibility along the chain is virtually zero. Without derivation we state that it can be estimated that this effect causes an increase of the bulk modulus by a factor of 2.25 compared to the modulus for actual van der Waals crystals (see Section 11.3). An estimate of Young's modulus E is thus given by

$$E = 2.25 \times 3(1-2\nu)K = 2.25 \times 3(1-2\nu)\frac{mn}{9}\frac{U_{\text{sub}}}{v_{\text{m}}} = 18\frac{U_{\text{sub}}}{v_{\text{m}}} \qquad (11.25)$$

where for the last step the Lennard-Jones potential ($m = 6$, $n = 12$) and $\nu = 1/3$ have been taken. Using typical values for $U_{\text{sub}} \cong 40$ kJ/mol and $v_{\text{m}} \cong 0.1$ nm^3 a value of about 10 GPa results, in reasonable agreement with experiments.

Oriented polymers

It is possible to orient the chains in the polymer resulting in macroscopically anisotropic material. Orientation can be introduced either by using non-flexible chains in a spinning process, e.g. aramide fibres, or by stretching via plastic deformation a material consisting of flexible chains extremely far, e.g. polyethylene (PE) fibres. In this case the contribution of the covalent bonds to the moduli is much larger. A

Fig. 11.7: Schematic of an oriented polymer.

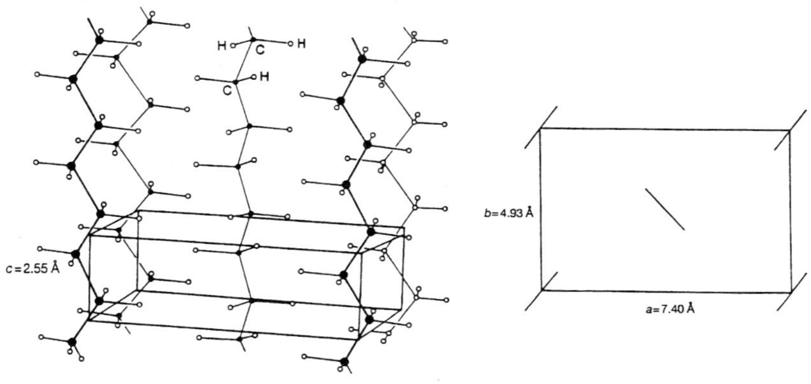

Fig. 11.8: The crystallographic structure of polyethylene.

schematic of the structure is given in Fig. 11.7. If the chains are long enough and the degree of orientation large enough, the secondary forces due to chain-chain interaction are sufficiently large to prevent sliding of the chains along each other. In that case the stiffness is determined by the covalent bonds.

The elastic properties of these fibres can be estimated by calculating the modulus for a single chain[m]. Let us consider PE fibres. We need the cross-section of the fibre as well as the spring constant. The cross-section can be determined from the orthorhombic, crystallographic unit cell structure (Fig. 11.8) with cell dimensions of $a = 0.740$, $b = 0.493$ and $c = 0.255$ nm. The length r of the C–C bonds is $r = 0.153$ nm while the valence angle θ between two C–C bonds is $\theta = 112°$. Hence the angle $\alpha = 34°$, see Fig. 11.9. In a plane perpendicular to the chains there are two chains per area $A = ab$. The cross-section per chain therefore is 0.182 nm^2. The fibre crystallises in a zigzag way and thus for considering the force along the chain direction we have to deal with the force constants k_r and k_θ. An estimate of these spring constants can be made by infrared spectroscopy and resulted in $k_r = 4.4$ N/cm and $k_\theta/r^2 = 0.35$ N/cm, respectively.

For the deformation of a single C–C bond (Fig. 11.9) it holds that

$$F_1 = F\cos\alpha = k_r\,\mathrm{d}r \qquad \text{or} \qquad \mathrm{d}r = (F\cos\alpha)/k_r$$

For the deformation of a bond angle we use the moment $\frac{1}{2}rF_2 = \frac{1}{2}rF\sin\alpha$ and $\mathrm{d}\alpha = -\mathrm{d}\theta/2$ to obtain

$$\tfrac{1}{2}rF\sin\alpha = k_\theta\,\mathrm{d}\theta = -2k_\theta\,\mathrm{d}\alpha \qquad \text{or} \qquad \mathrm{d}\alpha = -(Fr\sin\alpha)/4k_\theta$$

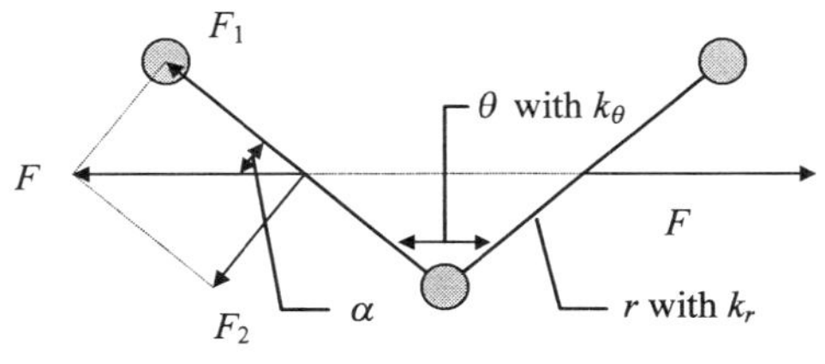

Fig. 11.9: The deformation of a single molecule.

[m] Treloar, L.R.G. (1960), Polymer **1**, 95.

From the length of the chain $L = Nr\cos\alpha$, where N denotes the number of bonds, the length change dL is given by

$$\mathrm{d}L = N\,\mathrm{d}(r\cos\alpha) = N[(\cos\alpha)\mathrm{d}r - r(\sin\alpha)\mathrm{d}\alpha]$$

and thus for the relative length change dL/L one obtains

$$\frac{\mathrm{d}L}{L} = \left(\frac{\cos^2\alpha}{k_r} + \frac{r^2\sin^2\alpha}{4k_\theta}\right)\frac{F}{r\cos\alpha} \tag{11.26}$$

Finally for the modulus E, using the data as given before, we obtain

$$E = \frac{F}{A}\bigg/\frac{\mathrm{d}L}{L} = \frac{r\cos\alpha / A}{[(\cos^2\alpha)/k_r] + [(r^2\sin^2\alpha)/4k_\theta]} \tag{11.27}$$

Substituting the numerical values for the parameters results in E = 182 GPa. This calculation for PE fibres and similar ones for other fibres shows that the moduli of stretched fibres can be quite high and this has stimulated the research of stretching of polymers. Using a more sophisticated modelling that includes secondary interactions and bond torsion, results in even higher numbers[n], e.g. up to 300 GPa. Experimentally fibres with moduli of 220 GPa have been realised in the mean time.

The main reason for the stiffness of the PE fibre is that they crystallise in a zigzag way in planes so that only the stiff bond stretch and bond angle deformation are activated when a force is applied. If the chain crystallises in a helix, e.g. as for polypropylene, also deformation of the torsion angle is possible. In that case the overall stiffness is much lower. The fibre stiffness thus is dependent on both the orientation and the chain conformation in the crystal. Finally in Table 11.6 estimates for various other fibres are given based on similar, but more sophisticated calculations as outlined above. As is clear from this table, quite high moduli can be obtained. Finally it should be noted that these high values can only be realised for sufficiently long molecules when only van der Waals interactions between the molecules exist. For the case of hydrogen bonding between chains, the theoretical values are already approached for much shorter chain lengths.

Table 11.6: Theoretical moduli for various polymers.

Material	PE	PTFE	PVA	PA	PET	i-PP
E (GPa)	300	200	250	200	150	75

PE: polyethylene, PTFE: polytetrafluorethylene, PVA: poly(vinyl alcohol), PA: polyamide, PET: poly(ethylene teraphtalate), i-PP polyisopropylene

Problem 11.10

Show that if one considers the C–C bond as infinitely stiff, $E = \dfrac{8r\,k_\theta\cos\alpha}{Ar^2\sin^2\alpha}$ and that thus the main contribution to the compliance $1/E$ is due to bond angle bending.

[n] Tashiro, K., Kobayashi, M. and Tadakoro, H. (1978), Macromolecules **13**, 914; Odajima, A., Madea, T.J. (1966) Polym. Sci. **C34**, 55.

11.7 Rubber elasticity

Many polymers are also capable of showing entropy elasticity. These materials are generally known as *rubbers* (or *elastomers*). The classical example is an amorphous thermoplastic material with a glass temperature T_g below room temperature, which is chemically cross-linked at a number of places along the chains so that a network results. Each part of the molecule between cross-links can undergo random coiling since a large number of conformational sequences is still possible. Hence one can describe it as a viscous (or visco-elastic) liquid, incapable of flow without force due to the cross-links. Cross-linking, which in the jargon of rubbers is denoted as *vulcanisation*, was originally often realised by sulphur but other agents, e.g. peroxide, are used as well nowadays. Also a partially crystallised thermoplastic material at a

Table 11.7: Some typical rubbers.

	Composition	T_g (°C)
$-CH_2-C(CH_3)=CH-CH_2-$	Polyisoprene (natural rubber)	–72
$-CH_2-C(Cl)=CH-CH_2-$	Polychloroprene (neoprene)	–50
$-CH_2-C(CH_3)_2-CH_2-C(CH_3)_2-$	Polyisobutylene (butyl rubber)	–
$-CH_2-CH=CH-CH_2-$	Polybutadiene	–85
$-CH_2-CH=CH-CH_2-CH_2-CH(C_6H_5)-$	Butadiene-styrene	–50
$-O-Si(CH_3)_2-$	Silicone rubber	–120

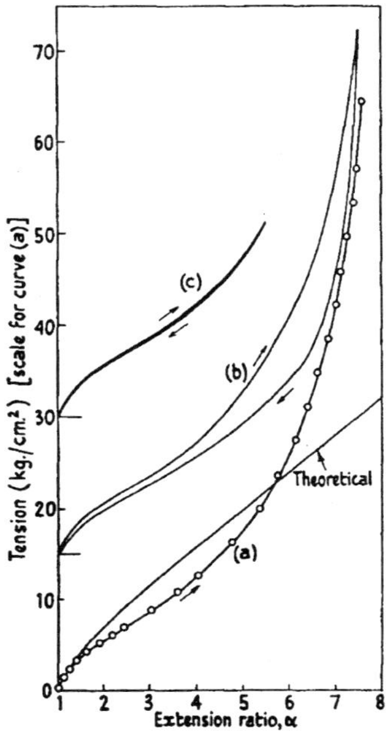

Fig. 11.10: Theoretical and experimental uni-axial stress-stretch curve for 8% sulphur vulcanised natural rubber. The theoretical curve is $s_1 = 4.0(\lambda-\lambda^{-2})$.

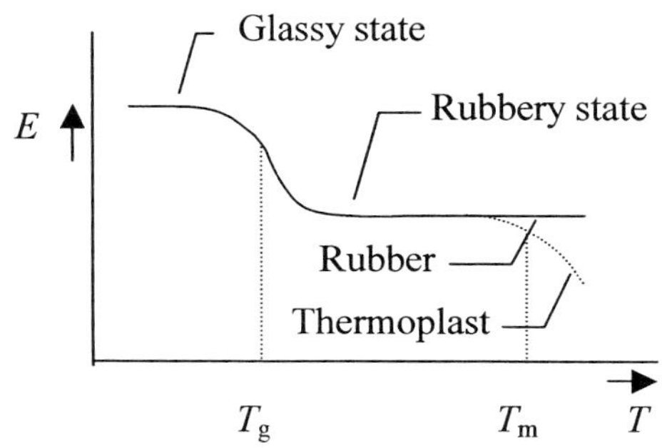

Fig. 11.11: Schematic of E versus T.

temperature between the glass transition temperature T_g and the melting point T_m is in the *rubbery state*. In the latter case the not yet melted crystals act as physical cross-links. It thus also can be described as a polymeric liquid since the secondary intra- as well as intermolecular bonds are fully broken allowing the chains to exhibit thermal motion, but, again, not to flow without force due to the cross-links. Some typical examples of rubbers are given in Table 11.7. Rubbers[o] have typically a low Young's modulus and a high reversible strain ranging up several hundreds of percent, dependent on the degree of cross-linking (see Fig. 11.10, curve c). After that they show hysteresis (see Fig. 11.10, curve b). A schematic figure of the modulus E versus temperature T is given in Fig. 11.11. Rubbers show a significant thermo-elastic effect, i.e. a stretched rubber shrinks upon heating and produces heat upon stretching. During short time deformation their volume is constant to a high degree of accuracy. In practice often (inorganic) fillers are used, either as a means to stiffen or to reinforce the material or as a (physical) cross-linker. A well-known application is the use of rubber filled with carbon black for tires of cars.

Experiments carried out as indicated in Section 11.1 (see as an example Fig. 11.12) clearly showed that the relative contribution of the internal energy u to the modulus for rubbers is relatively small (about 0.1) and can be ignored in a first approximation. The expression for the stress thus reduces to

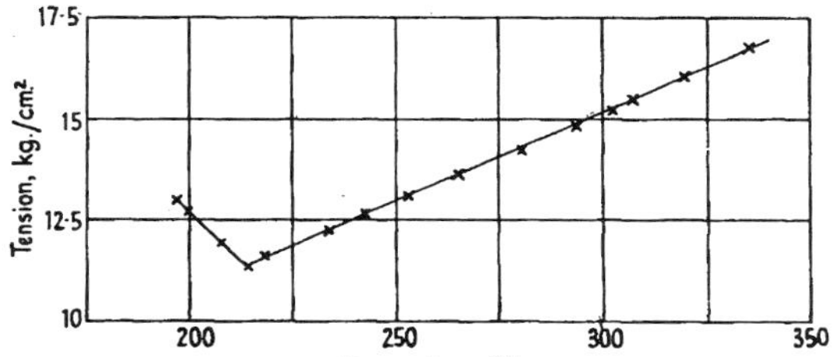

Fig. 11.12: Force at constant $\lambda = 4.5$ for 8% sulphur vulcanised natural rubber as a function of temperature. The stress-temperature relationship is linear from 70 °C to –50 °C. The change in slope at –50 °C indicates the glass transition temperature where below the energy contribution becomes dominant[p].

[o] Treloar, L.R.G. (1949), *The physics of rubber elasticity*, Clarendon, Oxford.

[p] Data from Meyer and Ferri (1935), Helv. Chim. Acta **18**, 570 and Rubb. Chem. Tech. **8**, 319 as represented by Treloar, L.R.G. (1949), *The physics of rubber elasticity*, Clarendon, Oxford.

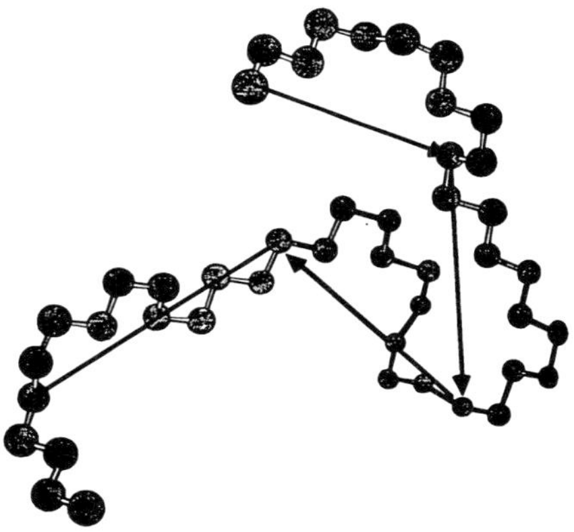

Fig. 11.13: The equivalent chain for polyethylene. A freely jointed chain with segments containing ~ 1/10 of the bonds of the polyethylene molecule and each segment length ~ 8 times that of a C–C bond length with the same average dimension $\langle r^2 \rangle$ and contour length as the polyethylene molecule.

$$\sigma_{ij} = \rho\frac{\partial f}{\partial \varepsilon_{ij}} = \rho\frac{\partial(u - Ts)}{\partial \varepsilon_{ij}} \cong -\rho\frac{\partial Ts}{\partial \varepsilon_{ij}}$$

However, since rubbers can show a large strain it is more convenient to use, instead of the strains ε_{ij}, the principal stretches $\lambda_I = L_I/L_{I0} \cong 1+\varepsilon_I$ along the co-ordinate axes. Here the initial length along the co-ordinate axis is denoted by L_{I0} while the final length is L_I. The index I runs from 1 to 3. Instead of the stress σ_{ij} the (principal) forces f_I, corresponding to the direction I, are used. The work done during deformation is $f_I\lambda_I$ and the above equation becomes

$$f_I \cong -T\partial s/\partial L_I = -(T/L_{I0})\partial s/\partial \lambda_I$$

The main task is now to estimate the entropy s as a function of the stretches λ_I in order to obtain the forces forces f_I.

For an estimate of the entropy derivative, the polymer is assumed to be a network of cross-linked chains of which the parts between cross-links are referred to as sub-chains. Since for a real polymer the various parts of the chain are constrained with respect to each other in lieu of bond angles and internal rotation along a backbone bond, the calculation of the conformational behaviour is complex. Therefore it is customary to use an *equivalent chain*, containing n segments of length l, for which we require that the average size of the molecular coil and the contour length are the same as for the real molecule but in which the segments can move completely freely with respect to each other apart from being connected. The equivalent chain for polyethylene is illustrated in Fig. 11.13.

We now introduce the *affine network model*, essentially introduced by Kuhn in 1936 and somewhat extended and modified by Treloar in 1943, for which the basic assumptions are:

- The deformation at the molecular level is the same as that on the macroscopic level, e.g. the deformation is affine.
- All conformational states have the same energy.
- The unstressed and stressed network is isotropic.
- The volume remains constant during deformation, e.g. $\lambda_{\text{I}}\lambda_{\text{II}}\lambda_{\text{III}} = 1$.

- The network consists of N mole of sub-chains, each sub-chain containing n segments with length l. The entropy of the network is the sum of the entropies of the individual sub-chains.
- The sub-chains between the cross-links can be represented by Gaussian statistics of unperturbed equivalent chains of the same length[q].

The probability distribution of the end-to-end vectors for a freely jointed chain is described by the random walk model[r] resulting in

$$P(n,\mathbf{r})\mathrm{d}\mathbf{r} = \left(\frac{3}{2\pi nl^2}\right)^{3/2} \exp\left[-\frac{3r^2}{2nl^2}\right]\mathrm{d}\mathbf{r} \tag{11.28}$$

From this distribution one obtains

$$\langle r^2\rangle_0 = \int r^2 P(n,\mathbf{r})\mathrm{d}\mathbf{r} = 4\pi\int_0^\infty r^4 P(n,\mathbf{r})\mathrm{d}r = nl^2$$

with $\langle r^2\rangle_0 = \langle x^2\rangle_0 + \langle y^2\rangle_0 + \langle z^2\rangle_0$ the mean square end-to-end distance of the sub-chains.

Justification 11.1

A simple justification for the distribution function $P(n,\mathbf{r})$ runs as follows. We consider a freely jointed (ideal) chain of n segments of length l. If we add to the chain an extra segment with vector $\mathbf{l}$ (components l_i) we have $P(n+1,\mathbf{r}) = P(n,\mathbf{r}+\mathbf{l})$. Assuming that $l \equiv \|\mathbf{l}\| \ll r \equiv \|\mathbf{r}\|$, we may expand $P(n+1,\mathbf{r})$ as a Taylor series in $\mathbf{l}$ obtaining

$$P(n,\mathbf{r}+\mathbf{l}) = P(n,\mathbf{r}) + \nabla P(n,\mathbf{r})\cdot\mathbf{l} + \tfrac{1}{2}\mathbf{l}^{\mathrm{T}}\cdot\nabla^2 P(n,\mathbf{r})\cdot\mathbf{l} + \cdots$$

Averaging over all possible orientations of $\mathbf{l}$, taking into account that $\langle\mathbf{l}\rangle = \mathbf{0}$, the result is

$$\langle P(n,\mathbf{r}+\mathbf{l})\rangle = \langle P(n,\mathbf{r})\rangle + (l^2/6)\,\langle\nabla^2 P(n,\mathbf{r})|_{\mathbf{l}=\mathbf{0}}\rangle + \cdots$$

where $\langle l_i l_j\rangle = \tfrac{1}{3}l^2\delta_{ij}$ has been used. We further write $\langle P(n,\mathbf{r})\rangle = P(n,r)$ for the average over all orientations of $\mathbf{r}$. Using $\langle P(n+1,\mathbf{r})\rangle = P(n+1,r)$ we also have for $n \gg 1$

$$P(n+1,r) = P(n,r) + \langle\partial P(n,\mathbf{r})/\partial n\rangle + \cdots$$

and equating $P(n+1,r)$ with $\langle P(n,\mathbf{r}+\mathbf{l})\rangle$ we have approximately

$$\partial P(n,\mathbf{r})/\partial n = (l^2/6)\nabla^2 P(n,\mathbf{r})$$

The relevant solution depends only on r since the undisturbed molecule has a spherical shape and reads $P \sim n^{-3/2}\exp(-3r^2/2nl^2)$. Using the normalisation condition $\int P(n,\mathbf{r})4\pi r^2\,\mathrm{d}r = 1$, we have for the complete solution

$$P(n,\mathbf{r}) = (3/2\pi nl^2)^{3/2}\exp(-3r^2/2nl^2)$$

Note that for $r > nl$, $P(n,r) \neq 0$, though this condition should be obeyed for a realistic solution. However, for $n \gg 1$, $P(n,r) \cong 0$ when $r > nl$.

---

[q] We recall that for the equivalent chain under theta conditions $\langle r^2\rangle = nl^2$ where we have omitted the prime and write n and l for the number and length of the segments, respectively. See Chapter 8.

[r] For a derivation, see e.g. Gedde (1995) or Weiner (1983).

The entropy of an individual sub-chain can be calculated from $s = k \ln W$, with W the possible number of conformations and where k denotes Boltzmann's constant, if we assume that W is proportional[s] to $P(\mathbf{r})\mathrm{d}\mathbf{r}$. After simplification the result becomes

$$s = C - \frac{3kr^2}{2\langle r^2\rangle_0} \qquad C = \text{constant} \tag{11.29}$$

The initial state is characterised by the end-to-end vector $\mathbf{r}_0 = (x_0, y_0, z_0)$ and for affine stretches the end-to-end vector is $\mathbf{r} = (x, y, z) = (\lambda_I x_0, \lambda_{II} y_0, \lambda_{III} z_0)$. The entropy difference $\Delta s = s - s_0$ for a single subchain between the initial, unstressed state and the final, stressed state thus becomes

$$\Delta s = s - s_0 = -\frac{3k}{2\langle r^2\rangle_0}\left[(\lambda_I^2 - 1)x_0^2 + (\lambda_{II}^2 - 1)y_0^2 + (\lambda_{III}^2 - 1)z_0^2\right] \tag{11.30}$$

and the total entropy $\Delta S = \Sigma\Delta s$ where the summation is over the number of sub-chains. Remembering that we assumed isotropic behaviour we have

$$\Sigma x_0^2 + \Sigma y_0^2 + \Sigma z_0^2 = \Sigma r_0^2 = NN_A \langle r^2\rangle_0$$

$$\Sigma x_0^2 = \Sigma y_0^2 = \Sigma z_0^2 = \Sigma r_0^2/3 = NN_A\langle r^2\rangle_0/3$$

where N_A is Avogadro's constant (remember that N is the number of moles of sub-chains). Inserting this result in Eq. (11.30) results in the total entropy expression

▶ $$\Delta S = -\frac{NR}{2}\left[\lambda_I^2 + \lambda_{II}^2 + \lambda_{III}^2 - 3\right] \tag{11.31}$$

where $R = kN_A$ is the gas constant. The corresponding Helmholtz energy is given by $\Delta F = -T\Delta S$. The expression for the entropy is valid for all stress states. For an uniaxial stress state $\lambda_I = \lambda$ and, since $\lambda_{II} = \lambda_{III}$, we obtain from the incompressibity equation $\lambda_I\lambda_{II}\lambda_{III} = 1$, $\lambda_{II} = \lambda_{III} = \lambda^{-1/2}$. The Helmholtz energy and force thus become

$$\Delta F = -T\Delta S = \frac{1}{2}NRT\left(\lambda^2 + \frac{2}{\lambda} - 3\right) \quad \text{and}$$
$$f_I = \frac{1}{L_{10}}\frac{\partial \Delta F}{\partial \lambda} = \frac{NRT}{L_{10}}\left(\lambda - \frac{1}{\lambda^2}\right) \tag{11.32}$$

To obtain the engineering stress s_I and natural stress σ_I we divide by the area A_0, respectively A resulting in

$$s_I = \frac{NRT}{V_0}\left(\lambda - \frac{1}{\lambda^2}\right) \quad \text{and}$$
$$\sigma_I = \frac{A_0}{A}s_I = \lambda s_I = \frac{NRT}{V}\left(\lambda^2 - \frac{1}{\lambda}\right) \tag{11.33}$$

Finally, for small deformations we can write the stretches in terms of (small displacement gradient) strains by taking the limit $\lambda \to 1$ resulting in $\lambda - 1/\lambda^2 \cong \lambda^2 - 1/\lambda \cong 3\varepsilon_I$ and Young's modulus is thus given by

---

[s] The justification of why the use of $s = k \ln W$ is allowed is complex and we refer to Weiner (1983).

$$E = \frac{3NRT}{V} = \frac{3\rho RT}{M_{\text{sub}}} \tag{11.34}$$

where $\rho = NM_{\text{sub}}/V$ is the density and M_{sub} the average molecular weight between the cross-links (molecular weight of the sub-chains).

Before comparing these results with experimental data a few words must be said on network defects such as loose ends, loops and entanglements. Of these defects loose ends and loops lower the stiffness of the network since these parts are not 'active', i.e. carry no load, while entanglements increase the stiffness since they act similarly as a cross-link. For loose ends Flory in 1952 noted that if there are N' primary molecules cross-linked by $N/2$ chemical reactions, a total of N chains will be formed. Of these there are $2N'$ inactive being loose ends. Hence the number of active chains as corrected for loose ends is $N-2N'$ or $N(1-2N'/N) = N(1-2M_{\text{sub}}/M)$, where M is the molecular weight of the primary molecule. The effect of entanglements (and loops) is incorporated by an empirical factor g. The final expression then becomes

$$E = g\frac{3\rho RT}{M_{\text{sub}}}\left(1-\frac{2M_{\text{sub}}}{M}\right)$$

It appears that this functional dependency on molecular weight M is well supported by experiments. For high molecular weight materials, e.g. M a few hundred thousands, with a reasonable degree of cross-linking, e.g. M_{sub} a few ten thousands, this correction is small but it obviously grows in importance for lower M and higher M_{sub}, i.e. for a lower degree of cross-linking. However, the variation with the degree of cross-linking (or with M_{sub}) does not conform so well to experiment. Using the above expression using independent estimates for M_{sub} and with g as a parameter generally results in g-values between 2 and 4 so that it can be concluded that for the usual low cross-link densities entanglement is dominant and accounts for the larger part of the stiffness. It is, however, not clear whether entanglements and loops are the only explanation for this factor (see Section 11.8). Therefore, in many cases M_{sub} is considered as an empirical parameter although for some well-defined systems the linear dependence of modulus E on the sub-chain density N/V can be described reasonably well using a somewhat more advanced network model (see Section 11.8).

Experimentally the functional relationship Eq. (11.33) is reasonably well obeyed

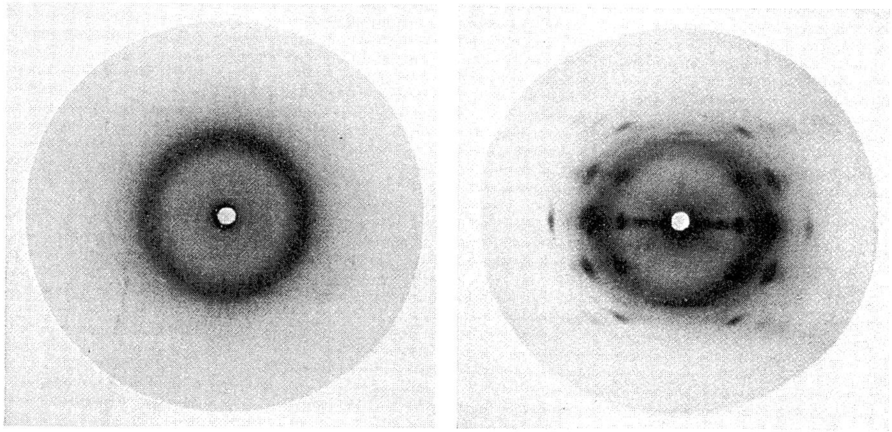

Fig. 11.14: X-ray micrograph of unstretched isoprene rubber and rubber stretched to $\lambda = 7$.

for stretch values from about 0.4 up to about 1.5 (for $\lambda > 1$, see Fig. 11.10, curve a). Moreover the stress-stretch relation for other deformations, e.g. (simple and pure) shear and bi-axial extension, as derived from the Helmholtz expression can be describe the deformation behaviour of a certain rubber with the same modulus. This provides a significant amount of trust in the model. For larger stretches the experimental data increase at first less rapidly but at large stretches more rapidly than theory predicts.

The deviations at lower stretch values ($2 \le \lambda \le 5$) are at present not very well understood but are probably caused by the non-affine deformation of the cross-linked network (Flory, 1989). The discrepancies for $\lambda \cong 5$ and higher are due to the fact that the chains do not behave any longer in Gaussian way (Treloar, 1975) and that stretch induced crystallisation can occur (Flory, 1989). This is illustrated by early micrographs for natural rubber due to Bragg[t] in Fig. 11.14[u]. The non-Gaussian behaviour is dominant while the crystallisation modifies the details of the stress-stretch curve. The 'upswing' is thus not caused by bond elongation and bond angle increase. This effect becomes only important when the chain has been extended to about 97% of its fully extended length of the unstressed chain, the so-called *hydrodynamic length* (Treloar, 1975). Finally, it is interesting to note that the same expression for the stress as a function of stretch has been obtained from pure continuum mechanics considerations (Section 7.7). This emphasises the basic nature of these results. However, for higher order approximations this correspondence is lost.

Problem 11.11

Prove Eq. (11.34) by expanding the factor $\lambda - 1/\lambda^2$ or $\lambda^2 - 1/\lambda$ and approximating the result for small deformations.

Problem 11.12

Show that the force for a single chain f is given by $f = 3kTr/\langle r^2\rangle_0$. Discuss the result.

11.8 Rubber refinements*

In this section we discuss briefly some refinements for the description of the elastic behaviour of rubbers.

First, we note that an exact statistical mechanical model for a polymer chain, including correlated, hindered rotation around the backbone C-C bonds, can be constructed. The solution of that model can be given only in terms of a moment expansion of the probability distribution for the end-to-end distance. Using only the first moment the Gaussian approximation is obtained. However, even the calculation of the first moment is somewhat complex and therefore the equivalent chain is introduced (see Chapter 8). The equivalent chain model is used throughout in polymer science. Smith (1972) presented a clear overview of all this.

---

[t] William Henry Bragg (1862-1942) and William Lawrence Bragg (1890-1971). Father and son English physicists who received the Nobel Prize for physics in 1915 for their study of crystal structures by means of X-rays.

[u] Bragg, W.L. (1933), *The crystalline state*, vol. 1, *A general survey*, G. Bell and Sons, London.

Second, we deal briefly with a more realistic model for the deformation of the network. In the affine network model it is assumed that the cross-links displace during deformation conform the macroscopic deformation. In reality, however, this needs not to be so and for a homogeneous macroscopic deformation only the cross-links at the surface of the deformed volume comply with this requirement. This was recognised as early as 1943 by James and Guth, who introduced the *phantom network model*, in which the cross-links take part in the thermal motion. This corresponds much more closely to the physical reality of the rubbers. The analysis involved is somewhat complex and we omit it here (see e.g. Boyd and Phillips, 1993) but the main conclusions are that:

- The distribution of the positions of the cross-links is a Gauss distribution, of which the width is independent of the strain.
- The average force between cross-links is the same as if they were fixed at their most probable position and behaves as a classical spring. Therefore, for a homogeneous strain the average position of the cross-links will be displaced as if they were embedded in an elastic continuum. This provides a basis for the affine assumption made in the affine network model.
- The average force as exerted by the network is the same whether cross-links are considered free or fixed at their most probable position.

The final result is that the functional relationship for ΔF is similar to the one for the affine network model. However, using the same Gaussian chain assumption as in the affine network model the pre-factor in the expression for ΔF is explicitly dependent on the network structure, contrary to the affine network model. If the pre-factor in the affine network model is given by $\frac{1}{2}NRT$, for the phantom network model the results is $\frac{1}{2}(1-C/N)NRT$, with C the molar number of cross-links. For a perfect network $C = N/2$ and the pre-factor of the phantom network model is thus half of the value of the affine network model. The reduction in ΔF is essentially due to the fluctuation in cross-link positions allowed in the phantom network model. The factor ½ shows up in the expression for force and modulus as well. It is for this model (including the factor ½) that for some systems good agreement between the experimental and theoretical values for E as a function of cross-link density is observed. However, it should be noted that various authors do not agree upon the pro's and con's of both models, see e.g. Smith (1972) and Boyd and Phillips (1993).

Third, let us consider the reference state for the network. In the derivations given in Section 11.7 the end-to-end distance of the uncross-linked state and cross-linked state was taken the same and equal to $\langle r^2\rangle_0^{1/2}$. However, it was noticed by Tobolsky that this is not necessarily the case and a 'front factor' $\eta = \langle r^2\rangle/\langle r^2\rangle_0$ can be introduced to correct for this. Introducing this factor throughout the calculation leads to

$$\Delta F = \tfrac{1}{2}NRT\left[\eta\left(\lambda_{\mathrm{I}}^2 + \lambda_{\mathrm{II}}^2 + \lambda_{\mathrm{III}}^2\right) - 3\right] - NRT\ln\eta^{3/2}\lambda_{\mathrm{I}}\lambda_{\mathrm{II}}\lambda_{\mathrm{III}}$$

and prevents ΔF from going to zero when $\lambda_{\mathrm{I}}\lambda_{\mathrm{II}}\lambda_{\mathrm{III}} \to 1$ unless the network experiences the same T and V conditions as present during network formation. In fact it introduces an energy component. The force f_{I} for uniaxial tension becomes

$$f_{\mathrm{I}} = \frac{1}{L_{10}}\frac{\partial \Delta F}{\partial \lambda} = \frac{NRT\eta}{L_{10}}\left(\lambda - \frac{1}{\lambda^2}\right)$$

However, the factor $\eta = \langle r^2\rangle/\langle r^2\rangle_0$ is, say 1.5, so that its influence on the force and therefore Young's modulus E is limited.

Fourth, we recall that for intermediate values of stretch the agreement with experiment is not so good. Two possible approaches are given in the literature. The first theory considers the fact that at intermediate values of stretch the deformation will not be longer affine. The distribution of the cross-links is no longer an isotropic Gauss function but becomes an anisotropic function. Since the shape of the distribution function of the cross-links becomes elongated in the direction of the applied force, the force increases less rapidly as for the Gauss distribution. The second theory includes explicitly the entanglements between the cross-links. The entanglements have a less permanent character than the cross-links so that with increasing stretch some of them will slip, again leading to a less rapidly increase in force. Boyd and Phillips (1993) provide a brief introduction to both these theories.

A fifth factor deals with the upswing at large values for λ and is related to the statistics. The statistics of the possible configurations was based on a Gaussian distribution of the end-to-end distance r. However, this distribution is only valid if $r = \langle \mathbf{r}^2 \rangle^{1/2}$ is considerably smaller than the contour length nl of the chain. For high values of r/nl Kuhn showed that a parameter β becomes important. This parameter is defined by

$$r/nl = L(\beta) \equiv \coth\beta - \beta^{-1} \qquad \text{or equivalently} \qquad \beta = L^{-1}(r/nl)$$

where $L(\beta)$ is the *Langevin function*. A useful approximation to $L^{-1}(r/nl)$ is given by[v]

$$L^{-1}\left(\frac{r}{nl}\right) = \frac{3r}{l} \bigg/ \left(1 - \frac{r}{nl}\right)^2$$

The probability distribution P for a single sub-chain is given in this approximation by

$$\ln P(r) = C - n\left(\frac{r\beta}{nl} + \ln\frac{\beta}{\sinh\beta}\right) \qquad C = \text{constant}$$

$$= C - n\left[\frac{3}{2}\left(\frac{r}{nl}\right)^2 + \frac{9}{20}\left(\frac{r}{nl}\right)^4 + \frac{99}{350}\left(\frac{r}{nl}\right)^6 + \cdots\right]$$

where the second step represents a series expansion. The Gaussian approximation is actually the first term in the expansion and is thus valid as long as the second term is small as compared to the first or if $(3/10)(r/nl)^2$ is small as compared to 1. At r/nl = 0.3 and 0.5 the Gaussian error is about 3% and 8%, respectively. Estimating the total entropy difference ΔS as before from the contributions of the sub-chains Δs we have[w]

$$\Delta S = \sum \Delta s = \sum \left\{ s\left[\left(\lambda_{\mathrm{I}}^2 x_0^2 + \lambda_{\mathrm{II}}^2 y_0^2 + \lambda_{\mathrm{III}}^2 z_0^2\right)^{1/2}\right] - s\left[\left(x_0^2 + y_0^2 + z_0^2\right)^{1/2}\right]\right\}$$

and the force from

$$f_{\mathrm{I}} = \frac{1}{L_{I0}}\frac{\partial \Delta F}{\partial \lambda_{\mathrm{I}}} = -\frac{1}{L_{I0}}\frac{\partial T\Delta S}{\partial \lambda_{\mathrm{I}}}$$

---

[v] Warner, H.R. (1972), Ind. Eng. Chem. Fund. **11**, 379.

[w] Since for Gaussian statistics the contribution of the x, y and z directions are independent they can be summed independently. This is no longer true for Langevin statistics where a deformation in one direction influences the others. Therefore additivity is an additional approximation. Moreover, the cross-links do not deform affinely any more in view of the non-linearity of the Langevin distribution. Hence this is another approximation.

The details are left as an exercise. Since the Gaussian approximation is quite reasonable, the correction the correction for intermediate stretch values is relatively small and often ignored. Moreover, this method essentially assumes freely jointed chains so that the procedure can only be applied to the equivalent chain model. The above-mentioned results are based on the statistics of a large number of bonds. Treloar has shown that the approximation is in fact quite reasonable down to 25 bonds between the cross-links. Further corrections for non-Gaussian behaviour can be made but we refer to the extensive review by Treloar (1975) for further details.

The last factor to be discussed is related to the morphology. Isotropic behaviour was assumed throughout the deformation but crystallisation introduces anisotropy. As an example the stretch induced crystallisation is illustrated for natural rubber (polymerised cis-isoprene, C_5H_8) in Fig. 11.14. While the unstretched rubber shows an X-ray micrograph with only rings, like that of a liquid or amorphous material and indicating a random distribution of the diffracting units, the stretched sample shows clearly shows discrete diffraction spots, indicating a certain amount of crystallinity for both raw and vulcanised rubber. The pattern begins to appear at $\lambda \cong 2$-3 and remains unaltered in position but grows in intensity as the rubber is stretched. At $\lambda \cong 7$ the degree of crystallinity is estimated as about 30%. The process is reversible, the spots disappearing again when the stress is relaxed. Apparently the extension of the rubber pulls out the entangled long chains until they assume a more or less parallel position, fitting into each other laterally as well as having a regular repetition longitudinally. A monoclinic unit cell results with lattice constants $a = 1.25$ nm, $b = 0.89$ nm and $c = 0.81$ nm and an included angle $\beta = 92°$ (between the b and c axes) containing 4 molecules and 8 isoprene units. An estimate of the 'crystallites' from the size of the diffuse spots is subject to considerable uncertainty but leads to dimensions ranging from 30×20×5 nm to 60×50×20 nm. The detailed incorporation of crystallisation phenomena in the stress-stretch behaviour is complex. Smith (1972) has presented a brief summary. Some authors suggest that crystallisation is relatively unimportant in general (Ward, 1983).

This concludes our brief extension of rubbers. For a clear review of chain behaviour we refer to Smith (1972). Boyd and Phillips (1993) provide a concise introduction to rubbers. The early work is well described by the extensive book by Treloar (1975) while Erman and Mark (1997) provide a more recent treatment.

Problem 11.13

Show that for the inverse Langevin approximation the force f on a single chain is given by

$$f = T\partial s/\partial r = (kT/l)\, L^{-1}(\beta) = (kT/l)\, L^{-1}(r/nl) \qquad \text{and thus that}$$

$$f = (kT/l)\, [3(r/nl) + 9(r/nl)^3/5 + 297(r/nl)^5/175 + \cdots]$$

Note that $r/nl = \lambda/n^{1/2}$ so that the entropy s is given by

$$s = \tfrac{1}{3}NkT[n^{1/2}L^{-1}(r/nl) - \lambda^{3/2}L^{-1}(\lambda^{-1/2}\, n^{-1/2}]$$

11.9 Thermal effects

So far we have discussed only the potential energy of atoms and neglected the kinetic energy. Taking into account the thermal vibrations is generally complex and we restrict ourselves to the thermal expansion coefficient α and the bulk modulus K. We will use only the simplest ideas on vibrations in this section, i.e. independent atomic vibrations, and follow a similar discussion as given by Born and Huang (1954).

So, we have a solid dependent on temperature T and volume V. We assume that the vibrational behaviour of solids is similar to each other and that this behaviour scales as $f(T,V) = Tg(T/\theta)$, where g is a function of the single variable T/θ and θ a parameter, often denoted as *characteristic temperature*, only dependent on the volume, i.e. $\theta = \theta(V)$. Using

$$\left(\frac{\partial g}{\partial \ln\theta}\right)_V = -\frac{T}{\theta}\frac{\partial g}{\partial(T/\theta)} = -\left(\frac{\partial g}{\partial \ln T}\right)_V \tag{11.35}$$

one can show that

$$\frac{\partial f}{\partial V} = \frac{\gamma}{V}\left(T\frac{\partial f}{\partial T} - f\right) \qquad \text{where} \quad \gamma = -\frac{\mathrm{d}\ln\theta}{\mathrm{d}\ln V} \tag{11.36}$$

which is supposed to be a constant, characteristic for the material.

In our simple approach to the thermal behaviour of solids we have two contributions to the Helmholtz energy F. First, let us consider the lattice energy contribution Φ. We assume that we may expand Φ in terms of $V - V_0$, where V_0 is the volume at 0 K, and thus that we may write

$$\begin{aligned}\Phi &= \Phi_0 + \frac{1}{2!}\frac{\partial^2\Phi}{\partial V^2}\bigg|_{V=V_0}(V-V_0)^2 + \frac{1}{3!}\frac{\partial^3\Phi}{\partial V^3}\bigg|_{V=V_0}(V-V_0)^3 + \cdots \\ &= \Phi_0 + \frac{1}{2!}\Phi_0''(V-V_0)^2 + \frac{1}{3!}\Phi_0'''(V-V_0)^3 + \cdots\end{aligned} \tag{11.37}$$

Omitting all terms after Φ_0'', the lattice potential is called *harmonic*. The term Φ_0''' and the higher ones denote *anharmonic terms*. At 0 K the lattice contribution to the bulk modulus $K = V(\partial^2F/\partial V^2)$ is given by $V_0(\partial^2\Phi_0/\partial V^2) = V_0\Phi_0''$ while at any T we have $\Phi'' = \Phi_0'' + \Phi_0'''(V-V_0)$, neglecting higher order terms.

The second contribution stems from the thermal vibrations. We assume that each atom vibrates independently of the others at a *characteristic frequency* $\omega = (\Phi''/m)^{1/2}$, where m is the mass of the atom. This characteristic frequency is equivalent with the *characteristic temperature* θ via $k\theta = \hbar\omega$. For the vibrational contribution F_{vib} to the Helmholtz energy we can write, according to Eq. (11.36),

$$\frac{\partial F_{\text{vib}}}{\partial V} = \frac{\gamma}{V}\left(T\frac{\partial F_{\text{vib}}}{\partial T} - F_{\text{vib}}\right) = \frac{\gamma}{V}\left(-TS_{\text{vib}} - F_{\text{vib}}\right) = -\frac{\gamma}{V}U_{\text{vib}}$$

The equation of state can be obtained from $p = -\partial F/\partial V$. Because $F = \Phi + F_{\text{vib}}$ we have

► $$p + \frac{\partial\Phi}{\partial V} = -\frac{\partial F_{\text{vib}}}{\partial V} = \frac{\gamma}{V}U_{\text{vib}} \tag{11.38}$$

This is the *Mie-Grüneisen equation of state*[x] with γ the *Grüneisen* parameter, the interpretation of which we will come to later. For solids at atmospheric pressure we may neglect p as compared to the other terms and obtain

$$\frac{\partial \Phi}{\partial V} = \frac{\gamma}{V} U_{\text{vib}}$$

Expanding Φ delivers $\partial\Phi/\partial V = \Phi' = \Phi_0' + \Phi_0''(V-V_0) + \cdots = \Phi_0''(V-V_0) + \cdots$ and retaining only linear terms in $(V-V_0)$ we find, meanwhile using $K_0 = V_0\Phi_0''$,

$$\frac{V - V_0}{V_0} = \frac{\gamma U_{\text{vib}}}{V_0 K_0} \qquad (11.39)$$

The (linear) thermal expansion coefficient α is then given by

► $$3\alpha \equiv \frac{1}{V}\frac{\partial V}{\partial T} \cong \frac{\gamma C_V}{V_0 K_0}, \qquad (11.40)$$

an expression often referred to as the *Grüneisen equation.* In this approximation α is thus proportional to C_V. Since for $T \to 0$, $C_V \to 0$ and thus $\alpha \to 0$, conform the third law.

To obtain the isothermal bulk modulus K_T we multiply the equation of state with V and differentiate with respect to V resulting in

$$p + V\frac{\partial p}{\partial V} + \frac{\partial \Phi}{\partial V} + V\frac{\partial^2 \Phi}{\partial V^2} = \gamma \frac{\partial U_{\text{vib}}}{\partial V} \left[= \frac{\gamma^2}{V}\left(TC_V - U_{\text{vib}}\right) \right]$$

where in the last step again use has been made of Eq. (11.36). Using $K_T = -V\partial p/\partial V$ we find

$$K_T = p + \frac{\partial \Phi}{\partial V} + V\frac{\partial^2 \Phi}{\partial V^2} - \frac{\gamma^2}{V}\left(TC_V - U_{\text{vib}}\right) \qquad (11.41)$$

At $T = 0$ K, $C_V = 0$ J/K·mol and $U_{\text{vib}} = 3R\Theta/2$ J/mol (zero-point energy per mole atoms) resulting in $K_T = K_0 + 3(\gamma+\gamma^2)R\Theta/2V_0 \cong K_0$, so that even at 0 K there is in principle a vibrational contribution to the bulk modulus, albeit a small one. In fact for any temperature there are three contributions. The term $V\Phi''$ contains the lattice contribution and an expansion contribution. The term Φ' is also an expansion contribution while the last tem is due to vibrations. To make this clearer an expansion of Φ' and Φ'' in $V - V_0$ to first order is made yielding, again neglecting p,

$$K_T = K_0 + K_0\left(1 + \frac{V_0\Phi_0'''}{\Phi_0''}\right)\frac{V - V_0}{V_0} - \frac{\gamma^2}{V_0}\left(TC_V - U_{\text{vib}}\right) \qquad \text{or using Eq. (11.39)}$$

► $$K_T = K_0 + \left(1 + \frac{V_0\Phi_0'''}{\Phi_0''}\right)\frac{\gamma U_{\text{vib}}}{V_0} - \frac{\gamma^2}{V_0}\left(TC_V - U_{\text{vib}}\right) \qquad (11.42)$$

Here we recognise in the first term a pure lattice term, the second represents the thermal expansion contribution and the third is a pure vibrational contribution.

[x] Grüneisen, E. (1926), Geiger-Scheel's Handbuch der Physik, **10**, 1, Springer, Berlin.

The link between expansion and vibrational contributions occurs via γ. To see that we recall that $\Phi'' = \Phi_0'' + \Phi_0'''(V-V_0) + \cdots$. From this expression and $\omega = (\Phi''/m)^{1/2}$ we find that

$$\frac{\mathrm{d}\omega}{\mathrm{d}V} = \omega\frac{\Phi_0'''}{2\Phi''} \cong \omega\frac{\Phi_0'''}{2\Phi_0''}$$

and for the Grüneisen parameter γ

$$\blacktriangleright \quad \gamma = -\frac{\mathrm{d}\ln\theta}{\mathrm{d}\ln V} = -\frac{V}{\theta}\frac{\mathrm{d}\theta}{\mathrm{d}V} = -\frac{V}{\omega}\frac{\mathrm{d}\omega}{\mathrm{d}V} = -\frac{V}{2}\frac{\Phi_0'''}{\Phi_0''} \tag{11.43}$$

Hence if the vibrations are harmonic, i.e. $\Phi_0''' = 0$, the thermal expansion coefficient α as well as the temperature dependence of the bulk modulus K vanish.

Remains to be discussed the typical values for θ and γ. The characteristic temperature θ ranges from about 100 K for soft metals (like Pb) via 400 to 500 K for engineering metals (like Al and Fe) and 1000 K for ceramic materials (like Al_2O_3) to about 2000 K for diamond. The value for the Grüneisen parameter γ is typically 1 to 2 for many materials.

In conclusion, in this model the thermal expansion coefficient α is proportional to the heat capacity C_V and the bulk modulus K contains a lattice contribution, an expansion contribution and a vibrational contribution. Given the values for θ, γ, V_0 and K_0, one can estimate the temperature dependence of α and K using the proper harmonic oscillator expressions. Even this simple model for the thermal behaviour of solids already leads to somewhat complex equations. The model can, however, explain the qualitative behaviour well.

Problem 11.14

Prove Eqs. (11.35) and (11.36).

11.10 Lattice dynamics*

In Section 11.9 thermal effects were explained in the simplest possible terms. In particular the correlation between atomic vibrations was neglected entirely. In this section a brief introduction into the coupled dynamics of atoms in crystals, usually denoted as *lattice dynamics*, is given which deals with this correlations. For a more elaborate introduction, see Cochran[y] (1973) or Ziman (1972), while Donovan and Angress[z] (1971) present a more complete treatment. We also recommend Born and Huang (1954).

Dispersion relations and density of states

We recall that the potential energy Φ is given as a function of the co-ordinates of all atoms indicated by $\mathbf{r}_{\mathbf{n}\alpha}$ (or in components by $r_{\mathbf{n}\alpha i}$), i.e. $\Phi = \Phi(\mathbf{r}_{\mathbf{n}\alpha})$. As before the label $\mathbf{n}$ indicates one of the N unit cells and the label α one of the r atoms in the unit

---

[y] Cochran, W. (1973), *The dynamics of atoms in crystals*, Edward Arnold, London.
[z] Donovan, B. and Angress, J.F. (1971), *Lattice vibrations*, Chapman and Hall, London.

cell. We now expand Φ for small displacement $\mathbf{u}_{\mathbf{n}\alpha}$ in a Taylor series around the equilibrium position $\mathbf{r}_{\mathbf{n}\alpha}$.

$$\begin{aligned}\Phi(\mathbf{r}_{\mathbf{n}\alpha}+\mathbf{u}_{\mathbf{n}\alpha}) &= \Phi(\mathbf{r}_{\mathbf{n}\alpha})+\tfrac{1}{2}\sum\nolimits_{\mathbf{m}\beta}\mathbf{u}_{\mathbf{n}\alpha}\cdot\frac{\partial^2\Phi(\mathbf{r}_{\mathbf{n}\alpha})}{\partial\mathbf{r}_{\mathbf{n}\alpha}\partial\mathbf{r}_{\mathbf{m}\beta}}\cdot\mathbf{u}_{\mathbf{m}\beta}+\cdots \\ &= \Phi(\mathbf{r}_{\mathbf{n}\alpha})+\tfrac{1}{2}\sum\nolimits_{\mathbf{m}\beta}\mathbf{u}_{\mathbf{n}\alpha}\cdot\Phi_{\mathbf{n}\alpha}^{\mathbf{m}\beta}\cdot\mathbf{u}_{\mathbf{m}\beta}+\cdots\end{aligned} \tag{11.44}$$

The linear terms are absent since the expansion is around the equilibrium, minimum energy position. Higher order terms are neglected for the time being. As before the summation runs over all atoms $\mathbf{n}\alpha$ (and components i). Eq. (11.44) indicates the potential energy for a 'generalised' harmonic oscillator and therefore the approximation is called the *harmonic approximation.* The second derivatives of the potential energy $\Phi_{\mathbf{n}\alpha}^{\mathbf{m}\beta}$, the so-called *coupling constants*, are the generalised spring constants.

The next step is to assume that Newton's laws are valid and this leads to

$$\begin{aligned}&m_\alpha\ddot{\mathbf{u}}_{\mathbf{n}\alpha}+\sum\nolimits_{\mathbf{m}\beta}\Phi_{\mathbf{n}\alpha}^{\mathbf{m}\beta}\mathbf{u}_{\mathbf{m}\beta}=0 \qquad \text{or in components}\\ &m_\alpha\ddot{u}_{\mathbf{n}\alpha i}+\sum\nolimits_{\mathbf{m}\beta j}\Phi_{\mathbf{n}\alpha i}^{\mathbf{m}\beta j}u_{\mathbf{m}\beta j}=0\end{aligned} \tag{11.45}$$

where m_α denotes the mass of the atoms associated with the label α. For rN atoms in total this leads to $3rN$ coupled differential equations. For the coupling constants several relations apply of which the most important one is due to the translational invariance. This invariance implies that the coupling constant can be only dependent on the difference between $\mathbf{n}$ and $\mathbf{m}$ and the relation reads

$$\Phi_{\mathbf{n}\alpha i}^{\mathbf{m}\beta j}=\Phi_{0\alpha i}^{(\mathbf{m}-\mathbf{n})\beta j} \tag{11.46}$$

If we assume periodic boundary conditions the displacements $u_{\mathbf{n}\alpha i}$ can be written as plane waves with wave vector $\mathbf{q}$, frequency ω and amplitude $U_{\alpha i}(\mathbf{q})$ given by

$$u_{\mathbf{n}\alpha i}=(m_\alpha)^{-1/2}U_{\alpha i}(\mathbf{q})\exp[i(\mathbf{q}.\mathbf{r}_\mathbf{n}-\omega t)] \tag{11.47}$$

and substitution in Eq. (11.45) results in

$$\begin{aligned}&-\omega^2U_{\alpha i}(\mathbf{q})+\sum\nolimits_{\beta j}\left\{\sum\nolimits_{\mathbf{m}}(m_\alpha m_\beta)^{-1/2}\Phi_{\mathbf{n}\alpha i}^{\mathbf{m}\beta j}\exp[i\mathbf{q}.(\mathbf{r}_\mathbf{m}-\mathbf{r}_\mathbf{n})]\right\}U_{\beta j}(\mathbf{q})=0\\ &\text{or}\quad -\omega^2U_{\alpha i}(\mathbf{q})+\sum\nolimits_{\beta j}D_{\alpha i,\beta j}(\mathbf{q})U_{\beta j}(\mathbf{q})=0\end{aligned} \tag{11.48}$$

The elements of the matrix $D_{\alpha i,\beta j}$, the so-called *dynamical matrix*, depend only on the difference $\mathbf{r}_\mathbf{m}-\mathbf{r}_\mathbf{n}$ due to the translational invariance. Eq. (11.48) is a linear homogeneous equation of order $3r$. In case we have a primitive unit cell, the summation over atoms $r = 1$ and for every wave vector $\mathbf{q}$ we have to solve a system of three equations. In general such a system has only solutions if the determinant vanishes and thus if

$$\blacktriangleright \qquad \det\left[D_{\alpha i,\beta j}(\mathbf{q})-\omega^2\delta_{\alpha i,\beta j}\right]=0 \tag{11.49}$$

For every wave vector $\mathbf{q}$ this equation has $3r$ solutions $\omega(\mathbf{q},p)$, denoted as *branches* and indicated by the branch label p. Since the dynamical matrix is Hermitian, the eigenvalues $\omega(\mathbf{q},p)$ are real. The dependence $\omega(\mathbf{q},p)$ is known as the *dispersion relation*. The branches for which $\omega(\mathbf{q},p) = 0$ at $\mathbf{q} = 0$ are called the *acoustic branches*, while for $\omega(\mathbf{q},p) \neq 0$ at $\mathbf{q} = 0$ the term *optical branches* is used (not necessarily indicating optical activity). In crystals there are always three acoustic branches while the number of optical branches depends on the number of atoms per unit cell. The eigenvectors, denoted $\mathbf{e}(\mathbf{q},p)$ and labelled by the wave vector $\mathbf{q}$ and the branch index p, are combinations of the waves represented by Eq. (11.47), can be considered as quasi-particles and are addressed as *phonons*.

The requirement that the properties of the lattice should repeat after every $N^{1/3}$ unit cells implies that the displacements $\mathbf{u}_{\mathbf{n}\alpha}$ also must repeat and leads to

$$\exp[iN^{1/3}\mathbf{q}\cdot(\mathbf{a}_1+\mathbf{a}_2+\mathbf{a}_3)] = 1$$

where $\mathbf{a}_1$, $\mathbf{a}_2$ and $\mathbf{a}_3$ denote the lattice vectors. This equation can only be satisfied if $\mathbf{q} = \mathbf{n}N^{-1/3}$ indicating the allowed $\mathbf{q}$-values or *states*. For large N the states are so densely packed that they form a quasi-continuous distribution. The number of states $Z(\omega)$ in the frequency range $\mathrm{d}\omega$ is then given by

$$Z(\omega)\mathrm{d}\omega = \frac{V}{(2\pi)^3}\int_{\omega}^{\omega+\mathrm{d}\omega}\mathrm{d}\mathbf{q} \tag{11.50}$$

and the function $Z(\omega)$ is called the *density of states*. The volume of the unit cell V is given by $V = \mathbf{a}_1\cdot\mathbf{a}_2\times\mathbf{a}_3$. If we separate the wave vector increment $\mathrm{d}\mathbf{q}$ in a length element $\mathrm{d}q$ perpendicular to the surface $\omega(\mathbf{q})$ = constant and an element of surface area on that surface $\mathrm{d}S_\omega$, we can write $\mathrm{d}\mathbf{q} = \mathrm{d}S_\omega \mathrm{d}q$. With $\mathrm{d}\omega = |\nabla_\mathbf{q}\omega|\mathrm{d}q$ one obtains

$$Z(\omega)\mathrm{d}\omega = \frac{V}{(2\pi)^3}\mathrm{d}\omega\int_{\omega=\mathrm{const.}}\frac{\mathrm{d}S_\omega}{|\nabla_\mathbf{q}\omega|} \tag{11.51}$$

from which it follows that a flat region in the dispersion relation corresponds to a singularity in the density of states (*van Hove singularity*).

For the calculation of the elements of the dynamical matrix $\Phi_{\alpha i,\beta j}$ one needs the second derivatives $\Phi_{\mathbf{n}\alpha}^{\mathbf{m}\beta}$ of the potential energy $\Phi(\mathbf{r}_{\mathbf{n}\alpha}+\mathbf{u}_{\mathbf{n}\alpha})$. The latter can be calculated in various ways:

- Empirically using atomic potentials like the Lennard-Jones potential or more elaborate expressions. Sometimes three body contributions are added in order to improve the accuracy and reliability. The parameters in the potentials are typically fitted to various bulk properties like heat of sublimation, bulk modulus, etc.
- Quantum-mechanically using either semi-empirical or ab-initio methods. While the former approach relies on (more or less) simplified quantum schemes with parameters fitted to atomic and/or molecular properties, the latter approach relies fully on theory. The importance of ab-initio methods is increasing due to increasing calculational computer power resulting from faster processors and more extensive memory.

From the solution of the characteristic equation of the dynamical matrix for each of the allowed $\mathbf{q}$-values, the density of states (DOS) can be calculated. This is obviously a non-trivial task and the accuracy of the DOS depends, apart from the quality of the potential used, also on the sampling scheme for $\mathbf{q}$. Generally the resulting dispersion relations are plotted only for a few selected directions in the Brillouin zone. A typical set of theoretical dispersion curves and the accompanying density of states for AlN

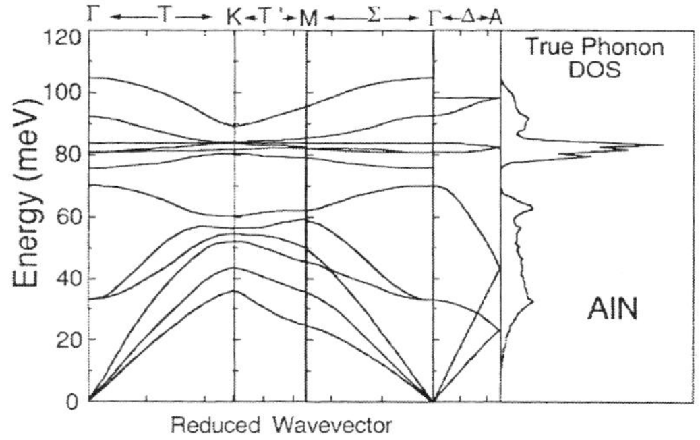

Fig. 11.15: The dispersion relations for AlN for a few directions and the corresponding density of states[aa].

are given in Fig. 11.15. Note the high density of states for the flat part of the dispersion relations around 80 meV, indicating the van Hove singularity. Since for each material a somewhat elaborate calculation is necessary to evaluate the density of states, simple models for $Z(\omega)$ are frequently used. Two such models are presented below.

Albert Einstein (1879-1955)

Born in Ulm, Germany, he was educated at the Zürich Polytechnic. In 1902 he became a clerk at the patent office in Berne and received his doctorate degree at the University of Zurich in 1905 on a new estimation of molecular dimensions. Independently of Gibbs he formulated the basis of statistical mechanics in 1902. Also in 1905 he published the paper *Zur Elektrodynamik bewegter Körper*, which formed the basis for the special theory of relativity. In 1907 he became Privatdocent at the university of Bern, in 1909 assistant professor at the University of Zurich, leaving two years later for the German University at Prague and returning to Zurich in 1912. In 1913 he accepted, courted by Planck, Nernst and Max von Laue (1879-1960) a research position at the Kaiser Wilhelm Institute for Physics and a professorship at the University of Berlin. When in the United States in 1933 he said that he would not return to Germany because the country's citizens no longer enjoyed "civil liberty, tolerance, and equality". This statement was considered slander against the fatherland by the secretary of the Berlin Academy of Sciences and Einstein immediately resigned both his membership of the academy and his German citizenship. He remained in the United States where he held a position at the Princeton Institute for Advanced Study until his retirement in 1948. Einstein could never accept the statistical interpretation of quantum mechanics, vividly worded by *God does not play dice*. He received the Nobel Prize in 1921 for the discovery of the law of the photoelectric effect.

[aa] Nipko, J.C. and Loong, C.-K. (1999), Phys. Rev. **B57**, 10550.

Example 11.1: The Einstein model

The simplest vibrational model one can take, first proposed by Einstein in 1907, assumes independent vibrations with frequency ω_E for all atoms. This implies a density of states $Z(\omega)\,d\omega = 3rN\delta(\omega - \omega_E)\,d\omega$, where $\delta(x)$ indicates the Dirac delta function. The frequency ω_E is denoted as the *Einstein frequency* and the model as such as the *Einstein model*. In fact this is the model as used in Section 11.8.

Example 11.2: The Debye model

Another simple but frequently used model, first proposed by Debye[bb] in 1912, assumes an isotropic elastic continuum with sound velocity v_L for the longitudinal waves and v_T for the two (degenerated) transverse waves. For each branch the surface $\omega(\mathbf{q},p)$ = constant is a sphere. Hence $|\nabla_\mathbf{q}\omega|$ is the sound velocity v_i of branch i, independent of the wave vector $\mathbf{q}$ and the dispersion relation is $\omega = v_i q$. The integration in Eq. (11.51) results then in the surface of the sphere $4\pi q^2$. Consequently, the density of states for each branch becomes

$$Z_i(\omega)\mathrm{d}\omega = \frac{V}{(2\pi)^3}\mathrm{d}\omega\frac{4\pi q^2}{v_i} = \frac{V}{2\pi^2}\frac{\omega^2}{v_i^3}\mathrm{d}\omega \qquad (11.52)$$

and the total density of states becomes

$$Z(\omega)\mathrm{d}\omega = \frac{V}{2\pi^2}\omega^2\left(\frac{1}{v_L^3}+\frac{2}{v_T^3}\right)\mathrm{d}\omega \qquad (11.53)$$

This $Z(\omega)$ not only applies to isotropic elastic continua but also to isotropic crystals at small frequency ω. Since the total number of states is $3rN$, we can normalise $Z(\omega)$ by taking a (common) maximum frequency ω_D to be determined by

$$3rN = \frac{V}{2\pi^2}\left(\frac{1}{v_L^3}+\frac{2}{v_T^3}\right)\int_0^{\omega_D}\omega^2\mathrm{d}\omega = \frac{V}{2\pi^2}\left(\frac{1}{v_L^3}+\frac{2}{v_T^3}\right)\frac{1}{3}\omega_D^3 \quad \text{or} \qquad (11.54)$$

$$\omega_D = \left(\frac{18\pi^2 rN}{V}\right)^{1/3}\left(\frac{1}{v_L^3}+\frac{2}{v_T^3}\right)^{-1/3} \qquad (11.55)$$

The cut-off frequency ω_D is denoted as the *Debye frequency* and the model as such as the *Debye model*. The density of states thus becomes $Z(\omega) = 9rN\omega^2/\omega_D^3$. We note that this model of the vibrational behaviour obeys the scaling assumption $f(T,V) = Tg(T/\theta)$ with $\theta = \theta(V)$ as discussed in Section 11.9, so that Eq. (11.36) and its consequences are also valid for the Debye model.

[bb] Pieter Josephus Wilhelmus Debye (1884-1966). Dutch born American scientist who received the Nobel Prize for chemistry in 1936 for his contribution to our knowledge of molecular structure through his investigations on dipole moments and on the diffraction of X-rays and electrons in gases.

Heat capacity

In order to estimate the energy involved in the dynamics we need the energy expression for a single oscillator. Quantum mechanics tells us that the energy is quantised and given by $u_n = \hbar\omega(n+½)$, where the quantum number $n = 0, 1, 2, \ldots$ and $\hbar = h/2\pi$ where h is Planck's constant. It can be shown further that by evaluating a Boltzmann distribution over the energy levels that the mean energy of an oscillator

$$\varepsilon(\omega,T) = \hbar\omega\left(\frac{1}{\exp(\hbar\omega/kT)-1} + ½\right) \equiv \hbar\omega(\langle n\rangle + ½) \tag{11.56}$$

where $\langle n\rangle$ is the expectation value for n for an oscillator in thermal equilibrium at temperature T. As usual in this context, k is Boltzmann's constant. The specific internal energy contribution $u(T)$ due to the dynamics may now be written as

$$u(T) = (\rho V)^{-1}\int_0^\infty Z(\omega)\varepsilon(\omega,T)\,\mathrm{d}\omega \tag{11.57}$$

while the specific heat capacity at constant volume c_V is given by

$$c_V = \frac{\partial u(T)}{\partial T} = (\rho V)^{-1}\int_0^\infty Z(\omega)\frac{\partial\varepsilon(\omega,T)}{\partial T}\,\mathrm{d}\omega \tag{11.58}$$

We now need the expression for $Z(\omega)$ and we take the Debye model so that

$$c_V = \frac{9rN}{\rho V}\frac{1}{\omega_\mathrm{D}^3}\int_0^{\omega_\mathrm{D}}\omega^2\frac{\partial}{\partial T}\left(\frac{\hbar\omega}{\exp(\hbar\omega/kT)-1}\right)\mathrm{d}\omega \tag{11.59}$$

Introducing the *Debye temperature* θ_D by $\hbar\omega_\mathrm{D} = k\theta_\mathrm{D}$ we obtain with $y = \hbar\omega/kT$

► $$c_V = \frac{3rNk}{\rho V}\left[\frac{3}{(\theta_\mathrm{D}/T)^3}\int_0^{\theta_\mathrm{D}/T}\frac{y^4\mathrm{e}^y}{(\mathrm{e}^y-1)^2}\mathrm{d}y\right] \tag{11.60}$$

and where the term in brackets is typically referred to as a *Debye function*. For $kT >> \hbar\omega_\mathrm{D}$, the upper limit is small, the integrand may be expanded and c_V reduces to

$$c_V = \frac{3rNk}{\rho V} \qquad (T >> \theta_\mathrm{D}) \tag{11.61}$$

the well-known *law of Dulong and Petit*. For $kT << \hbar\omega_\mathrm{D}$, one obtains by extending the upper limit of the integration from θ_D/T to ∞

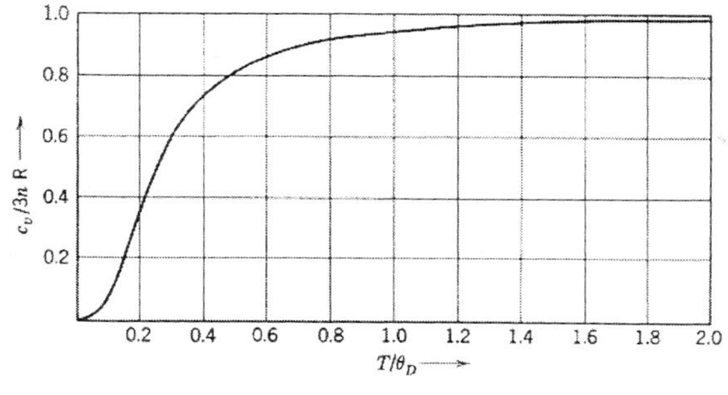

Fig. 11.16: The normalised heat capacity c_V as a function of T/θ_D for the Debye model.

Table 11.8: Debye temperatures θ_D (K) for various compounds.

Ar	95	Fe	474
Pb	105	Cr	630
Au	165	Si	640
KCl	235	$CaTiO_3$	750
Pt	240	Al_2O_3	1030
Nb	275	$MgSiO_3$	1094
NaCl	321	SiO_2	1217
Cu	343	Be	1440
Al	428	C	2230

$$c_V = \frac{9rNk}{\rho V}\frac{4\pi^4}{15}\left(\frac{T}{\theta_D}\right)^3 \qquad (T << \theta_D) \tag{11.62}$$

also experimentally well established. The overall behaviour is shown in Fig. 11.16. Table 11.8 provides a selected set of Debye temperatures[cc]. From Fig. 11.15 it is clear that the Debye density of states is well obeyed at low **k**. In this particular case Debye-like behaviour observed only below about 30 meV while the whole spectrum extends to about 100 meV. In spite of the fact that the Debye density of states is far from realistic, the predictions and experimental values for the heat capacity match closely if the material contains only one kind of atoms (metals) or atoms of similar mass and bonding characteristics. This is due to the fact that the heat capacity, being a property averaged over the density of states, is rather insensitive to the precise distribution of states. If the bond type, co-ordination and mass of the various atoms in the unit cell become more and more different, the Debye approximation becomes increasingly inadequate. However, the Debye approximation is frequently used in view of its (relative) simplicity, in spite of the relative large errors involved[dd].

Problem 11.15

Consider the Einstein density of states. Show that the heat capacity is given by

$$c_V = 3rNk\left[\frac{y^2}{\left(e^y - 1\right)^2}\right]$$

where $y = \hbar\omega_E / kT$.

a) Show that in the high temperature limit the expansion for c_V reduces also to the law of Dulong and Petit.
b) Show that in the low temperature limit the expansion for c_V reduces to the expression $c_V \approx T^{-2}\exp(\hbar\omega_E/kT)$.
c) Estimate the Einstein temperature, defined by $\hbar\omega_E = k\theta_E$, as a fraction of the Debye temperature θ_D taking the same values of c_V at θ_D, $\theta_D/2$ and $\theta_D/4$.

[cc] It must be said that various definitions of θ_D are possible. For an overview, see Grimvall (1986).
[dd] van der Laag, N.J., Fang, C.M. and de With, G. (2003), unpublised note.

Thermal expansion

So far we discussed the thermal behaviour within the harmonic approximation. Since the movement of atoms in this approximation occurs in a symmetrical potential well, the equilibrium position is independent of T. To describe thermal expansion we thus need higher order terms. However, while the calculation of the second derivatives is already somewhat involved, the calculation of the third order terms is even more complex and incorporating higher order terms complicates the solution of the dynamical equation tremendously. In fact the principal axes transformation to obtain the dispersion relations cannot be applied any longer successfully. The usual procedure is therefore to use the harmonic solutions as a basis for perturbation theory to obtain more accurate answers. The approach is complex and we refer to the literature[ee].

Here we will use another approach generally referred to as the *quasi-harmonic approximation.* To that purpose we need the partition function $Q = \Sigma\exp(-u_i/kT)$ (where the sum runs over all quantum states i) and the corresponding Helmholtz energy $F = -kT \ln Z$. Consider for a moment a single oscillator of which the partition function Z_{vib} is given by

$$Z_{\text{vib}} = \sum_n \exp\left(-\frac{u_n}{kT}\right) = \sum_n \exp\left[\frac{-\hbar\omega(n+½)}{kT}\right] = \frac{e^{-y/2}}{1-e^{-y}} \tag{11.63}$$

where $y = \hbar\omega/kT$ has been used again, and the vibrational Helmholtz energy F_{vib} by

$$F_{\text{vib}} = -kT\ln Z_{\text{vib}} = ½\hbar\omega + kT\ln\left[1-\exp\left(-\frac{\hbar\omega}{kT}\right)\right] \tag{11.64}$$

The vibrational Helmholtz energy is independent of the displacement u from the equilibrium position and we confirm that there is no thermal expansion. In the quasi-harmonic approximation the vibrational energy of the oscillator is still given by $u_n = \hbar\omega(n+½)$ but the frequency ω is taken a function of the time averaged position a. For the potential energy we take

$$\phi = \phi_0 + ½f(a-a_0)^2 \tag{11.65}$$

where a_0 denotes the position of the potential energy minimum and f the force constant. The total Helmholtz energy $F = F_{\text{vib}}+\phi$ should be minimal and thus $\partial F/\partial a = 0$. Inserting Eq. (11.64) and Eq. (11.65) results in

$$f(a-a_0) + \frac{1}{\omega}\frac{\partial\omega}{\partial a}\varepsilon(\omega,T) = 0 \qquad \text{or} \qquad a = a_0 - \frac{1}{f\omega}\frac{\partial\omega}{\partial a}\varepsilon(\omega,T) \tag{11.66}$$

The (linear) thermal expansion coefficient α is thus

$$\alpha \equiv \frac{1}{a}\frac{\partial a}{\partial T} = -\frac{1}{fa^2}\frac{\partial\ln\omega}{\partial\ln a}\frac{\partial\varepsilon(\omega,T)}{\partial T} \tag{11.67}$$

For a 3D solid, containing many of these oscillators, we only have to substitute VK_T for a^2f (where the bulk modulus K_T is given by $K_T = -V(\partial p/\partial V)_T$ and sum over all wave vectors $\mathbf{q}$ and branches p to obtain

[ee] Leibfried, G. (1961), Solid State Physics **12**, F. Seitz and D. Turnbull, eds., Academic Press, New York, page 276. See also Wallace, D.C. (1972), *Thermodynamics of crystals*, Wiley, New York (also Dover, 1998).

► $$\alpha = -\frac{1}{3K_T V}\sum_{\mathbf{q}p}\frac{\partial \ln\omega}{\partial \ln V}\frac{\partial}{\partial T}\varepsilon[\omega(\mathbf{q},p),T] \tag{11.68}$$

where the factor 3 results from $\mathrm{d}(\ln V) = 3\mathrm{d}(\ln a)$. Obviously the expansion coefficient α behaves (generally but not always) similarly to c_V due to the factor $\partial\varepsilon/\partial T$. The quantities $\gamma(\mathbf{q},p) = -\partial\ln\omega(\mathbf{q},p)/\partial\ln V$ are the *Grüneisen parameters*. They show typically a weak dependence on $\omega(\mathbf{q},p)$. For the Debye solid there is only one Grüneisen parameter $\gamma = -\partial\ln\theta_D/\partial\ln V$. Also for real solids often a single value for γ is used of which the value for many solids is typically between 1 and 2. In case a single γ describes the material behaviour we obtain

$$\alpha = \frac{\gamma c_V}{3K_T V} \qquad \text{or using } \frac{K_S}{K_T} = \frac{c_P}{c_V} \text{ (see Chapter 6)} \qquad \alpha = \frac{\gamma c_P}{3K_S V} \tag{11.69}$$

a relation generally referred to as the *Grüneisen relation*. The parameter γ is in this case addressed as the *thermodynamic Grüneisen parameter*. It should be noted that, dependent on the details of the derivation, expressions with V_0K, VK_0 or V_0K_0 also are given in the literature.

Within the quasi-harmonic approximation the heat capacity at constant pressure c_p can be expressed explicitly in terms of c_V. To that purpose we note first that from thermodynamics we have

$$c_p = T\left(\frac{\partial S}{\partial T}\right)_p = T\left(\frac{\partial S}{\partial T}\right)_V + T\left(\frac{\partial S}{\partial V}\right)_T\left(\frac{\partial V}{\partial T}\right)_p = c_V + T\left(\frac{\partial S}{\partial V}\right)_T 3\alpha V$$

Second, we want to introduce γ and therefore we rewrite $3\alpha = \gamma c_V/VK_T$ as $\gamma c_V = 3\alpha VK_T$ and consider the product $3\alpha K_T$

$$3\alpha K_T = -\frac{1}{V}\left(\frac{\partial V}{\partial T}\right)_p V\left(\frac{\partial p}{\partial V}\right)_T$$

$$= -\left(\frac{\partial V}{\partial T}\right)_p\left(\frac{\partial p}{\partial V}\right)_T\left(\frac{\partial T}{\partial p}\right)_V\left(\frac{\partial p}{\partial T}\right)_V = \left(\frac{\partial p}{\partial T}\right)_V = \left(\frac{\partial S}{\partial V}\right)_T$$

Here in the third step the −1-rule for p, V and T, i.e. $(\partial V/\partial T)_p(\partial p/\partial V)_T(\partial T/\partial p)_V = -1$, is used while in the last step the Maxwell relation $(\partial S/\partial V)_T = (\partial p/\partial T)_V$ (as obtained from $\mathrm{d}F = -S\mathrm{d}T - p\mathrm{d}V$) is inserted. Therefore, we have $\gamma c_V = 3\alpha VK_T = V(\partial S/\partial V)_T$ and the expression for c_p becomes

$$c_p = c_V(1 + 3\alpha\gamma T)$$

Elastic constants

Another effect of anharmonicity is the temperature (and pressure) dependence of the elastic constants. Typically the value of Young's modulus E for crystalline solids depends only slightly on temperature but decreases with increasing temperature. Near absolute zero the temperature derivative approaches zero. Above about half the Debye temperature the temperature derivative of E becomes a constant. It has been shown experimentally that Young's modulus of many oxides can be reasonably well described[ff] by

[ff] Wachtman, J.B., Tefft, W.E., Lam, D.G. and Apstein, C.S. (1961), Phys. Rev. **122**, 1754.

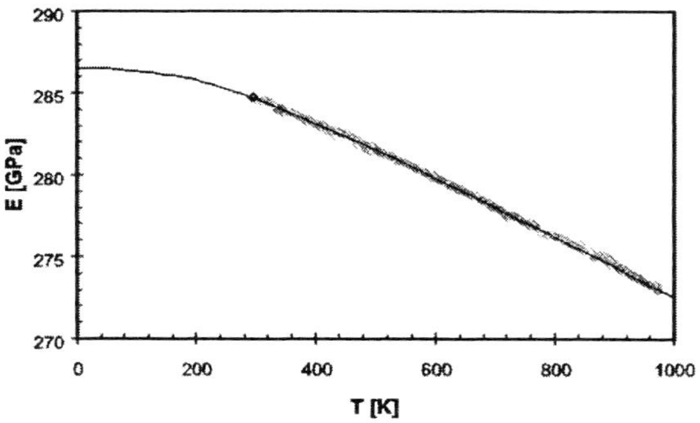

Fig. 11.17: The temperature dependence of Young's modulus E for polycrystalline $MgSiN_2$ as measured between 293 and 973 K and the Wachtman fit $E = 286.6 - 0.0195T\exp(-326/T)$ (solid line).

$$E = E_0 - bT\exp(-T_0 / T) \tag{11.70}$$

In fact the empirical relationship Eq. (11.70) was rationalised by Anderson[gg] using considerations based on the Mie-Grüneisen equation of state, the Mie potential and Debye vibrational behaviour. His final result reads

$$\blacktriangleright \quad K_S = K_0 - \frac{\gamma\delta}{V_0}U_{\text{vib}} \tag{11.71}$$

As before γ denotes the Grüneisen constant and the subscript refers to the properties at 0 K. The parameter δ, which is often referred to as the *Grüneisen-Anderson parameter*, is, like γ, approximately constant. The full derivation is given in Justification 11.2. The vibrational energy U_{vib} is evaluated in the Debye approximation. From Eq. (11.57) we obtain with $y = \hbar\omega/kT$

Table 11.9: Several values of E, $E^{-1}(\mathrm{d}E/\mathrm{d}T)$, α, γ and θ_D.

Material	E (GPa)	$E^{-1}(\mathrm{d}E/\mathrm{d}T)$ (10^{-4}/K)	α (10^{-6}/K)	γ (–)	θ_D (K)
Alumina	410	–1.4	6.2	1.3	1030
Magnesia	320	–1.9	12.0	1.6	900
Thoria	260	–1.4	8.9	–	560
Yttria	140	–1.1	8.3	–	735
Fused silica	74	–	0.41	0.03	470
SiC	415	–	4.7	1.6	–
Si_3N_4	310	–	2.7	0.6	1248
Aluminium	70	–4.4	23.1	2.1	428
Copper	110	–3.4	16.5	2.0	343
Iron	207	–2.7	11.8	1.8	467
Lead	17.9	–10.4	28.9	2.4	105
Titanium	103	–6.7	8.6	1.2	420
Nickel	207	–3.3	12.4	1.6	450
PMMA	2.9	–	50	0.51	–
PS	3.3	–	50	–	–
PP	1.3	–	81	–	–

[gg] Anderson, O.L. (1966), Phys. Rev. **144**, 553.

$$U_{\text{vib}}(T) = \int_0^\infty Z(\omega)\varepsilon(\omega,T)\mathrm{d}\omega = 3rNkT\left[\frac{3}{(\theta_{\text{D}}/T)^3}\int_0^{\theta_{\text{D}}/T}\frac{y^3}{\mathrm{e}^y-1}\mathrm{d}y\right] \qquad (11.72)$$

where $Nk = R$ is the gas constant and the term in brackets is again referred to as a Debye function. We now compare this expression with the empirical relationship $K = K_0 - bT\exp(-T_0/T)$ and conclude that the exponential is an approximation to the vibrational energy contribution. Therefore $b \cong 3r\gamma\delta R/V_0$. To interpret T_0 we note that for high temperature the Debye function may be approximated by $y/[\exp(y)-1]$ which on its turn can be expanded as

$$\frac{y}{\mathrm{e}^y-1} = \left[1+\left(\frac{\theta_{\text{D}}/2}{T}\right)+\frac{2}{3}\left(\frac{\theta_{\text{D}}/2}{T}\right)^2+\cdots\right]^{-1}$$

Now also expanding the exponential $\exp(-T_0/T)$ the result is

$$\mathrm{e}^{-T_0/T} = \left[1+\frac{T_0}{T}+\frac{1}{2}\left(\frac{T_0}{T}\right)^2+\cdots\right]^{-1}$$

and we conclude that $T_0 \cong \theta_{\text{D}}/2$, explaining the empirical finding mentioned before. For MgO Anderson obtained good agreement with the experimental data using the independently determined values $\gamma = 1.53$, $\delta = 3.1$, $V_0 = 5.56$ g/cm^3 and $\theta_{\text{D}} = 930$ K but using K_0 as an empirical parameter with value $K_0 = 166$ GPa.

Justification 11.2

Since the Debye model, like the Einstein model, obeys the scaling relation $f(T,V) = Tg(T/\theta)$ with $\theta = \theta(V)$ we can use the results of Section 11.9 and in particular the expression for the isothermal bulk modulus K_T given by Eq. (11.41). However, in most cases ultrasonic techniques are used which measure the adiabatic bulk modulus K_S instead of K_T and therefore we use the thermodynamic relation $K_S/K_T = c_p/c_V$ (see Chapter 6) which in the quasi-harmonic approximation reads

$$K_S/K_T = c_p/c_V = (1+3\alpha\gamma T) \qquad \text{with} \quad 3\alpha = \gamma c_V/V_0K_0$$

We thus have

$$K_S = K_T + 3\alpha\gamma TK_T = p + \Phi' + V\Phi'' - \frac{\gamma^2}{V}(Tc_V - U_{\text{vib}}) + \frac{\gamma^2 Tc_V K_T}{V_0K_0}$$

Approximating K_T/V_0K_0 by $1/V$ and using the equation of state $\gamma U_{\text{vib}}/V = \Phi' + p$ the result is

$$K_S = (1+\gamma)p + (1+\gamma)\Phi' + V\Phi'' \cong (1+\gamma)\Phi' + V\Phi''$$

a result that Grüneisen obtained by direct calculation from the equation state by differentiation at constant entropy. In the last step the contribution $(1+\gamma)p$ is neglected in comparison with the other terms, which is allowed for the experimental accessible pressure range.

So far the result is general and we now specialise by using the Mie potential

$$\Phi = \frac{N_A}{2}\left(-\frac{C_m}{r^m}+\frac{B_n}{r^n}\right) \equiv -\frac{C}{r^m}+\frac{B}{r^n} = -\frac{C}{V^{m/3}}+\frac{B}{V^{n/3}}$$

and we easily obtain

$$\Phi' = \frac{m}{3}\frac{C}{V^{\frac{m}{3}+1}} - \frac{n}{3}\frac{B}{V^{\frac{n}{3}+1}} \quad \text{and} \quad V\Phi'' = -\frac{m}{3}\left(\frac{m}{3}+1\right)\frac{C}{V^{\frac{m}{3}+1}} + \frac{n}{3}\left(\frac{n}{3}+1\right)\frac{B}{V^{\frac{n}{3}+1}}$$

From the equilibrium condition at 0 K, $\Phi' = \partial\Phi/\partial V = 0$, one derives that $B = mV_0^{(n-m)/3}C/n$. Introducing further the modulus at 0 K, $K_0 = m(n-m)C/9V_0^{(m/3)+1}$, the result for K_S after some calculation becomes

$$K_S = K_0\left[\frac{3\gamma - m}{n-m}\left(\frac{V_0}{V}\right)^{\frac{m}{3}+1} - \frac{3\gamma - n}{n-m}\left(\frac{V_0}{V}\right)^{\frac{n}{3}+1}\right]$$

This expression cannot be simplified unless a further approximation is made. Because $\Delta V = V - V_0 << V$ we may write $(V_0/V)^x = (1-\Delta V/V)^x \cong (1 - x\Delta V/V)$ and using this approximation for both power terms we obtain after some further calculation[hh]

$$K_S = K_0\left[1-\left(\frac{n+m+3}{3}-\gamma\right)\frac{\Delta V}{V}\right] \equiv K_0\left[1-\delta\frac{\Delta V}{V}\right]$$

which, using the same approximation in reverse, may be written as

$$K_S = K_0\left[1-\delta\frac{\Delta V}{V}\right] \cong K_0\left[1-\frac{\Delta V}{V}\right]^\delta = K_0\left(\frac{V_0}{V}\right)^\delta$$

an expression first given by Grüneisen. From this last expression we derive

$$\frac{\mathrm{d}K}{K} = -\delta\frac{\mathrm{d}V}{V} \quad \text{or} \quad \frac{\mathrm{d}\ln K/\mathrm{d}T}{\mathrm{d}\ln V/\mathrm{d}T} = \frac{(1/K)\mathrm{d}K/\mathrm{d}T}{(1/V)\mathrm{d}V/\mathrm{d}T} = -\delta \qquad (11.73)$$

identifying the meaning of the Grüneisen-Anderson parameter δ as the ratio of the relative temperature coefficients of K and V. Since γ is assumed (and appears approximately) to be constant, δ is also a constant. Although using a slightly different procedure, in essence Anderson used $K_S = K_0(1-\delta\Delta V/V)$, replaced $\Delta V/V$ by $\Delta V/V_0$ and substituted $\Delta V/V_0 = \gamma U_{\text{vib}}/V_0K_0$ (Eq. (11.39) so that

$$K_S = K_0 - \frac{\gamma\delta}{V_0}U_{\text{vib}}$$

This concludes our excursion to the temperature dependence of the bulk modulus for inorganic materials. It will have become clear that even in this approximate treatment the argument is somewhat complex but we note again that the overall

[hh] Grüneisen (1912, Ann. Physik **39**, 257) defined the bulk modulus as $K^* = -V_0(\partial p/\partial V)$, while we now generally use $K = -V(\partial p/\partial V)$ and this leads to an extra factor V_0/V with as result that the associated parameter becomes $\delta^* = (m+n+6)/3-\gamma$. Anderson inconsistently used δ^* with K.

behaviour can be explained qualitatively very well, if not semi-quantitatively. However, it appears difficult to proceed further with this general discussion and detailed calculations for specific materials are based on complex calculations (see e.g. Example 8.3).

11.11 Bibliography

Born, M. and Huang, K. (1954), *Dynamical theory of crystal lattices*, Oxford University Press, Oxford.

Boyd, R.H. and Phillips, P.J. (1993), *The science of polymer molecules*, Cambridge University Press, Cambridge.

Erman, B. and Mark, J.E. (1997), *Structure and properties of rubberlike networks*, Oxford University Press, New York.

Flory, P.J (1989), *Statistical mechanics of chain molecules*, Hanser, New York.

Gedde, U.W. (1995), *Polymer physics*, Chapman and Hall, London.

Grimvall, G. (1986), *Thermophysical properties of materials*, North-Holland, Amsterdam.

Treloar, L.R.G. (1975), *The physics of rubber elasticity*, 3rd ed., Clarendon, Oxford.

Smith, K.T. (1972), p. 323 in *Polymer science*, A.D. Jenkins, ed., North-Holland, Amsterdam.

Strobl, G. (1997), *The physics of polymers*, Springer, Berlin.

Ward, I.M. (1983), *Mechanical properties of solid polymers*, 2nd ed., Wiley, Chichester.

Weiner, J.H. (1983), *Statistical mechanics of elasticity*, Wiley, New York.

Ziman, J.M. (1972), *Principles of the theory of solids*, Cambridge University Press, London.

12

Microstructural aspects of elasticity

After having discussed the molecular basis of elastic behaviour in the previous chapter, in this chapter the microstructural aspects of elastic behaviour are dealt with. Again the division into four categories: inorganics, metals, polymers and composites, is kept and we present some general considerations first. We conclude with a brief discussion of some first principles aspects.

12.1 Basic models

Both metals and inorganic materials generally are polycrystalline. In a polycrystalline material the anisotropy as present in single crystals can be averaged out to yield an isotropic material although also polycrystalline materials with a preferred orientation can be made (Grimvall, 1986). Since the individual grains remain anisotropic the calculation of the average is not straightforward. In an exact calculation the compatibility between the grains has to be maintained which is far from trivial. Therefore, a number of approximate and bounding procedures have been developed which nevertheless can yield quite accurate predictions for polycrystals.

The elastic equations for isotropic materials can be treated decoupled when using the isotropic and deviatoric parts of the strain tensor. The appropriate elastic constants are the bulk modulus K and shear modulus G. We can therefore expect that the calculation of these quantities is the simplest. For cubic crystals the relevant elastic stiffness constants are C_{11}, C_{12} and C_{44}. We then have to calculate from these values the value of the effective bulk and shear modulus.

Voigt in 1910 assumed that the strain is uniform throughout the polycrystal (Fig. 12.1). The bulk and shear modulus for cubic crystals then become

$$K_V = (C_{11} + 2C_{12})/3 \quad \text{and} \quad G_V = (C_{11} - C_{12} + 3C_{44})/5$$

Similarly, Reuss in 1929 assumed that the stress is uniform throughout the polycrystal (Fig. 12.1). In this case the bulk and shear modulus become

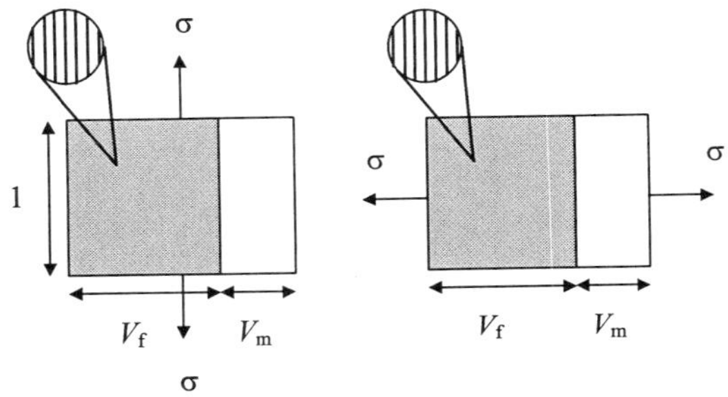

Fig. 12.1: Voigt or parallel model and Reuss or series model.

$K_R = (C_{11} + 2C_{12})/3$ and $G_R = 5(C_{11} - C_{12})C_{44}/[4C_{44} + 3(C_{11} - C_{12})]$

For cubic lattices it thus holds that $K_V = K_R$. Hill[a] has shown that the *Voigt estimate* and *Reuss estimate* provide upper and lower bounds on K and G for any crystal structure (see Justification 12.1). Hence

$$K_R \leq K \leq K_V \quad \text{and} \quad G_R \leq G \leq G_V \tag{12.1}$$

From the general relation between E, G and K

$$\frac{1}{E} = \frac{1}{3G} + \frac{1}{9K} \tag{12.2}$$

it is clear that the smallest possible value for E is given with G_R and K_R. This value is denoted by E_R. Similarly, the largest possible value is obtained by using G_V and K_V and it is denoted by E_V. It follows that

$$E_R \leq E \leq E_V$$

Improved bounds are available, in particular through the work of Hashin and Shtrikman[b]. For cubic crystals the moduli C_{11}, C_{12} and C_{44} are used. Alternatively one may define

$$K = (C_{11} + 2C_{12})/3 \qquad \mu = (C_{11} - C_{12})/2 \qquad \text{and} \qquad \mu' = C_{44}$$

where K represents the bulk modulus and μ and μ' are two different shear moduli. The bounds for the shear modulus read, when $\mu' < \mu$,

$$\mu^{(-)} = \mu + 3\left[\frac{5}{\mu' - \mu} + \frac{12(K + 2\mu)}{5\mu(3K + 4\mu)}\right]^{-1} \quad \text{and}$$

$$\mu^{(+)} = \mu + 2\left[\frac{5}{\mu - \mu'} + \frac{18(K + 2\mu')}{5\mu'(3K + 4\mu')}\right]^{-1}$$

When $\mu' > \mu$, the lower bound becomes the upper bound and vice versa. These bounds provide a considerable improvement over the Reuss-Voigt bounds. Also bounds for non-cubic crystals are available. Generally the expressions are somewhat complicated and we do not discuss them here and refer to Mura (1987).

Hill has suggested to combine the Voigt and Reuss values in the arithmetic average for an estimate of the moduli for polycrystalline materials. Usually these averages are referred to as the *Voigt-Reuss-Hill estimates*:

$$K_{VRH} = (K_V + K_R)/2 \quad \text{and} \quad G_{VRH} = (G_V + G_R)/2$$

Clearly the estimate E_{VRH} as obtained from K_{VRH} and G_{VRH} using Eq. (12.2) is not the same as the estimate $(E_R + E_V)/2$.

Example 12.1

For single crystalline MgO the elastic stiffness constants are given by:

[a] Hill, R. (1952), Proc. Phys. Soc. A**65**, 344.

[b] Hashin, Z. and Shtrikman, S. (1963), J. Mech. Phys. Solids **11**, 127.

$C_{11} = 289.2$ GPa $\qquad C_{12} = 88.0$ GPa $\qquad C_{44} = 154.6$ GPa

Therefore, we calculate

$$K_V = K_R = (C_{11} + 2C_{12})/3 = 155.1 \text{ GPa}$$

$$G_V = (C_{11} - C_{12} + 3C_{44})/5 = 133.0 \text{ GPa}$$

$$G_R = 5(C_{11} - C_{12})\, C_{44}/[4C_{44} + 3(C_{11} - C_{12})] = 127.3 \text{ GPa}$$

resulting in $K_{VRH} = (K_V + K_R)/2 = 155.1$ and $G_{VRH} = (G_V + G_R)/2 = 130.2$ GPa. From $1/E = 1/3G + 1/9K$ it follows that $E_{VRH} = 305.2$. Similarly $E_R = 300.0$ GPa and $E_V = 310.3$ GPa and thus that $(E_R + E_V)/2 = 305.2$ GPa. In this particular case the estimates do not differ since the elastic anisotropy of single crystalline MgO as expressed by $A = 2C_{44}/(C_{11}–C_{12}) = 1.54$ is relatively small. The experimental value for Young's modulus for fully dense polycrystalline MgO is 305 GPa.

The approach discussed above can also be used for composites. In that case the Voigt and Reuss estimates for the bulk modulus are given by

► $K_V = f_1 K_1 + f_2 K_2$ $\qquad$ and $\qquad 1/K_R = f_1/K_1 + f_2/K_2$

where f_1 and $f_2 = 1 - f_1$ denote the volume fractions of phase 1 and 2, respectively, and K_1 and K_2 their bulk moduli. For the shear modulus corresponding equations hold.

► $G_V = f_1 G_1 + f_2 G_2$ $\qquad$ and $\qquad 1/G_R = f_1/G_1 + f_2/G_2$

These estimates also yield upper and lower bounds[c]. The Voigt expression is an example of the *rule-of-mixtures*, which for any property X reads

► $$X = f_1 X_1 + f_2 X_2 \tag{12.3}$$

From the general relation between E, G and K we obtain

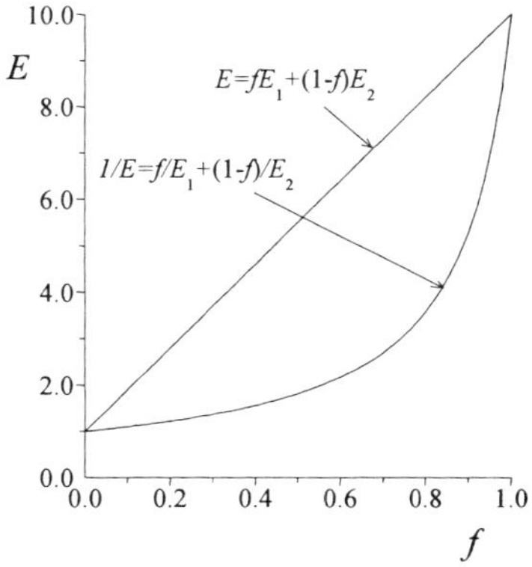

Fig. 12.2: Young's modulus E for a composite with $E_1 = 1$, $E_2 = 10$ and $\nu_1 = \nu_2$.

---

[c] Paul, B. (1960), Trans. ASME **218**, 36.

$$1/E_R = 1/(3G_R) + 1/(9K_R) \quad \text{and} \quad 1/E_V = 1/(3G_V) + 1/(9K_V)$$

resulting again in

$$E_R \le E \le E_V$$

One always has $1/E_R = f_1/E_1 + f_2/E_2$ but generally $E_V \neq f_1 E_1 + f_2 E_2$, except when $G_1K_2 = K_1G_2$ (cf. Problem 12.2) which includes the special case $\nu_1 = \nu_2$ shown in Fig. 12.2. For fibre composites, however, one often uses $E_V = f_1 E_1 + f_2 E_2$, even if the condition $G_1K_2 = K_1G_2$ is not met. The above method yields satisfactory results in case the properties of the constituting phases are not too widely different.

For the calculation of the elastic properties of polycrystals composites several other semi-empirical and first principle models are available in the literature. The method of solving the equations for the latter models is complex and the final equations are somewhat lengthy. For some considerations about effective behaviour and details on first principles and semi-empirical methods, see Sections 12.6 and 12.7.

Woldemar Voigt (1850-1919)

Born in Leipzig, Germany, and educated at the Leipzig University. In 1874 he prepared his thesis on the elastic properties of rock salt and in 1875 he became an assistant of Neumann. In 1893 he was elected to a chair in Göttingen where he lectured on theoretical physics and set up a laboratory where work on elasticity played an important role. For example, he solved the old controversy on whether elastic isotropy is to be described by one or two constants. Dealing with theoretical problems in torsion and bending of prisms cut out of crystals he extended St. Venant's theory. He was the first to introduce tensor notation in the theory of elasticity. His numerous publications in this area were assembled in his *Lehrbuch der Kristallphysik*, published in 1910 and still referred to today. He also worked on the ultimate strength of materials and observed that that the tensile strength greatly depends upon the orientation of the axis of the specimen with respect to the crystal's axes. He finally came to the conclusion that the question of strength is too complicated.

Problem 12.1

Calculate E_{VRH} and $(E_R + E_V)/2$ for polycrystalline $Mn_{0.45}Zn_{0.50}Fe_{2.05}O_4$, using the data for single crystals as given in Chapter 7, and compare these estimates with the experimental result[d] E = 178 GPa. Comment on the results.

[d] de With, G. and Damen, J.P.M. (1981), J. Mater. Sci. **16**, 838.

Problem 12.2

Show that $E_V = f_1E_1 + f_2E_2 + \dfrac{27f_1f_2(G_1K_2 - G_2K_1)^2}{(3K_V + G_V)(3K_1 + G_1)(3K_2 + G_2)}$. Calculate the relative difference $\delta = [E_V - (f_1E_1 + f_2E_2]/(f_1E_1 + f_2E_2)$ for a metal-ceramic (metal $\nu_1 = 0.33$ ceramic $\nu_1 = 0.2$) composite as a function of the volume fraction f_1 of the metal and $x = E_1/E_2$. Determine the maximum of δ for the three material combinations Al_2O_3-Al, Al_2O_3-Fe and Al_2O_3-Cu.

12.2 Inorganics

As discussed in Chapter 1 the microstructure of an inorganic material contains several features. Inorganic materials are very often polycrystalline so that grains with different orientations are present. Often the material is not fully dense but contains some porosity. Moreover microcracks, due to differential thermal expansion, may be present and the grains might be oriented so that the material becomes anisotropy. These features are the most important ones, the effect of which we discuss in the following.

A typical 'clean' microstructure of a nearly fully dense ceramic material is shown in Fig. 12.3. The polycrystalline approach outlined in Section 12.1 can be directly applied to fully dense ceramics. The calculated estimates for the elastic moduli generally agree quite well with the experimental results. We quote two examples (see Table 12.1). For cubic polycrystalline $Y_3Al_5O_{12}$, using the elastic constants as given in

Table 12.1: Estimates for the Young's modulus for two polycrystalline materials.

	$Y_3Al_5O_{12}$ [e]	Al_2O_3 [f]
E_{upper} (GPa)	282.73	408.4
E_{lower} (GPa)	282.66	397.3
E (GPa)	282.7	402.9
E_{exp} (GPa)	290	404, 406

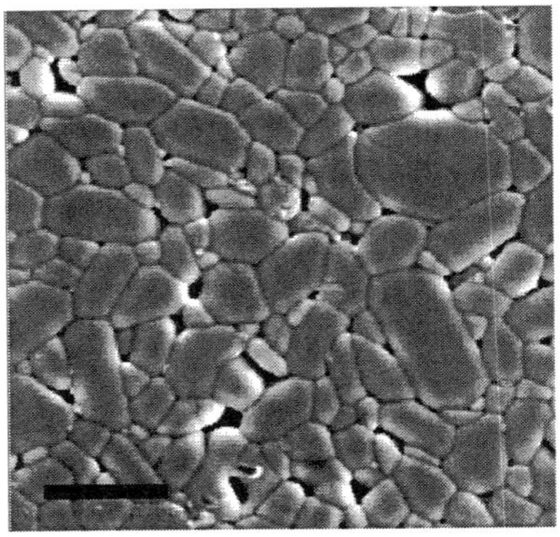

Fig. 12.3: Microstructure of nearly fully dense alumina. The scale represents 5 μm.

[e] de With, G. and Parren, J.E.D. (1985), Solid State Ionics **16**, 87.

[f] Dörre, E. and Hübner, H. (1984), *Alumina*, Springer, Berlin; de With, G. (1984), J. Mater. Sci. **19**, 2195.

Table 7.3, there is a small but significant discrepancy between the theoretical estimate and the experimental value due to alumina inclusions originating from a slight Al-excess. Moreover, note the closeness of the upper and lower bound estimates due to the limited anisotropy of this garnet. For tetragonal, fully dense Al_2O_3 the agreement between estimate and experiment is excellent. It must be said, though, that also in this case the anisotropy in the single crystal is relatively small leading to only a small difference in upper and lower bounds.

Polycrystalline ceramics typically have another characteristic feature, apart from their polycrystalline nature. This feature, more frequently encountered for these materials than for either metals or polymers, is their porosity. This porosity arises either as a result from a sintering process that has not densified the material completely or has been included on purpose, e.g. in membranes. The porosity may range from virtually zero, e.g. in modern engineering ceramics, to about ten percent, e.g. in refractory materials, to several tens of percent, e.g. in high temperature thermal insulating materials.

For porous materials with a pore fraction or *porosity* P the composite approach using voids for one phase can be used. These voids have zero G and K because the Young's modulus of a void is taken to be zero. The results, however, are not very satisfactory because the properties of the composing phases are rather different and because a void is actually a region without material, which is not the same as a material with vanishing modulus of elasticity.

Many other approaches are presented in the literature, empirical data fitting on a range of polycrystalline materials with varying porosity being one of the more frequently encountered procedures. Typical equations are

$$E = E_0(1 - aP + bP^2) \quad \text{and} \quad E = a\exp(-bP)$$

The former equation can also be derived for a continuous matrix with dispersed, spherical shaped pores[g], strictly speaking for $\nu = 0.3$. In that case $a \cong 1.9$ and $b \cong 0.9$ which has been verified experimentally for a number of cases. In the latter equation the parameter a provides an estimate for Young's modulus of the fully dense material. Note that the limiting behaviour towards $P = 1$ is incorrect for the exponential

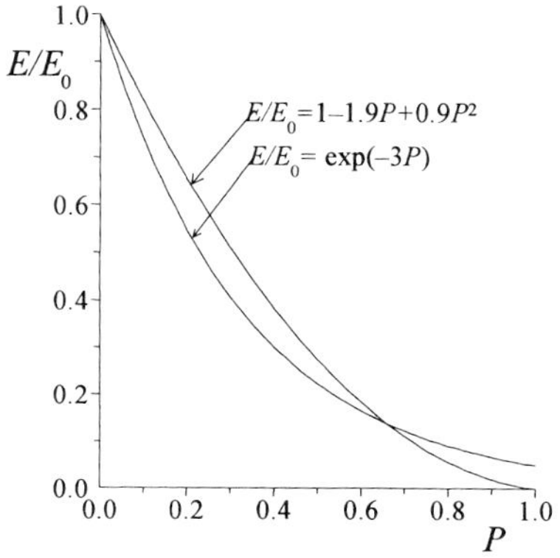

Fig. 12.4: Typical expressions for the porosity dependence of Young's modulus.

[g] MacKenzie, J.K. (1950), Proc. Phys. Soc. **B63**, 2.

expression (Fig. 12.4). Rice has provided an extensive review[h].

Another important effect on Young's modulus is the presence of microcracks, which generally leads to a decrease in stored energy, a lower modulus and in a non-linear stress-strain relationship. The microcracks typically originate from differential shrinkage during cooling after sintering due to anisotropy in either thermal expansion coefficient or elastic modulus or both.

In this case also some models are available to estimate the influence on Young's modulus. Defining $A_0 = 16(10-3\nu_0)(1-\nu_0^2)/45(2-\nu_0)$ one model[i] yields

$$\frac{E}{E_0} = \left(1 + A_0 N a^3\right)^{-1} \tag{12.4}$$

where E_0 and ν_0 are Young's modulus and Poisson's ratio, respectively, for the uncracked material, N the volume density of the number of microcracks and a the mean crack radius. The factor A_0 varies between 1.77 and 1.50 for ν_0 between 0 and 0.5, so that in first approximation the expression simplifies to

$$\frac{E}{E_0} = \left(1 + 1.63 N a^3\right)^{-1}$$

Another model[j] uses Poisson's ratio ν for the cracked material and defining A similar as A_0, but with ν_0 replaced with ν, the final expressions read

$$\frac{E}{E_0} = 1 - A N a^3 \quad \text{and} \quad \frac{\nu}{\nu_0} = 1 - \frac{16 N a^3}{9} \tag{12.5}$$

A similar reasoning for A as before for A_0 leads to

$$\frac{E}{E_0} = 1 - 1.63 N a^3$$

For small values of Na^3 both models yield approximately the same answers but at larger values the results differ significantly.

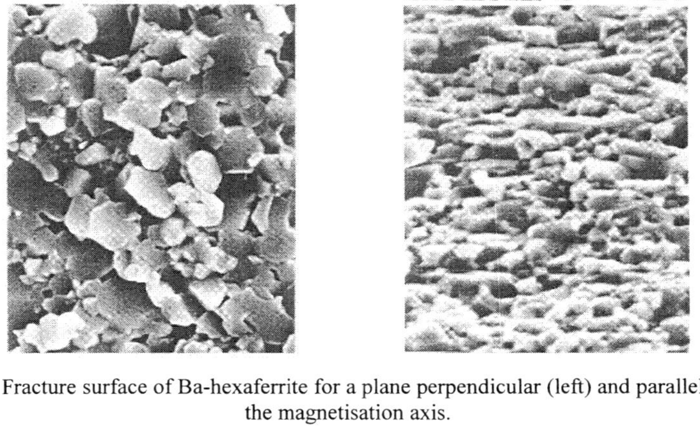

Fig. 12.5: Fracture surface of Ba-hexaferrite for a plane perpendicular (left) and parallel (right) to the magnetisation axis.

---

[h] Rice, R.W. (1977), Treatise Mater. Sci. Tech. **11**, 200, Academic Press, New York.

[i] Sagalnik, R.C. (1973), Izv. Akad. Nauk. SSR Mekh. Tverd. Tila **8**, 149.

[j] O'Connell, R.J. and Budiansky, B. (1974), J. Geol. Res. **79**, 5412.

A final aspect to mention also is anisotropy. In Chapter 9 it was indicated that even cubic crystals can be highly anisotropic. If a preferred orientation is present, the polycrystal may also be anisotropic to a considerable extent. This may be due to either the morphology of the powder used or due to some orientation process applied during manufacturing (or both). Since the degree of anisotropy that can be achieved depends strongly on the size of piece of material produced, the anisotropy depends typically on the size of the billet of material produced. The amount of anisotropy can be quantified by X-ray diffraction but is, if sufficiently large, also visible in the microstructure. For non-cubic crystals the effect can be quite large, e.g. for Ba-hexaferrite[k], (hexagonal structure, $BaO{\cdot}6Fe_2O_3$), oriented magnetically during pressing, the elastic modulus parallel and perpendicular to the magnetisation axis for a ceramic of 95% density is 154 and 317 GPa, respectively. The plate-like grain shape is evident from the fracture surfaces (Fig. 12.5).

Problem 12.3

The Reuss approach does not make sense for a porous material. What is the basic reason for this?

12.3 Metals

For metals the same considerations as for inorganics apply. Polycrystalline metals are normally (nearly) fully dense, although the grain size is typically larger as compared with inorganic materials. An example of the microstructure of Zr metal is given in Fig. 12.6. For most metals the anisotropy is small rendering the Voigt-Reuss-Hill satisfactory. Sometimes porous metals are used, e.g. in filters. Also in this case similar considerations as for inorganic materials apply. Since most metals are much more ductile than inorganic materials microcracking is only important for brittle metals like the refractory metals Mo and W and fully hardened steels.

While for inorganic materials the polycrystal generally is isotropic, polycrystalline metals are frequently anisotropic. This is generally due to the metal processing, in particular rolling and drawing. The amount of anisotropy introduced in this way is far from negligible. In Fig. 12.7 the microstructure of a steel containing 0.5% C is shown.

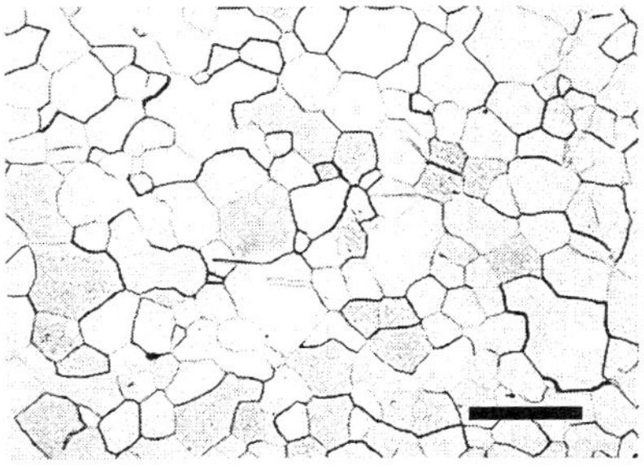

Fig. 12.6: Microstructure of fully dense polycrystalline Zr metal annealed at 750 °C. The bar indicates 100 μm.

[k] Iwasa, M., Liang, C.E., Bradt, R.C. and Nakamura, Y. (1981), J. Am. Ceram. Soc. **64**, 390.

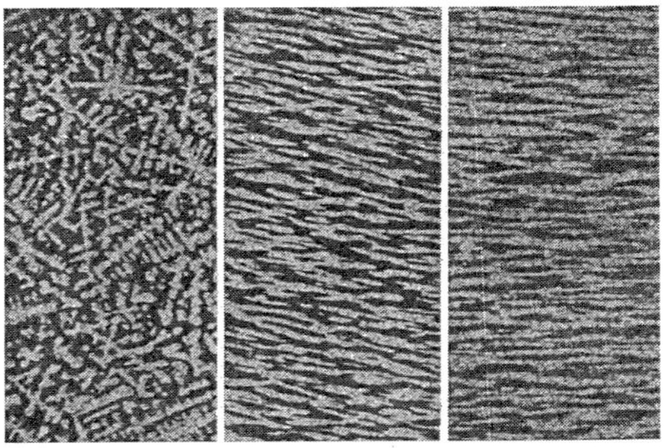

Fig. 12.7: Microstructure of 0.5% C steel as cast (left), after thickness reduction to 1/5 (middle) and 1/30 (right).

While after casting a dendritic structure is observed, hot working introduces a clear anisotropy, showing a banded structure. The final microstructure is heavily dependent on the thickness reduction. A similar anisotropy originates from wire drawing where the higher the draw ratio[1], the more texture results. The increase in preferred orientation as obtained after wire drawing of Cu and as determined by X-ray diffraction using the Laue flat film technique with monochromatic radiation is shown in Fig. 12.8. Young's modulus for Cu varies with direction (E_{111} = 200 GPa and E_{100} = 70 GPa). Since a preferred orientation of about 50% of ideal with the [111] direction as the preferred orientation can be realised, an increase of the modulus to about 150 GPa can be observed, to be compared with a value for a random polycrystalline material of 110 GPa. Similarly an increase in yield strength from 170 to 300 MPa and an increase in fracture strength from 220 to 380 MPa is observed for a draw ratio of 2.

12.4 Polymers

Similar to inorganics and metals, polymers do have a microstructure, although polymer researchers generally refer to it as morphology. As mentioned briefly in Chapter 1, many different features can be present. The most common one is probably the spherulitic structure, as illustrated in Fig. 12.9. In each spherulite fibrils, consisting of folded chain crystallites, are oriented radially from it centre. Thus, although similar

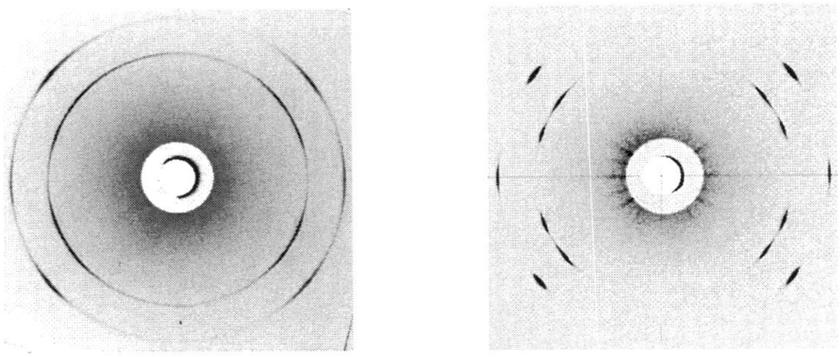

Fig. 12.8: X-ray patterns of a Cu wire after a draw ratio of 2 (left) and 6 (right), showing sharper diffraction spots with increased deformation indicating increased texture.

[1] The draw ratio is defined as the length of a line in the draw direction after and before drawing. Assuming volume conservation this ratio can be measured as the ratio of initial and final diameter.

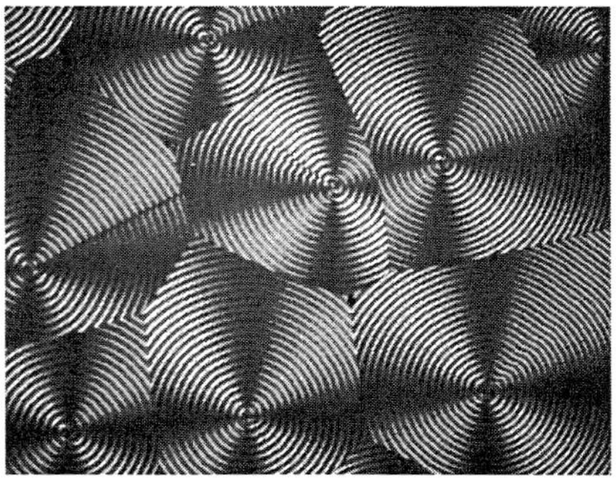

Fig. 12.9: Spherulitic structure.

in appearance to a grain in a polycrystalline material, their nature is entirely different. For this type of material the elastic modulus can be estimated reasonably from molecular considerations, as has been discussed in Chapter 11.

Although in many cases polymers do show isotropic behaviour, anisotropy may be present. This is generally due to the chain-like nature of the molecules which can lead to ordering of the molecules due to the fabrication process, e.g. in the case of injection moulded parts or in drawn wires. As an example we show in Fig. 12.10 X-ray photographs for isotactic polypropylene at various draw ratios, clearly showing the increased orientation. This increased orientation leads to increased values for several mechanical properties. For polymers the microstructural aspects generally are highly interwoven with the molecular aspects, so that a separated discussion is somewhat artificial. However, one can distinguish one-dimensional, two-dimensional and three-dimensional aspects.

1D considerations

Starting with the one-dimensional aspects, the first thing that comes to the mind is a fibre. Two major production routes[m] can be discerned, which use either flexible molecules or rigid molecules as a base material. The most important example of rigid chain polymers are the aromatic polyamides (aramids), notably poly(*p*-phenylene

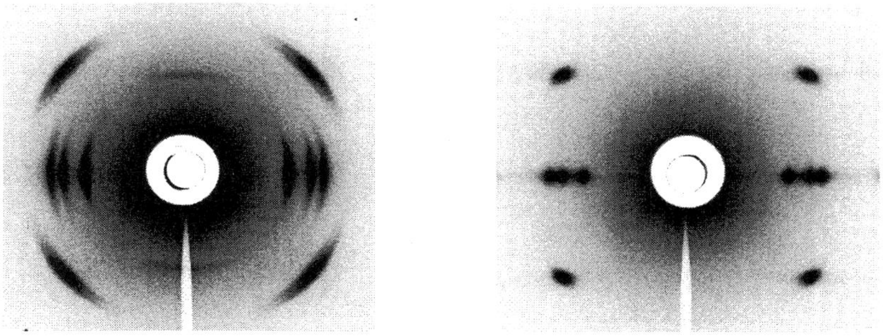

Fig. 12.10: X-ray patterns of polypropylene wire after a draw ratio of 4 (left) and 15 (right). The sharper diffraction spots with increased deformation indicate increased orientation of the molecules. This results in an increase in E from 2 to 9.5 GPa and an increase in strength from 200 to 500 MPa.

[m] Lemstra, P.J., Kirschbaum, R., Ohta, T. and Yasuda, H. (1987), page 39 in *Developments in oriente polymers*-2, I.M. Ward, ed., Elsevier Applied Science, London.

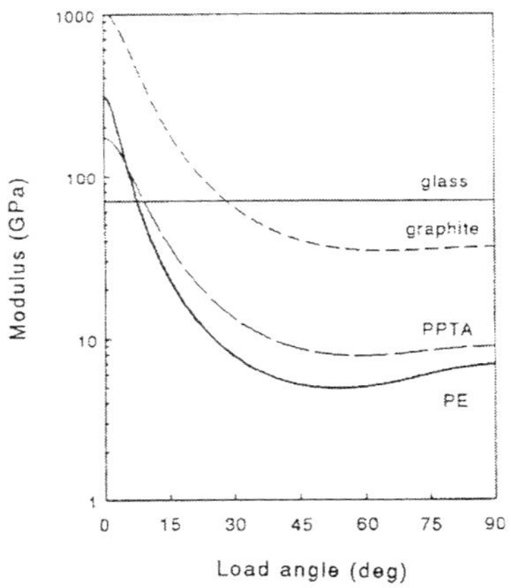

Fig. 12.11: Modulus of PE and some other materials versus orientation angle θ.

terephtalamide, PPTA), commercially known as Kevlar® (Du Pont) and Twaron® (Akzo Nobel). More recent developments include poly-(phenylene benzobisoxazole), (PBO), commercially known as Zylon® (Toyoba) and the experimental fibre denoted as M-5 (Akzo-Nobel), based on PIPD (polypyridobisimidazole). The latter has a significant better compressive strength. Here use is made of the intrinsic stiffness of the chains to align them.

The most important flexible fibre is based on polyethylene (PE). We have seen in Chapter 9 that PE as a single crystal shows considerable anisotropy and that this property is related to the small cross-section and packing of the molecules in the unit cell. Using the single crystal data as given in Chapter 9 we can estimate the elastic modulus along any direction different from the axis[n]. Young's modulus in the [100] direction is given by $E = S_{11}^{-1}$ (Nye, 1957) where, as before, the elastic compliance constant is denoted by S_{ij}. Here we need the elastic modulus for a direction with angle θ between the test direction and the molecular axis in the *a*-*c* plane[o]. Evaluating Nye's relationship for this situation, the result is given by the expression

$$E(\theta) = S_{33}^{-1}(\theta) = \begin{pmatrix} S_{11}\sin^4\theta - S_{15}\sin^3\theta\cos\theta \\ + (2S_{13} + S_{55})\sin^2\theta\cos^2\theta - 2S_{35}\sin\theta\cos^3\theta + S_{33}\cos^4\theta \end{pmatrix}^{-1} \quad (12.6)$$

For $\theta = 0$ this results in 312 GPa. This relationship is shown in Fig. 12.11 for PE together with that of some other materials and illustrates that the elastic behaviour of PE single crystals is indeed highly anisotropic. For a load direction more or less in the fibre direction a high stiffness results due to the intrinsic stiffness of the chains. As soon as the load direction deviates more than a few degrees the stiffness decreases significantly due to the low shear resistance of the PE crystals originating from the van der Waals interactions between the chains.

---

[n] Bastiaansen, C.W.M., Leblans, P.J.R. and Smith, P. (1990), Macromolecules **23**, 236.

[o] The standard result as provided by Nye applies for Young's modulus in the *x*-direction while here we need the expression for the *z*-direction. Therefore the indices 1, 3, 4 and 6 become 3, 1, 6 and 4, respectively. Moreover, since we consider the *a*-*c* plane the direction cosine $l_2 = 0$.

Table 12.2: Estimated Young's modulus E for flexible chain molecules.

Material	E (GPa)	Material	E (GPa)
Polyethylene	235	Poly(ethylene teraphtalate)	110
Poly(vinyl alcohol)	230	Polypropylene	40

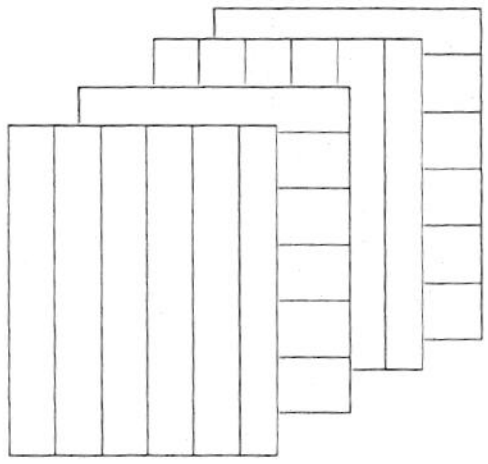

Fig. 12.12: Laminate of crystalline layers.

The problem is thus to align the molecules in the fibre in such a way that use can be made of the axial stiffness of the molecules. This alignment is realised, e.g. in ultra-high molecular weight polyethylene (UHMW-PE) by the gel-spinning process. Commercial brands are Dyneema® (DSM) and Spectra® (Allied Signal). For several other molecules also high Young's moduli can be reached, see Table 12.2.

*2D considerations**

Although in one dimension polymeric chain molecules show a considerable stiffness (and strength), their behaviour is significantly modified in two dimensions. To that purpose we consider a laminate of single crystals, again for the case of polyethylene. The simplest model assumes macroscopic axial symmetry with defect-free, 100% crystalline layers oriented along the c-axis (Fig. 12.12). Since polyethylene is orthorhombic either the a-axis or the b-axis can be perpendicular to the plane of the layer. Moreover, we have to decide on the adhesion between the layers. As extreme cases we have perfect adhesion or no adhesion at all. Assuming that the b-axis is perpendicular to the layer the modulus E_b is obtained by averaging using Eq. (12.6). This leads to

$$E_b = (2\pi)^{-1} \int_{-\pi}^{+\pi} E(\theta)\,\mathrm{d}\theta \tag{12.7}$$

A similar estimate can be made for E_a, assuming that the a-axis is perpendicular to the layer, or for E_{ab}, where it is assumed that both the a-axis and b-axis are random in the plane perpendicular to the c-axis. The values of E_a, E_b and E_{ab} calculated from the single crystal constants are given in Table 12.3, both for the case of full adhesion (Voigt average) and no adhesion (uniform strain). From this table it is clear that the estimates for the moduli depend to a large extent on the structure assumed. However, a significant drop as compared with the uniaxial case occurs. Moreover, experiments

Table 12.3: Theoretical moduli for equi-biaxially oriented polyethylene single crystal laminates.

	E_a (GPa)	E_b (GPa)	E_{ab} (GPa)
Uniform strain in 1-D	34.9	25.0	29.0
Voigt average	111	109	110

show that the structure assumed does not occur in reality for biaxially oriented films but that a more realistic model assumes a random distribution of aggregates of single crystals in the plane of films. 3D effects thus have to be invoked.

*3D considerations**

For a three-dimensional model similar assumptions as before lead directly to the Voigt and Reuss averages, as discussed in Section 12.1. In Table 12.4 the results for these calculations with the *c*-axis oriented in the plane are given, again assuming either the *a*-axis or the *b*-axis perpendicular to the layer and with a random distribution of *a*- and *b*-axes. Again it can be seen that the moduli depend to a large extent on the structure assumed. From experiments it became clear that the fibrils in biaxially oriented UHMW-PE films contain crystallites of limited size and are connected by an amorphous phase. The modulus of this phase is so low that the strain in each crystal depends only on the crystal orientation with respect to the test direction. Consequently, this results in equal stresses in the fibres for which the Reuss average is the most appropriate. Experimentally a modulus of about 5 GPa is observed for UHMW-PE ($M_w = 5\times10^3$ kg/mol) with a draw ratio of 15×15, in agreement with the microstructural considerations. Other polymers yield similar theoretical values. The extreme stiffness as present in the fibres itself is thus largely lowered by the morphology of the layers.

As a summary, these fibres although intrinsically stiff, loose their stiffness largely in not fully oriented configurations due to their low shear resistance. It will thus not come as a surprise that about three quarters of the fibre production (1999) find their application in cables, ropes, nets or in ballistics where essentially only uniaxial loading is present.

Table 12.4: Theoretical moduli for equi-biaxially oriented single crystal aggregates.

	E_a (GPa)	E_b (GPa)	E_{ab} (GPa)
Voigt average PE	111	109	110
Reuss average PE	11.9	7.5	9.6
Reuss average PVA	18.7	8.7	12.6
Reuss average Nylon-6	7.1	5.6	9.7

12.5 Composites

The influence of the microstructure is rather important for composites. Generally we distinguish between metal, inorganic and polymer matrices and between inorganic and metal fillers. Rubbers are used only in polymers as filler. The filler shape can be a particle, a platelet or a fibre. Also the connectivity of the filler phase (see Fig. 1.10) is important. First, we show some examples of composites and indicate their application. Some of materials are generally used and some of them are experimental. Thereafter we deal with their elastic behaviour.

For metals generally inorganic particle reinforcement is used. Because the properties of the constituting phases generally are in the same range, the simple estimates, as discussed in Section 12.1, yield reasonable results for isotropic composites. One example is provided by *hard metals* (Co-bonded WC polycrystals, Fig. 12.13), which are generally used as cutting tool materials. Typically they contain 5 to 25 vol% of Co while the remainder is WC. The grain size of the WC phase

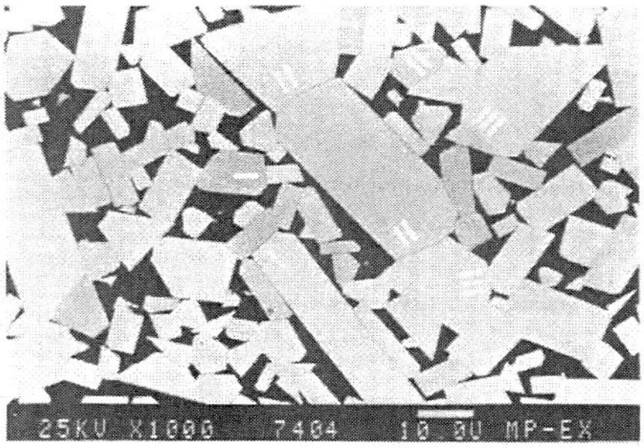

Fig. 12.13: Microstructure of hard metal (WC - 9% Co) annealed for 25 h at 1400 °C showing different type of grain boundaries. I straight, II faceted, III curved.

typically ranges from 0.5 to 20 μm. The elastic modulus of WC is 696 GPa and the value for Co is 207 GPa. Estimates for the elastic modulus E of 1 to 2 μm grain size composites based on averaging the series and parallel estimate together with the experimental data[p] are shown in Fig. 12.14. Although the trend is well predicted, quantitative agreement obviously is absent.

Al metal reinforced with Al_2O_3 whiskers provides another example[q]. Fig. 12.15 shows Al reinforced with 23 vol% Al_2O_3 whiskers that have been coated with 0.4 μm thick Ni-Ti to improve the adherence. The average cross-section area of the whiskers is 4 μm^2 while their average strength is about 5.5 GPa. The elastic modulus of alumina is 400 GPa while for Al the value is 70 GPa. The rule-of-mixtures estimates, together with the experimental data are given in Table 12.5. The experimental uncertainty was determined to be about 3 GPa. Obviously the elastic modulus is

Table 12.5: Elastic modulus of alumina whisker reinforced aluminium.

Vol. %	E_{exp} (GPa)	E_{the} (GPa)
8.3	85	97
10.6	95	105
16.9	106	126

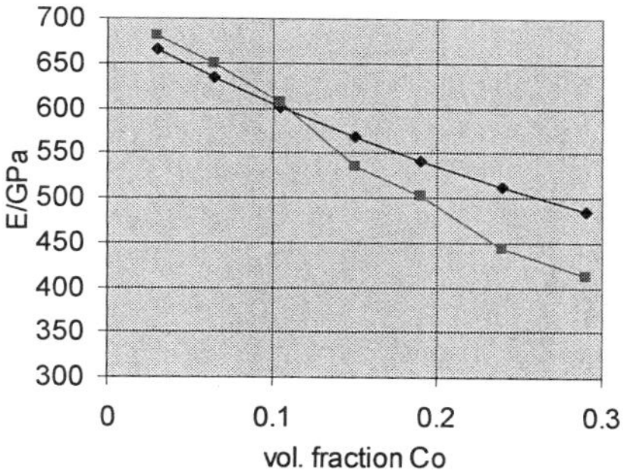

Fig. 12.14: Experimental (■) and estimated (♦) values for the elastic modulus of hard metals as a function of volume content Co.

[p] Chermant, J.L., Iost, A. and Osterstock, F. (1975), Proc. Brit. Ceram. Soc. **25**, 197.
[q] Mehan, R.L. (1967), page 29 in *Metal matrix composites*, ASTM-STP 438, ASTM, Philadelphia.

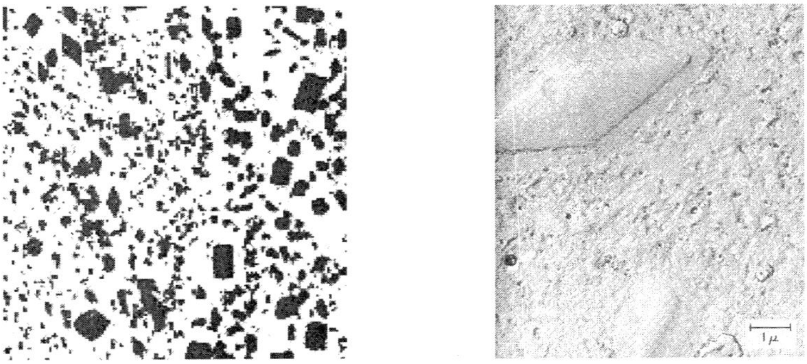

Fig. 12.15: Overall microstructure of Al reinforced with 23 vol% Al_2O_3 fibres (left) and a detail showing the Ni-Ti coating on an Al_2O_3 fibre (right).

overestimated. In this particular case the reason may be the incomplete infiltration of the preform with liquid aluminium during processing.

Not only inorganic materials are used to reinforce metals, also metals can be used. A fracture surface of an Al matrix uniaxially reinforced with steel fibres[r] is shown in Fig. 12.16. Since the elastic modulus of steel is 200 GPa, in this case the rule-of-mixtures for a 25 vol% composite yields 103 GPa, in good agreement with the experimental value.

With inorganic matrices both metal and inorganic particles are used. For example, attempts have been made to toughen hydroxy-apatite ceramics with Ag particles. The mineral part of bone is also hydroxy-apatite and this material shows excellent biocompatibility. A recent development is provided by ZrO_2-Al_2O_3 composites. These materials are highly wear resistant and are used as die material for wire drawing.

Also glass-ceramics can be considered as composites. As briefly mentioned in Chapter 1, these materials are made using glass technology. After shaping the materials are partially crystallised so that crystal of different properties appear in a glass matrix. An example is a glass of composition SiO_2-Al_2O_3-MgO-CaO, which after an addition of LiO_2 as crystallising agent shows crystallisation into an MgO and

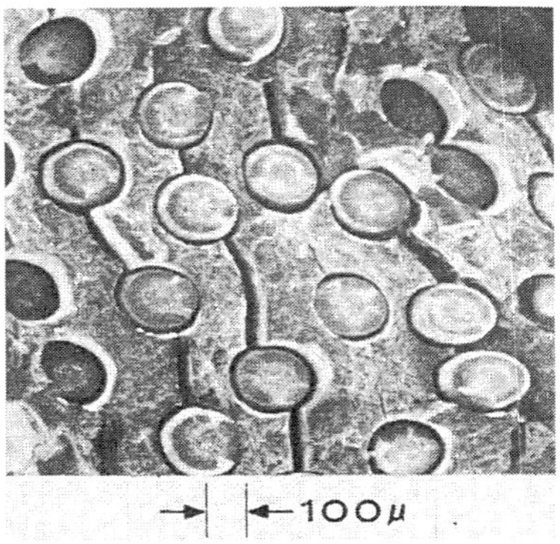

Fig. 12.16: Fracture surface of Al reinforced with 25 vol% steel fibres.

[r] Jones, R.C. (1967), *Deformation of wire reinforced metal matrix composites*, page 183 in Metal matrix composites, ASTM-STP 438, ASTM, Philadelphia.

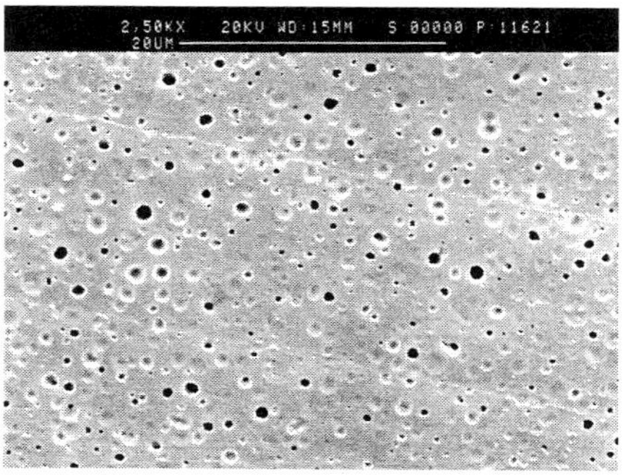

Fig. 12.17: Fracture surface of rubber-toughened polystyrene as used, e.g. for coffee cups.

Al_2O_3 rich phase, cordierite, with an elastic modulus of 200 GPa. This material shows a Young's modulus of about 140 GPa at a crystal volume fraction $f = 0.7$.

Polymer matrices can be reinforced with inorganic particles. Many polymer objects in use in daily life are filled with glass particles, mainly for economical reasons. Also rubber particles are used. In this case the main purpose is to toughen the material (Fig. 12.17). Fibres, either oriented more or less uniaxially or oriented in planes, are also frequently used to do reinforce polymer matrices. Fig. 12.18 shows the fracture surface of a glass fibre reinforced epoxy matrix. In the latter case often various layers are stacked and bonded together using different directions for the various plates. These composites are referred to as (cross-ply) *laminates*.

Estimating the elastic modulus of composites is frequently done by the rule-of-mixtures. From the examples discussed it is clear that significant deviations can occur using this rule. Moreover, the rule-of-mixtures assumes isotropic elastic behaviour while many composites are anisotropic, elastically as well as otherwise. A first guess for Young's modulus of a uniaxially oriented fibre composite in the direction of the fibres is still given by the rule-of-mixtures, although, as has been stated, before, the Voigt expression for E_V should be used. An empirical expression frequently used is the *Halpin-Tsai equation*[s] given by

$$E = E_2[E_1 + \xi(f_1E_1 + f_2E_2)]/(f_1E_2 + f_2E_1 + \xi E_2)$$

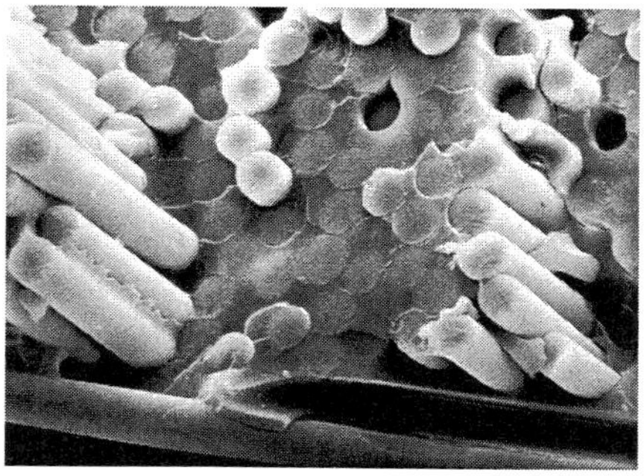

Fig. 12.18: Fracture surface of a composite of glass fibres in an epoxy matrix as used, e.g. in boats. Courtesy dr. Peijs, Imperial College, London, UK.

[s] Halpin, J.C. and Kardos, J.L. (1976), Pol. Eng. Sci. **16**, 344-352.

where ξ is an adjustable parameter that results in series coupling for $\xi = 0$ and parallel coupling for $\xi \to \infty$. As before f_1 and f_2 denote the volume fractions of phase 1 and phase 2, respectively.

12.6 Effective properties*

After the experimental survey preceded by some general considerations we discuss in this section more fully the effective properties of composite materials. We follow in broad terms the discussion as given by Francois et al. (1998). As has been emphasized in Chapter 1, in the description of materials a meso-level is generally important. Atomic/molecular aspects determine the properties of the meso-level and the meso-level itself co-determines the macro-behaviour of a material. Due to the heterogeneity of a real material, we must replace it by an equivalent, homogeneous material in such a way that the overall properties are the same for both. In order to be able to do this a representative volume element (RVE) or meso-cell must be defined with a content that is statistically the same as that for the real, heterogeneous material. Although we require that the response of the meso-cell must be the same as for the real material, i.e. the global properties must be the same, the stress and strain distribution within the meso-cell will vary, i.e the local properties are different. Denoting the structure elements of the meso-cell symbolically by Ω we have for linear elastic behaviour

$$\boldsymbol{\varepsilon}(\mathbf{r}) = \mathbb{A}(\mathbf{r},\Omega,...):\mathrm{E} \qquad \text{and} \qquad \boldsymbol{\sigma}(\mathbf{r}) = \mathbb{B}(\mathbf{r},\Omega,...):\Sigma \tag{12.8}$$

where $\boldsymbol{\varepsilon}$ and $\boldsymbol{\sigma}$ denote the local strain and stress, respectively, dependent on the position $\mathbf{r}$ in the meso-cell, and E and Σ the global strain and stress, constant throughout the meso-cell. The fourth order tensors $\mathbb{A}$ and $\mathbb{B}$ relate the local values of strain and stress to the global values. Now we must remember that we have only control over the boundary conditions and not over the interior of the meso-cell. For load control we write

$$t_i = \sigma_{ij} n_j = \Sigma_{ij} n_j \qquad \text{on } A_t \text{ with } \sigma_{ij,j} = 0 \text{ in } V \text{ (neglecting body forces)}$$

and we find[t]

$$\int \sigma_{ij}\,\mathrm{d}V = \int \sigma_{ik} x_{j,k}\,\mathrm{d}V = \int (\sigma_{ik} x_j)_{,k}\,\mathrm{d}V = \int \sigma_{ik} x_j n_k\,\mathrm{d}A$$
$$= \Sigma_{ik} \int x_j n_k\,\mathrm{d}A = \Sigma_{ik} \int x_{j,k}\,\mathrm{d}V = \Sigma_{ik} \int \delta_{jk}\,\mathrm{d}V = \Sigma_{ik}\delta_{jk} V = \Sigma_{ij} V$$

For displacement control on the other hand

$$u_i = E_{ik} x_k \qquad \text{on } A_u$$

and we have

$$\int \varepsilon_{ij}\,\mathrm{d}V = \int u_{i,j}\,\mathrm{d}V = \int u_i n_j\,\mathrm{d}A = \int E_{ik} x_k n_j\,\mathrm{d}A$$
$$= E_{ik} \int x_k n_j\,\mathrm{d}A = E_{ik} \int x_{k,j}\,\mathrm{d}V = E_{ik} \int \delta_{kj}\,\mathrm{d}V = E_{ik}\delta_{kj} V = E_{ij} V$$

Therefore the equality of global properties means that

---

[t] The following relations hold: $\int\sigma_{ij}\,\mathrm{d}V = \int\sigma_{ik}\delta_{jk}\,\mathrm{d}V = \int\sigma_{ik}x_{j,k}\,\mathrm{d}V = \int(\sigma_{ik}x_j)_{,k}\,\mathrm{d}V - \int\sigma_{ik,k}x_j\,\mathrm{d}V = \int(\sigma_{ik}x_j)_{,k}\,\mathrm{d}V = \int\sigma_{ik}x_jn_k\,\mathrm{d}A$, where the equilibrium condition $\sigma_{ik,k} = 0$ and Gauss' theorem $\int\alpha_{i,j}\,\mathrm{d}V = \int\alpha_i n_j\,\mathrm{d}A$ are used.

$$\blacktriangleright \quad \mathrm{E} = E_{ij} = \frac{1}{V}\int \varepsilon_{ij}\,\mathrm{d}V \equiv \langle \varepsilon_{ij} \rangle \quad \text{and} \quad \Sigma = \Sigma_{ij} = \frac{1}{V}\int \sigma_{ij}\,\mathrm{d}V \equiv \langle \sigma_{ij} \rangle \tag{12.9}$$

where $\langle \ldots \rangle$ is an obvious short-hand notation.

Now assume a statically admissible stress field $\sigma_{ij}{}^{*}$ and a kinematically admissible strain field $\varepsilon_{ij}{}^{**}$ that obey the afore-mentioned boundary conditions. Then

$$V\langle \boldsymbol{\sigma}^{*}:\boldsymbol{\varepsilon}^{**}\rangle = \int \boldsymbol{\sigma}^{*}:\boldsymbol{\varepsilon}^{**}\,\mathrm{d}V = \int \sigma_{ij}^{*}u_{i,j}^{**}\,\mathrm{d}V = \int (\sigma_{ij}^{*}u_{i}^{**})_{,j}\,\mathrm{d}V = \int \sigma_{ij}^{*}u_{i}^{**}n_{j}\,\mathrm{d}A$$

and thus if $\sigma_{ij}{}^{*}n_j = \Sigma_{ij}\, n_j$ on A_t

$$V\langle \sigma_{ij}^{*}\varepsilon_{ij}^{**}\rangle = \Sigma_{ij}\int u_{i}^{**}n_{j}\,\mathrm{d}A = \Sigma_{ij}\int \varepsilon_{ij}^{**}\,\mathrm{d}V = \Sigma_{ij}E_{ij}V = \langle \sigma_{ij}^{*}\rangle\langle \varepsilon_{ij}^{**}\rangle V$$

or if $u_i{}^{**} = E_{ik}\,x_k$ on A_u

$$V\langle \sigma_{ij}^{*}\varepsilon_{ij}^{**}\rangle = \int \sigma_{ij}^{*}u_{i,j}^{**}\,\mathrm{d}V = \int \sigma_{ij}^{*}u_{i}^{**}n_{j}\,\mathrm{d}A = E_{ik}\int \sigma_{ij}^{*}x_{k}n_{j}\,\mathrm{d}A$$
$$= E_{ik}\int (\sigma_{ij}^{*}x_{k})_{,j}\,\mathrm{d}A = E_{ik}\int \sigma_{ik}^{*}\,\mathrm{d}V = \langle \varepsilon_{ik}^{**}\rangle\langle \sigma_{ik}^{*}\rangle V$$

We thus always have

$$\blacktriangleright \quad \langle \boldsymbol{\sigma}^{*}:\boldsymbol{\varepsilon}^{**}\rangle = \langle \boldsymbol{\sigma}^{*}\rangle:\langle \boldsymbol{\varepsilon}^{**}\rangle \tag{12.10}$$

a result usually referred to as *Hill's theorem*.

After these preliminaries we can relate the global variables to the local variables. Using $\boldsymbol{\varepsilon} = \mathbb{A}:\mathrm{E} = \mathrm{E}:\mathbb{A}^{\mathrm{T}}$ (since $\boldsymbol{\varepsilon} = \boldsymbol{\varepsilon}^{\mathrm{T}}$ and $\mathrm{E} = \mathrm{E}^{\mathrm{T}}$) or $\varepsilon_{ij} = A_{ijkl}E_{kl}$ we obtain

$$\langle \boldsymbol{\varepsilon}\rangle = \langle \mathbb{A}:\mathrm{E}\rangle = \langle \mathbb{A}\rangle:\mathrm{E} = \langle \mathbb{A}\rangle:\langle \boldsymbol{\varepsilon}\rangle \qquad \text{or} \qquad \langle \mathbb{A}\rangle = \mathbb{I}$$

and from $\boldsymbol{\sigma} = \mathbb{B}:\Sigma$ similarly

$$\langle \boldsymbol{\sigma}\rangle = \langle \mathbb{B}:\Sigma\rangle = \langle \mathbb{B}\rangle:\Sigma = \langle \mathbb{B}\rangle:\langle \boldsymbol{\sigma}\rangle \qquad \text{or} \qquad \langle \mathbb{B}\rangle = \mathbb{I}$$

If we have the local relation $\boldsymbol{\sigma} = \mathbf{f}(\boldsymbol{\varepsilon})$ we have on the one hand

$$\langle \boldsymbol{\sigma}:\boldsymbol{\varepsilon}\rangle = \langle \boldsymbol{\varepsilon}:\mathbf{f}(\boldsymbol{\varepsilon})\rangle = \langle \mathrm{E}:\mathbb{A}^{\mathrm{T}}:\mathbf{f}(\mathbb{A}:\mathrm{E})\rangle = \mathrm{E}:\langle \mathbb{A}^{\mathrm{T}}:\mathbf{f}(\mathbb{A}:\mathrm{E})\rangle$$

and on the other hand

$$\langle \boldsymbol{\sigma}:\boldsymbol{\varepsilon}\rangle = \langle \boldsymbol{\sigma}\rangle:\langle \boldsymbol{\varepsilon}\rangle = \langle \mathbf{f}(\mathbb{A}:\mathrm{E})\rangle:\langle \mathbb{A}:\mathrm{E}\rangle = \langle \mathbf{f}(\mathbb{A}:\mathrm{E})\rangle:\langle \mathbb{A}\rangle:\mathrm{E} = \langle \mathbf{f}(\mathbb{A}:\mathrm{E})\rangle:\mathrm{E}$$

Since $\langle \boldsymbol{\sigma}:\boldsymbol{\varepsilon}\rangle = \Sigma:\mathrm{E}$ and $\langle \boldsymbol{\sigma}\rangle = \Sigma = \mathbf{f}_{\mathrm{eff}}(\boldsymbol{\varepsilon})$ we find

$$\blacktriangleright \quad \mathbf{f}_{\mathrm{eff}}(\boldsymbol{\varepsilon}) = \langle \mathbf{f}(\mathbb{A}:\mathrm{E})\rangle = \langle \mathbb{A}^{\mathrm{T}}:\mathbf{f}(\mathbb{A}:\mathrm{E})\rangle \tag{12.11}$$

Similarly from the local relation $\boldsymbol{\varepsilon} = \mathbf{g}(\boldsymbol{\sigma})$ we obtain

$$\blacktriangleright \quad \mathbf{g}_{\mathrm{eff}}(\boldsymbol{\sigma}) = \langle \mathbf{g}(\mathbb{B}:\Sigma)\rangle = \langle \mathbb{B}^{\mathrm{T}}:\mathbf{g}(\mathbb{B}:\Sigma)\rangle \tag{12.12}$$

For elastic behaviour these expressions reduce to $\langle \boldsymbol{\sigma}\rangle = \mathbb{C}_{\mathrm{eff}}:\langle \boldsymbol{\varepsilon}\rangle$ and $\langle \boldsymbol{\varepsilon}\rangle = \mathbb{S}_{\mathrm{eff}}:\langle \boldsymbol{\sigma}\rangle$, where $\mathbb{C}$ and $\mathbb{S}$ denote the elastic constants and compliances, respectively.

Mean values, energy and dissipation

In many cases the response of a heterogeneous material will not only be due to the elastic response of the phases involved but also due to an internal strain, e.g. resulting from local plastic deformation, a phase transformation or thermal strain. In this case we have that if the global stress $\Sigma = \mathbf{0}$, nevertheless a local stress $\boldsymbol{\sigma}$ can be present. For concreteness we can think of local plastic deformation and describe the total local strain $\boldsymbol{\varepsilon}$ as the sum of an elastic part $\boldsymbol{\varepsilon}^{(e)}$ and a plastic part $\boldsymbol{\varepsilon}^{(p)}$, i.e. $\boldsymbol{\varepsilon} = \boldsymbol{\varepsilon}^{(e)} + \boldsymbol{\varepsilon}^{(p)}$. Moreover we have $\boldsymbol{\sigma} = \boldsymbol{\sigma}^* + \boldsymbol{\sigma}^{(r)}$, where $\boldsymbol{\sigma}^*$ denotes the stress field that would be present if the material remained elastic under the same loading and $\boldsymbol{\sigma}^{(r)}$ the residual stress. In this case the global stress and strain and their rates of change are functions of the total local values of stress and strain but this is not true for the elastic and plastic parts separately since they are generally incompatible. We have

$$\Sigma = \langle\boldsymbol{\sigma}\rangle = \langle\boldsymbol{\sigma}^*+\boldsymbol{\sigma}^{(r)}\rangle \quad \text{or} \quad \langle\boldsymbol{\sigma}^{(r)}\rangle = \mathbf{0} \quad \text{since if } \Sigma = \mathbf{0},\ \langle\boldsymbol{\sigma}^*\rangle = \mathbf{0}$$

and according to Eq. (12.8) also $\boldsymbol{\sigma}^* = \mathbb{B}{:}\Sigma$. We obtain for elastic behaviour with $\boldsymbol{\varepsilon}^{(e)} = \mathbf{S}{:}\boldsymbol{\sigma}$ the following relations

$$\begin{aligned}\langle\mathbb{B}^{\mathrm{T}}{:}\boldsymbol{\varepsilon}\rangle &= \langle\mathbb{B}^{\mathrm{T}}{:}\boldsymbol{\varepsilon}^{(e)}\rangle + \langle\mathbb{B}^{\mathrm{T}}{:}\,\boldsymbol{\varepsilon}^{(p)}\rangle = \langle\mathbb{B}^{\mathrm{T}}{:}\mathbf{S}{:}\boldsymbol{\sigma}\rangle + \langle\mathbb{B}^{\mathrm{T}}{:}\,\boldsymbol{\varepsilon}^{(p)}\rangle\\ &= \langle\boldsymbol{\sigma}{:}\mathbf{S}{:}\mathbb{B}\rangle + \langle\mathbb{B}^{\mathrm{T}}{:}\,\boldsymbol{\varepsilon}^{(p)}\rangle \quad (\text{since } \mathbf{S} = \mathbf{S}^{\mathrm{T}})\\ &= \langle\boldsymbol{\sigma}\rangle{:}\langle\mathbf{S}{:}\mathbb{B}\rangle + \langle\mathbb{B}^{\mathrm{T}}{:}\,\boldsymbol{\varepsilon}^{(p)}\rangle\end{aligned}$$

by Hills's theorem since $\mathbf{S}{:}\mathbb{B}$ and $\boldsymbol{\sigma}$ are kinematically and statically admissible and thus

$$\langle\mathbb{B}^{\mathrm{T}}{:}\boldsymbol{\varepsilon}\rangle = \Sigma{:}\mathbf{S}_{\mathrm{eff}} + \langle\mathbb{B}^{\mathrm{T}}{:}\,\boldsymbol{\varepsilon}^{(p)}\rangle = \mathbf{E}^{(e)} + \mathbf{E}^{(p)}$$

Also we have

$$\langle\mathbb{B}^{\mathrm{T}}{:}\boldsymbol{\varepsilon}\rangle = \langle\mathbb{B}^{\mathrm{T}}\rangle{:}\langle\boldsymbol{\varepsilon}\rangle = \mathbb{I}{:}\mathbf{E} = \mathbf{E}$$

by Hills's theorem since $\boldsymbol{\varepsilon}$ and $\mathbb{B}^{\mathrm{T}}$ are kinematically and statically admissible so that

$$▶ \quad \mathbf{E} = \langle\boldsymbol{\varepsilon}\rangle \quad \mathbf{E}^{(e)} = \langle\mathbb{B}^{\mathrm{T}}{:}\boldsymbol{\varepsilon}^{(e)}\rangle \neq \langle\boldsymbol{\varepsilon}^{(e)}\rangle \quad \text{and} \quad \mathbf{E}^{(p)} = \langle\mathbb{B}^{\mathrm{T}}{:}\boldsymbol{\varepsilon}^{(p)}\rangle \neq \langle\boldsymbol{\varepsilon}^{(p)}\rangle \qquad (12.13)$$

Let us now consider the energy. For the average stress energy we find

$$\begin{aligned}\tfrac{1}{2}\langle\boldsymbol{\sigma}{:}\mathbb{S}{:}\boldsymbol{\sigma}\rangle &= \tfrac{1}{2}\langle(\boldsymbol{\sigma}^*+\boldsymbol{\sigma}^{(r)}){:}\mathbb{S}{:}(\boldsymbol{\sigma}^*+\boldsymbol{\sigma}^{(r)})\rangle\\ &= \tfrac{1}{2}\langle\boldsymbol{\sigma}^*{:}\mathbb{S}{:}\boldsymbol{\sigma}^*\rangle + \langle\boldsymbol{\sigma}^{(r)}{:}\mathbb{S}{:}\boldsymbol{\sigma}^*\rangle + \tfrac{1}{2}\langle\boldsymbol{\sigma}^{(r)}{:}\mathbb{S}{:}\boldsymbol{\sigma}^{(r)}\rangle\end{aligned}$$

Using $\boldsymbol{\sigma}^* = \mathbb{B}{:}\Sigma$ the first term on the right-hand side of this equation becomes

$$\tfrac{1}{2}\langle\boldsymbol{\sigma}^*{:}\mathbb{S}{:}\boldsymbol{\sigma}^*\rangle = \tfrac{1}{2}\Sigma{:}\langle\mathbb{B}^{\mathrm{T}}{:}\mathbb{S}{:}\mathbb{B}\rangle{:}\Sigma = \tfrac{1}{2}\Sigma{:}\mathbb{S}_{\mathrm{eff}}{:}\Sigma$$

The second term results in

$$\langle\boldsymbol{\sigma}^{(r)}{:}\mathbb{S}{:}\boldsymbol{\sigma}^*\rangle = \langle\boldsymbol{\sigma}^{(r)}\rangle{:}\langle\mathbb{S}{:}\boldsymbol{\sigma}^*\rangle = 0$$

by Hill's theorem and the relation $\langle\boldsymbol{\sigma}^{(r)}\rangle = \mathbf{0}$. The total result is therefore

$$▶ \quad \tfrac{1}{2}\langle\boldsymbol{\sigma}{:}\mathbb{S}{:}\boldsymbol{\sigma}\rangle = \tfrac{1}{2}\Sigma{:}\mathbb{S}_{\mathrm{eff}}{:}\Sigma + \tfrac{1}{2}\langle\boldsymbol{\sigma}^{(r)}{:}\mathbb{S}{:}\boldsymbol{\sigma}^{(r)}\rangle \qquad (12.14)$$

The total internal stress energy is thus the sum of the macroscopic stress energy and the energy stored due to the residual stress. The interaction energy between the macroscopic stress Σ and the residual stress $\boldsymbol{\sigma}^{(r)}$ vanishes, in agreement with Collonetti's theorem.

In a similar way we can derive an expression for the dissipation. The local dissipation is equal to the local plastic power $\boldsymbol{\sigma}:\dot{\boldsymbol{\varepsilon}}^{(p)}$. To obtain the global dissipation we note that

$$\langle\boldsymbol{\sigma}:\dot{\boldsymbol{\varepsilon}}\rangle = \langle\boldsymbol{\sigma}\rangle:\langle\dot{\boldsymbol{\varepsilon}}\rangle = \Sigma:\dot{\mathbf{E}} = \Sigma:\dot{\mathbf{E}}^{(e)} + \Sigma:\dot{\mathbf{E}}^{(p)} \qquad \text{and that}$$

$$\langle\sigma:\dot{\varepsilon}\rangle=\langle\sigma:\dot{\varepsilon}^{(e)}\rangle+\langle\sigma:\dot{\varepsilon}^{(p)}\rangle=\langle\sigma:\mathbb{S}:\dot{\sigma}\rangle+\langle\sigma:\dot{\varepsilon}^{(p)}\rangle$$

Using Eq. (12.14) we obtain

$$\langle\sigma:\mathbb{S}:\dot{\sigma}\rangle=\Sigma:\mathbb{S}_{\text{eff}}:\dot{\Sigma}+\langle\sigma^{(r)}:\mathbb{S}:\dot{\sigma}^{(r)}\rangle=\Sigma:\dot{E}^{(e)}+\langle\sigma^{(r)}:\mathbb{S}:\dot{\sigma}^{(r)}\rangle$$

and thus for the macroscopic dissipation

► $$\langle\sigma:\dot{\varepsilon}^{(p)}\rangle=\Sigma:\dot{E}^{(p)}-\langle\sigma^{(r)}:\mathbb{S}:\dot{\sigma}^{(r)}\rangle \quad (12.15)$$

The macroscopic plastic power $\Sigma:\dot{E}^{(p)}$ is thus not completely dissipated but also partially stored in the residual stress field.

In the following we use these concepts for the calculation of Voigt and Reuss bounds.

Justification 12.1: Voigt and Reuss bounds

As before we denote the average stress and strain in a meso-cell representing either a polycrystalline or composite material by Σ and E. Moreover we have $\Sigma = \mathbb{C}_{\text{eff}}:E$, $E = \mathbb{S}_{\text{eff}}:\Sigma$, the traction $\mathbf{t} = \sigma\cdot\mathbf{n} = \Sigma\cdot\mathbf{n}$ on A_t and the displacement $\mathbf{u} = \mathbf{u}^{\text{o}}$ on A_u. Let us define the stress $\sigma_{ij}{}^{\text{o}}$ as the stress that would exist in a grain having the strain E_{ij}. Similarly the strain $\varepsilon_{ij}{}^{\text{o}}$ is the strain that would be produced in the grain by the stress Σ. Therefore

$$\sigma^{\text{o}} = \mathbb{C}:E \quad \text{and} \quad \varepsilon^{\text{o}} = \mathbb{S}:\Sigma$$

Since we also have $\sigma = \mathbb{C}:\varepsilon$ and $\varepsilon = \mathbb{S}:E$ we obtain

$$\sigma^{\text{o}}:\varepsilon = \mathbb{C}:E:\varepsilon = \sigma:E \quad \text{and} \quad \sigma:\varepsilon^{\text{o}} = \sigma:\mathbb{S}:\Sigma = \varepsilon:\Sigma$$

Therefore, we have

$$\sigma:\varepsilon + (\sigma-\sigma^{\text{o}}):(\varepsilon-E) = \sigma^{\text{o}}:E + 2(\varepsilon-E):\sigma \quad \text{and}$$

$$\sigma:\varepsilon + (\sigma-\Sigma):(\varepsilon-\varepsilon^{\text{o}}) = \Sigma:\varepsilon^{\text{o}} + 2(\sigma-\Sigma):\varepsilon$$

Since $(\sigma-\sigma^{\text{o}}):(\varepsilon-E)$ can be written as $\mathbb{C}:(\varepsilon-E):(\varepsilon-E)$, this term is positive. Moreover we have $\int(\varepsilon-\mathbf{E}):\sigma\,\mathrm{d}V = \int(\mathbf{u}^{\text{o}}-\mathbf{u}^{\text{o}})\cdot\mathbf{t}\,\mathrm{d}A = 0$. Consequently

$$\int\sigma:\varepsilon\,\mathrm{d}V \equiv \Sigma:\mathbf{E}V = \mathbb{C}_{\text{eff}}:\mathbf{E}:\mathbf{E}V \le \int\sigma^{\text{o}}:\mathbf{E}\,\mathrm{d}V$$

$$= \mathbf{E}:\int\sigma^{\text{o}}\,\mathrm{d}V = \mathbf{E}:\int\mathbb{C}:\mathbf{E}\,\mathrm{d}V = \mathbf{E}:\mathbf{E}:\int\mathbb{C}\,\mathrm{d}V$$

$$\text{or} \quad \mathbb{C}_{\text{eff}} \le V^{-1}\int\mathbb{C}\,\mathrm{d}V \quad \text{for every } \mathbf{E}$$

In the same way we can write $(\sigma-\Sigma):(\varepsilon-\varepsilon^{\text{o}}) = \mathbb{S}:(\sigma-\Sigma):(\sigma-\Sigma) > 0$ and $\int(\sigma-\Sigma):\varepsilon\,\mathrm{d}V = \int(\mathbf{t}-\mathbf{t})\cdot\mathbf{u}^{\text{o}}\mathbf{t}\,\mathrm{d}A = 0$. Hence

$$\int\sigma:\varepsilon\,\mathrm{d}V \equiv \Sigma:\mathbf{E}V = \mathbb{S}_{\text{eff}}:\Sigma:\Sigma V \le \int\Sigma:\varepsilon^{\text{o}}\,\mathrm{d}V$$

$$= \Sigma:\int\varepsilon^{\text{o}}\,\mathrm{d}V = \Sigma:\int\mathbb{S}:\Sigma\,\mathrm{d}V = \Sigma:\Sigma:\int\mathbb{S}\,\mathrm{d}V$$

or $\quad \mathbb{S}_{\text{eff}} \le V^{-1}\int \mathbb{S}\,\mathrm{d}V \quad$ for every Σ

Altogether we may write $\langle \mathbb{S}\rangle^{-1} \le \mathbb{S}_{\text{eff}}{}^{-1} = \mathbb{C}_{\text{eff}} \le \langle \mathbb{C}\rangle$

Problem 12.4

Prove Eq. (12.12).

12.7 Improved estimates*

For the detailed calculation of the elastic properties of composites several first principle methods and semi-empirical methods are available in the literature. We discuss briefly some aspects of the first principle approach and deal thereafter with a semi-empirical method used for polymers as well as inorganics.

First principle methods

For particulate composites with low volume fraction inclusions a certain volume of matrix material containing a single spherical inclusion represents a simple meso-cell. In the previous chapter the elastic state of such a spherical inclusion in a matrix was discussed. Basically the effects discussed are due to the non-fitting of the free inclusion in the hole in the matrix. This is called the *size effect*. However, if a matrix material containing inclusions of a fitting size but with different elastic constants is loaded externally, there is still an effect, which is usually addressed as the *inhomogeneity effect*. Obviously such an inclusion is called an inhomogeneity. The description of that effect leads to an estimate of the elastic constants of a composite containing a dilute concentration of inclusions. We limit ourselves to entirely to isotropic matrices with spherical inclusions and refer for cylindrical, lamellar or anisotropic inclusions to e.g. Christensen (1979) and Mura (1987). An extensive theoretical reference is Nemat-Nasser and Hori (1993).

Let us now assume that a homogeneous material with elastic constants $\mathbb{C}$ is subjected to external forces which we limit for convenience to surface tractions $\mathbf{t}$. At certain positions we change the elastic constants to $\mathbb{C}^*$, e.g. due to the introduction on an inclusion. Keeping the tractions constant, the change in elastic energy is given[u] by

$$\Delta W = \tfrac{1}{2}\int (\sigma_{ij}^*\varepsilon_{ij}^* - \sigma_{ij}\varepsilon_{ij})\,\mathrm{d}V = \tfrac{1}{2}\int \sigma_{ij}(u_i^* - u_i)n_j\,\mathrm{d}A \tag{12.16}$$

or equivalently

$$\Delta W = \tfrac{1}{2}\int (\sigma_{ij}\varepsilon_{ij}^* - \sigma_{ij}^*\varepsilon_{ij})\,\mathrm{d}V = \tfrac{1}{2}\int (C_{ijkl}^* - C_{ijkl})\varepsilon_{ij}^*\varepsilon_{kl}\,\mathrm{d}V \tag{12.17}$$

since the difference term can be transformed to

---

[u] Similar as before we note that for body of volume V and surface area A often expressions like $\int\sigma_{ij}\varepsilon_{ij}\,\mathrm{d}V = \int\sigma_{ij}u_{i,j}\,\mathrm{d}V = \int(\sigma_{ij}u_i)_{,j}\,\mathrm{d}V - \int\sigma_{ij,j}u_i\,\mathrm{d}V = \int\sigma_{ij}n_ju_i\,\mathrm{d}A = \int t_iu_i\,\mathrm{d}A$ are encountered in which the (reduced) equilibrium condition $\sigma_{ij,j} = 0$ and the Gauss theorem $\int\alpha_{i,j}\,\mathrm{d}V = \int\alpha_in_j\,\mathrm{d}A$ have been used.

$$\tfrac{1}{2}\int(\sigma_{ij}^* - \sigma_{ij})(u_i^* - u_i)n_j\,\mathrm{d}A$$

which vanishes since the traction $t_i = \sigma_{ij}n_j = \sigma_{ij}^*n_j$ is kept constant. For isotropic materials $C_{ijkl} = \lambda\delta_{ij}\delta_{kl} + \mu(\delta_{ik}\delta_{jl} + \delta_{il}\delta_{jk})$ and Eq. (12.17) reduces to

$$\Delta W = \tfrac{1}{2}\int\left[(\lambda^* - \lambda)\varepsilon_{ii}^*\varepsilon_{jj} + 2(\mu^* - \mu)\varepsilon_{ij}^*\varepsilon_{ij}\right]\mathrm{d}V \tag{12.18}$$

Introducing the deviator $\varepsilon_{ij}' = \varepsilon_{ij} - \tfrac{1}{3}\varepsilon_{kk}\delta_{ij}$ we obtain

$$\Delta W = \tfrac{1}{2}\int\left[(K^* - K)\varepsilon_{ii}^*\varepsilon_{jj} + 2(\mu^* - \mu)\varepsilon_{ij}^{*\prime}\varepsilon_{ij}'\right]\mathrm{d}V \tag{12.19}$$

where the bulk modulus $K = \lambda + \tfrac{2}{3}\mu$ has been used. Now in case the strain external to the inclusion $\boldsymbol{\varepsilon}$ is uniform, i.e. does not depend on $\mathbf{r}$, one can show[v] that the strain $\boldsymbol{\varepsilon}^*$ inside a spherical inhomogeneity is also uniform and given by

$$\mathrm{tr}\,\boldsymbol{\varepsilon}^* = (A+1)\mathrm{tr}\,\boldsymbol{\varepsilon} \qquad \text{and} \qquad \boldsymbol{\varepsilon}^{*\prime} = (B+1)\boldsymbol{\varepsilon}' \qquad \text{where}$$

$$A = \frac{K^* - K}{(K^* - K)\alpha - K} \qquad B = \frac{\mu^* - \mu}{(\mu^* - \mu)\beta - \mu}$$

$$\alpha = \frac{1}{3}\frac{1+\nu}{1-\nu} \qquad \beta = \frac{2}{15}\frac{4-5\nu}{1-\nu}$$

Introducing these results in Eq. (12.19) we obtain

$$\begin{aligned}\Delta W &= -\tfrac{1}{2}\Omega\int\left[AK\varepsilon_{ii}\varepsilon_{jj} + 2\mu B\varepsilon_{ij}'\varepsilon_{ij}'\right]\mathrm{d}V \\ &= -\tfrac{1}{2}\Omega\int\left[\frac{A}{9K}\sigma_{ii}\sigma_{jj} + \frac{B}{2\mu}\sigma_{ij}'\sigma_{ij}'\right]\mathrm{d}V\end{aligned} \tag{12.20}$$

where σ_{ij}' is the deviator of the stress outside the inhomogeneity of volume Ω. If n denotes the number of inhomogeneities per unit volume, the total change in elastic energy is $n\Delta W$. On the other hand the elastic energy in the absence of inhomogeneities

$$W = \tfrac{1}{2}\int\left[K\varepsilon_{ii}\varepsilon_{jj} + 2\mu\varepsilon_{ij}'\varepsilon_{ij}'\right]\mathrm{d}V \tag{12.21}$$

so that the apparent elastic constants are

$$K_{\mathrm{app}} = (1-\phi A)K \qquad \text{and} \qquad \mu_{\mathrm{app}} = (1-\phi B)\mu \tag{12.22}$$

where $\phi = n\Omega$ is the volume faction of inhomogeneities. Obviously these equations hold only for small volume fractions of inhomogeneities since any interaction between them has been neglected. Frequently the parameters A and B are considered as parameters, to be determined experimentally.

There are various other possible approaches for higher concentrations of which we discuss briefly here two models: the composite spheres model and the three-phase model. In both models the inclusion is modelled as a single sphere in a spherically shaped piece of second phase material and this unit, the *composite sphere*, is

---

[v] Eshelby, J.D. (1961), Prog. Solid Mech. II, 89.

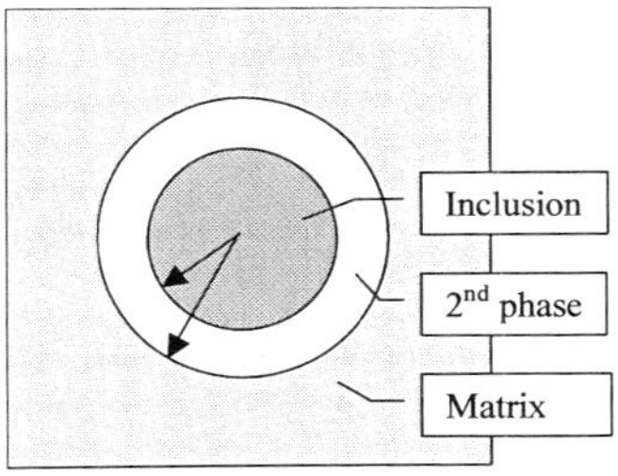

Fig. 12.19: The composite sphere embedded in a matrix, the properties of which are either determined in a self-consistent way or from an assembly of space filling composite spheres of varying dimensions.

embedded in a matrix material (Fig. 12.19). The ratio of inclusion radius a and second phase radius b is chosen according to $a^3/b^3 = \phi$, where ϕ is the volume fraction inclusions. In the *composite spheres model* the matrix consists of composite sphere units of varying size in such a way that the matrix is entirely filled. In the *three-phase model* the matrix is an effective material with unknown K and G. These parameters are determined by requiring that the strain energy stored in the material is the same as that for a fully homogeneous material under conditions of identical average strain. In both cases the solution is somewhat complex and we discuss here only the solution. For details we refer to the literature, e.g. Christensen (1979).

For the composite spheres model the bulk modulus K can be calculated by bounding procedures. It appears that the upper and lower bounds coincide. They represent therefore the exact solution, which is given by

$$\frac{K-K_{\mathrm{m}}}{K_{\mathrm{i}}-K_{\mathrm{m}}}=\frac{\phi}{1+\left[(1-\phi)(K_{\mathrm{i}}-K_{\mathrm{m}})/(K_{\mathrm{m}}-4G_{\mathrm{m}}/3)\right]} \tag{12.23}$$

where the indices m and i indicate the matrix and inclusion, respectively. However, for the shear modulus G the bounds do not coincide except for very small or very large volume fractions and in these cases the expressions reduce to the dilute solution case. An exact solution is not known.

For the three-phase model the expression for the bulk modulus appears to be the same as for the composite spheres model. The value for the shear modulus is given by the quadratic equation

$$A\left(\frac{G}{G_{\mathrm{m}}}\right)^2+2B\left(\frac{G}{G_{\mathrm{m}}}\right)+C=0 \tag{12.24}$$

where A, B and C are complex expressions given by

$$A=8\left(\frac{G_{\mathrm{i}}}{G_{\mathrm{m}}}-1\right)(4-5\nu_{\mathrm{m}})\eta_1\phi^{10/3}-2\left[63\left(\frac{G_{\mathrm{i}}}{G_{\mathrm{m}}}-1\right)\eta_2+2\eta_1\eta_3\right]\phi^{7/3}$$

$$+252\left(\frac{G_{\mathrm{i}}}{G_{\mathrm{m}}}-1\right)\eta_2\phi^{5/3}-50\left(\frac{G_{\mathrm{i}}}{G_{\mathrm{m}}}-1\right)(7-12\nu_{\mathrm{m}}+8\nu_{\mathrm{m}}^2)\eta_2\phi+4(7-10\nu_{\mathrm{m}})\eta_2\eta_3$$

$$B = -2\left(\frac{G_i}{G_m} - 1\right)(1 - 5\nu_m)\eta_1\phi^{10/3} + 2\left[63\left(\frac{G_i}{G_m} - 1\right)\eta_2 + 2\eta_1\eta_3\right]\phi^{7/3}$$
$$- 252\left(\frac{G_i}{G_m} - 1\right)\eta_2\phi^{5/3} + 75\left(\frac{G_i}{G_m} - 1\right)(3 - \nu_m)\eta_2\nu_m\phi + \frac{3}{2}(15\nu_m - 7)\eta_2\eta_3$$

$$C = 4\left(\frac{G_i}{G_m} - 1\right)(5\nu_m - 7)\eta_1\phi^{10/3} - 2\left[63\left(\frac{G_i}{G_m} - 1\right)\eta_2 + 2\eta_1\eta_3\right]\phi^{7/3}$$
$$+ 252\left(\frac{G_i}{G_m} - 1\right)\eta_2\phi^{5/3} + 25\left(\frac{G_i}{G_m} - 1\right)(\nu_m^2 - 7)\eta_2\phi - (7 + 5\nu_m)\eta_2\eta_3$$

with

$$\eta_1 = (49 - 50\nu_i\nu_m)\left(\frac{G_i}{G_m} - 1\right) + 35\frac{G_i}{G_m}(\nu_i - 2\nu_m) + 35(2\nu_i - \nu_m)$$

$$\eta_2 = 5\nu_i\left(\frac{\mu_i}{\mu_m} - 8\right) + 7\left(\frac{\mu_i}{\mu_m} + 4\right) \quad \text{and} \quad \eta_3 = \frac{G_i}{G_m}(8 - 10\nu_m) + (7 - \nu_m)$$

The final result reduces, as expected, to the dilute solution case for low concentrations. The three-phase model is capable of describing the volume dependence of the moduli for composites with a matrix and inclusion material of rather different moduli quite accurately.

Finally we note in the literature often use is made of a self-consistent scheme to estimate the moduli of composites. While for polycrystals this approach is considered reasonable, for porous materials its use is questionable. In particular it predicts zero modulus at a finite volume fraction of pores. This approach has been severely criticised by Christensen (1979) and we refer to his book for details.

Problem 12.5

Prove Eq. (12.17).

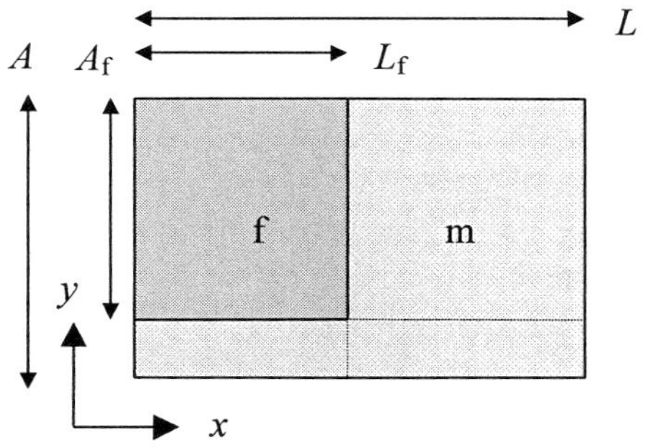

Fig. 12.20: A schematic view of a representative element.

Semi-empirical estimates

In Section 12.1 the basic series and parallel elements have been described. Two improved estimates are possible. The first one is by considering the parallel connection and after that the series connection of a block[w] of phase f to phase m. We refer to this as the P-S model. Second, the other way around, i.e. by considering the series connection and then the parallel connection of a block. Logically this is indicated as the S-P model. Fig. 12.20 shows these elements in a two-dimensional representation. In all cases we neglect effects due to Poisson's ratio. This way of estimating the moduli has been proposed for inorganic materials[x] (mainly for elastic properties of porous materials, hard metals, etc.) and polymers[y] (mainly for viscoelastic behaviour). We emphasize that many other semi-empirical approaches have been presented in the literature but limit ourselves to the model mentioned.

For the P-S model the parallel connection between f and m extends over L_f, the length of the f block (Fig. 12.20). For the parallel element we have

$$E_\mathrm{p} = V_\mathrm{f} E_\mathrm{f} + V_\mathrm{m} E_\mathrm{m} \tag{12.25}$$

where E_p denotes the parallel modulus of the two-phase material in the indicated direction, V_f the volume fraction of phase f, $V_\mathrm{m} = 1-V_\mathrm{f}$ the volume fraction of the matrix m, E_f the elastic constant of the phase f and E_m the elastic constant of the phase m. In this case

$$V_\mathrm{f} = A_\mathrm{f}L_\mathrm{f}/AL_\mathrm{f} = A_\mathrm{f}/A \quad \text{and} \quad V_\mathrm{m} = (A - A_\mathrm{f})L_\mathrm{f}/AL_\mathrm{f} = (A - A_\mathrm{f})/A$$

The part of phase m at the right-hand side of phase f can be considered to be in series connection with the left-hand part, so that

$$\frac{1}{E_\mathrm{PS}} = \frac{V_\mathrm{p}}{E_\mathrm{p}} + \frac{1-V_\mathrm{p}}{E_\mathrm{m}} \quad \text{or} \quad E_\mathrm{PS} = \frac{E_\mathrm{p}E_\mathrm{m}}{V_\mathrm{p}E_\mathrm{m} + (1-V_\mathrm{p})E_\mathrm{p}} \tag{12.26}$$

where E_PS denotes the effective modulus of the two-phase material. Here it holds that

$$V_\mathrm{p} = AL_\mathrm{f}/AL = L_\mathrm{f}/L \quad \text{and} \quad 1-V_\mathrm{p} = (L - L_\mathrm{f})A/AL = (L - L_\mathrm{f})/L$$

The final equation thus becomes

$$E_\mathrm{PS} = \frac{\left[\frac{A_\mathrm{f}}{A}E_\mathrm{f} + \left(1-\frac{A_\mathrm{f}}{A}\right)E_\mathrm{m}\right]E_\mathrm{m}}{\frac{L_\mathrm{f}}{L}\left[\frac{A_\mathrm{f}}{A}E_\mathrm{f} + \left(1-\frac{A_\mathrm{f}}{A}\right)E_\mathrm{m}\right] + \left(1-\frac{L_\mathrm{f}}{L}\right)E_\mathrm{m}} \tag{12.27}$$

In a similar way for the S-P model the series connection between f and m extends over the surface A_f. The part of m outside the interface A_f, which extends over the full length L, can be considered to be in parallel connection with the inner part. After similar evaluation the final result is

$$E_\mathrm{SP} = \frac{E_\mathrm{f}E_\mathrm{m}}{\left[\frac{L_\mathrm{f}}{L}E_\mathrm{f} + \left(1-\frac{L_\mathrm{f}}{L}\right)E_\mathrm{m}\right]}\frac{A_\mathrm{f}}{A} + \left(1-\frac{A_\mathrm{f}}{A}\right)E_\mathrm{m} \tag{12.28}$$

[w] Here 'block' is used as an alternative for 'parallellepiped'.
[x] Veldkamp, J.D.B. (1979), J. Phys. D: Appl. Phys. **12**, 1375.
[y] Takanayagi, M., Imada, K. and Kajayima, T. (1966), J. Pol. Sci. **C15**, 263.

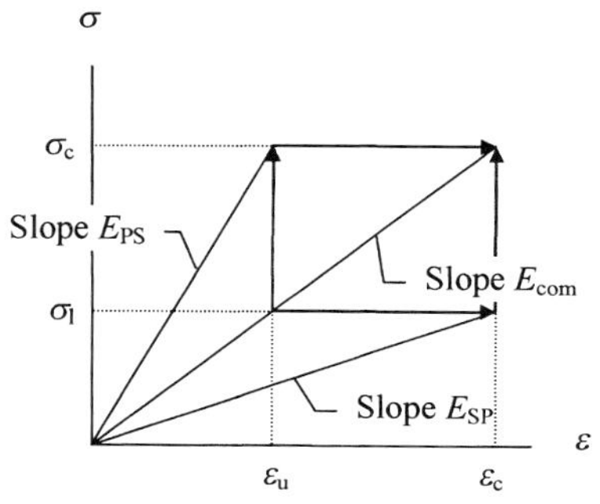

Fig. 12.21: Shear weakening and stiffening.

For an understanding of these two estimates it helps to visualise the elastic deformation, as shown in Fig. 12.21.

Using the P-S model with Eq. (12.27) one can imagine at $x = L_f$ (see Fig. 12.20) an infinitely stiff flat plane, which totally blocks lateral displacements within the representative element with respect to the outer surface. In reality this plane is absent, thus for a real deformation of the representative element, lateral displacements take place due to shear deformation. This shear deformation can be imagined to deform the plane at L_f at constant average stress σ_c at $x = L$ from ε_u up to the moment of the real deformation (ε_c in Fig. 12.21). This shear deformation is accompanied by shear work (here called the shear-weakening energy) and leads to a stiffness E_{com}, lower than E_{PS}.

On the other hand using the S-P model with Eq. (12.28) one assumes a non-bonded horizontal interface between the series part and the matrix (see Fig. 12.20). This implies a maximum lateral displacement within the element with respect to the outer surface. In reality a bond at the interface is present and these lateral displacements are much smaller due to shear deformation within the element. The bond formation can here be imagined to gradually improve the shear resistance at the horizontal interface at constant strain at $x = L$ up to the moment of the real deformation. This shear deformation, also accompanied by shear work (here called the shear-stiffening energy), leads to a stiffness E_{com}, higher than E_{SP}.

An estimate for the composite E_{com} thus would be the average of E_{SP} and E_{PS}, i.e.

► $$E_{com} = \tfrac{1}{2}(E_{SP} + E_{PS}) \tag{12.29}$$

When the shear weakening energy at constant stress σ_c equals the shear stiffening energy at constant strain ε_c, it can be easily shown that

► $$E_{com} = \sqrt{E_{PS}E_{SP}} \tag{12.30}$$

In case the estimates E_{SP} and E_{PS} do not differ too much, both ways of averaging yield reasonable values for E_{com}. Finite element calculations and measurements show that the average of Eqs. (12.27) and (12.28) indeed yields reasonably reliable estimates for the effective properties of a composite material.

Problem 12.6

Derive Eqs. (12.28) and (12.30).

The equivalent element

In both the first principle and empirical methods the shape of the inclusion was regular. In general this is not the case and we need a method to quantify the fractions. Below a method to do so is indicated.

For arbitrarily shaped phases f and m one must transform the shape of the inclusions of phase f into a block, meanwhile maintaining the block shape of the representative element. This transformation requires basically an integrated approach taking into account shapes and dimensions of the phases and the stress and strain fields. However, for simplicity the transformation is carried out only on the basis of shapes and dimensions. The centre of gravity of the new block of phase f must coincide with the centre of gravity of the representative cell of the two-phase material.

Let us consider a volume V of the phase f with a maximum length in the direction of loading in the representative element of L_f and a maximum cross-sectional area of A_f. The average length $\overline{L}_\text{f}$ and the average cross-section $\overline{A}_\text{f}$ are defined as V/A_f and V/L_f, respectively. The length and cross-section of the new particle block are L and A, respectively. If one assumes that for the transformation the following relation holds

$$V = LA = \overline{L}A_\text{f} = L_\text{f}\overline{A} \tag{12.31}$$

and thus that

$$\frac{A}{L} = \frac{\overline{A}}{\overline{L}} = \frac{A_\text{f}}{L_\text{f}} \tag{12.32}$$

one easily obtains

$$A^2 = \frac{A_\text{f}}{L_\text{f}}V \qquad \text{and} \qquad L^2 = \frac{L_\text{f}}{A_\text{f}}V \tag{12.33}$$

In this way the original representative element of our two-phase material is transformed into the equivalent element of Fig. 12.20, which is of equal outer dimensions and which contains a particle block with length L_f and cross-section A_f. Of course, assuming orthorhombic symmetry this transformation can be carried out for each of the three dimensions separately.

12.8 Laminates*

In Section 12.5 we briefly mentioned laminates. A laminate is a particular kind of composite material, which consists of a stack of layers, each with a different orientation and/or properties. A well-known example is plywood. For long-fibre composites the elastic and failure behaviour is quite anisotropic and therefore layers of this type of material are often combined in a laminate with different directions of the fibres to obtain a smaller degree of anisotropy (meanwhile keeping the improved elastic and/or failure behaviour). Since laminates have often the shape of plates we must be able to describe the deformation of plates. To that purpose one conventionally assumes that planar cross-sections remain planar during deformation, like in the theory of beams. This condition leads to the *Kirchhoff theory*[z] of plates. In this theory one neglects, although important, several other aspects, e.g. edge effects. In order to

---

[z] Gustav Robert Kirchhoff (1824-1887). German scientist and pupil of F. Neumann, in engineering theory well known for his theory of plates but who also contributed to physics, e.g. together with Robert Bunsen (1811-1899) on spectrum analysis as published in 1859.

apply the plate theory to a laminate the effective elastic constants of the laminate are required and this involves the response of each of the layers. Since the loading direction is usually not along the fibre direction, we need the transformation of the elastic constants of the layer to an arbitrary orientation of the axes system. In the next sections, we first derive the effective elastic constants of the laminate and thereafter briefly discuss the plate theory (Hull and Clyne, 1996).

Transformation rules

Let us consider the transformation of the elastic constants from one axes system to another. In general we have in tensor notation

$$\sigma_{ij} = C_{ijkl}\varepsilon_{kl} \qquad \text{or in pseudo-vector notation} \qquad \sigma_i = C_{ij}\varepsilon_j$$

The number of relevant elements varies with the symmetry properties of the material, as discussed in Chapter 9. We only recall here three important cases. The first is the cubic material with

$$C_{ij} = \begin{pmatrix} C_{11} & & & & & \\ C_{21} & C_{11} & & & \text{sym} & \\ C_{21} & C_{21} & C_{11} & & & \\ 0 & 0 & 0 & C_{44} & & \\ 0 & 0 & 0 & 0 & C_{44} & \\ 0 & 0 & 0 & 0 & 0 & C_{44} \end{pmatrix}$$

The second is the isotropic material for which the above equation holds with $C_{44} = ½(C_{11}–C_{12})$. The third is the orthorhombic (or orthotropic) material with

$$C_{ij} = \begin{pmatrix} C_{11} & & & & & \\ C_{21} & C_{22} & & & \text{sym} & \\ C_{31} & C_{32} & C_{33} & & & \\ 0 & 0 & 0 & C_{44} & & \\ 0 & 0 & 0 & 0 & C_{55} & \\ 0 & 0 & 0 & 0 & 0 & C_{66} \end{pmatrix}$$

Since the laminate is usually in plate-shape, it is conventional to assume that the material is in plane stress, i.e. σ_3 (= σ_{33}) = σ_4 (= σ_{23}) = σ_5 (= σ_{31}) = 0. Here we have taken the plane of the plate as the 1-2 or z-plane. This reduces the stress-strain relationship to

$$\begin{pmatrix} \sigma_1 \\ \sigma_2 \\ \sigma_6 \end{pmatrix} = \begin{pmatrix} Q_{11} & Q_{12} & 0 \\ Q_{21} & Q_{22} & 0 \\ 0 & 0 & Q_{66} \end{pmatrix} \begin{pmatrix} \varepsilon_1 \\ \varepsilon_2 \\ \varepsilon_6 \end{pmatrix} \qquad (12.34)$$

In Chapter 9 we only discussed the isotropic case for which $Q_{11} = Q_{22} = E/(1+\nu)(1-2\nu)$, $Q_{12} = Q_{21} = \nu Q_{11}$ and $Q_{66} = E/2(1+\nu) = \mu$, while $Q_{16} = Q_{26} = 0$. In this case the direction of loading is unimportant as far as the elastic constants are concerned. Moreover, since in this case $Q_{16} = Q_{26} = 0$, there is no coupling effect between the normal and shear stresses. For the orthorhombic case the expression for Q_{ij} reads

$$Q_{ij} = C_{ij} - \frac{C_{i3}C_{j3}}{C_{33}}$$

and the normal and shear stresses remain decoupled as long as the loading direction coincides with the 1-axis of the material, but this is generally no longer true for an off-axis loading. Therefore, we have to find the expressions for the elastic constants for arbitrary loading directions.

The general rule (see Chapter 6) for the transformation of the elastic constants is

$$C_{ijkl}^{*} = a_{ip}a_{jq}a_{kr}a_{ls}C_{pqrs}$$

We limit ourselves here to orthorhombic materials in plane stress since this is the most important case in this connection. For the special case where one symmetry plane is coincident with the 1-2 plane (as usual for lamina) the stress-strain relationship becomes

$$\begin{pmatrix} \sigma_1 \\ \sigma_2 \\ \sigma_6 \end{pmatrix} = \begin{pmatrix} Q_{11} & Q_{12} & Q_{13} \\ Q_{21} & Q_{22} & Q_{23} \\ Q_{31} & Q_{32} & Q_{33} \end{pmatrix} \begin{pmatrix} \varepsilon_1 \\ \varepsilon_2 \\ \varepsilon_6 \end{pmatrix} \tag{12.35}$$

again with

$$Q_{ij} = C_{ij} - \frac{C_{i3}C_{j3}}{C_{33}} \tag{12.36}$$

If now the new axes 1^*-2^*-3^* are rotated about the 3-axis of the 1-2-3-system, applying the transformation to the C_{ij} results in

$$C_{11}^{*} = c^4C_{11} + c^2s^2(2C_{12}+4C_{66}) + s^4C_{22}$$

$$C_{12}^{*} = (c^4+s^4)C_{12} + c^2s^2(C_{11}+C_{22}-4C_{66})$$

$$C_{22}^{*} = s^4C_{11} + c^2s^2(2C_{12}+4C_{66}) + c^4C_{22}$$

$$C_{16}^{*} = c^3s(C_{11}-C_{12}-2C_{66}) - cs^3(C_{22}-C_{12}-2C_{66})$$

$$C_{26}^{*} = cs^3(C_{11}-C_{12}-2C_{66}) - c^3s(C_{22}-C_{12}-2C_{66})$$

$$C_{66}^{*} = c^2s^2(C_{11}+C_{12}-2C_{12}) + (c^2-s^2)^2C_{66}$$

where $c = \cos\varphi$, $s = \sin\varphi$ and φ is the angle of rotation. From these expressions the Q_{ij}^{*} can be calculated using Eq. (12.36). It can be shown from these relations that the laminate responds as an isotropic material for in-plane loading if there are at least three or more identical lamina in the laminate at equal angles. However, for bending this is not true, as we will see in the next section, and therefore one refers to this behaviour usually as *pseudo-isotropic*.

Plate theory

Now we briefly discuss classical plate theory as applied to laminates. The laminate is thought to build up by several layers, each with a stress-strain relationship like Eq. (12.35) As stated before, we will assume that plane sections remain plane during bending. In this case the displacements u, v and w in the x, y and z directions, respectively, become

$$u = u_0(x,y) - z\frac{\partial w_0(x,y)}{\partial x} \qquad v = v_0(x,y) - z\frac{\partial w_0(x,y)}{\partial y} \qquad w = w_0(x,y)$$

where, as before, the z-coordinate is perpendicular to the undeformed centre plane of the laminate. The corresponding strains are

$$\varepsilon_{xx} = \frac{\partial u_0}{\partial x} - z\frac{\partial^2 w_0}{\partial x^2} \qquad \varepsilon_{yy} = \frac{\partial v_0}{\partial y} - z\frac{\partial^2 w_0}{\partial y^2} \qquad \varepsilon_{xy} = \frac{1}{2}\left(\frac{\partial u_0}{\partial y} + \frac{\partial v_0}{\partial x}\right) - z\frac{\partial^2 w_0}{\partial x \partial y}$$

which can be written as

$$\begin{pmatrix} \varepsilon_{xx} \\ \varepsilon_{yy} \\ \varepsilon_{xy} \end{pmatrix} = \begin{pmatrix} \varepsilon_{xx}^0 \\ \varepsilon_{yy}^0 \\ \varepsilon_{xy}^0 \end{pmatrix} + z\begin{pmatrix} \kappa_{xx} \\ \kappa_{yy} \\ \kappa_{xy} \end{pmatrix}$$

where

$$\varepsilon_{xx}^0 = \frac{\partial u_0}{\partial x} \qquad \varepsilon_{yy}^0 = \frac{\partial v_0}{\partial y} \qquad \varepsilon_{xy}^0 = \frac{1}{2}\left(\frac{\partial u_0}{\partial y} + \frac{\partial v_0}{\partial x}\right)$$

$$\kappa_{xx} = -\frac{\partial^2 w_0}{\partial x^2} \qquad \kappa_{yy} = -\frac{\partial^2 w_0}{\partial y^2} \qquad \kappa_{xy} = -\frac{\partial^2 w_0}{\partial x \partial y}$$

We now define the normal and the shear stresses resultants by

$$\begin{pmatrix} N_{xx} & N_{yy} & N_{xy} \end{pmatrix} = \int_{-h/2}^{+h/2} \begin{pmatrix} \sigma_{xx}^{(k)} & \sigma_{yy}^{(k)} & \sigma_{xy}^{(k)} \end{pmatrix} \mathrm{d}z \tag{12.37}$$

$$\begin{pmatrix} R_{xz} & R_{yz} \end{pmatrix} = \int_{-h/2}^{+h/2} \begin{pmatrix} \sigma_{xz}^{(k)} & \sigma_{yz}^{(k)} \end{pmatrix} \mathrm{d}z \tag{12.38}$$

where h is the total thickness, $z = 0$ is taken at the middle of the laminate and the superscript (k) indicates the kth layer. The bending moments are defined by

$$\begin{pmatrix} M_{xx} & M_{yy} & M_{xy} \end{pmatrix} = \int_{-h/2}^{+h/2} \begin{pmatrix} \sigma_{xx}^{(k)} & \sigma_{yy}^{(k)} & \sigma_{xy}^{(k)} \end{pmatrix} z\,\mathrm{d}z \tag{12.39}$$

Further the reduced equilibrium conditions are given by

$$\frac{\partial \sigma_{px}}{\partial x} + \frac{\partial \sigma_{py}}{\partial y} + \frac{\partial \sigma_{pz}}{\partial z} = 0$$

where p denotes either x, y or z. Integrating these expressions with respect to z and taking the shear stress at the top and bottom surface as vanishing, i.e. $\sigma_{xz}(\frac{1}{2}h) = \sigma_{xz}(-\frac{1}{2}h) = 0$, results in

$$\frac{\partial N_{xx}}{\partial x} + \frac{\partial N_{xy}}{\partial y} = 0 \qquad \frac{\partial N_{yx}}{\partial x} + \frac{\partial N_{yy}}{\partial y} = 0 \qquad \frac{\partial R_{zx}}{\partial x} + \frac{\partial R_{zy}}{\partial y} + q = 0 \tag{12.40}$$

where $q = \sigma_{zz}(\frac{1}{2}h) - \sigma_{zz}(-\frac{1}{2}h)$. Multiplying the first reduced equilibrium condition by z and integrating with respect to z we obtain

$$\frac{\partial M_{xx}}{\partial x} + \frac{\partial M_{xy}}{\partial y} + \int_{-h/2}^{+h/2} \frac{\partial \sigma_{xz}^{(k)}}{\partial z} z \,\mathrm{d}z = 0$$

Using $z\, \partial\sigma_{xz}^{(k)}/\partial z = \partial(z\sigma_{xz}^{(k)})/\partial z - \sigma_{xz}^{(k)}$ and $\sigma_{xz}(\frac{1}{2}h) = \sigma_{xz}(-\frac{1}{2}h) = 0$ finally results in

$$\frac{\partial M_{xx}}{\partial x} + \frac{\partial M_{xy}}{\partial y} - R_{xz} = 0 \tag{12.41}$$

In a similar way one can obtain

$$\frac{\partial M_{yx}}{\partial x} + \frac{\partial M_{yy}}{\partial y} - R_{yz} = 0 \tag{12.42}$$

By substituting the derivative of Eq. (12.41) with respect to x and of Eq. (12.42) with respect to y in the third of expressions (12.40) one obtains

$$\frac{\partial^2 M_{xx}}{\partial x^2} + 2\frac{\partial^2 M_{xy}}{\partial x \partial y} + \frac{\partial^2 M_{yy}}{\partial y^2} + q = 0$$

which is identical to the result of homogeneous plate theory, apart from the modified definitions of the stress resultants. The final result in terms of the strain field is obtained by substituting the stresses as given by Eq. (12.35) into the stress resultants given by Eqs. (12.37), (12.38) and (12.39) and reads

$$\begin{pmatrix} N_{xx} \\ N_{yy} \\ N_{xy} \\ M_{xx} \\ M_{yy} \\ M_{xy} \end{pmatrix} = \begin{pmatrix} A_{11} & A_{12} & A_{16} & B_{11} & B_{12} & B_{16} \\ A_{21} & A_{22} & A_{26} & B_{21} & B_{22} & B_{26} \\ A_{61} & A_{62} & A_{66} & B_{61} & B_{62} & B_{66} \\ B_{11} & B_{12} & B_{16} & D_{11} & D_{12} & D_{16} \\ B_{21} & B_{22} & B_{26} & D_{21} & D_{22} & D_{26} \\ B_{61} & B_{62} & B_{66} & D_{61} & D_{62} & D_{66} \end{pmatrix} \begin{pmatrix} \varepsilon_{xx}^0 \\ \varepsilon_{yy}^0 \\ \varepsilon_{xy}^0 \\ \kappa_{xx} \\ \kappa_{yy} \\ \kappa_{xy} \end{pmatrix} \tag{12.43}$$

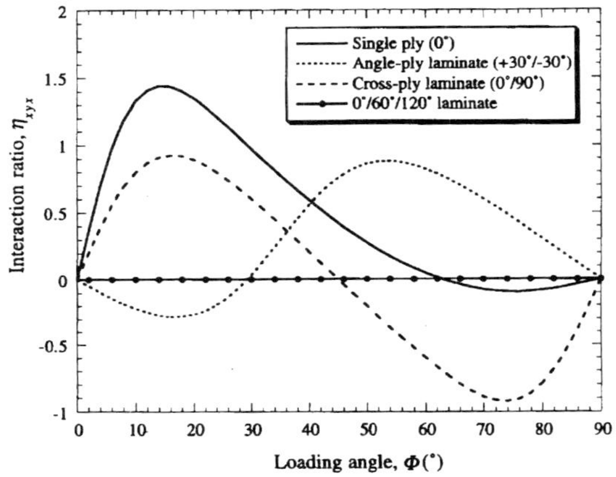

Fig. 12.22: The interaction ratio η_{xyx} for various glass fibre-epoxy laminates as a function of loading angle Φ.

where

$$(A_{ij} \quad B_{ij} \quad D_{ij}) = \int_{-h/2}^{h/2} Q_{ij}^{(k)} (1 \quad z \quad z^2) \mathrm{d}z \tag{12.44}$$

We see now that in general bending and stretching are coupled through the B-part of the matrix. In Fig. 12.22 the interaction ratio η_{xyx}, which represents the ratio of the induced shear stress σ_{xy} due an applied normal stress σ_{xx}, for a glass-epoxy composite as layer material is shown. From this plot it is clear that a significant shear stress can arise. However, the plot provides only a particular example with a relatively large difference in elastic moduli between the fibre and the matrix material. A smaller degree of anisotropy will lead to a smaller value of the interaction parameter. Also more layers in the layers in the laminate will reduce the value of the interaction parameter. The plot also shows the isotropy for the 0°/60°/120° laminate, conform the last remark of the previous section.

From Eq. (12.44) it follows that the B-part vanishes for a complete symmetry of the individual lamina thickness, properties and orientation. Sometimes this is referred to as a *symmetric laminate* (Fig. 12.23). A symmetric stacking of layers only is insufficient to guarantee decoupling. In practice one prefers symmetric laminates since according to the above theory coupling effects, which usually are a nuisance, are completely (and in reality largely) avoided. From the analysis it is clear that also for certain non-symmetric stacks the coupling effects can be largely avoided. In that case one refers to the laminate as a *balanced laminate*. If a fully symmetric laminate cannot be realised, a balanced laminate still is usually preferred above a non-balanced one. Apart from warping problems, also problems due to edge effects are less dominant. Although calculations along these lines are not particularly complex, they require a rather large number of numerical manipulations. Finally, it should be noted that this theory is far from a complete theory of laminates. Several assumptions have been made. To mention a few: plane stress, planar deformation and no edge effects present. More complex deformation fields have been incorporated in the theory and the effect of edge effects has been considered as well. For an introduction these topics we refer to the literature (e.g. Christensen, 1979). Standard software is available nowadays, not only for the classical theory but also for more advanced theory including non-linear, temperature and moisture effects.

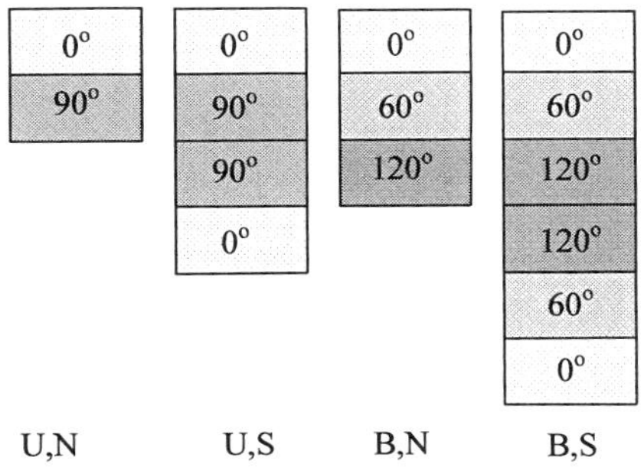

Fig. 12.23: Balanced and symmetric stacks. U unbalanced, B balanced, N non-symmetric, S symmetric.

Example 12.2

If there is only one isotropic lamina in the stack one regains the result of the homogeneous plate theory, which read

$$A_{11} = A_{22} = \frac{Eh}{1-\nu^2} \equiv A \qquad A_{12} = \nu A \qquad A_{16} = A_{26} = 0 \qquad A_{66} = (1-\nu)A$$

$$D_{11} = D_{22} = \frac{Eh^3}{12(1-\nu^2)} \equiv D \quad D_{12} = \nu D \quad D_{16} = D_{26} = 0 \quad D_{66} = (1-\nu)D$$

Moreover $B_{ij} = 0$, so that there is no coupling between bending and stretching. In this form plate theory is equivalent to the theory of homogeneous beams.

If there is only a single orthorhombic layer in the stack the result is

$$A_{ij} = Q_{ij}h \qquad B_{ij} = 0 \qquad D_{ij} = \frac{Q_{ij}h^3}{12}$$

Also in this case there is no coupling between bending and stretching.

12.9 Bibliography

Christensen, R.M. (1979), *Mechanics of composite materials*, Wiley, New York.

Francois, D., Pineau, A. and Zaoui, A. (1998), *Mechanical behaviour of materials*, vol. I: Elasticity and Plasticity, Kluwer, Dordrecht.

Grimvall, G. (1986), *Thermophysical properties of materials*, North-Holland, Amsterdam.

Hull, D. and Clyne, T.W. (1996), *An introduction to composite materials*, 2nd ed., Cambridge University Press, London.

Mura, T. (1987), *Micromechanics of defects in solids*, 2nd ed., Nijhoff, Dordrecht.

Nemat-Nasser, S. and Hori, M (1993), *Micromechanics: overall properties and heterogeneous materials*, North-Holland, Amsterdam.

Nye, J.F. (1957), *Physical properties of Crystals*, Oxford University Press, London.

Index

Q

R

S